Springer Series in Statistics

Advisors:
P. Bickel, P. Diggle, S. Fienberg,
U. Gather, I. Olkin, S. Zeger

To Shelby
with thanks and
warm regards
[signature]
Aug 18, 2005

Juha M. Alho and Bruce D. Spencer

Statistical Demography and Forecasting

With 33 Illustrations

Juha Alho
Department of Statistics
University of Joensuu
Joensuu, Finland

Bruce Spencer
Department of Statistics
Northwestern University
Evanston, IL 60208
USA

Library of Congress Control Number: 2005926699 (hard cover)
Library of Congress Control Number: 2005927649 (soft cover)

ISBN 10: 0-387-23530-2 (hard cover)
ISBN 13: 978-0387-23530-1 (hard cover)
ISBN 10: 0-387-22538-2 (soft cover)
ISBN 13: 978-0387-22538-8 (soft cover)

Printed on acid-free paper.

Printed in the United States of America. (TB/MVY)

9 8 7 6 5 4 3 2 1 SPIN 11011019 (hard cover) SPIN 11013662 (soft cover)

springeronline.com

To Irja and Donna

Preface

Statistics and demography share important common roots, yet as academic disciplines they have grown apart. Even a casual survey of leading journals shows that cross-references are rare. This is unfortunate, because many social problems call for a multi-disciplinary approach. Both statistics and demography are necessary ingredients in any serious analysis of the sustainability of pension or health care systems in the aging societies, in the assessment of potential inequities of formula-based allocations to local governments, in the estimation of the size of elusive populations such as drug users, in the investigation of the consequences of social ills such as unemployment, and so forth. This book was written to bring together much of the basic statistical theory and methodology for estimating and forecasting population growth and its components of births, deaths, and migration. Although relatively simple mathematical methods have traditionally been used to assess demographic trends and their role in the society, use of modern statistical methods offers significant advantages for more accurately measuring population and vital rates, for forecasting the future, and for assessing the uncertainty of the demographic estimates and forecasts.

For statisticians the book provides a unique introduction to demographic problems in a familiar language. For demographers, actuaries, epidemiologists, and professionals in related fields the book presents a unified statistical outlook on both classical methods of demography and recent developments. The book provides a self-contained introduction to the statistical theory of demographic rates (births, deaths, migration) in a multi-state setting. The book has a dual character. On the one hand, it is a monograph that can be consumed by a lone reader. There are many results that have appeared in journals or working papers only. Some appear here for the first time. The book is also useful as a classroom text, and includes exercises and complements to explore special topics in detail without interrupting the flow of the text. More than half of the book is readily accessible to undergraduates, but to fully benefit from the complete text may require more maturity.

Joensuu, Finland — *Juha M. Alho*
Evanston, Illinois, USA — *Bruce D. Spencer*

Acknowledgments

This book was some 15 years in the making. We are grateful to many colleagues and students for advice, encouragement and helpful comments, both specific and general. We thank Bill Bell, Katie Bench, Henry Bienen, Petra Can, Tom Espenshade, Steve Fienberg, Marty Frankel, Olavi Haimi, Joan Hill, Jan Hoem, Jeff Jenkins, Jay Kadane, Anne Kearney, Nico Keilman, Nathan Keyfitz, Donna Kostanich, Bill Kruskal, Esa Läärä, Jukka Lassila, Ron Lee, Risto Lehtonen, Chijien Lin, Lincoln Moses, Fred Mosteller, Tom Mule, Jukka Nyblom, Erkki Pahkinen, Päivi Partanen, Rita Petroni, Jiahe Qian, Dave Raglin, Chris Rhoads, Gregg Robinson, Mikko A. Salo, the late I. Richard Savage, Eric Schindler, Tom Severini, Eric Song, Richard Suzman, Shripad Tuljapurkar, Tarmo Valkonen, Jim Vaupel, Nic van de Walle, Larry Wu, Sandy Zabell. Shelby Haberman and Mary Mulry went above and beyond the call in close reading and advice. Responsibility for remaining errors, of course, remains with the authors.

During preparation of the book we received financial support from U.S. National Institute on Aging grant R01 AG10156-01A1 to Northwestern University; The Searle Fund grant on Limits of Empirical Social Science for Policy Analysis, to Northwestern University; U.S. Census Bureau contract 50-YABC-7-66020 with Abt, Associates; Academy of Finland Grants 8684, 41495, and 201408, Statistics Finland Grant 5012, and European Commission Grant HPSE-CT-2001-00095 to University of Joensuu; and European Commission Grant QLRT-2001-02500 to the Research Institute of the Finnish Economy.

Joensuu, Finland — *Juha M. Alho*
Evanston, Illinois, USA — *Bruce D. Spencer*

Contents

List of Examples

List of Figures

1
Introduction

1. Role of Statistical Demography

The world population exceeded six billion (6,000,000,000) in 1999. According to current United Nations projections, in 2050 the population is expected to be 9.3 billion, although under plausible scenarios it might be as low as 7.7 billion or as high as 10.9 billion. In all cases, the increase will intensify competition for arable land, clean water, and raw materials. Soil erosion and deforestation will continue in many parts of the world. The increased production of food, housing, and consumer goods will increase the production of greenhouse gases and, thus, contribute to climate change.

Underneath the global trends there is a great diversity. In the middle of the 19th century, European women gave birth to five children or more, on average. A newborn was expected to live 40 years or less. In a matter of a century the average number of children dropped to two and life expectancy rose to over 60 years. Many developing countries (notably China) have later followed a similar path, but a key factor in the uncertainty regarding global trends is whether all developing countries will go through a similar transition, and if so, at what pace.

Even within the industrialized world a great diversity persists. The average number of children per woman (as measured by the total fertility rate) varies from 1.2 children per woman in Italy and Spain, to 2.0 in the United States. The U.S. value is over 50% higher than that of the primarily catholic Mediterranean countries that have had a history of relatively high fertility! Yet, all values are below the level (approximately 2.1) that is needed for population replacement. Although births currently exceed deaths, this is a temporary phenomenon caused by an age-distribution that still has relatively many people in the child-bearing ages. In the near future the situation will change, and the age-distributions of the industrialized countries will be older than in any national population ever before on earth. This will put stress on the health care and retirement systems, a stress whose magnitude is not fully appreciated by decision makers, yet.

The "graying" of the industrialized populations will be accentuated by two factors. First, the large baby-boom cohorts born after World War II will be retiring in 2010–2020. This may prove to be a one time phenomenon, but no-one can say for certain that fertility fluctuations would have come to an end. The second factor is the continuing increase in longevity. Forecasters have repeatedly assumed that the decline in mortality cannot continue for more than a decade or two, only to have been proved wrong by the subsequent development.

Interestingly, populations can be quite heterogeneous with respect to life expectancy, as well. Women live longer than men, the rich and the well-educated live longer than the poor and the less-educated, and those in marriage live longer than those divorced, for example. The elderly are in many ways disadvantaged in the current industrialized societies. A happier future may lay ahead, if only by selection: it is possible that we will see a well-educated, healthy and wealthy retired population that is capable of exercising political power for its own benefit.

Since the rate of population growth in the developing countries far exceeds that of the industrialized countries, the geographic distribution of the world population will change. For example, the combined population of Europe and North America is currently 17% of the world population, but since the combined population is not expected to change by 2050, its share is expected to drop to 11%. A key social policy issue is to what extent the declining trend is counterbalanced by immigration from the less developed regions. An influx of immigrants would probably be advantageous to the elderly, since the immigrants could keep the economies growing and the "pay-as-you-go" retirement systems solvent. However, those in working age may reasonably see immigrants as competing in the same labor market, so racism and xenophobia may also gain ground.

Apart from global issues, demographics has an important role in the day-to-day decision making of national and local governments. Ever since the biblical times demographic data have served as a basis of taxation, military conscription, apportionment of political representation, and allocation of funds. Systematic biases in data may cause inequities across ethnic domains or geographic regions. When small areas are considered, random variations may cause inequalities in treatment. Lack of timeliness is always a potential source of systematic bias, but the remedy of frequent adjustments adds an element of unpredictability in the planning by local units.

Relatively simple mathematical methods have traditionally been used to assess demographic trends and their role in the society. The methods have typically been based on the measurement of demographic rates by age and sex. Summary measures, such as total fertility rate and life expectancy can then be calculated. A substantive line of research tries to explain variation in the rates across social groups, regions, or time, in terms of sociological or economic concepts. Another, less ambitious line of research tries to elucidate the long-term implications of the current rates. Classical methods from matrix algebra and differential and integral equations are used in the latter.

Simple methods have served and, undoubtedly, will continue to serve demography well. However, there are three reasons for expanding a demographer's toolkit into a statistical direction. First, as noted above, there is considerable interest in exploring variations in demographic rates in ever finer subpopulations. For example, if we find that young widows have an elevated risk of death but numbers are small, how can we know that this is not due to chance? Or, if the duration of unemployment is associated with mortality, how can this be evaluated? Cross tabulations are a classical, but clumsy, way to study such issues. In epidemiology, cross tabulations have largely been replaced by statistical relative risk regression techniques. We believe the same will happen in demography. Apart from simply adding new techniques to a demographer's toolkit, a methodological consequence is that principles of statistical inference, in particular the assessment of estimation error, should become a standard part of demographic analysis.

Second, many of the issues mentioned above involve forecasting in one way or another. In econometrics, the standard way to handle forecasting problems is to use statistical time-series techniques. We believe demographers can also benefit from the time-series toolkit provided that it is judiciously applied, in a manner that respects the demographic context. Demographic forecasts can then be made using data driven techniques, in addition to the judgmental methods that are currently favored. A methodological consequence of the adaptation of such techniques is that forecast uncertainty can be handled probabilistically. For example, instead of merely saying that it is plausible that world population is between 7.7 and 10.9 billion in 2050, we may say that it is within such an interval with a specific probability. Empirical analyses based on the accuracy of earlier U.N. forecasts suggest that in this case the probability is roughly 95%.

Third, even though the quality of basic demographic data on population size is likely to continue to improve, more elusive populations have become of concern. For example, we need information on the spread of drug use to assess its cost to the society and to determine the success anti-drug policies. Direct enumeration is, clearly, out of the question. Or, we need estimates of populations by health status to anticipate future demands on institutional care and housing that are accessible to those physically impaired. Such populations present us with complex definitional challenges, and information concerning them must derived via statistical techniques that may suffer both from biases and sampling error.

After these remarks we are reminded of two characterizations of the demographic profession. Jim Vaupel has defined a demographer as "someone who knows Lexis". Earlier Joel Cohen defined a demographer as "someone who forecasts population wrong", and a mathematical demographer as "someone who uses mathematics to forecast population wrong". Perhaps we could define a statistical demographer as "someone who knows Lexis, forecasts population wrong, but can at least quantify the uncertainty".

We have written this book with two types of readers in mind. First, we have thought of a mathematically oriented demographer, who is interested in learning the statistical outlook on the familiar problems. We have tried to define all relevant concepts in the book. However, the exposition is necessarily brief, so previous, familiarity with basic mathematical statistics, regression analysis, and time-series analysis is probably necessary for a full understanding of many of the arguments. Second, we have thought of a statistician, who is interested in working with demographic problems. We have tried to present the central demographic concepts in the context of statistical models, and indicate conditions under which the classical demographic procedures are optimal. Empirical examples are provided to give a flavor of what makes demography interesting. In addition to demographers and statisticians, we have thought of, for example, economists interested in pension and health care problems, epidemiologists interested in risk assessment, and actuaries and public health people interested in gerontology as potential readers of the book.

The application of statistical models in demography is not always straight forward, however. Along the way we try to indicate how a blind application of statistics can lead to unacceptable results. In fact, a central virtue of demographic teaching is a kind of "source criticism", in which one examines, much like a historian does, the mechanisms that have produced the data being analyzed. The most fashionable statistical analysis is not worth much if it is applied to data that are not what they seem. The book points out such issues, so it may be of a more general methodological interest to statistical readers.

2. Guide for the Reader

The book was originally conceived as a monograph intended for a lone reader. There are many results that have appeared in journals or working papers only. Some appear here for the first time. Yet, we have included exercises and complements to permit the use of the book in classroom. Some of the technical material is useful for reference (e.g., formulas for estimators and variances), and may be skipped on a first reading. Guidance is provided throughout the book. Parts of the earlier versions of the book have been used at the Universities of Joensuu and Jyväskylä, Finland; Örebro University, Sweden; Max Planck Institute at Rostock, Germany; and Northwestern University, U.S.A., to teach advanced undergraduate and graduate students in statistics and demography. For a statistical audience, additional discussion of the demographic issues has often proved useful. For a demographic audience, we have spent more time on the basics of statistics.

At least three threads of thought can be distinguished within the book:

* Chapters 2 and 4–6 provide an introduction to Statistical Demography; a shorter course that might be called Biometrics is obtained from Chapters 2 and 4;
* Chapters 2–4, 10 and 12 provide an introduction the Demographic Data Sources and their Quality;
* Chapters 4, 6–9 and 11 provide an introduction to Demographic Forecasting; a shorter course concentrating on Demographics of Pensions and Public Finances is obtained from sections of Chapters 4, 8–9, and 11.

In each case, other chapters provide supporting material.

3. Statistical Notation and Preliminaries

The remainder of this chapter introduces some notation for random variables and their distributions emphasizing vector and matrix formulations. We also give a heuristic review of basic results from maximum likelihood estimation that we assume as known in the sequel. Additional reminders/results will appear interspersed in the text, where needed. Some references for this material, at the same general mathematical level of the text, include Rice (1995), DeGroot (1987), Lindsey (1996), Azzalini (1996) and, at a more advanced mathematical level, Rao (1973), Severini (2000), Bickel and Doksum (2001), and Williams (2001).

The probability of an event A will be denoted by $P(A)$. If X is a *random variable* (i.e., a function whose value is determined by a random experiment), its *distribution function* or *cumulative distribution function (c.d.f.)* is $F(x) = P(X \leq x)$. The probability that X exactly equals x is $P(X = x) = F(x) - \lim_{h \searrow 0} F(x - h)$. Note that whenever $F(.)$ is continuous this probability is zero. If $F(.)$ is differentiable, then $F'(.) = f(.)$ is the *density function* of X.

Example 3.1. Normal (Gaussian) Distributions. The *standard normal distribution* $N(0, 1)$ has the expectation 0 and variance 1. Its density is $f(x) = (2\pi)^{-1/2} \exp(-x^2/2)$. Suppose X has this distribution, or $X \sim N(0, 1)$, then $Y = \mu + \sigma X$ has the normal (Gaussian) distribution $N(\mu, \sigma^2)$ with mean μ and variance σ^2. The density of Y is $f(y) = (2\pi)^{-1/2}\sigma^{-1} \exp(-(y - \mu)^2/(2\sigma^2))$. ◊

Example 3.2. Bernoulli Distribution. If X takes the value 1 with probability p and 0 with probability $1 - p$, then X has a Bernoulli distribution with parameter p, or $X \sim \text{Ber}(p)$. In this case $P(X = x) = p^x(1 - p)^{1-x}$, where $0 \leq p \leq 1$ and $x \in \{0, 1\}$. ◊

In mathematical demography one typically considers $X \geq 0$ and it is often more convenient to work with *survival probabilities* $p(x) = P(X > x)$ than with c.d.f.'s. If $p(.)$ is differentiable, then $f(x) = -p'(x)$.

The joint probability of events $A_1, \ldots, A_n$ is $P(A_1 \cap \ldots \cap A_n)$, but we sometimes write $P(A_1, \ldots, A_n)$ for short. The *conditional probability* of one event given another is defined as $P(A_1|A_2) = P(A_1 \cap A_2)/P(A_2)$, when $P(A_2) > 0$. If $X_1, \ldots, X_n$ are random variables, their *joint distribution function* is $F(x_1, x_2, .., x_n) = P(X_1 \leq x_1, X_2 \leq x_2, .., X_n \leq x_n)$. Writing column vectors $\mathbf{x} = (x_1, \ldots, x_n)^T$ and $\mathbf{X} = (X_1, \ldots, X_n)^T$, with ^{T} denoting transpose, we may also write $F(\mathbf{x}) = P(\mathbf{X} \leq \mathbf{x})$ where the inequality holds for each component.

The *expectation* of X is denoted by $E[X]$. If X has density $f(.)$, or if X takes discrete values $x_1, x_2, \ldots$, then

$$E[X] = \int_{-\infty}^{\infty} x f(x)\, dx \quad \text{or} \quad E[X] = \sum_i x_i\, P(X_i = x_i), \tag{3.1}$$

respectively. If X and Y are random variables and a and b are scalars, then we have the linearity property $E[aX + bY] = aE[X] + bE[Y]$. The *variance* of X is defined as $\text{Var}(X) = E[(X - E[X])^2]$. It has the property $\text{Var}(a + bX) = b^2$ $\text{Var}(X)$.

The expectation of a random vector $\mathbf{X}$ is defined componentwise, $E[\mathbf{X}] = (E[X_1], \ldots, E[X_n])^T$. If $\mathbf{a}$ is a vector and $\mathbf{B}$ is a matrix such that $\mathbf{a} + \mathbf{B}\mathbf{X}$ is well-defined, then $E[\mathbf{a} + \mathbf{B}\mathbf{X}] = \mathbf{a} + \mathbf{B}E[\mathbf{X}]$. The *covariance* between X_1 and

X_2 is defined as $\text{Cov}(X_1, X_2) = E[(X_1 - E[X_1])(X_2 - E[X_2])]$. The covariance matrix of $\mathbf{X} = (X_1, \ldots, X_n)^T$ is an $n \times n$ matrix $\text{Cov}(\mathbf{X})$ whose (i, j) element is $\text{Cov}(X_i, X_j)$. Using vector notation we may write $\text{Cov}(\mathbf{X}) = E[(\mathbf{X} - E[\mathbf{X}])(\mathbf{X} - E[\mathbf{X}])^T]$. It has the property $\text{Cov}(\mathbf{a} + \mathbf{B}\mathbf{X}) = \mathbf{B}\text{Cov}(\mathbf{X})\mathbf{B}^T$.

The *conditional expectation* of X_1 given X_2 is denoted by $E[X_1|X_2]$. It has the linearity property of the usual expectation. It may be shown that, when the moments exist, $E[X_1] = E[E[X_1|X_2]]$. The *conditional variance* is $\text{Var}(X_1|X_2) = E[X_1^2|X_2] - E[X_1|X_2]^2$. It has the property, $\text{Var}(X_1) = E[\text{Var}(X_1|X_2)] + \text{Var}(E[X_1|X_2])$. Similarly, the *conditional covariance* is defined as $\text{Cov}(X_1, X_2|X_3) = E[X_1X_2|X_3] - E[X_1|X_3]E[X_2|X_3]$ and has the property $\text{Cov}(X_1, X_2) = E[\text{Cov}(X_1, X_2|X_3)] + \text{Cov}(E[X_1|X_3], E[X_2|X_3])$.

Example 3.3. Multivariate Normal Distribution. Suppose a $k \times 1$ vector $\mathbf{X}$ has $E[\mathbf{X}] = \boldsymbol{\mu}$ and $\text{Cov}(\mathbf{X}) = \boldsymbol{\Sigma}$. It has a multivariate normal distribution, $\mathbf{X} \sim N(\boldsymbol{\mu}, \boldsymbol{\Sigma})$, if $\mathbf{a}^T\mathbf{X} \sim N(\mathbf{a}^T\boldsymbol{\mu}, \mathbf{a}^T\boldsymbol{\Sigma}\mathbf{a})$ for any $k \times 1$ vector $\mathbf{a}$. If $\boldsymbol{\mu} = \mathbf{0}$ and $\boldsymbol{\Sigma} = \mathbf{I}$, the identity matrix, then $\mathbf{X}^T\mathbf{X} \sim \chi^2$ *distribution* with $k \geq 1$ degrees of freedom. $\diamondsuit$

The multivariate normal distribution is an example of a parametric family of distributions. Consider n independent observations X_i coming from densities $f_i(x_i; \boldsymbol{\theta})$, $i = 1, \ldots, n$, where $\boldsymbol{\theta}$ is, say, a $k \times 1$ vector of parameters belonging to some set $\boldsymbol{\Theta} \subset \mathbb{R}^k$. We do not assume here that the observations are necessarily identically distributed, because in regression applications of interest they typically are not. For example, in normal theory regression, if X_i would be the dependent variable and $\mathbf{z}_i$ would be a vector of explanatory variables, we would have the density $f_i(x_i; \boldsymbol{\theta}) = (2\pi)^{-1/2}\sigma^{-1}\exp(-(x_i - \mathbf{z}_i^T\boldsymbol{\beta})^2/(2\sigma^2))$, where $\boldsymbol{\theta} = (\boldsymbol{\beta}^T, \sigma^2)^T$.

When viewed as a function of $\boldsymbol{\theta}$ the probability of the observed data is called the *likelihood function*, $L(\boldsymbol{\theta}) = f_1(x_1; \boldsymbol{\theta}) \cdots f_n(x_n; \boldsymbol{\theta})$. The natural logarithm of the likelihood function is the *loglikelihood function* $\ell(\boldsymbol{\theta}) = \log L(\boldsymbol{\theta})$. The principle of maximum likelihood means that we try to determine a value of $\boldsymbol{\theta}$ that maximizes $L(\boldsymbol{\theta})$, or equivalently $\ell(\boldsymbol{\theta})$. The maximizing value (if one exists) is called a *maximum likelihood estimator (MLE)*. Define a $k \times 1$ vector of partial derivatives $\mathbf{S}_i(\boldsymbol{\theta}) = \partial/\partial\boldsymbol{\theta} \log(f_i(x_i; \boldsymbol{\theta}))$ for each $i = 1, \ldots, n$. Their sum $\mathbf{S}(\boldsymbol{\theta}) = \mathbf{S}_1(\boldsymbol{\theta}) + \cdots + \mathbf{S}_n(\boldsymbol{\theta})$ is called the *score* (e.g., Rao 1973, 367), and the MLE solves the system of k equations $\mathbf{S}(\boldsymbol{\theta}) = \mathbf{0}$.

Before the observations $X_i = x_i$ have been made, the score is a random variable, because its components are random: $\mathbf{S}_i(\boldsymbol{\theta}) = \partial/\partial\boldsymbol{\theta} \log(f_i(X_i; \boldsymbol{\theta}))$. Assuming that the order of differentiation and integration can be changed, we have that $E[\mathbf{S}_i(\boldsymbol{\theta})] = \partial/\partial\boldsymbol{\theta} \int f_i(x_i; \boldsymbol{\theta})\, dx_i = \mathbf{0}$. The latter equality holds because the integral equals 1 for all $\boldsymbol{\theta}$. Therefore, the expectation of the score is $E[\mathbf{S}(\boldsymbol{\theta})] = \mathbf{0}$. Write $\text{Cov}(\mathbf{S}_i(\boldsymbol{\theta})) = \mathcal{I}_i(\boldsymbol{\theta})$, $i = 1, \ldots, n$, and define $\mathcal{I}(\boldsymbol{\theta}) = \mathcal{I}_1(\boldsymbol{\theta}) + \cdots + \mathcal{I}_n(\boldsymbol{\theta})$. It follows that $\text{Cov}(\mathbf{S}(\boldsymbol{\theta})) = \mathcal{I}(\boldsymbol{\theta})$, because the observations are independent. This is one form of the so-called *Fisher information* of the sample. Subject to regularity conditions on densities $f_i(x_i; \boldsymbol{\theta})$ (that may involve conditions on both the range of values of possible explanatory variables and on the tails of the density), none of components of the score $\mathbf{S}_i(\boldsymbol{\theta})$ take too large a share of the variance of the score, so one can appeal to the central limit theorem to assert the asymptotic normality of the score. Therefore, we have that $\mathbf{S}(\boldsymbol{\theta}) \sim N(\mathbf{0}, \mathcal{I}(\boldsymbol{\theta}))$ asymptotically.

Example 3.4. Score tests. Consider a hypothesis $H_0 : \boldsymbol{\theta} = \boldsymbol{\theta}_0$. Under the null hypothesis, $\mathbf{a}^T\mathbf{S}(\boldsymbol{\theta}_0) \sim N(0, \mathbf{a}^T\mathcal{I}(\boldsymbol{\theta}_0)\mathbf{a})$ for any $k \times 1$ vector $\mathbf{a}$, so depending on the alternative hypothesis, a large number of the so-called *score tests* can be constructed. ◊

Define a $k \times k$ matrix $\mathbf{H}_i(\boldsymbol{\theta}) = \partial^2/\partial\boldsymbol{\theta}\partial\boldsymbol{\theta}^T \log(f_i(X_i;\boldsymbol{\theta}))$, for each $i = 1, \ldots, n$. I.e., this is a matrix whose (r, s) element is $\partial^2/\partial\boldsymbol{\theta}_r\partial\boldsymbol{\theta}_s \log(f_i(X_i;\boldsymbol{\theta}))$. Their sum $\mathbf{H}(\boldsymbol{\theta}) = \mathbf{H}_1(\boldsymbol{\theta}) + \cdots + \mathbf{H}_n(\boldsymbol{\theta})$ is called the *Hessian*. By a direct calculation one can show that $E[\mathbf{H}_i(\boldsymbol{\theta})] = \partial^2/\partial\boldsymbol{\theta}\partial\boldsymbol{\theta}^T \int f_i(x_i;\boldsymbol{\theta})\,dx_i - E[\mathbf{S}_i(\boldsymbol{\theta})\mathbf{S}_i(\boldsymbol{\theta})^T]$. As in the case of the score, the first term on the right hand side is zero. Using the result, $E[S_i(\boldsymbol{\theta})\mathbf{S}_i(\boldsymbol{\theta})^T] = \mathrm{Cov}(\mathbf{S}_i(\boldsymbol{\theta})) = \mathcal{I}_i(\boldsymbol{\theta})$, we find an alternative expression for Fisher information, $-E[\mathbf{H}(\boldsymbol{\theta})] = \mathcal{I}(\boldsymbol{\theta})$.

Example 3.5. Fisher Information for Normal Distribution. Consider the normal distribution $N(\mu, \sigma^2)$. Let $\boldsymbol{\theta} = (\mu, \sigma^2)^T$. The Fisher information $\mathcal{I}(\boldsymbol{\theta})$ is given by the matrix

$$\begin{bmatrix} 1/\sigma^2 & 0 \\ 0 & 1/(2\sigma^4) \end{bmatrix}. \tag{3.2}$$

If instead we take $\boldsymbol{\theta} = (\mu, \sigma)^T$ then the lower diagonal entry of $\mathcal{I}(\boldsymbol{\theta})$ changes to $2/\sigma^2$. ◊

Suppose $\hat{\boldsymbol{\theta}}$ is the MLE. By Taylor's theorem there is vector $\boldsymbol{\theta}'$ between the MLE and the true value $\boldsymbol{\theta}$ such that $\mathbf{S}(\hat{\boldsymbol{\theta}}) = \mathbf{S}(\boldsymbol{\theta}) + \mathbf{H}(\boldsymbol{\theta}')(\hat{\boldsymbol{\theta}} - \boldsymbol{\theta})$. We get from this that $\hat{\boldsymbol{\theta}} - \boldsymbol{\theta} = -\mathbf{H}(\boldsymbol{\theta}')^{-1}\,\mathbf{S}(\boldsymbol{\theta})$ provided that the inverse exists. Subject to regularity conditions $\mathbf{S}(\boldsymbol{\theta})/n \to \mathbf{0}$,[1] as $n \to \infty$, and $\mathbf{H}(\boldsymbol{\theta})/n$ has a limit $\mathbf{H}^*(\boldsymbol{\theta})$ that is a continuous function of $\boldsymbol{\theta}$ at least in the neighborhood of the true parameter value. In this case the MLE also converges to $\boldsymbol{\theta}$, so it is *consistent*. Being essentially a linear function of the score, the MLE inherits the multivariate normal distribution from the score and asymptotically $\mathrm{Cov}(\hat{\boldsymbol{\theta}}) = \mathcal{I}(\boldsymbol{\theta})^{-1}$. For practical inferential purposes we may assume, for large n, that $\hat{\boldsymbol{\theta}} \sim N(\boldsymbol{\theta}, -\mathbf{H}(\hat{\boldsymbol{\theta}})^{-1})$. This leads to the so-called *Wald tests*.

There is yet a third type of test that naturally arises from the above theory. Consider a hypothesis $H_0 : \boldsymbol{\theta} = \boldsymbol{\theta}_0$. Using a second order Taylor series development for $\ell(\boldsymbol{\theta})$ around $\hat{\boldsymbol{\theta}}$ and noting that $\mathbf{S}(\hat{\boldsymbol{\theta}}) = \mathbf{0}$, we get that

$$2(\ell(\hat{\boldsymbol{\theta}} - \ell(\boldsymbol{\theta}_0)) = -(\hat{\boldsymbol{\theta}} - \boldsymbol{\theta}_0)^T\,\mathbf{H}(\boldsymbol{\theta}')(\hat{\boldsymbol{\theta}} - \boldsymbol{\theta}_0), \tag{3.3}$$

where $\boldsymbol{\theta}'$ is a point between $\boldsymbol{\theta}$ and $\hat{\boldsymbol{\theta}}$. The asymptotic result given for the Wald tests shows that the right hand side has a approximate χ^2 distribution with k degrees of freedom. This is one form of the so-called *likelihood ratio test*. The three tests are

[1] This can mean either convergence in probability or almost sure convergence (Rice 1995, 164).

asymptotically equivalent, but their small sample characteristics may differ (Rao 1973, 415–418).

We conclude with definition of $o(.)$ and $O(.)$ notation. Let $\{a_n\}_{n=1}^{\infty}$ and $\{b_n\}_{n=1}^{\infty}$ be two sequences of numbers. We say that a_n is $o(b_n)$ if $\lim_n |a_n/b_n| = 0$, and $a_n = O(b_n)$ if $|a_n/b_n|$ is bounded when n is large. To allow continuous arguments we say that $a(x)$ is $o(b(x))$ or $O(b(x))$ as $x \to L$ if $a(x_n)$ is $o(b(x_n))$ or $O(b(x_n))$ for any sequence $\{x_n\}_{n=1}^{\infty}$ with $x_n \to L$. For example, $6x^4$ is $O(x^4)$ and $o(x^5)$ as $x \to \infty$, and $6x^4$ is $O(x^4)$ and $o(x^3)$ as $x \to 0$.

2
Sources of Demographic Data

1. Populations: Open and Closed

We can think of a population size as a *process*. At any given time t a set of individuals satisfy the membership criterion of the population. In the case of a geographic area, for example, the criterion is "being in the area". The population can increase via births and in-migration. It can decrease via deaths and out-migration.[1] Thus, births, deaths, and migration form the relevant *vital processes*.

Traditionally, the term *vital event* is used for births, deaths, marriages and divorces but not for migration (cf., Shryock and Siegel 1976, 20). Although this usage has an origin in civil registration, the distinction is not useful in statistical demography and we consider vital processes to include migration. Changes of marital status can be vital processes, if the population of interest has been defined in terms of marital status, but so can be such processes as getting a job or becoming unemployed, if the population is defined in terms of employment status.

In a limiting case we define a population as *closed* if it has no vital processes. A closed population is simply a set of individuals. (In demography it is common to call a population closed even if it experiences births and deaths. We take here a broader view.) In most demographic applications a population is open in some respects. For example, in a follow-up study of a fixed set of individuals, the population is closed with respect to births and in-migration, but it is open with respect to deaths. Annoyingly from the researcher's point of view, such a population may, in practice, be open to out-migration and other forms of attrition or loss from follow-up, as well.

As discussed below, the distinction between closed and open populations is important in the design of the data collection for demographic studies. However, in most parts of this book we have the prototype of national population in mind. National populations are open to births, deaths, migration etc.

[1] A population can also change when its definition changes, e.g., when a country, state, or city annexes or de-annexes an area. Such changes do not involve vital processes, and analysis of past data on population change should make allowance for any significant boundary changes that occurred.

At first thought nothing seems simpler than to define a population. National identity is so ingrained that a special effort is required to appreciate the conventional aspects of the membership criterion. Therefore, consider the following two examples.

Example 1.1. Who Counts in the U.S. Census? The United States Constitution (Article I, sec. 2) stipulates that "Representatives and direct Taxes shall be apportioned among the several States which may be included within this Union, according to their respective Numbers, which shall be determined by adding to the whole Number of free Persons, including those bound to Service for a Term of Years, and excluding Indians not taxed, three fifths of all other Persons." Since nontaxed Indians were not included in these numbers, their coverage in historical censuses (that started in 1790) is dubious. Slaves were to be counted in a separate category in censuses prior to 1870. It seems that slaves were to be counted in full in the census and then their numbers reduced by two fifths for Federal apportionment – slaves did not figure into population counts for apportionment of state legislatures by southern states (cf., Shryock and Siegel 1976, 14–16; Savage 1982; Anderson and Fienberg 1999, 13). ◊

Example 1.2. Who Belongs to the Sami Population? In the mid-1990's considerable controversy was caused in Northern Finland by the question of who belongs to the *Sami* (*Lapp*) population of Lapland. Some advocated a definition emphasizing the role of Sami language, others the length of family history in the area. Different cultures had mixed in Lapland over the centuries, so no clear-cut distinction between the families could be given. Fueling the controversy was the thought that the original people of the area may be treated preferentially in future legislation. In the Law on the Sami Cultural Self-Government from 1995 the following (freely translated) definition was given:

> A person belongs to the Sami population, if he considers himself to be Lapp, provided that (1) he himself or at least one of his parents or grandparents has spoken Sami as his mother tongue; or (2) he is a descendant of a person who has been marked as mountain, forest, or fisher Lapp in the books of land or taxation; or (3) at least one of his parents has been marked or could have been marked as having the right to vote in the election of Sami representatives.

In addition, a map of the area within which this definition was to be applied, was published. ◊

These examples display many of the problems that one encounters in trying to define a membership criterion for a human population. Economic, cultural, and administrative considerations are typically involved. Even subjective factors ("... if he considers himself to be Lapp ...") were involved in the very definition of the Sami population. How can or ought one define the "true size" of the Sami population at a given point in time? Not only is the definition subjective, but so is its measurement: a person's self-identification may vary over time as well as how the question asking for self-identification is presented.

A similar issue arises forcefully in the definition and assignment of racial classifications. The American Anthropological Association concluded that "The concept of race is a social and cultural construction, with no basis in human biology – race can simply not be tested or proven scientifically."[2] In the U.S., ever since the 1970 census a person's race is based on self-identification. Since some people identify with more than one group, the United States began in the 2000 Census to allow for "multi-race" categories: 63 racial classifications with 6 categories[3] for single-race only and 57 for combinations of races (U.S. Census Bureau 2000). Analysis of time series statistics for racial groups in the U.S. requires care for allowing for definition changes pre- and post-2000.

Below, we briefly discuss some aspects of the operational definition of national and sub-national populations and relate these to the coverage and classification errors that frequently occur. We next discuss censuses and population registers as sources of population data. We pay attention to historical aspects of the registration of the vital events, because analysis of past time series of statistics on vital events will help us understand the accuracy of forecasts. Similarly we introduce the concept of the Lexis diagram for insight into the complexities of using grouped data to estimate vital rates in open populations. After that we consider registers and cohort and case-control study designs as prototypes of data collection for specific demographic (or epidemiological) problems. We conclude the chapter by discussing the role of statistical sampling in population estimation. Sampling more generally will be discussed in Chapter 3.

2. *De Facto* and *De Jure* Populations

At any moment in time any specific geographic area has a *de facto* population, which consists of all individuals who are present in the area. This concept is unequivocal but may not always be highly relevant. Consider the following groups mentioned in the "Recommendations for the 1990 Censuses of Population and Housing in the ECE Region" (United Nations 1987, 9–10):

(1) persons usually resident and present;
(2) persons usually resident but absent;
(3) persons temporarily present but usually resident elsewhere.

The *de facto* population comprises (1) and (3), but excludes (2). Often one is interested in the usually resident, or *de jure*, population consisting of (1) and (2). The distinction may seem simple until one considers the cases frequently encountered in practice:

---

[2] American Anthropological Association, Press Release/OMB 15, Sept. 8, 1997.
[3] American Indian and Alaska Native, Asian, Black or African American, Native Hawaiian and Other Pacific Islander, Some Other Race, White.

(a) persons maintaining more than one residence;
(b) students not living with parents;
(c) persons living away from home during work week;
(d) persons in military service;
(e) military personnel who maintain a home elsewhere;
(f) institutional populations such as hospitals, or prisons;
(g) persons intending to return to a former home place;
(h) persons who have arrived a short time ago who consider some other place as their home;
(i) persons expected to return soon from elsewhere.

Categories (g)–(i) may consist of illegal aliens, nomads, vagrants, military, naval, or diplomatic personnel and their families. They may include merchant seamen, fishermen, transients in ships, trains, cars, or airplanes, refugees etc. For different purposes different choices can reasonably be made concerning which of these groups are included into the population. In many countries and many subnational areas these categories may be small and so their operational definitions may not matter in practice. Sometimes these groups do matter, however.

Example 2.1. Accident Rates in Nordic Countries. A comparison of the rate of traffic accidents in the cities of Gothenburg, Helsinki, Oslo, and Stockholm from 1990–1994 (Nieminen 1996, 22) shows that Helsinki has had a lower rate of accidents involving passengers inside vehicles (about 1 passenger accident per 1,000 inhabitants in a year) than the other cities (1.5–2.5 per 1,000), but a higher rate of accidents involving pedestrians (about 0.5 per 1,000) than the other cities (0.35–0.5 per 1,000). There can be many causes for such differences, including possible variations in the completeness of the registration. However, a map of the locations of the accidents in Helsinki (Nieminen 1996, 13) shows that accidents concentrate near the central railway station, a major gateway for commuters to work. Although we cannot determine whether this explains the differences between the cities, it is clear that while the accidents are tabulated according to the place of occurrence, the denominator population is the *de jure* population. This is a mismatch. A proper denominator for the risk rate would be the *de facto* population because many accidents occur to individuals who commute to work. ◊

In the industrialized countries, the official population figures typically rely on some form of *de jure* definition (Shryock and Siegel 1976, 50). Once the definition of the population is agreed upon, it is important to consider the quality of demographic information. If the analysis of time trends is of interest, have the definitions remained the same over time? If comparisons between different areas are of interest, are the definitions the same in the different countries? Finally, if the definitions are comparable, are the counts and classifications accurate?

Example 2.2. Undercount in U.S. Censuses. Consider the population sizes reported by U.S. censuses of 1940–2000. The "net undercount" – true size minus census count – can be estimated by several methods (cf., Chapter 11). To appreciate the order of magnitude, consider the following estimates of the undercount (in %) by race based on "demographic analysis" (Robinson et al. 1993, 1065, and Robinson, Adlakha, and West 2002, 26):

	Non-Black		Black	
year	male	female	male	female
1940	5.2	4.9	10.9	6.0
1950	3.8	3.7	9.7	5.4
1960	2.9	2.4	8.8	4.4
1970	2.7	1.7	9.1	4.0
1980	1.5	0.1	7.5	1.7
1990	1.6	0.6	8.1	3.1
2000	0.2	– 0.8	5.1	0.5

We see that Blacks have higher undercount rates than Non-Blacks, and males have higher undercount rates than females. Note that the rates show the *net* effect of both census misses and census duplications or other erroneous enumerations. By and large the net undercount rates declined from 1940 to 1980, and increased in 1990. It is possible that attempts to obtain a complete count may lead to increased erroneous enumerations, and the 2000 census appears to have overcounted non-black females. Demographic analysis also shows that net undercount varies markedly by age. For example, in 1990 Black males in ages 25–60 had the lowest probabilities of being enumerated in the census whereas non-blacks in ages 15–25 may even have been overcounted. Clearly, census numbers suffer from problems of comparability across sex, age, race or ethnic group, and time. ◊

Migration can also lead to surprising conceptual problems. In the case of international geographic migration most countries are unable to keep track of emigration, and many countries have difficulty in keeping track of (especially illegal) immigration. The United States, for example, does not have any statistics concerning emigration, and while it has annual statistics of legal immigration, only indirect estimates (e.g., Muller and Espenshade 1985, Espenshade 1997) are available for the much larger illegal immigration. In Europe, the quality of migration data varies considerably. The Nordic countries with well-functioning population registers have relatively good data on people moving in, because typically many aspects of daily life (health care, child care, opening of bank accounts, access to subsidized public transportation etc.) depend directly or indirectly on their being registered. It is somewhat harder to keep track of people moving out, unless the out-movers go to a country with a good register that agrees to supply information about new migrants received. The European countries that rely on censuses face problems similar to those of the United States. A practical problem in compiling statistics on migration is caused by the fact that the countries do not adhere to the same definition as to who is a (long term) *migrant* (Poulain 1993, 354). The U.N. has recommended that an intention of staying at least a year in a country (after an absence of at least a year) would be required to consider a person a migrant, but this is not followed by most European countries (Poulain 1993, 355; Eurostat 2004, 151–153). The use of different definitions of migrants implies that a person may be counted as belonging to the population of two countries at the same time, for example. Thus, even if the practices of census taking would agree between two countries, the definition of the population during intercensal years need not be the same across countries.

A further problem in published population statistics arises from possible misclassifications by age, race, marital status, place of residence etc. Although age is nowadays accurately known for inhabitants of most industrialized countries, a self-reported age may be in error. In non-industrialized countries age may have been less important, especially in the past. For example in the population of Philippines in 1960 showed remarkable *digit preference* (or *age heaping*) for multiples of 5 years. For example, the counts in ages 59, 60, and 61 were 72,206; 275,436; and 31,299, respectively (cf., Shryock and Siegel 1976, 116).[4,5] Where feasible, such reporting problems may be mitigated by recording year and date of birth as well as age (to cross-check).

Although demographic methods typically are applied to human populations, demographic concepts have methodological value more broadly. Some notions that are basic for the study of human populations can be usefully extended to populations consisting of other types of elements. Populations of types of consumer goods (cars, refrigerators, ...) or species of animals (rabbits, fish, insects, ...) are obvious examples experiencing births, deaths and migration, and having a changing age structure. In addition, one can also study interesting populations consisting of human aggregates such as households and enterprises. Their definition often has an administrative, *de jure* basis, but for application one is typically interested in the *de facto* numbers.

Example 2.3. What Is a Household? *Households* can be defined in terms of housekeeping, or one or more persons live in a housing unit and provide themselves with food and possibly other necessities of life (cf., Van Imhoff and Keilman 1991, 10). Housing units often have not only *de jure* residents but *de facto* residents as well. Therefore, the composition of a household may only be revealed by special surveys. Note that no aspect of kinship is usually involved in the definition of a household even though many households are familial units also. In addition to births and deaths, households may also *split.* ◊

Example 2.4. Corporate Demography. In *enterprise* or *corporate demography* (cf., Ilmakunnas, Laaksonen, and Maliranta 1999; Carroll and Hannan 2000, 51) data often are available for individual *establishments*, such as factories, warehouses, restaurants, or stores. In some cases, data may exist for departments within establishments, such as different production lines in a factory. Enterprises, corporations

[4] The age heaping was still present to a lesser extent in the 1990 census, where the numbers for the three ages were 275,560; 322,233; and 205,177, respectively (Hobbs 2004, 137).

[5] Similar phenomena occur in other statistics. For example, Breslow and Day (1987, 163) presented data on smoking from the so-called British Doctors' study (cf., Example 5.1 below). Smoking status was classified into classes 0, –4, 5–9, ..., 30–34, 35–40 cigarettes/day. An estimate of the average number of cigarettes is also given for each class. The averages are quite close to lower limits of the classes suggesting that the respondents have had a clear digit preference of multiples of five.

and other economic organizations with a legally defined (*de jure*) status may consist of several establishments. Finally, conglomerates consisting of legally separate corporations may form a unit of analysis. Data on enterprises are usually collected for some administrative purpose such as taxation or occupational health. Enterprises with low level of economic activity may be inadequately surveyed or even completely omitted by the legal definitions in use. Therefore, the size of the enterprise population may be underestimated in official statistics at the same time that total employee population statistic is relatively accurate. In addition to births, deaths, and splits, enterprises may also *merge*. ◊

3. Censuses and Population Registers

In statistics it has become customary to contrast censuses and samples. A *census* is a study comprising the whole population of interest, whereas a sample involves only a part. A population census refers more specifically to a complete count of the population of an area at a given time. Censuses may be combined with samples in various ways. Some data (e.g., age) may be collected for 100% of the population and other data (e.g., income) collected from, say, every 100th unit. A census can be *de facto* or *de jure* based and typically collects such basic information as age, sex, and, perhaps on a sample basis, marital status, literacy, educational attainment, occupation, industry, place of usual residence, place of birth (cf., Shryock and Siegel 1976, 32; United Nations 1987, 5–7). Most countries of the world (including the Unites States, England, France, China, and India) rely on censuses as the basic source of population data. In practice, censuses are carried out via mail questionnaires and door-to-door interviewing. Since population counts are often used to apportion political power, for military conscription, or for taxation, a census may not always be an innocuous operation.

Example 3.1. Nigerian Censuses. Prior to the 1991 census the population of Nigeria was estimated to be 95.7 million in 1985 by the United Nations, 110 million in 1988 by the World Bank, and 112 million in 1987 by the Nigerian government. Estimates for the year 1991 were in the range 112–123 million (*Population Today*, June 1992, No. 6). The history of the Nigerian censuses goes back to the 1860's but apparently the quality of the results, including that of the previous census, in 1973, has been less than satisfactory. Presumably, the ethnic diversity of the country has played a part in this. With this background it was quite a shock that the 1991 census count was 88.5 million, or more than 20% less than the estimates. Evidently, any attempt at a statistical analysis of the population of African countries must somehow account for the uncertainty of the census results. ◊

In countries using censuses a separate system has been in place for the estimation of births, deaths, marriages, migration etc. For example, in the United States death registration became fairly complete in Massachusetts around 1865 (Shryock and Siegel 1976, 21). In the year 1900 a "death registration area" was established comprising the District of Columbia and ten states. A "birth registration area" was established in 1915 with the same area included. Complete geographic coverage was achieved in 1933 although only 90% registration was required for the admission of a state into the area (Shryock and Siegel 1976, 274). We see that even in the industrialized world one cannot expect long time-series of known statistical quality, on vital events.

In contrast to the statistics usage, in demography censuses typically are contrasted with population registers. Registers provide continuous information about all members of the (typically *de jure*) population. The Nordic countries, Japan, and Russia are examples of countries with population registers. Although nowadays population registers are maintained as computerized databases in many countries, they have a long history. Finland and Sweden have continuous, register based population statistics from the year 1749 onwards. The registers were kept by the church based on an ecclesiastic law of 1686. Each parish would keep track of the vital processes of births, deaths, marriages, and changes of parish. Initially, these registers developed out of books that were maintained since the 1500's for the follow-up of parishioners' progress in the knowledge of reading, writing, and the Bible (Nieminen and Markelin 1974). The establishment of the population statistics around 1750 seems to have occurred in part because estimates compiled by the Royal Academy in Stockholm showed that the true population was only about 2 million instead of the generally believed figure of 3 million (Teräsvirta 1987, 3), a situation not unlike the one that occurred much later in Nigeria!

The reliability of the Finnish vital statistics has been studied using parish level data by Pitkänen (1977), for example. He has shown that many infant deaths were omitted from the registers during the 18th century, because unbaptized children were recorded as stillborn, and baptized infants who died young were deliberately omitted. Pitkänen (1986) also shows that a curious increase in the mortality of the middle-aged and older men during the first decades of the 20th century may have been an artifact caused by migration to the United States. Apparently a fairly large number of deaths that occurred overseas were recorded in the parish registers, even though the persons themselves had been marked as emigrated. The mis-match of the numerator and denominator (as in Example 2.1) could have caused an artificial increase of a few percent in the estimated mortality (Pitkänen and Laakso 1999).

Countries with population registers do conduct censuses every five or ten years to provide occupational and educational details that are not included in the population register itself. The situation varies between countries but for example in Finland this involves the linking of computerized databases rather than door-to-door activities (Harala and Tammilehto-Luode 1999).

4. Lexis Diagram and Classification of Events

A formal aspect of the recording of the vital events is their classification by age and time. Much the same way as with defining populations, initially nothing seems simpler. However, since it is customary to compile statistics on vital events by discrete time, rather annoying complications arise. To appreciate the problem, we

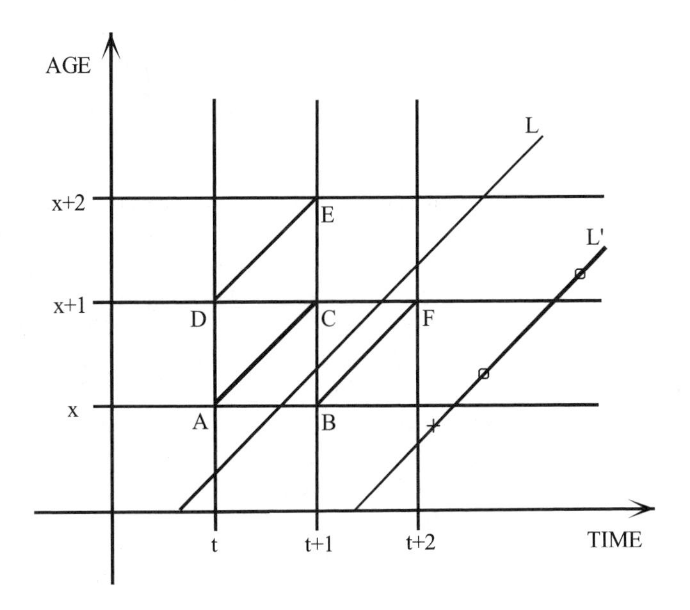

FIGURE 1. Lexis Diagram.

introduce the concept of a Lexis diagram, one of the most useful technical devices of demography.[6]

We let horizontal axis refer to time t and vertical axis to age x in Figure 1. For each person in a well-defined population we may draw a *life line* that starts at a time and age when the person enters the population and ends at the time and age when the person exits the population. Typically the entry would occur at birth and the exit at death, but entries or exists due to other vital processes (e.g., migration) may occur at other ages. The line L of Figure 1 is an example of a life line.

The complications referred to above arise from the following. Suppose we are interested in describing the mortality of the population in age x during year t. We have three options. (1) We may take those who were in age $\in [x, x+1)$ at exact time t, and observe their mortality experience during year t. The life lines of these individuals touch or cross the line AD and the deaths among them occur in the parallelogram ACED. The problem is that these individuals have their $(x+1)^{\text{st}}$ birthday during the year, so the deaths occur to both x and $x+1$ year-olds. (2) We may take those whose x^{th} birthday occurs during year t. Their life lines cross the line AB and their deaths occur in the parallelogram ABFC. The obvious problem is that the deaths occur in part during year $t+1$. (3) We may consider those who are present in the population in age x during any part of the year t. Their life lines cross either AB or AD, and the deaths are recorded in the rectangle ABCD. One problem

[6] Wilhelm Lexis (1837–1914) was a German statistician and economist who was among the first users of the diagram in Lexis (1875). Others (e.g., Gustav Zeuner, Karl Becker) had used similar graphics in the 1870's also.

in this approach is that it mixes deaths from two *birth cohorts*: life lines crossing AD belong to those born during calendar year $t - x - 1$; life lines crossing AB belong to those born during year $t - x$. Also, unlike the other two approaches, it is less directly applicable to forecasting because forecasts are typically formulated in terms of cohorts.

Many countries routinely compile their vital statistics based on the rectangles. They give rise to *period measures* (i.e., measures relating to a particular observation period such as a calendar year) of life expectancy, for example. Since such calculations combine data concerning different cohorts (mortality experience of the $x + 1$ years olds is recorded from the rectangle above DC, for example), one often thinks of them as referring to *synthetic cohorts*, whose experience corresponds to those alive during any part of the year t.

A more refined analysis is feasible if continuous-time data are available. Consider the lifeline L' of Figure 1. Suppose it refers to a woman, whose marriage is marked by '+', whose first and second children were born at mark '∘'. The analysis of the "waiting times" between the marks is called *event history analysis*. Statistical techniques for such analyses will be discussed in Chapters 4 and 5.

In general, the follow-up of cohorts requires that events are classified by the year of occurrence, age, and birth year. With the *triple classification of vital events*, the events of interest can be divided into the *triangles* of Figure 1, so any of the above approaches could be implemented. In modern computerized registration systems triple classification poses no particular problems. However, one should note that in all countries of the world demographic statistics have earlier been based on separate tabulations that have been extracted from the primary source materials by hand. In many countries they still are. In non-automated tabulations the requirement of triple classification is an additional burden. Consequently, one cannot expect long time-series based on triple classification in any country in the world.

There is an even more fundamental problem in some demographic and related statistics. Above, we have taken for granted that the events are classified by the *year of occurrence*. However, sometimes events are tabulated by the *year of reporting*. This seemingly illogical practice may sometimes be followed because it is desired to published statistics in a timely fashion. One can argue that if the number of missed reports during year t equals the number of those reports that actually relate to events from earlier years, but come in during t, then no error occurs. This argument is misleading, however, since much of the interest in official statistics is in changes of trends, and the trends will be distorted if tabulations are made by the year of reporting.

The timeliness requirement does produce a problem for all statistics, even those based on the most modern computer systems. For example, it is typical that information about deaths occurring abroad come into the registration system months, or years, after the event. For this reason, statistical agencies establish rules as to how long they wait for reports of the events. Statistics compiled in this manner may sometimes have to be revised, if the missing events are numerically important. The historical Finnish parish registers discussed above are a case in point.

One should also note that there are events of demographic interest for which the time of occurrence is not easily observable. For example, the onset time of many cancers, or that of HIV infection, is not directly observable, and the presence of a disease may only become known when the disease has progressed sufficiently. In other cases, such as noise-induced hearing loss, the impairment may progress gradually, and no clear-cut definition is feasible. In such cases the reporting of the events may depend crucially on the severity of the symptoms and the efficiency of medical screening. In these cases there may not exist any estimates of actual onset times, and tabulation by year of reporting is the only practical possibility. Nevertheless, we caution that the statistics thus obtained may misrepresent actual trends.

5. Register Data and Epidemiologic Studies

5.1. Event Histories from Registers

Much of demography deals with data classified by age group, time period etc. With modern computing power, the analysis of data sets consisting of individual level data has become feasible. Computerized population registers contain life histories of all individuals in a population (cf., Harala and Tammilehto-Luode 1999). These have been supplemented by information from other registers, or from censuses, to analyze mortality, for example (Valkonen and Martelin 1999). Census data are entered into databases, and historical parish records have been available in computerized form (e.g., the Umeå Demographic Database at http://www.ddbumu.se, or the Scanian Demographic Database at http://ddss.nu/Ldd/fortext.htm, both in Sweden). Social security systems or insurance companies often have highly detailed work histories that are continually updated.

In addition to the administrative data sources mentioned above, computerized data bases have been created for specific research tasks. For example, cancer incidence data are available in many countries from specific cancer registries (e.g., Teppo and Hakulinen 1999). Some countries, such as Finland, maintain a large number of other special purpose databases on births, congenital malformations, occupational diseases, causes of death, abortions, sterilizations, implants, visual impairments, intellectual disabilities, diabetes, infectious diseases etc. (Gissler 1999)

The strength of the continuously operating administrative and special purpose registers is their ability, in principle, to provide information on trends. However, their usefulness may be limited by narrow data content and their information may be biased for specific research uses because they cover only certain groups of persons.

5.2. Cohort and Case-Control Studies

Complementing census or register based information, we have increasingly available databases from large epidemiological studies and from social surveys. These databases have the advantage that they have been created with specific research hypotheses in mind, so, in general, they can be expected to provide superior data sources for certain kinds of causal research.

In Section 4, we used "cohort" to refer to those born in a given year. More generally, a *cohort* consists of those individuals that have experienced a given event at the same time. Strictly speaking, one can then think of a cohort as a closed population. In practice, the term is often used in a way that allows for the possibility that a cohort is depleted by deaths. Or, a cohort can be open with respect to deaths.

In addition to birth cohorts, those entering college during a given semester form a cohort, women who have given birth on the same day form a cohort, etc. In response to the increased public interest in effects of environment and individuals' behavior on health, governments have funded increasingly many follow-up studies to try to unravel the causal chains involved. As a result, there is an increasing number of high quality data sets containing individual-level information on cohorts.

An alternative, *case-control* (or *case-referent*) study design in epidemiology tries to assess relative risk by comparing those who have fallen ill ("cases") to those who could have fallen ill, but have not ("controls" or "referents"). Case-control data typically are collected from an open population by sampling, so its study design is quite different from that of a cohort study.[7]

Both designs are much used in epidemiology, and they are both well-suited to demographic studies. We briefly introduce their basic logic and point out some possible pitfalls. For a more detailed discussion, Breslow and Day (1980, 1987), Kleinbaum, Kupper and Morgenstern (1982), Woodward (1999) or dos Santos Silva (1999) may be consulted.

5.3. Advantages and Disadvantages

A cohort study is based on the idea that one follows a cohort over time, records the exposures or the occurrence of other potential causal agents, and estimates the extent to which the subsequent illnesses among the members of the cohort vary by exposure history. Since specific illnesses typically are rare and may have a long latency time, cohort studies can be both costly and time consuming.

Example 5.1. British Doctors' Study. In the famous *British Doctors' Study* (Doll and Peto 1976) the primary objective was to study the lung cancer risk caused by smoking. In October 1951, all men and women in the British Medical Register who were believed to be resident in the U.K. were sent a questionnaire. The first analyses related to the men only. A total of 34,440 men (or 69% of the men alive at the time) gave their name, address, age, and sufficient information about their smoking habits to be included in the study. Follow-up started in November 1, 1951, and

---

[7] Increasingly, case-control studies are conducted within cohorts, i.e., both cases and controls are restricted to members of a predefined cohort. The cohort is followed and controls are selected over time as cases appear. These hybrid designs are called *nested case-control, case-cohort,* or *case-base designs* (cf., Prentice, Self and Mason 1986; Flanders, Dersimonian and Rhodes 1990).

continued until October 31, 1971. Repeat questionnaires were sent in 1957, 1966, and 1972 to collect current information on smoking. The numbers of respondents (as proportion of those alive in parenthesis) were 31,318 (98.4%), 26,163 (96.4%), and 23,299 (97.9%), respectively. A total of 10,072 deaths were observed during the follow-up, with 441 caused by lung cancer. In addition, much information was obtained concerning other cancers, cardio-vascular diseases and other diseases. Among the results, one may note that the age-standardized death rate (Section 3.3 of Chapter 5) due to lung cancer was 0.1 per 1,000 person years among the non-smokers and 1.4 among the cigarette smokers – the relative risk of the smokers is about 14-fold. Among the latter, the risk increased from 0.78 for those smoking 1–14 cigarettes/day, to 1.27 for those smoking 15–24 cigarettes/day, to 2.51 for those smoking over 25 cigarettes/day. The evidence on increasing dose-response was clear. ◊

A case-control study is based on the idea that if we find a group of people with a specific illness, and select a group of those who could have the illness (i.e., are at risk) but do not have the illness, then any differences in the earlier exposures of the two groups may be causally related to the illness. The difficulty in carrying out the study centers on the investigator's ability to find controls that can be validly compared to the cases (Feinstein 1985). No exact rules are available, but if one can identify the population out of which the cases arose, then a random sample of the same population are eligible for being controls. (For a lively debate on the matter, see the 1985 contributions of O. Miettinen, J. Schlesselman, A. Feinstein and O. Axelsson in *Journal of Chronic Disease* 38, 543–558.)

Example 5.2. Doll and Hill Study. Prior to the British Doctors' Study, Doll and Hill (1950) had used the case-control design to investigate the role of smoking and atmospheric pollution as risk factors for lung cancer. The study was planned in 1947. Twenty London hospitals were asked to notify the investigators of all carcinomas of the lung, stomach, colon, or rectum. The latter three cancers were investigated to provide a possible contrast to lung cancer. Although complete notification was not achieved, the authors believe that omissions could not bias the inquiry by being a select group, since the hospitals did not know the detailed hypotheses being studied. Between April 1948 and October 1949 a total of 2,370 cancers were reported. It had been decided beforehand that patients 75 years of age and older would not be admitted, so 150 cases were excluded from the study. In 80 cases the cancer diagnosis was found to be erroneous, so 2,140 patients were left. Of these, 408 could not be interviewed due to early discharge (189), being too ill (116), death (67), deafness (24), being unable to speak English clearly (11). One case was excluded due to "wholly unreliable" replies. Thus, 1,732 cancer cases remained. Of these, 709 were lung cancer cases. Despite the exclusions, the authors claimed that the cases were "a representative sample of the lung-carcinoma patients attending selected London hospitals". As controls for the lung cancer cases, the investigators chose 709 patients at the same hospitals who had come there for some other illness. For each case, the control had to be of the same sex, within the same 5-year age-group, and have come to the same hospital at about the same time.

In other words, the controls were individually *matched* to the cases. Somewhat more of the cases turned out to live outside London than of the controls, but again the authors believe that this can hardly influence the results. As one indication of the excess risk they mention that the odds of never smoking were 2:647 among the male lung carcinoma patients, whereas the odds were 27:622 among the male controls. Alternatively, one could say that the odds of cancer were 2:27 among the non-smokers and 647:622 among the smokers. (I.e., there were 29 non-smokers in the data set with 2 lung cancers, and 1,269 smokers with 647 lung cancers.) The resulting odds ratio for cancer is 647:622/2:27 = 14 indicating a similar relative risk as the one later found in the British Doctors' Study. (This analysis does not allow for the matching that was used in the study, however, and the analysis would now be done in a different way, see Example 7.5 of Chapter 5). ◊

Examples 5.1 and 5.2 suggest the following, simplified characterization of the merits of the two approaches. The cohort study is often relatively *slow and costly*, especially if the illness is rare and the latency time of the illness is long, but the *results are more trustworthy*. The case-control study typically is *quicker and less expensive* but it may be *less reliable* if the choice of controls is biased in some way. We will come back to this issue in Section 2.3 of Chapter 5. Moreover, when cohort studies are carried out prospectively, the exposures and illnesses both occur after the study has been initiated.[8] In contrast, often a case-control study is retrospective, so that information on exposures must be obtained from remaining records, or it must be remembered by the subjects or by other people who have known them.[9] Therefore, the exposure information is typically weaker, and possibly biased, and imperfect controls may also cause bias.

However, the potential gains in efficiency are often seen to outweigh the risk of bias, and the case-control design has become a standard tool of epidemiologic investigation. With this background it is surprising that in demography, most investigations with causal goals have cohort designs.

A very large number of demographic studies are cross-sectional, so they follow neither paradigm. Since the time element is missing from those designs, they often lack credibility for causal inferences.

5.4. *Confounding*

A defining feature of experimental research is that the researcher can manipulate and control the causal factors of interest. For example, in a study of drug efficiency, groups with precise dosage are formed and subjects are randomized into them. In many epidemiologic studies, such as those discussed in Examples 5.1 and 5.2, ethical considerations prohibit manipulation of exposures. Similarly, in most demographic studies (e.g., when investigating the determinants of fertility) the

[8] Logically, a retrospective cohort study is also a possibility. In this case one defines a historical cohort and collects information on it from existing records.

[9] In nested case-control studies data collection is usually prospective.

researcher has no choice but to observe what happens, and to try to make comparisons in as valid a manner as possible. We call such studies *observational*. The validity of an observational study with causal aims can sometimes be compromised by unobserved interdependencies of the variables being studied.

Two variables are said to be *confounded* in a study if their separate effects cannot be distinguished from each other (Moses 1986, 9–10).[10] If one variable has negligible effects then the possible confounding may not be important (cf., Bailey 1982). There are also a multitude of other ways in which a comparative study may fail. Yet, possible confounding is often a major concern.

Confounding may be present in an observational study when those subjects who receive a treatment differ systematically from those who do not. For example, when the large-scale randomized (and double-blind placebo-controlled) Salk vaccine trials were conducted, an observational study was also done to compare (i) polio incidence rates for second-grade students who were vaccinated and whose parents gave permission for vaccination with (ii) the rates for first-grade and third-grade students in the same schools. Comparison with a randomized controlled experiment showed the risk of contracting polio was confounded with parental permission – higher income children more readily received permission but had lower immunity from the disease.

Confounding may also be present even in a randomized controlled experiment when subjects leave the study or otherwise do not follow protocol for reasons related to the assignment of the treatment. For example, subjects assigned a placebo or a treatment may perceive it as inferior and leave the study to pursue other treatment.

For an illustration, consider the artificial data of Figure 2. The aim of the study is to understand what might explain variations in Y. Two groups are involved: there are 24 individuals marked with a '+' and 36 individuals marked with a '∘', and there is one continuous explanatory variable X. Define $G = 1$, for the individuals of type '+', and let $G = 0$ otherwise. The data are well described by the estimated regression equation

$$Y_i = 1.47 + 6.65G_i + 0.915X_i + e_i, \tag{5.1}$$

where the estimated residuals $e_i, i = 1, \ldots, 50$, have the variance 2.19^2. The coefficient of G has a t-value $= 10.27$ and the coefficient of X has a t-value $= 5.90$. With P-values < 0.001, both effects appear highly significantly different from 0.

Suppose now that an investigator has no knowledge of the two types of individuals, and fits a simple linear regression with X alone as an explanatory variable. The estimated equation is

$$Y_i = 9.94 + 0.192X_i + e_i, \tag{5.2}$$

---

[10] This is a rather general characterization. In particular, it does not include specific assumptions concerning the causal roles of the variables. For a review of the many complexities that arise when the concept is operationalized in an epidemiologic context, see Geng, Guo and Fung (2002).

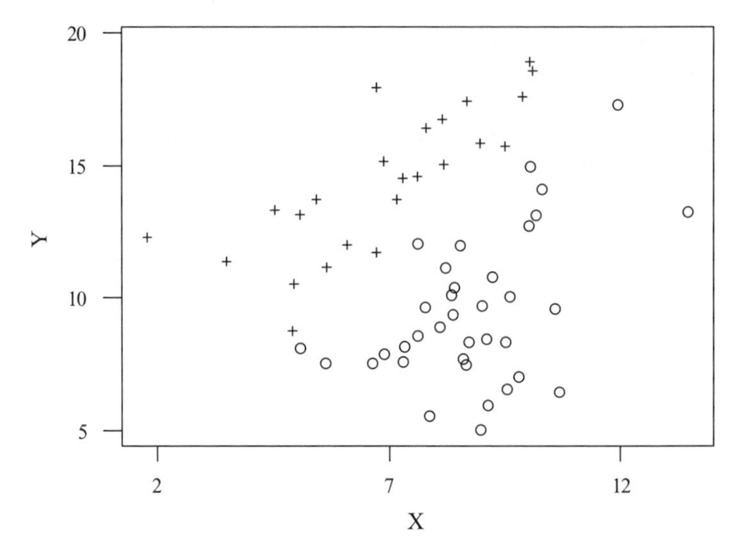

FIGURE 2. Example of Confounding.

where the residual variance is 3.67^2. The coefficient of X has a t-value $= 0.83$ and a P-value $= 0.41$, suggesting that X has no influence on Y. The estimated effect of X is tangled up with the unmeasured group indicator, and the conclusion of the study is incorrect.

Note that had the researcher restricted his or her study to those of type '+' only, and regressed Y on X, the estimated slope would have been 0.83 with a P-value of 0.003, so the correct conclusion would have been reached. The same is true if only those of type '∘' had been studied (resulting in the estimated slope $= 0.99$, and P-value < 0.001). This suggests that restricting the scope of the study by controlling a variable is one way to avoid confounding.

On the other hand, suppose the investigator was interested in comparing the two groups, and did not measure X. Using a two-sample t-test, he or she would have found that a 95% confidence interval for the mean of those of type '+' minus the mean of those of type '∘' is (3.47, 6.37). The conclusion that those with a '+' have a higher mean would have been correct, but the difference would have been underestimated by approximately a half due to the confounding of G and X.

Both cohort and case-control designs often give rise to contingency tables whose analysis can be invalid, if confounding is present. In complements we indicate some classical procedures for handling suspected confounding via stratified analysis. In Chapter 5 we show how regression techniques can be used to do the same.

6. Sampling in Censuses and Dual System Estimation

If it were not for the need of geographic detail (for municipalities, city neighborhoods or blocks, etc.), sample surveys would probably have replaced censuses a long time ago. Samples would be less expensive to carry out and they reduce the burden of respondents because only a fraction is included. More extensive information can be collected by well-trained personnel in a sample survey than in a census that has to rely on temporary work force. In addition, being based on deliberate randomization, the precision of statistical sampling can be assessed based on the sample itself (Chapter 3), whereas errors in a census cannot be evaluated based on the census only. These advantages have been used to complement census information in various ways.[11]

Sampling has been used in the U.S. decennial censuses since 1940 to collect part of the information. The so-called long form requesting detailed data on income and other characteristics is given to approximately 10% of the respondents, the fraction being larger in smaller areas and smaller in larger areas. Major savings in response burden are achieved by this without unduly compromising data quality.

Sampling has also been used in the United States to evaluate the accuracy of the decennial censuses. The "demographic analysis" estimates of Example 2.2 are essentially based on consistency checks between the current census, earlier censuses, and the recorded vital events. A problem in such estimates is that they rely on the assumption that such other pieces of earlier information are trustworthy, an uncertain proposition at best, and they depend on consistency in definitions (e.g., racial classification) among the various data sources.

A direct statistical evaluation of the census can be made by redoing the census on a sample basis in different parts of the country. Suppose the unknown population of an area is N, with n_1 individuals counted in the census. Suppose the second census count is n_2, and one can verify that m individuals were counted in both censuses. A more refined analysis will be given in Section 5 of Chapter 5, but let us condition here on n_1 and n_2. Assume that the two counts are independent, and that individuals are equally likely to be counted during either occasion. The probability of counting m individuals in both censuses is equal to the number of ways of choosing m from the n_1 in the first census times the number of ways of choosing $n_2 - m$ from the $N - n_1$ not counted in the first census, divided by the number of ways of choosing n_2 from N. The resulting probability of observing m can be written as $P(m; n_1, N - n_1, n_2)$, when we first define

$$P(x; \alpha, \beta, \gamma) \equiv \binom{\alpha}{x}\binom{\beta}{\gamma - x} \Big/ \binom{\alpha+\beta}{\gamma}. \tag{6.1}$$

Here $\max\{0, \gamma - \beta\} \leq x \leq \min\{\alpha, \gamma\}$ and $P(x; \alpha, \beta, \gamma) = 0$ otherwise (Exercise 8). This probability distribution is called the *hypergeometric distribution* (DeGroot 1987, 247–250) and we can use it to calculate the probability of observing m when we know N (and n_1 and n_2). In the census context, we observe values of n_1, n_2, and m but we do not know N. One way to formulate a guess

[11] The existence of censuses is very important for many sample surveys, because the census can provide a frame or list from which a probability sample can be drawn. The census can also provide information adjusting a sample or calibrating estimates based on the sample to agree with observations on the whole population. We will not pursue these aspects, however.

(or estimate) of N is to choose the value that makes the observed data as likely as possible. We view (6.1) as a function of N (a likelihood function) and choose the value of N that maximizes (6.1) (cf., Feller, 1968, 45–46). The maximizing N is the maximum likelihood estimator. Here, the MLE is essentially $\hat{N} = n_1 n_2 / m$ (Exercises 9, 10).

Example 6.1. Underreporting of Occupational Diseases. The Finnish Register of Occupational Diseases obtains its information from two sources. A suspected case of occupational disease must be reported by the examining physician to authorities (first capture). The case must also be reported to the insurance institution responsible for compensation (second capture). The following data were obtained in 1980: $n_1 = 3{,}769$, $n_2 = 3{,}053$, and $m = 1{,}591$. The total number of cases reported was $M = 3{,}769 + 3{,}053 - 1{,}591 = 5{,}231$. In this case $\hat{N} = 3{,}769 \times 3{,}053 / 1{,}591 = 7{,}232$, or the ratio between the estimated cases to the reported cases would appear to be $c \equiv \hat{N}/M = 1.38$. However, it was suspected that the likelihood of reporting would depend on the diagnosis. The main diagnostic groups were (a) noise-induced hearing loss with $M = 1{,}856$ and $c = 1.20$, (b) diseases caused by repetitive or monotonous work with $M = 1{,}448$ and $c = 2.47$, (c) skin diseases with $M = 1{,}171$ and $c = 1.23$, (d) other diseases with $M = 756$ and $c = 1.34$. Adding the disease specific estimates leads to an overall estimate of 8,258 cases in 1980. The fact that diseases in category (b) are poorly reported is understandable, because the connection between working conditions and the disease is particularly hard to establish for them. ◊

Some populations are especially hard to estimate, because their membership criterion involves illegal activities. Drug use is an example in which users are expected to be reluctant to reveal their user status (cf., Turner, Lessler and Gfroerer 1992). Yet, a drug user may end up being registered in several administrative registers. This provides a basis for population estimation.

Example 6.2. Numbers of Drug Users. In Finland, information about heavy drug use is available through several registers. The most important ones are the Hospital Discharge Register and the Criminal Report Register. In 2001 there were $n_1 = 446$ reports from the former, $n_2 = 825$ reports from the latter, and $m = 53$ reports from both registers, for heavy drug use in the Helsinki Region (Helsinki, Espoo, Vantaa, Kauniainen). This yields the estimate $\hat{N} = 446 \times 825/53 = 6{,}942$. We will come back to this in Example 3.7 of Chapter 5. ◊

A form of this *capture-recapture* method was used by Sir Francis Bacon in the study of wildlife populations around 1650 (Cormack 1968). Laplace applied it to human populations in the 1780's. The method has been reinvented many times, whence the names "Petersen's method" or "Lincoln index" in ecology. Its modern use in demography is usually accredited to Chandra Sekar and Deming (1949). In demography it is often called *dual systems estimation (DSE)* (Marks, Seltzer, and Krotki 1974).

Simple as $\hat{N} = n_1 n_2 / m$ may seem, in practice the application of dual systems estimation to the study of the census is complicated by several factors. First, the population may be heterogeneous with respect to the probability of being captured. If the heterogeneity is observable, it can be modeled by stratification (Chandra Sekar and Deming 1949) as we did in Example 6.1 or by logistic regression (Huggins 1989, Alho 1990b). Second, error in n_1, n_2, and m may arise from data errors (names, addresses etc.) that should be corrected. Third, actual human populations are typically open, so the *de facto* population of an area may not be the same during the two counts (cf., Alho et al. 1993). Nevertheless, the dual systems approach provides a practical way to analyze the coverage of a census (cf., Mulry and Spencer 1993; Kostanich 2003a,b; U.S. Census Bureau 2004). A more detailed discussion of population heterogeneity will be taken up in Section 5 of Chapter 5, and Chapter 10 presents an overview of the whole problem of census evaluation using dual systems techniques in the U.S. context.

Exercises and Complements (*)

1. Consider (a) your own country, (b) the city you live in. Which is bigger, the *de jure* or the *de facto* population?
2. Digit preference has been quantified in demography using statistics that are based on comparing the size of the enumerated population to the population one would expect to see in the absence of digit preference. Define V_x = enumerated population in age x. *Whipple's index* (for digit preference of ages 25, 30, ..., 60) is defined as,

$$\sum_{y=1}^{8} V_{20+5y} \Big/ \frac{1}{5} \sum_{x=23}^{62} V_x.$$

   This is of the observed/expected form if in reality all V_x's are equal. Give some more general conditions, under which this index still works. (Hint: Consider 5-year intervals [23, 27], [28, 32], ... and assume that V_x is (a) linear in each interval, (b) an odd function around the center of the interval: $V_{25-x} - V_{25} = -(V_{25+x} - V_{25})$ for $x = 1, 2$, etc.) For more information about quantifying digit preference, see Shryock and Siegel (1976, 116–118).
3. Consider Example 5.2, where an odds ratio for disease (among smokers and non-smokers has been calculated as 647:622/2:27. (a) Show that the odds ratio for smoking (among those diseased and non-diseased) has the same value. Therefore, the value of the odds-ratio does not depend on whether the data come from a case-control, or a cohort study. (b) Given that the data come from a case-control study, can one say that the risk of cancer is 2/29 for the non-smokers and 647/1269 among the smokers?

*4. Suppose the results of either a cohort or a case-control study are presented as a 2 × 2 table,

	Ill	Not	Total
Exposed	a	b	n_1
Not	c	d	n_2
Total	m_1	m_2	N

Here $N = n_1 + n_2 = m_1 + m_2$ is the total number of subjects. The odds ratio is estimated as $OR = ad/(bc)$ under both study designs. Condition on all the margins n_1, n_2, m_1, m_2. Then, any one element of the matrix defines the others. Denote the upper left hand corner of the matrix by A and its value in a particular experiment by a. Under the null hypothesis that the true odds ratio is $= 1$, the probability of having a exposed who are ill is $P(a; n_1, n_2, m_1)$ as defined in (6.1). Thus, $E[A] = m_1 n_1/N$ and $\mathrm{Var}(A) = m_1(n_1/N)(n_2/N)(N - m_1)/(N-1)$ (e.g., DeGroot 1987, 247–250). As discussed by Feller (1968, 194) the variable $X = (A - E[A])/\mathrm{Var}(A)^{1/2} \sim N(0, 1)$ asymptotically, so $X^2 \sim \chi^2$ distribution with one degree of freedom. Thus, the null hypothesis is rejected at risk level α, if $X^2 \geq k_{1-\alpha}$, where $k_{1-\alpha}$ is the $1-\alpha$ fractile of the χ^2 distribution. Show that the observed value of the test statistic can be written as

$$X^2 = (N-1)\frac{(ad - bc)^2}{n_1 n_2 m_1 m_2}.$$

*5. Continuation. When one wants to control for the values of a potentially confounding third variable with values, say, $k = 1, \ldots, K$, then we have K independent strata with

	Ill	Not	Total
Exposed	a_k	b_k	n_{1k}
Not	c_k	d_k	n_{2k}
Total	m_{1k}	m_{2k}	N_k

Denote the true odds ratio in stratum k by θ_k. Consider the situation in which $\theta_k \equiv \theta$ for $k = 1, \ldots, K$. Now test $H_0\colon \theta = 1$ against $H_A\colon \theta \neq 1$. The famous *Cochran-Mantel-Haenszel statistic* for this hypothesis is

$$X^2 = \left\{\sum_{k=1}^{K}(A_k - E[A_k])\right\}^2 \Big/ \sum_{k=1}^{K}\mathrm{Var}(A_k),$$

where the expectation and variance are calculated as in Complement 4 for each table $k = 1, \ldots, K$ (Cochran 1954, Mantel and Haenszel 1959). The remarkable fact is that asymptotically $X^2 \sim \chi^2$ distribution with *one degree of freedom* even if the strata are very small (e.g., $N_k = 2$), as long as K is large. (For large strata the result is obvious.) Show that the observed value of the test statistic can be written as

$$X^2 = \left\{\sum_{k=1}^{K}\left(\frac{a_k d_k - b_k c_k}{N_k}\right)\right\}^2 \Big/ \sum_{k=1}^{K}\left(\frac{n_1 n_2 m_1 m_2}{N_k^2(N_k - 1)}\right).$$

*6. Continuation. In the setting of Complement 5, the so-called *Mantel-Haenszel estimator of the common odds ratio* is defined (Mantel and Haenszel 1959) as

$$\hat{\theta} = \sum_{k=1}^{K}\left(\frac{a_k\, d_k}{N_k}\right) \Big/ \sum_{k=1}^{K}\left(\frac{b_k\, c_k}{N_k}\right).$$

Show that if $b_k c_k > 0$ for all $k = 1, \ldots, K$, then we can write

$$\hat{\theta} = \sum_{k=1}^{K} w_k \hat{\theta}_k,$$

where $\hat{\theta}_k = a_k d_k/(b_k c_k)$, and $w_k = (b_k c_k/N_k)/\sum_j b_j c_j/N_j$.

*7 Continuation. In a matched case-control study in which one case is matched with one control, each pair forms a stratum $k = 1, \ldots, K$ because the matching criteria may correspond to possible confounders. The results of such a study are often represented as 2×2 table as follows:

		Control	
		Exposed	Not
Case	Exposed	a	b
	Not	c	d

This table is a sum of the K stratum specific tables of the type considered in Complement 5. In this case $N_k = 2$ for all $k = 1, \ldots, K$ because there is one case and one control in each stratum. There are $N = 2K$ individuals in all. There are four types of tables: (i) a tables with both the case and the control exposed, (ii) b tables with the case exposed but the control is not, (iii) c tables with the case not exposed but the control is, (iv) d tables with neither the case nor the control exposed. In case (i), for example, the table is of the form

	Ill	Not	Total
Exposed	1	1	2
Not	0	0	0
Total	1	1	2

(a) Verify that in cases (i) and (iv) we have $a_k d_k = b_k c_k = 0$, in case (ii) $a_k d_k = 1$, $b_k c_k = 0$, and in case (iii) $a_k d_k = 0$, $b_k c_k = 1$. (b) Show that the Cochran-Mantel-Haenszel test statistic is then of the form $X^2 = (b - c)^2/(b + c)$. This is also known as the *McNemar test statistic*. (c) Show that the Mantel-Haenszel estimator of the common odds ratio is $\hat{\theta} = b/c$. Thus, in both statistics only the "discordant pairs" matter.

8. Consider the capture-recapture case in which n_1 is the number of first captures, n_2 recaptures, and m is the number caught both times. (The traditional notation used in capture-recapture literature does not follow the usual conventions of statistics; note that these symbols have here a meaning different from the one in the previous examples!) Show that the labeling of the censuses as first or second in Section 6 does not matter, so that $P(m; n_1, N - n_1, n_2) = P(m; n_2, N - n_2, n_1)$, as defined in (6.1).
9. Show that by equating m to its expected value (that is given in Complement 4) one obtains the classical estimator, $\hat{N} = n_1 n_2/m$.
10. Show that the MLE based on (6.1) is essentially the same as $\hat{N}$ defined above. (Hint: show that $P(m; n_1, N - n_1, n_2)/P(m; n_1, N - 1 - n_1, n_2) = (N - n_1)(N - n_2)/(N - n_1 - n_2 + m)$, which is increasing when $n_1 n_2/m > N$, so

that (6.1) is increasing for $N < n_1 n_2/m$ and decreasing for $N > n_1 n_2/m$. Conclude that the exact MLE is $= \lfloor n_1 n_2/m \rfloor$, where $\lfloor x \rfloor$ is the largest integer $\leq x$.)

11. To estimate Var($\hat{N}$) under the hypergeometric model in which n_1 and n_2 are fixed, note first that $E[m] = n_1 n_2/N$ and $\text{Var}(m) = n_2(n_1/N)((N - n_1)/N)(N - n_2)/(N - 1)$. Since $\hat{N}$ is a nonlinear function of m we linearize the statistic at $E[m]$ using a Taylor series, or $\hat{N} \approx n_1 n_2/E[m] - (n_1 n_2/E[m]^2)(m - E[m])$. This yields the approximate variance, $\text{Var}(\hat{N}) \approx (n_1 n_2/E[m]^2)^2 \text{Var}(m)$. Assume that N is large enough so that $N - 1$ can be replaced by N in Var(m). Show that by plugging in the estimator $\hat{N}$ the approximate *variance of the capture-recapture estimator* can be estimated by $n_1 n_2 u_1 u_2/m^3$, where $u_j = n_j - m$, $j = 1, 2$. This is an example of the so-called delta method that will be discussed in more detail in Section 7.2 of Chapter 3.

3
Sampling Designs and Inference

Cohort and case-control studies are usually restricted to a carefully selected subset of the total population, because the possibility of confounding is an overriding concern. For example, in cohort studies of carcinogenicity one tries to find groups that differ from each other as much as possible in terms of the exposures of interest but that are otherwise similar. There is no attempt to cover the population at large, the assumption being that the causal effect found in the groups under study will be similar for persons outside the groups. Even with that assumption, the complementary task of assessing the risk caused by the exposures at population level requires a "representative sample" from which to estimate the actual pattern of exposures. The concept of representative sampling is more slippery than might first appear (Kruskal and Mosteller 1979a–c, 1980), but will be explicitly defined below. For most studies, we hope to generalize either to the population from which the sample members (or study subjects) were selected or even more generally to a larger population, sometimes called a "superpopulation". Much of the data used for social, economic, demographic, or epidemiologic analyses comes from samples.

Although sampling theory is not always viewed as part of demography, we present selected aspects of the theory here because it plays a central role in the production of some basic population data. For example, the Current Population Survey is a stratified multi-stage survey of U.S. households that provides important data on economic and social activities. As another example, U.S. Post Enumeration Surveys (PES's) are conducted after the decennial censuses to assess their accuracy. Poststratification plays an important role in their analysis. In the 1970s and early 1980s, the World Fertility Survey was carried out in 41 nations in Africa, the Americas, Asia, and Europe. Our goal is to give enough details of the theory so that the reader can appreciate the complexities of the relevant large scale surveys and make inferences appropriately from the survey data. In particular, Section 7 discusses principles of statistical inference in a sampling context.

A *sampling design* (or *sampling procedure* or *selection procedure*) is a rule for choosing a single sample from the set of possible samples. An individual element is selected if the chosen sample contains the element. If the rule assigns probabilities to the possible samples such that each element in the population has a non-zero probability of being selected, we say the sample resulting from the rule is a *random sample* or a *probability sample*.

Samples in which nature provides the randomization do not necessarily satisfy the definition of random sampling. *A fortiori*, this holds for *purposive samples* in which the researcher handpicks "representative elements" (cf., Cochran 1977, 10-11), and for *self-selected samples*, such as the popular internet surveys in which any individual with access to internet may have his or her view about a particular issue recorded. Although inferences can be made from nonrandom samples, the strength of the inferences can be assessed internally – from the sample itself – only if additional assumptions are invoked; see Smith (1983). In contrast, if each element in the sample has a positive selection probability and the selection probability is known for each element in the sample, then an unbiased estimator of the population total is available (Section 4.2) – such samples will be called *representative samples*. Moreover, if the inclusion probability of every pair of elements is known for every sampled pair and is positive for every pair in the population, the standard error of the total can be estimated from the sample (Section 5.3) and the sample is called a *measurable sample*.

We take the view that in analyzing data from a sample one should generally acknowledge the method used to select the sample. Point estimates may be adjusted for probabilities of selection, and variance estimates should account both for unequal probabilities and for dependencies in sample selection. Exceptions to this rule may be made for certain analyses of well-specified models (Sections 4.4, 7.3) and for analyses in which one is willing to accept bias as a compensation for reduced variance (Section 4.4). We review some major types of sampling designs underlying demographic data and discuss how the designs affect analyses of the data. Basic references include Cochran (1977), Lohr (1999), and Levy and Lemeshow (1999). More recent and very practical references are Korn and Graubard (1999) and Lehtonen and Pahkinen (2004). More advanced theoretical treatments include Särndal, Swensson, and Wretman (1992), Thompson (1997), Skinner, Holt, and Smith (1989), Chambers and Skinner (2003), and the classic Kish (1965), which provides much practical advice for large scale survey design. A concise and accessible overview is provided by Frankel (1983).

In past years, only a few specialized software packages were available for carrying out statistical analyses that took the sampling design into account. Currently a number of strong packages are available. Descriptions and links to reviews are available from the Survey Research Methods Section of the American Statistical Association[1].

1. Simple Random Sampling

The most elementary kind of sample selection is *simple random sampling* (*SRS*), in which each possible sample of *n* elements from a population of size *N* elements has an equal chance of selection. The *selection probability* for an element of the

[1] http://www.fas.harvard.edu/%7Estats/survey-soft/survey-soft.html

population is the probability that the element is contained in the sample. In simple random sampling, each individual has the same selection probability, which equals the *sampling fraction* $f = n/N$. In *without-replacement sampling*, no element is selected more than once, and in *with-replacement sampling* an element may be selected more than once (up to n times).

To select a SRS of n units from a population of size N we need a listing of the population units, called a *sampling frame*. Construction and maintenance of a sampling frame is an important practical matter (e.g., Kish 1965, 53–59), with attention required to ensure completeness and detect duplications and erroneous inclusions. A sample of the population can be based on random digit dialing, so the frame is implicitly formed by the list of all phone numbers. Multi-stage area samples can be based on maps and database listings of housing units. In both cases, the frame represents the ideal target population in an approximate sense only. Population counts can be used for controls for ratio estimates of totals (Sections 4.2, 5.4), and those counts may be based on censuses or on postcensal estimates (Chapter 10). In countries that have a population register, the register can be a flexible source of sampling frames for many uses. However, when the target population of the sample is defined by some social, economic or educational criteria that are only available for census years, the register becomes gradually outdated, as time from the census elapses. Errors caused by the mismatch of the frame and the ideal target population are typically not assessed in surveys. It would involve completely different methods - methods of the type that are used in statistical forecasting.

A way to think of drawing a simple random sample of size n is to take a list of the N elements in the population and randomly permute their order, and then to take the first n. Forming a random permutation requires care, however.

Example 1.1. The 1970 Draft Lottery in the U.S. During the U.S. participation in the Vietnam War, concerns about the unfairness of the military draft led to a decision to randomize the selections. A random permutation of birth dates in the year would be formed, and those young men who would end up first on the list would be chosen first, and so forth. In practice, capsules labeled with dates were put into a bowl to be chosen at random one at a time, so that the date on the ith selected capsule was assigned draft number i. The capsules were not well mixed in the bowl, however, which led to a significant negative correlation between birth date and draft number (Fienberg 1971). A recent analysis of deaths recorded on the Vietnam Memorial in Washington (Sommers 2003) found a similar negative association between death rate and draft number. An improved randomization method, relying on random number tables and physical randomization, was later used (Rosenblatt and Filliben 1971). $\diamond$

Consider using the sample to estimate the population mean for some numerical characteristic, or *variable*. We denote the population values by $y_1, \ldots, y_N$ and the sample values by $y_1, \ldots, y_n$. (Other symbols than y may be used as well.) Although y_i in the sample is not the same as y_i in the population, it will be clear from the context which is which. We will use upper-case letters to refer to population characteristics or summaries and lower-case for sample values of the variable. The population mean is denoted by $\bar{Y} = (y_1 + \cdots + y_N)/N$ and the sample mean is denoted by $\bar{y} = (y_1 + \cdots + y_n)/n$. The population total will be denoted by $T_Y = N\bar{Y}$. The "finite-population variance" S^2 and the sample variance s^2 are defined as

$$S^2 = \sum_{k=1}^{N}(y_k - \bar{Y})^2/(N-1), \quad s^2 = \sum_{k=1}^{n}(y_k - \bar{y})^2/(n-1). \tag{1.1}$$

Example 1.2. Child Stunting. Burgard (2002) uses household surveys of women in Brazil and South Africa to analyze child stunting (i.e., stunted or checked growth in children). For example, if y_i is the number of stunted-growth children of women in household i, the mean of the y_i's is then the average number of stunted growth children per household containing a woman. The population total of the y_i's is the number of stunted-growth children in households containing women. That total can be divided by the total number of children in households (say, from census records) to estimate the proportion of stunted growth children. ◊

Both $\bar{y}$ and s^2 are examples of *statistics* – functions of the data – with probability distributions that depend on the population and on the sample design used, which here is a SRS of size n from the population of size N. In later chapters we may view population characteristics themselves (e.g., vital rates) as random variables, as random even though there is no sampling from the population – the population itself is viewed as stochastic. In this chapter we are conditioning on the population at hand and regarding the data collection as a random process. We refer to the probability distribution for a statistic as its *sampling distribution.*

For without-replacement simple random sampling one can show (Exercise 1) that

$$E[\bar{y}] = \bar{Y}, \tag{1.2}$$

$$\mathrm{Var}(\bar{y}) = (1 - n/N)S^2/n, \tag{1.3}$$

$$E[s^2] = S^2. \tag{1.4}$$

In other words, the mean of the sampling distribution of $\bar{y}$ is $\bar{Y}$, the mean of the sampling distribution of s^2 is S^2, and the variance of the sampling distribution of $\bar{y}$ is $(1 - n/N)S^2/n$. The standard deviation of a statistic is called the *standard error* (SE), for a non-negative variable the ratio of the standard error to the mean (or to the population value being estimated, which may be different if the statistic is biased) is called the *coefficient of variation* (CV), and the square of the coefficient of variation is called the *relative variance.*

Thus, the sample mean $\bar{y}$ is an *unbiased estimator* of the population mean. Its variance is the product of three factors: the fraction not sampled, the heterogeneity in the population, and the reciprocal of the sample size. The first factor $1 - f$, called the *finite population correction*, explains why a large sampling fraction is not needed to obtain high precision. A large sampling fraction helps reduce variance, but a small sampling fraction does not hurt. Often the sampling fraction

f is small enough to ignore. Plugging in s^2 for S^2 in the formula in (1.3) yields an unbiased estimator of $\text{Var}(\bar{y})$

$$\hat{\text{V}}\text{ar}(\bar{y}) = (1 - f)s^2/n. \tag{1.5}$$

To estimate the population total T_Y we use $\hat{T}_Y = N\bar{y}$. In general the population total may be estimated by the sum of the sample values divided by their selection probabilities. As an illustration notice that $\hat{T}_Y = \breve{y}_1 + \cdots + \breve{y}_n$, with $\breve{y}_i = y_i/f$. The variance may be estimated by

$$\hat{\text{V}}\text{ar}(\hat{T}_Y) = (1 - f)\frac{n}{n-1}\sum_{k=1}^{n}(\breve{y}_k - \bar{\breve{y}})^2 \tag{1.6}$$

where $\bar{\breve{y}} = (\breve{y}_1 + \cdots + \breve{y}_n)/n$. Although we have emphasized the unbiasedness property, we do not regard exact unbiasedness as a critical property. Many useful statistics are not exactly unbiased. For example, although (1.4) holds, we have that $E[s] \neq S$. Yet, the bias in s does not affect the development of confidence intervals based on t distribution under the usual normal-theory assumptions. In many other cases, what is important is that the bias becomes small as the sample size increases, so that the bias is small relative to the standard error. For example, the coverage of 95% normal-theory two-sided confidence intervals for the mean is still close to 95% if the ratio of the absolute value of the bias of the estimate of the mean to standard error of the estimate of the mean is 0.1 or less (Cochran 1977, 14).

2. Subgroups and Ratios

The simplest important nonlinear statistic arises in estimating the population ratio R of the totals of two variables in a population, say $R = T_Y/T_X$. If measurements y_i and x_i are made for each element in a simple random sample of size n, we may estimate R by

$$\hat{R} = \sum_{i=1}^{n} y_i \Big/ \sum_{i=1}^{n} x_i. \tag{2.1}$$

This statistic is the ratio of two random variables. In Example 1.2, if we wanted to estimate the proportion of children who were stunted but did not know the total number of children, we could use (2.1) with y_i the number of stunted children in household i and x_i the total number of children in household i. The expected value of $\hat{R}$ is not exactly equal to R. The "ratio-estimator bias" does not arise from problems in the sample but rather from non-linearity of the ratio estimator. To analyze the mean and variance we will approximate $\hat{R}$ by a linear statistic. Define $\varepsilon_i = y_i - Rx_i$ and notice that

$$\hat{R} - R = (\bar{y} - R\bar{x})/\bar{x} \approx (\bar{y} - R\bar{x})/\bar{X} = \bar{\varepsilon}/\bar{X}, \tag{2.2}$$

if the sample size is large enough that $\bar{x}$ will be close to $\bar{X}$. The approximation will work well provided the CV of the denominator of (2.1) is small (Complement 3).

The right hand side of (2.2) is a linear function of the observations, and we use the mean and variance of the right hand side to approximate the mean and variance of the left hand side (Cochran 1977, David and Sukhatme 1974). Because the expectation of the right hand side of (2.2) is zero, we say that the ratio estimator is *approximately unbiased* or *asymptotically unbiased* (for large n). The variable ε_i is the residual of y_i from the line through the origin with slope R. We estimate it by $e_i = y_i - \hat{R}x_i$ and use

$$\hat{\text{V}}\text{ar}(\hat{R}) = \frac{1-f}{\bar{x}^2 n} s_e^2 \tag{2.3}$$

as an estimator of the variance. Here, s_e^2 is given by s^2 in (1.1) with e_i substituted for y_i. If the population mean $\bar{X}$ is known, it may be used in the denominator of (2.3), but whether the estimator of variance is improved depends on the relation of y and x in the population (Cochran 1977, 155–156; Rao and Rao 1971).

Ratio estimators are commonly used when estimating characteristics of a subgroup whose sample size is random. This occurs often in the context of small area estimation, or more generally in the estimation of small domains that can also be defined by criteria other than the geographic. For example, consider using a simple random sample of size n to estimate the mean and total of a variable for a subgroup G. Define the indicator variable $x_i = 1$ if element i belongs to the subgroup and $x_i = 0$ otherwise, $i = 1, \ldots, N$. Define y_i to equal the variable of interest if $x_i = 1$ and to equal 0 otherwise (or replace y_i by $x_i y_i$). The total for the subgroup is T_Y and the size of the subgroup is $N_G = T_X$. Therefore, the mean for the subgroup is equal to $R = T_Y/T_X$. If $n_G \equiv x_1 + \cdots + x_n > 0$, we can estimate R by $\hat{R}$ in (2.1), which equals the mean of interest in the sample. If we consider the conditional sampling distribution with samples of a fixed size $n_G > 0$ from the subgroup, $\hat{R}$ is an unbiased estimator of R with (conditional) variance $(1 - n_G/N_G)S_G^2/n_G$, which may be estimated by $(1 - n_G/N_G)s_G^2/n_G$, with

$$S_G^2 = \sum_{i=1}^{N} x_i(y_i - R)^2/(N_G - 1) \quad \text{and} \quad s_G^2 = \sum_{i=1}^{n} x_i(y_i - \hat{R})^2/(n_G - 1). \tag{2.4}$$

3. Stratified Sampling

3.1. Introduction

In stratified sampling, the population is divided into some number H of non-overlapping strata, with $N_h > 0$ units in stratum $h = 1, \ldots, H$. Note that $N = N_1 + \cdots + N_H$. Samples are taken *independently* from each of the strata. In fact, completely different sampling methods may be used in different strata. Stratified sampling is used for a variety of purposes, including (i) reducing sampling variance, (ii) ensuring that sample sizes from certain strata do not fall below thresholds, (iii) controlling cost.

In stratified simple random sampling, a SRS of size $n_h \geq 1$ is selected from stratum $h = 1, \ldots, H$. To estimate the overall population mean, form an estimate of the mean of each stratum and then take a weighted average of those stratum means, with weights proportional to N_h. This weighted estimate will not be the same as the unweighted mean unless the sampling fractions $f_h = n_h/N_h$ are all equal. The total sample size is $n = n_1 + \cdots + n_H$.

Example 3.1. NELS:88 Base-Year School Sample. The National Educational Longitudinal Study of 1988 (NELS:88) was a survey conducted to provide data on a cohort of students who were in eight grade in 1988. The purpose was to provide data to inform policy research on schooling and later behavior and choices by the students. The base-year sample was taken from schools in the U.S. enrolling eighth grade students in 1988 (Spencer et al. 1990). Subsamples of the students were surveyed in successive years in follow-up surveys, allowing for estimation of growth and change in student attributes (Example 5.2, below). A list of public and private schools was obtained and used for a sampling frame; the schools in the frame were believed to contain 99% of the eighth grade students. Strata were developed in two steps, in order to group schools that were relatively similar in terms of variables deemed relevant to the survey's objectives. Superstrata were formed by cross-classification of school type (for public, private religious, and other private schools) by geographic region (8 regions for public and 4 aggregate regions for other schools). Substrata were formed within each superstratum by urban/suburban/rural location of school and, for public schools only, cross-classified by percentage of students who were black or Hispanic. The schools were selected independently from the different superstrata with unequal probabilities set roughly proportional to the estimated size of the eighth grade class. Within each superstratum, schools were sorted by stratum and within stratum by size (estimated eighth grade enrollment) and selected with systematic sampling (Section 6). For public schools, a sample of 817 out of 22,818 in the frame were selected and participated, compared to 240 out of 16,048 nonpublic private schools; although the sampling rate for the public schools appears to be larger than for nonpublic schools, the public schools tended to be much larger than the nonpublic schools, and the size-weighted sampling fractions were much larger for nonpublic schools. The latter, especially "other private" schools, were oversampled – selected with greater (size-weighted) sampling fractions than schools as a whole – to provide sufficient sample sizes for separate analyses and for comparison of public, private religious, and other private schools. The number of participating students in the base-year sample was 24,599. ◊

3.2. Stratified Simple Random Sampling

Denote the population value for unit i in stratum $h = 1, \ldots, H$ by Y_{hi}, $i = 1, \ldots, N_h$, and denote the sample values by $y_{hi}, i = 1, \ldots, n_h$. The population mean for stratum h is denoted by $\bar{Y}_h = (Y_{h1} + \cdots + Y_{hN_h})/N_h$ and the sample mean is denoted by $\bar{y}_h = (y_{h1} + \cdots + y_{hn_h})/n_h$. The corresponding variances S_h^2 and s_h^2 are obtained from (1.1) by substituting y_{hi} for y_i, $\bar{Y}_h$ for $\bar{Y}$, $\bar{y}_h$ for $\bar{y}$, N_h for N, and n_h for n. The overall population mean is a weighted sum of the stratum means, $\bar{Y} = W_1\bar{Y}_1 + \cdots + W_H\bar{Y}_h$ with the stratum weights defined as $W_h = N_h/N$. Since the sample mean in stratum h is an unbiased estimator of the population mean for the stratum, the weighted mean $\bar{y}_w = W_1\bar{y}_1 + \cdots + W_H\bar{y}_h$ is an unbiased estimator of $\bar{Y}$.

The variance of $\bar{y}_w$ is

$$\text{Var}(\bar{y}_w) = \sum_{h=1}^{H} W_h^2(1 - f_h)S_h^2/n_h. \tag{3.1}$$

Notice that the variance depends only on variability within strata. It will be small if the strata are internally homogeneous. Thus, in the design stage of the survey one may use prior information about the variability in deciding how to define strata.

If $n_h \geq 2$ we may unbiasedly estimate S_h^2 by s_h^2, leading to the variance estimator

$$\hat{\text{V}}\text{ar}(\bar{y}_w) = \sum_{h=1}^{H} W_h^2(1 - f_h)\frac{s_h^2}{n_h}. \tag{3.2}$$

If $n_h = 1$, unbiased estimation of variance is not possible. A common fix is to combine (or "collapse") strata that are adjacent or similar in some sense and pretend that the sampling used larger sample sizes in fewer strata (Wolter 1985).

Sample sizes sometimes are chosen proportional to N_h, leading to a sample distribution across strata identical to the population distribution. This *proportional allocation* of a sample typically reduces sampling variance relative to SRS with the same sample size. The sample allocation may also be chosen to minimize variance (for a particular statistic) for fixed sample size or (if costs vary across strata) for fixed cost. Then, one speaks of the so-called *optimal allocation* or *Neyman allocation*. The optimal allocation for one statistic may not be optimal for another, however, and Neyman allocation can lead to variances greater than SRS for some statistics. Allocating the sample to achieve thresholds ($n_h \geq \tau_h$ for thresholds τ_h) is sometimes called *oversampling* when the resulting stratum sample sizes exceed what they would be under proportional allocation (fN_h). Compared to proportional allocation, oversampling may increase the variance for statistics such as $\bar{y}_w$ that weight each stratum proportional to size (N_h).

3.3. Design Effect for Stratified Simple Random Sampling[2]

The *design effect* (*deff*) for a statistic under a given sampling design is defined as the ratio of its variance to what the variance would be for a comparable statistic under simple random sampling (Kish 1965, 258). For example, the design effect for the estimate of the mean under stratified sampling is the ratio of (3.1) to (1.3).

---

[2] This is a specialized topic and may be skipped without loss of continuity.

Although the numerator (3.1) may be estimated by (3.2), a proper estimator of deff is not immediately obvious, because s^2 in (1.1) typically is not an unbiased estimate of S^2 of (1.1) for sampling designs other than SRS.

Matters are simpler when estimating proportions, however, because if each y_i is 0 or 1, then $S^2 = \bar{Y}(1 - \bar{Y})N/(N-1)$ so (1.3) may be unbiasedly estimated by $(1-f)\bar{y}_w(1-\bar{y}_w)/(n-1)$. In this case the *estimated deff* is

$$\frac{\sum_{h=1}^{H} W_h^2(1-f_h)\bar{y}_h(1-\bar{y}_h)/(n_h-1)}{(1-f)\bar{y}_w(1-\bar{y}_w)/(n-1)}. \tag{3.3}$$

If proportional allocation is used, the design effect typically is less than 1 (Exercise 6), but the design effect can well exceed 1 if oversampling is used or if optimal allocation is used to minimize variance for a different statistic than the one we are analyzing.

Estimates of deff are useful both as summaries of efficiencies (or inefficiencies) of sample designs and for approximating the sampling variance of a statistic. For example, suppose that design effects are calculated for a variety of estimated proportions and have a median value of c, and we have estimated another proportion by p from a sample of size n. A quick estimate of the sampling variance of p is $c(1-f)p(1-p)/(n-1)$. This estimate could be off, however, as different statistics may have quite different design effects, and examination of not just the median design effect (or average) but also their spread is appropriate.

Example 3.2. Design Effects for NELS:88. Design Effects were calculated for a large number of base-year questionnaire items in NELS:88. The mean design effects for school questionnaire items were 1.82 for all schools, 2.23 for public schools, and 1.40 for private schools (Spencer et al. 1990, 52). The design effects were greater than 1.0 because the schools were selected with unequal probabilities across strata (private schools were oversampled) and, more important, within strata schools were selected with probabilities proportional to estimated eighth grade enrollment, which is efficient for surveying students but not efficient for estimating school characteristics based on equal treatment of large versus small schools. ◊

3.4. Poststratification

If an SRS is selected and stratum sizes N_h are known, the sample may be stratified after the fact and analyzed as if it were stratified initially; this practice is called *poststratification*. Poststratification does not cause bias in the estimate of a population mean or total if the sample means for the poststrata are conditionally unbiased (given the sample sizes from the poststrata). Poststratification can cause bias if the choice of poststrata depends on the observed values of the means, which can be avoided if the poststrata are chosen prior to analysis of the sample data. Poststratification improves variance nearly as much as proportional allocation provided the sample sizes within strata are not too small – Cochran (1977) recommends $n_h > 20$.

Example 3.3. Poststratification in the 1990 U.S. Post Enumeration Survey (PES). Post Enumeration Surveys are used to estimate undercounts and overcounts in censuses. The rates are known to vary among subgroups defined by variables such as age, sex, race, geographic location, family type, and housing type. As discussed in Example 6.1 of Chapter 2, overall estimates will be biased if the subgroup membership is not taken into account. In the 1990 PES, the U.S. Census Bureau initially used 1,392 poststrata in calculating the estimates (Example 4.1, Chapter 10). Excessive sampling variance due to small sample sizes for some of the poststrata led to a "revised" poststratification using 357 poststrata (Hogan 1993). The latter poststratification was based in part on analysis of the data. Also, the PES used cluster sampling and because sample elements from the same cluster could fall into different poststrata, statistics calculated for different poststrata are not independent. We continue the discussion in Examples 4.4 and 7.2, below. ◊

The term "poststratification" is used not only to describe stratification after the fact, but also for calibration of sample weights to sum to known totals (Section 4.2), to reduce non-response bias (Section 4.3), and to adjust for survey undercoverage or overcoverage (Chapter 10, Section 5.2). In these other applications, independence of selections in different poststrata is not assumed.

4. Sampling Weights

4.1. Why Weight?

In many applications, one has a sample of elements that appear in the sample with unequal probabilities. Sometimes the unequal probabilities occur by design, other times as a result of nonresponse or nonparticipation (Kish 1965, 425; Kish 1992). Define indicator random variables $I_k = 1$ if element k is selected in the sample and $I_k = 0$ if it is not. Define the *first-order inclusion probability* $\pi_k = E[I_k]$ to be the probability that element k is in the sample. The unweighted sample mean $\bar{y}$ typically will be biased. To see this, first reexpress $\bar{y}$ as

$$\bar{y} = \frac{1}{n} \sum_{k=1}^{N} y_k I_k \tag{4.1}$$

and notice that the y_k's are fixed (but unknown except for the sample) and the I_k's are random. Take expected values to obtain

$$E[\bar{y}] = \frac{1}{n} \sum_{k=1}^{N} y_k \pi_k. \tag{4.2}$$

Define the population covariance between π and Y as

$$\sigma_{Y\pi} = \sum_{k=1}^{N} y_k \pi_k / N - \bar{Y} \sum_{k=1}^{N} \pi_k / N. \tag{4.3}$$

Note that $\pi_1 + \cdots + \pi_N = n$ (because $I_1 + \cdots + I_N = n$) and so $E[\bar{y}] = (N/n)\sigma_{Y\pi} + \bar{Y}$. This shows that the unweighted sample mean has bias $(N/n)\sigma_{Y\pi}$, and hence $E[\bar{y}] = \bar{Y}$ if and only if the correlation between the selection probabilities and the variable is zero.

In general, if elements are selected with unequal probabilities, there can be no assurance that unweighted estimates will be approximately unbiased. For example, the weighted mean in stratified sampling may be written as

$$\bar{y}_w = \frac{1}{N} \sum_{h=1}^{H} \sum_{i=1}^{n_h} w_{hi} y_{hi} \tag{4.4}$$

with $w_{hi} = N_h/n_h = 1/f_h$ = the reciprocal of selection probability. Suppose $H = 2$, and we had a sample of 10% from stratum $h = 1$ and 25% from stratum $h = 2$, where the two strata were each half the population $(N_1 = N_2 = N/2)$. The unweighted mean would be biased unless the means of the two groups were exactly equal.

4.2. Forming Weights

The basic principle of weighting (as, e.g., in (4.4)) is to set a unit's weight equal to the reciprocal of its selection probability. The weights often are called either *sample weights* or *case weights*. If the weights are $w_k = 1/\pi_k$, the *Horvitz-Thompson estimator* of the population total is defined as the weighted sum

$$\hat{T}_{HT} = \sum_{k=1}^{n} w_k y_k, \tag{4.5}$$

and is unbiased for the population total T_Y (Exercise 7). Consider the case when each $y_k \equiv 1$ and notice that the sum of the weights is an unbiased estimator of N. Correspondingly, if $y_k = 1$ when element k is in a subgroup G and $y_k = 0$ otherwise, then $\hat{T}_{HT}$ is the sum of the weights for the members of G in the sample, so it is an unbiased estimator of N_G. In stratified SRS, the sum of the weights in stratum h is exactly N_h and the sum of the weights for all sampled elements is N.

Example 4.1. NELS:88 First Followup Schools. In 1990, two years after the NELS:88 base-year survey, the sampled students were surveyed again (actually, to save money, subsamples of the more than 24,000 base-year students were surveyed). Most of the students were in tenth grade, and most of the students were in different schools. For analyses of the schools in the first follow-up survey, school-level sampling weights were needed. The weights were set inversely proportional to the probability that a school was in the first follow-up survey. A school had a positive probability of being selected in the first follow-up if it enrolled at least one student who was eligible for selection in the base year, and in general that probability was a function of the numbers of students in the school who in 1988 were eighth grade students, their base-year selection probabilities, and how they were clustered in different schools in 1988. The probabilities could be estimated from specially collected data on what 1988 schools contributed students to the school in question in 1990 (Spencer and Foran 1991), and weights were set equal to the reciprocals of those probabilities. ◊

Although the expected value of the sum of the sample weights ($w_k = 1/\pi_k$) is always N, in many applications the sum of the weights – for the population as a whole and especially for subgroups – is random. When the sum of the weights is a random variable (or when non-response or population undercoverage or overcoverage is present), adjustments may be imposed so that the weights sum to a known total or the weights for subgroups sum to known sets of totals. A widely used adjustment forces the weights to sum to the population size N. Or the *calibrated weight* equals sample weight × adjustment factor, that is,

$$\tilde{w}_k = \frac{1}{\pi_k} \frac{N}{\sum_{i=1}^{n} 1/\pi_i}. \tag{4.6}$$

The analytical properties of estimators using such calibrated weights are more complex, but their use does confer some advantages in usual practice. (The complexity arises because the weight $\tilde{w}_k$ for unit k depends on which other units are in the sample, a dependence not affecting w_k.) If one is estimating a proportion by a weighted mean, using the weights w_k could lead to an estimate greater than 1, but weights $\tilde{w}_k$ always lead to estimates between 0 and 1. In many cases estimators based on $\tilde{w}_k$ will have smaller variance than those based on w_k (Särndal, Swensson, and Wretman 1992, 182–184). Statistical agencies sometimes make additional adjustments to weights to force various linear statistics to equal population values or other control values, via raking (Deming 1964) and its extensions (Haberman 1984) or regression models (Deville and Särndal 1992, Deville, Särndal, and Sautory 1993). A concise discussion is given by Rao (2003, 13–15, 20–21).

Advanced techniques (not recommended for casual use, but often carefully implemented in public use data files for large-scale surveys) modify the weights to reduce sampling variance of estimators though at the cost of introducing bias. Such techniques may involve "trimming" the largest weights or by shrinking all of the weights (averaging the vector of weights with a vector of constants); see Kish (1992), Potter (1990), Qian and Spencer (1994), and Kalton and Flores-Cervantes (2003).

Example 4.2. Extreme Weights in the 1990 U.S. PES. The 1990 PES was a sample survey conducted to provide data for estimating the gross overcount and gross undercount in the 1990 U.S. census. The sample consisted of a stratified sample of more than 5,000 small areas, called clusters. (See Chapter 10, Example 4.1 for further details.) Within each cluster, the census was essentially redone, and data were collected to allow for dual-system estimation as described in Section 6 of Chapter 2 and Section 5 of Chapter 5. The clusters were selected with unequal probabilities, so that areas with small numbers of households (as estimated from pre-census listings of housing units) had very small selection probabilities, and densely populated city blocks had larger selection probabilities. Some clusters, however, had large numbers of housing units but, as a result of errors in the pre-census listings, were selected with small probabilities and, when they appeared in the sample, received very large weights. One such cluster in the sample contributed 0.75 million to the estimate of undercount. The problem arose from a combination of a large weight and an outlier data value. Zaslavsky, Schenker, and Belin (2001) discuss the problem and discuss the use of robust methods for treating it. $\diamond$

When using statistical software that accommodates unequal probability samples, one should be aware that the software may assume that the weights are of the form w_k rather than $\tilde{w}_k$. Although one can use variants of (2.3) to estimate the variance when (4.6) is used, estimating variance when more complex weighting adjustments are used requires special software or procedures, e.g., Stukel, Hidiroglou, and Särndal (1996). Unless we construct the adjusted weights ourselves, we may not have the data to account for the variances in the weights. The effect on variance estimates of ignoring the complexity in the weights often is not severe in practice unless the differences between w_k and $\tilde{w}_k$ are large.

4.3. Non-Response Adjustments

Non-response is a common problem in demographic surveys: targeted respondents may not be located, may be located but not contacted, may be contacted but not provide usable data. Lohr (1999, Chapter 8) gives an accessible overview of nonresponse and a recent extensive treatment is provided by Groves et al. (2001). *Unit non-response* is said to occur when virtually no data are provided by the targeted respondent. Often, the unit non-respondents are not treated as part of the data file and a weighting adjustment is used to allocate the sampling weight for the unit non-respondent to one or more respondents. Some adjustments are based on a model that the survey participants are the result of two stages of random selection, first is probability of selection into the sample and second is a response *propensity* or conditional probability of responding given selection. The propensities are estimated with statistical models for estimating probabilities or rates (e.g., Section 5 of Chapter 5) and may be used directly (e.g., Alho et al. 1991) or to define weighting cells. Weighting cells are analogous to poststrata, except that the counts for weighting cells are based not on the whole population but on sample-weighted numbers of sample selections falling into the cells. The response propensity for a weighting cell is calculated as the ratio of the sample-weighted number of respondents to the sample-weighted number of selections in the cell. The *non-response adjustment factor* for a respondent is the reciprocal of the estimated propensity for the respondent. The assumptions or model behind the weighting will be incorrect to one degree or another, and bias may result. To assess the degree of error from imperfect non-response weighting adjustments, alternative weighting methods sometimes are used, but how well the resulting spread of estimates reflects the error will vary from situation to situation.

Example 4.3. Nonparticipation in a Survey in an STD Clinic. An extreme case of error from unit response occurred in blood testing of patients at a clinic for treating sexually transmitted diseases (STD's). Everyone in the group had given a blood sample, and the samples without identifying information were tested for HIV, with 17 positives found. In a survey of the patients, 82 percent agreed to participate, but only 8 tested positive. Had the survey been able to test the remaining 18% (who were nonparticipants), an additional 9 would have tested positive. Nonparticipation caused the survey estimate to be biased downward by a factor of 0.57 (Hull et al. 1988). ◊

Example 4.4. The Dual System Estimator as a Propensity-Weighted Census. Section 6, Chapter 2 presented a model-based estimate of population size based on a census with n_1 enumerations and a second, sample survey with n_2 enumerations, of which m were counted in both. The dual-system estimator (DSE) was $n_1 n_2/m$. A person not being counted in the census can be viewed as non-response, and we can consider an individual i to have a response propensity, which we will view as an enumeration probability π_i. If we view m/n_2 as an estimate of π_i, we can interpret the MLE as a Horvitz-Thompson estimator with estimated weights,

$$\sum_{i=1}^{n_1} y_i/\hat{\pi}_i, \tag{4.7}$$

with $\hat{\pi}_i = m/n_2$ and $y_i = 1$. Example 6.1 of Chapter 2 showed how unequal probabilities of enumeration could lead to bias in the DSE, but if the estimates could be poststratified so that the probabilities were homogeneous within poststrata, the bias could be corrected. ◊

Item non-response occurs when the targeted respondent's data are included in the data file but a variable is missing because the response to one or more questionnaire items is not available or not usable. A common practice that facilitates data analysis in the presence of item non-response is to use imputation to predict or fill-in the missing data item or items. Using imputed values as if they are actual observed values carries two risks. First, the imputations may be systematically wrong, e.g., if people with extremely high or extremely low incomes are more prone to non-report income data (even when other observed characteristics are taken into account), using reported values to impute non-reported values might bias the median up and the mean down. Second, variances computed from imputed values treated as actual observations tend to be too small. For example if a sample of size n includes some imputations that are used in estimating a mean, s^2 may be smaller in expected value than S^2 (depending on how imputations are made) and n will be larger than the actual number of observations, with the result that s^2/n may tend to underestimate the sampling variance. Methods for estimating the variance with allowance for randomness in the imputations include multiple imputation and jackknife methods and is an active area of research; see Rubin (1987, 1996), Fay (1996), Rao (1996), Rao and Shao (1992) and, for overview, Korn and Graubard (1999, 211–218). These methods might not be applicable in secondary analysis of a data file unless details on the imputation are available, including which cases were used to impute for other cases.

4.4. *Effect of Weighting on Precision*

As noted in Section 4.1, unless the covariance $\sigma_{Y\pi}$ is zero, weighting is needed to ensure unbiasedness or approximate unbiasedness of estimates of population means or totals. If the covariance is zero, or sufficiently small, more accuracy may be attainable without weights. If the covariance is zero, so weighting is unnecessary, but the weights w_k or $\tilde{w}_k$ are nevertheless used to estimate the population mean, the weighting multiplies the variance of the estimator by a factor of $g = (n/N)\bar{W} \geq 1$, where $\bar{W}$ is the population mean of the w_k's (Kish 1965, Section 11.7; Gabler, Haeder, and Lahiri 1999; Spencer 2000a). The factor g may be estimated from the sample by the formula, "one plus the relative variance of the weights" in the sample as recommended by Kish (1965, 1992). The factor g is often called the *design effect from weighting* or the *variance inflation factor* (Kalton and Flores-Cervantes 2003).

Given the increase in variance from unnecessary weighting, how can one decide whether weighting is necessary? It is possible to compare weighted and unweighted estimates to see if they have the same expected values, and if they do then it is not unreasonable to use unweighted estimates. DuMouchel and Duncan (1983) and Fuller (1984) describe hypothesis tests for linear models. Nordberg (1989) provides tests for generalized linear models to compare weighted versus unweighted coefficients. Pfefferman (1993) describes use of the Hausman specification test for additional models. He makes the important points, however, that the null hypothesis in all of those tests asserts that the expected values are the same with and without weighting, and lack of power in a test can lead one to incorrectly fail to reject the null hypothesis. Furthermore, even if expected values are equal, any probability statements could still be incorrect if the error structure is more complicated than specified under the null hypothesis.

What should one do if the weighted and unweighted estimates appear to have different expected values? The answer depends on one's goals and the standard errors of the estimates. It is possible for weighted estimates to have smaller standard errors, although often weighted estimates have higher standard errors. If the difference is caused by outliers that have large weights due to their small sampling or response probability, we would consider trimming or shrinking the weights, as mentioned in Section 4.2. In model-based analyses (including many studies with causal aims), a large difference in estimates of expected values suggests that some aspects of the models being entertained may be incorrectly specified – in that case, one can try to revise the model or use weighted estimates, which at least have the property of estimating the population-level parameters of the model one has specified.

On the other hand, if the design effect from weighting is quite large despite weight trimming or shrinkage, some compromise strategy might be appropriate, even in descriptive studies. For such cases, Korn and Graubard (1999, 1995) recommend modifying the estimand to include variables strongly related to the weights (or stratum definitions) and using unweighted point estimates or reducing the variability of the weights as discussed in Section 4.2.

Example 4.5. Extreme Weights in the Survey of Consumer Finance. The Survey of Consumer Finances (SCF) collects data on household finances, income, assets, debts, demographics, attitudes, employment, and other activities. The sample is selected from two frames. One sample is selected with area-based cluster sampling and provides data for the population generally. A second sample is selected from lists of persons who filed individual income tax returns. An index of wealth is constructed from the tax return data, and individuals are stratified by that index (Frankel and Kennickell 1995). The second sample provides most of the data on high-income and high-wealth individuals. In the 1983 SCF, a single respondent in the list sample had an unusually low selection probability but reported ownership of a \$200 million business; the sample-weighted wealth for the individual "represented \$1 trillion, or about 10 percent of total wealth" (Avery, Elliehausen, and Kennickell 1986, 20). Later, a reinterview showed that the \$200 million datum was an interviewer error – the business should have been recorded as \$2 million. This underscores the critical importance of data quality in addition to correct statistical methods (cf., U.S. Federal Committee on Statistical Methodology 2001). ◊

5. Cluster Sampling

5.1. Introduction

Selecting a SRS or stratified SRS may be difficult in practice. A listing of individual population elements with contact information (e.g., for sample of the national population) may not be available. Field costs can be high if the sample is spread out geographically and administrative costs can be high in sampling individuals from institutions if many institutions (such as hospitals or schools) are in the sample. A solution to these problems is to group individual elements into clusters and sample the clusters. Clusters may be geographic or institutional or derived in other ways. For example, Roberts et al. (2004) applied cluster sampling to estimate mortality related to the 2003 Iraq war.

In single stage cluster sampling a sample of clusters is chosen (the clusters are "primary" sampling units, or PSUs) and all elements within the sampled clusters form the final sample. In a two-stage cluster sample a sample of clusters is first selected, and then a sample of the elements of the chosen clusters is selected ("secondary" sampling units). These form the final sample. This readily generalizes to hierarchical multistage sampling with more than two stages of selection (Kish 1965, 155).

Often, the design effect for a statistic from a cluster sample is greater than 1, indicating less precision than a SRS of the same size. Indeed, it is possible for the design effect to be vastly greater than 1, implying that if the clustering is not taken into account in the variance estimation, the estimated variances could be the wrong order of magnitude. However, cluster sampling often is more cost-effective than element sampling, so that the sample may include a larger number of elements with cluster sampling than with SRS. Thus, the ratio of precision to cost *may* be lower even if deff $>$ 1.

We first consider single stage sampling with replacement. This will turn out to be of practical importance as an approximation for estimation under more complicated designs.

5.2. Single Stage Sampling with Replacement

Suppose a sample of size 1 is selected from among A clusters so that cluster α has probability $z_\alpha > 0, \alpha = 1, \ldots, A$, of being selected. Note that $z_1 + \cdots + z_A = 1$. Suppose y_α is the cluster total of the variable of interest. Let $I_\alpha = 1$ if cluster α is selected, and $I_\alpha = 0$ otherwise. The Horvitz-Thompson estimator of the population total is then $I_1 y_1/z_1 + \cdots + I_A y_A/z_A$. Since $P(I_\alpha = 1) = z_\alpha$, this is unbiased. Suppose we independently repeat the selection a times. Let the estimate obtained in the ith selection be y_i/z_i, $i = 1, \ldots, a$. Averaging the estimates obtained in this manner yields the *Hansen-Hurwitz estimator*

$$\hat{T}_{HH} = \frac{1}{a} \sum_{i=1}^{a} y_i/z_i. \tag{5.1}$$

As an average of unbiased estimators, this is also unbiased for the population total. To estimate the population mean, simply divide the estimator of the total by N (or by an estimate of N). The variance of (5.1) is unbiasedly estimated (Exercise 8) by

$$\hat{\text{V}}\text{ar}(\hat{T}_{HH}) = \frac{a}{a-1} \sum_{i-1}^{a} (\breve{y}_i - \bar{\breve{y}})^2 \tag{5.2}$$

with $\breve{y}_i = y_i/(az_i)$, the value of y_i inversely weighted by the expected number of times it appears in the sample, and $\bar{\breve{y}} = \breve{y}_1/a + \cdots + \breve{y}_n/a$.

5.3. Single Stage Sampling without Replacement

Consider now that a of the A units are selected without replacement, with π_α the probability that unit α is selected into the sample and $\pi_{\alpha\alpha'}$ the probability that units α and α' are both selected in the sample. In this case a cluster can only be sampled once, so we index the sampled clusters by α. Again, the Horvitz-Thompson estimator

$$\hat{T}_{HT} = \sum_{\alpha=1}^{a} \breve{y}_\alpha \tag{5.3}$$

with $\breve{y}_\alpha = y_\alpha/\pi_\alpha$ is unbiased for the population total. (This is really the same setup as (4.5), if we recognize that each y_i in (4.5) is now the total for PSU i.) However, its variance depends not just on first-order selection probabilities π_α but also on joint selection probabilities $\pi_{\alpha\alpha'}$ for PSUs α and α'. Without additional assumptions, unbiased estimation of the sampling variance is possible only when $\pi_{\alpha\alpha'} > 0$ for all pairs of PSUs. This condition is not satisfied by many sample designs in which the PSUs are selected with systematic sampling (Section 6). In addition the $\pi_{\alpha\alpha'}$'s must be known for all pairs of PSUs in the sample, which may not be the case in secondary analysis of data collected by others.

A practical expedient is to estimate the variance as if the sample were selected with replacement in a independent draws with draw-by-draw selection probabilities $z_\alpha = \pi_\alpha / a$. With these specified probabilities, the calculation of $\hat{T}_{HT}$ and $\hat{T}_{HH}$ yield the same results as point estimates, and (5.2) provides a serviceable approximation for the variance. It is reasonable to suppose that the estimated variance will be conservative (tend to be too large in expected value) because the without-replacement aspect of the sampling is ignored (Durbin 1953), although just how conservative depends on the sampling rates.

To estimate the mean for the population or for a subgroup more generally, we can divide the estimator of the total by the size of the population or subgroup (if known) or by an estimate. Define $x_{\alpha\beta} = 1$ if element β in PSU α is in the subgroup and $x_{\alpha\beta} = 0$ otherwise and define $y_{\alpha\beta}$ to equal the variable of interest if element β in PSU α is in the subgroup and $y_{\alpha\beta} = 0$ otherwise (or redefine $y_{\alpha\beta}$ as $x_{\alpha\beta} y_{\alpha\beta}$). The total for PSU α is $y_\alpha = y_{\alpha 1} + \cdots + y_{\alpha B}$ and the size of the subgroup in PSU α is $x_\alpha = x_{\alpha 1} + \cdots + x_{\alpha B}$. Define weighted PSU sample totals by $\breve{y}_\alpha = w_\alpha y_\alpha$ and $\breve{x}_\alpha = w_\alpha x_\alpha$ with $w_\alpha = 1/\pi_\alpha$ and estimate the mean by the ratio of the weighted totals,

$$\hat{R} = \sum_{\alpha=1}^{a} \breve{y}_\alpha \Big/ \sum_{\alpha=1}^{a} \breve{x}_\alpha . \tag{5.4}$$

From the linearization argument of Section 2 we know that this is approximately unbiased and its variance may be estimated (under the with-replacement assumption) by

$$\hat{\text{V}}\text{ar}(\hat{R}) = \frac{a}{a-1} \sum_{\alpha=1}^{a} \breve{e}_\alpha^2 \Big/ \left(\sum_{\alpha=1}^{a} \breve{x}_\alpha \right)^2 \tag{5.5}$$

with $\breve{e}_\alpha = \breve{y}_\alpha - \hat{R}\breve{x}_\alpha$.

Example 5.1. Survey of the Homeless in Chicago. In 1985 and 1986 two sample surveys were conducted to estimate the number of homeless people in Chicago and their characteristics. An operational definition of homeless was needed, and was based on where a person needed to spend the night at the time the survey was fielded. Homeless people were divided into two groups, those in public shelters and those "on the street". A list of public shelters was obtained, stratified by number of beds, and sampled. (Within shelters, residents were sampled, which is a form of multi-stage sampling as discussed in the next section.) To sample the homeless on the street, PSUs were defined as "census blocks, usually identical to residential or commercial blocks as conventionally understood, but also including open places, parks, railroad yards, or vacant land. Census blocks are divisions of the entire area of a city, including all land, whatever the use to which that land may be dedicated. For the city of Chicago, the 1980 Census defined approximately 19,400 blocks" (Rossi, Fisher, and Willis 1986, 11). A SRS of the blocks would yield few homeless, as the homeless tended to concentrate in certain areas. Stratification of the blocks was based on the subjective ratings of those members of the Chicago Police who were closely familiar with the blocks, and disproportionate sampling was used to minimize variance (based on prior assumptions). Each sampled block was included in the survey, and a professional interviewer and an off-duty Chicago policeman as a pair visited each face of the block at a time between midnight and 4 A.M. and attempted to find each person on the street (or parked car, or unlocked entryway, etc.). The surveys were run for two two-week periods, September 22–October 4, 1985 and February 22–March 7, 1986. The estimated average daily numbers of homeless in those periods were 2,344 (735) and 2,020 (275), with estimated standard errors shown in parentheses. ◊

5.4. *Multi-Stage Sampling*[3]

For efficiency purposes, it is common to choose a random subsample of elements from the sampled PSUs. The subsamples need not be selected by simple random sampling themselves; e.g., they may be drawn in one or more stages, e.g., in the U.S., counties or groups of counties may be the PSUs, then cities (or areas outside cities) may be selected at the second stage, then blocks may be selected at the third stage, and then housing units may be selected at the fourth stage. Stratified or systematic sampling may be used as well. Using the "ultimate cluster" method of variance estimation, we do not need to keep track of all stages of sampling, but only which selections came from each PSU (or "ultimate cluster"). Let $w_{\alpha\beta}$ denote the sampling weight for element $\alpha\beta$; e.g., if PSU α was selected with probability π_α and the conditional probability that element β was selected given that the PSU was selected is $\pi_{\beta|\alpha}$, then the weight is $w_{\alpha\beta} = 1/(\pi_\alpha \pi_{\beta|\alpha})$. Let $y_{\alpha\beta}$ denote the value of the variable of interest for element $\alpha\beta$ if it is in the subgroup and $y_{\alpha\beta} = 0$ otherwise, and let $x_{\alpha\beta} = 1$ if element $\alpha\beta$ is in the subgroup of interest and $x_{\alpha\beta} = 0$ otherwise. Form weighted values $\breve{y}_{\alpha\beta} = w_{\alpha\beta} y_{\alpha\beta}$ and $\breve{x}_{\alpha\beta} = w_{\alpha\beta} x_{\alpha\beta}$ and form weighted PSU sample totals as

$$\breve{y}_\alpha = \sum_{\beta=1}^{b_\alpha} \breve{y}_{\alpha\beta} \quad \text{and} \quad \breve{x}_\alpha = \sum_{\beta=1}^{b_\alpha} \breve{x}_{\alpha\beta}, \tag{5.6}$$

with b_α the number of elements subsampled from PSU α.

To estimate the total for the subgroup we can use the Horvitz-Thompson estimator (5.1) and we can estimate its variance by (5.2) (Complement 9). The variance estimation method is called the *ultimate cluster method.*

Alternatively, if the size of the subgroup is known to be, say, T_X, we can estimate the total by the "ratio-estimator of the total", $\hat{R}T_X$. An estimate of its variance is provided by T_X^2 times (5.5). For practical purposes, we may compute both estimates

[3] This is a specialized topic and may be skipped without loss of continuity.

and their variance estimates and choose the simpler estimate (Horvitz-Thompson) unless the ratio estimate appears to have appreciably smaller variance.

5.5. *Stratified Samples*[4]

Stratification and multistage sampling are often used together. We review here some of the complexities that arise. Often, the PSUs are stratified, and it is also possible that cluster sampling will be used in some strata and not others. In some cases, even if stratification is not explicitly used, some large PSUs may be selected with certainty, and then the analysis should proceed as each certainty PSU comprised a separate stratum (and then the secondary sampling units are treated as PSUs within the stratum) and the remaining sample selections were in another stratum (or strata, as the case may be).

To estimate the population total T, one may separately estimate the total for each stratum and then sum the estimates, using say $\hat{T}_1 + \cdots + \hat{T}_H$, with $\hat{T}_h$ an estimate of the total for stratum h. The latter may be Horvitz-Thompson estimates or ratio-estimates. The variance of the estimator of the total is estimated as the sum of the variances of the individual $\hat{T}_h$'s, namely $\text{Var}(\hat{T}_1) + \cdots + \text{Var}(\hat{T}_H)$. Specifically, consider sampled element $h\alpha\beta$, i.e., subsampled element β in sampled PSU α from stratum h. Denote its sampling weight by $w_{h\alpha\beta}$, let $y_{h\alpha\beta}$ denote the value of the variable of interest for element $h\alpha\beta$ if it is in the subgroup and $y_{h\alpha\beta} = 0$ otherwise, and let $x_{h\alpha\beta} = 1$ if element $h\alpha\beta$ is in the subgroup of interest and $x_{h\alpha\beta} = 0$ otherwise. Form weighted values $\breve{y}_{h\alpha\beta} = w_{h\alpha\beta} y_{h\alpha\beta}$ and $\breve{x}_{h\alpha\beta} = w_{h\alpha\beta} x_{h\alpha\beta}$ and define weighted PSU sample totals by

$$\breve{y}_{h\alpha} = \sum_{\beta=1}^{b_{h\alpha}} \breve{y}_{h\alpha\beta} \quad \text{and} \quad \breve{x}_{h\alpha} = \sum_{\beta=1}^{b_{h\alpha}} \breve{x}_{h\alpha\beta}, \tag{5.7}$$

with $b_{h\alpha}$ the number of elements subsampled from PSU α in stratum h. The Horvitz-Thompson estimator of the population total is then

$$\hat{T}_{Y,st} = \sum_{h=1}^{H} \sum_{\alpha=1}^{a_h} \breve{y}_{h\alpha}. \tag{5.8}$$

If we use the with-replacement estimator of variance from (5.2), we have

$$\sum_{h=1}^{H} \frac{a_h}{a_h - 1} \sum_{\alpha=1}^{a_h} (\breve{y}_{h\alpha} - \bar{\breve{y}}_h)^2 \tag{5.9}$$

as an estimator of variance of (5.8), where $\bar{\breve{y}}_h = \sum_{\alpha=1}^{a_h} \breve{y}_{h\alpha}$.

To estimate the mean, one may use either $\hat{T}_{Y,st}/T_X$, if T_X is known, or the ratio mean

$$\hat{R}_c = \hat{T}_{Y,st}/\hat{T}_{X,st}, \tag{5.10}$$

---

[4] The topic is rather specialized so the section may be skipped without loss of continuity.

with $\hat{T}_{X,st}$ defined analogously to $\hat{T}_{Y,st}$ in (5.8). A linear approximation to the error in $\hat{R}_c$ is

$$\hat{R}_c - R \approx (\hat{T}_{Y,st} - R\hat{T}_{X,st})/T_X = \sum_{h=1}^{H}\sum_{\alpha=1}^{a_h} \breve{\varepsilon}_{h\alpha}/T_X \tag{5.11}$$

with $\breve{\varepsilon}_{h\alpha} = \breve{y}_{h\alpha} - R\breve{x}_{h\alpha}$. To approximate the mean and variance of the left side of (5.11), we look at the mean and variance of the right hand side, which is $1/T_X$ times a Horvitz-Thompson "estimator" based on the unobservable $\breve{\varepsilon}_{h\alpha}$. To estimate the variance, we define $\breve{e}_{h\alpha} = \breve{y}_{h\alpha} - \hat{R}_c\breve{x}_{h\alpha}$ and use

$$\hat{\mathrm{V}}\mathrm{ar}(\hat{R}_c) = \sum_{h=1}^{H} \frac{a_h}{a_h - 1} \sum_{\alpha=1}^{a_h} (\breve{e}_{h\alpha} - \bar{\breve{e}}_h)^2 / \hat{T}_{X,st}^2 \tag{5.12}$$

with $\bar{\breve{e}}_h = \sum_{\alpha=1}^{a_h} \breve{e}_{h\alpha}/a_h$. The variance of the combined ratio estimate of the mean may be estimated by (5.12) as stated or with T_X^2 used in the denominator.

Example 5.2. NELS:88 Sample of Students. From each school in the base-year sample in NELS:88 (Example 3.1), a sample of eighth-grade students was selected. The schools were selected with probability proportional to the estimated number of eighth-grade students (based on information available for all schools in the frame), and for any given type of school the proportionality factor was constant, so that if a constant number of students were sampled in each school and the estimated numbers of students were correct, each student in a given type of school would have the same selection probability. The actual number of students selected per school varied slightly because within the sampled schools, oversamples of black and Hispanic students were selected with stratified sampling. The fact that stratified sampling was used within schools does not need to be taken into account in variance estimation if the collapsed stratum method is used.

A subsample of students in the base-year sample were surveyed again in follow-ups in 1990, 1992, 1994, and 2000. Students reported on school, work, and home experiences, activities, and attitudes, and achievement tests were administered as part of the survey in 1988–1992. Students' teachers, parents, and school administrators were also surveyed. (Determining selection probabilities for teachers is difficult, although if teacher data are analyzed as student attributes the student weights may be used.) For analysis of student growth over time, it is important to note that the original PSU – the eighth grade school – remains the PSU for variance estimation. ◊

Example 5.3. The U.S. Current Population Survey. The Current Population Survey (CPS) is a stratified multi-stage sample survey of the U.S. population, with a sample size on the order of 60,000 households per month (although budgetary fluctuations cause sample sizes to vary from one set of years to another). The sample overlaps heavily from one month to the next, in a deliberate design known as a *rotation sample*. A housing unit is in the sample for 4 consecutive months, is left out for the next 8, and then it returns into the sample for the following 4 months, after which it is replaced by a new selection. The rotation design is less expensive than sampling independently each month and improves the precision of estimates of monthly and annual change. Compared to a permanent panel, the rotation design gives more precise estimates of averages across years and eases response burden as well. Although its primary purpose is to provide employment data, in some months (or years) the CPS includes detailed questions on income, fertility, education, and other topics.

Since there is no list of people (and contact information) in the U.S., the CPS sampling frame is based on geographic areas. The U.S. is partitioned into about 2,000 PSUs, which typically consist of counties or groups of counties in the same state. Highly populated PSUs are selected with certainty ("self-representing PSUs") and each comprises its own stratum; the remaining PSUs are stratified based on number of male unemployed, number of female unemployed, and household demographics, for 432 strata in all (as of 1995). One PSU is selected from each stratum. Within each sample PSU, lists of *ultimate sampling units* (USUs, typically, clusters of 4 adjacent addresses) are prepared based on the previous census and a systematic sample (Section 6) is selected. In large USUs, further subsampling may be done. A sample of building permits supplements the list of USUs to account for recently constructed housing units. The design is quite sophisticated and has evolved over many years; a comprehensive reference is U.S. Census Bureau (2002). ◊

6. Systematic Sampling

We consider selecting a systematic sample of n units from a listing of N units such that each unit has the same selection probability. For simplicity, first suppose $k = N/n$ is an integer. A systematic sample consists of units $r, r+k, r+2k, \ldots, r+(n-1)k$ with r chosen to be an integer between 1 and k. Once r is randomly picked, the rest of the sample is determined. There are r possible systematic samples. Alternatively, the procedure may be viewed as choosing 1 of k possible clusters at random. If the list is in random order, the sampling is equivalent to random sampling, but more often the list is sorted by some criterion prior to sample selection. As we have described it, systematic sampling uses equal probabilities of selection, so the unweighted mean is unbiased.

It is perhaps slightly surprising that the variance of the estimator of the population total from a systematic sample *can be smaller* than that of a single random sample of the same size. This occurs if the variance of the y values within the systematic samples is larger than the population level variance of the y values, or equivalently when the *intracluster* (or *intraclass*) *correlation* within systematic samples is negative (Cochran 1977, 208–209). Another way of looking at systematic sampling is to see it as stratified sampling with dependent selections. In the sampling frame, the first k units are called the first *zone*, the next k units are the second zone, and so on until the nth zone consisting of the last k units. If we selected one unit from each zone, independently across zones, we would have a stratified sample. In systematic sampling, we select one unit from each zone but not independently: if we select the jth unit from the first zone, we select the jth unit from the every zone. For this reason, zones are often called *implicit strata.* The analogy with stratified sampling is helpful for variance estimation. A common method for estimating sampling variance when there is no replication (i.e., the sample consists of a single cluster) is to pretend the systematic sampling is equivalent to stratified sampling with one selection per stratum and to use the collapsed stratum method (Section 3.2) to estimate variance. The success of such variance estimation methods depends on the sort-order of the population list. Unlike the other methods of random sampling we have discussed, the sampling variance does not necessarily decrease as the sample size increases.

Example 6.1. Systematic Sampling of Private Schools in the National Assessment of Educational Progress. The National Assessment of Educational Progress (NAEP) is a test given to samples of students in several grades in the U.S. The main component is a public school sample, but a private school sample is also selected and is important for analyses comparing public and private student performance. The private school students are selected in two-stage sampling, rather similar to NELS:88 (Examples 3.1 and 5.1), with schools selected with systematic sampling with probabilities proportional to a measure of size of the school. In an investigation of the properties of variance estimators, Burke and Rust (1995) created a population of 105 schools that were selected in NAEP for 1994, and assigned a mean score to each school based on the observed mean from the 1994 student sample from the school (based on about 30 students per school). The schools were sorted using the characteristics underlying the NAEP private-school sample design and systematic samples of various sizes (numbers of schools) were selected. Analysis showed that the sampling variances (and mean square errors) did *not* decline monotonically with the sample size. The variance estimation methods performed well however, even with small sample sizes. ◊

Implementation of systematic sampling when $k = N/n$ is not an integer is discussed in texts such as Kish (1965, 115–116). One straightforward method is to randomly choose a number $r \in [0, k)$ and then randomly select units $\lfloor r + 1 + j \times k \rfloor$, $j = 0, \ldots, n-1$, with $\lfloor x \rfloor$ denoting the largest integer $\leq x$. The method extends to selection of units with unequal probabilities (Cochran 1977, 265–266).

7. Distribution Theory for Sampling

7.1. Central Limit Theorems

Central limit theorems apply to the weighted sample mean and Horvitz-Thompson estimators from many kinds of complex sample designs used in demographic surveys. The classical central limit theory assumes the sampling is with replacement, so that selections are made independently, meaning that $\pi_{ij} = \pi_i \pi_j$ for units i and j. Thus, if we select a simple random sample with replacement from a population of size N with mean $\bar{Y}$ and variance $0 < S^2 < \infty$, the distribution of the standardized sample mean is asymptotically normal $N(0, 1)$. If the units are selected with unequal probabilities z_i and with replacement, then the unbiased estimator of the total, $\hat{T}_{HH}$ given by (5.1), is the sample mean of the independent and identically distributed variates y_i/z_i and again the classical central limit theorem implies that the asymptotic distribution of $(\hat{T}_{HH} - T_Y)/\sqrt{\text{Var}(\hat{T}_{HH})}$ is $N(0, 1)$, where the variance $\text{Var}(\hat{T}_{HH})$ is shown in Exercise 8. A central limit theorem also applies to the weighted mean from a stratified simple random with-replacement sample with a fixed number of strata with increasing sample sizes (because a weighted average of normal random variables is normal) or with the number of strata increasing with N and the sample sizes in the strata fixed (Krewski and Rao 1981).

Those central limit theorems need modification to apply to sampling without replacement, because in that method the individual observations are not independent. If the sample is selected without replacement, or if number of strata increases with n, then the concept of n growing without limit requires us to consider N growing as well, for otherwise the sample would include the whole population (and keep growing!). Thus, we consider a sequence of sampling situations with increasing population sizes N and increasing sample sizes n such that $\lim n/N < 1$.

Versions of the central limit theorem have been proved for without-replacement sampling designs that are similar to simple random sampling in that either

$$\pi_{jk} \text{ is approximately proportional to } \pi_j \pi_k \tag{7.1}$$

for PSUs i and j or *successive sampling* is used (Complement 26). For example, Hájek (1960) and Erdös and Renyi (1959) showed that under some realistic conditions on the population, the standardized sample mean, $(\bar{y} - \bar{Y})/(S\sqrt{(1 - f)/n})$, is asymptotically normal $N(0, 1)$. The asymptotic normality of the Horvitz-Thompson estimator in unequal-probability sampling has been established for single-stage (Hájek 1964, Rosén 1972) and for multi-stage sampling designs whose PSU-selection probabilities satisfy (7.1) and whose weighted PSU sample totals in (5.1) satisfy certain moment-like conditions (Sen 1988). Additional conditions involve the PSU selection probabilities being too small for some units relative to others, the idea being that no single unit or small number or units contribute too much to the variance. Asymptotic normality has also been proved for the weighted mean in stratified simple random sampling when either the stratum sizes or the number of strata grow with the population sizes and $2 \leq n_h \leq N_h$ (Bickel and Freedman 1984). The results extend to stratified multistage sampling. The results do not apply to systematic sampling from a fixed population, where the limited number of possible systematic samples may be an impediment to normality, and where the variance can only be estimated under assumptions.

The central limit results mentioned above also apply to vectors of means or Horvitz-Thompson estimators, whose asymptotic distribution is multivariate normal.

We have not focused on the moment (or similar) conditions for the population that are required to formally prove the central limit theorem (Thompson 1997). When we are considering a finite population, practical considerations such as skewness and the presence of extreme values (or extreme sample-weighted values) – in relation to the sample size – become the most critical considerations. For example, a statistic computed from a sample of municipalities can have a highly skewed sampling distribution, if most units are small but some large cities belong to the list. Cochran (1977, 39–44) provides useful guidance concerning applicability of the theory to finite samples, and discusses how the minimum n for the normal approximation to work varies with the skewness in the underlying population.

7.2. The Delta Method

The delta method is a procedure for approximating random variables and especially their means, variances, and covariances. We have considered it already in Exercise 11 of Chapter 2 and we used it in a special case to approximate the ratio in (2.2). In this section we let T_n denote a general statistic (that may, but need not, be an estimator of a population total). Suppose the sequence $T_n, n = 1, 2, \ldots$ is such that T_n is asymptotically normal, specifically the limiting distribution of $\sqrt{n}(T_n - \theta)$ is $N(0, \sigma^2(\theta))$. If $g(.)$ is a function with a continuous non-zero derivative at θ, $g'(\theta) \neq 0$, and $\sigma(.)$ is continuous, then the distribution of

$$\frac{\sqrt{n}[g(T_n) - g(\theta)]}{g'(T_n)\sigma(T_n)} \tag{7.2}$$

approaches $N(0, 1)$ as $n \to \infty$. The basic idea is that $g(T_n) \approx g(\theta) + g'(\theta)(T_n - \theta)$ by Taylor's theorem. For smaller sample sizes, Student's t distribution may often provide a better approximation, although the appropriate number of degrees of freedom depends on the population, the sample design, and on the method used for variance estimation.

This result generalizes to k-variate statistics $\mathbf{T}_n = (T_{1n}, \ldots, T_{kn})^T$, for example vectors of weighted means or totals. Suppose we have a sequence of statistics $\mathbf{T}_n, n = 1, 2, \ldots$ such that the limiting distribution of $\sqrt{n}(\mathbf{T}_n - \boldsymbol{\theta})$ is multivariate normal $N(\mathbf{0}, \boldsymbol{\Sigma}(\boldsymbol{\theta}))$, with $\boldsymbol{\Sigma}(.)$ a continuous function of $\boldsymbol{\theta}$. Suppose further that we have a function $\mathbf{g} = (g_1, \ldots, g_q)^T$ from $\mathbb{R}^k$ to $\mathbb{R}^q$ such that the matrix of partial derivatives $\mathbf{G}(\boldsymbol{\theta}) = (\partial g_i / \partial \theta_j)$ is continuous. Then the distribution of $\sqrt{n}[\mathbf{g}(\mathbf{T}_n) - \mathbf{g}(\boldsymbol{\theta})]$ approaches $\mathrm{N}(\mathbf{0}, \mathbf{G}(\boldsymbol{\theta})\boldsymbol{\Sigma}(\boldsymbol{\theta})\mathbf{G}(\boldsymbol{\theta})^T)$ as $n \to \infty$. Furthermore, for inferential purposes we may approximate the limiting distribution by $N(\mathbf{0}, \mathbf{G}(\mathbf{T}_n)\boldsymbol{\Sigma}(\mathbf{T}_n)\mathbf{G}(\mathbf{T}_n)^T)$ (Rao 1973, 385–389). The latter covariance is called the *linearization estimate*, and for practical purposes we may use alternative estimates of covariance (as discussed in the Section 8) that are asymptotically equivalent.

When the limiting distribution is normal with mean zero, it is customary to say that the estimator is *asymptotically unbiased*. This does not necessarily mean that the bias of the estimator goes to zero. For example, consider $\bar{x}$ and $\bar{y}$ to be sample means and $g(\bar{x}, \bar{y}) = \bar{y}/\bar{x}$ to be the ratio estimator. If $\bar{x}$ and $\bar{y}$ are jointly normally distributed, then one can show that $E[g(\bar{x}, \bar{y})]$ does not exist[5], although as sample sizes get large and variances of $\bar{x}$ and $\bar{y}$ go to zero, the distribution of $g(\bar{x}, \bar{y}) - g(E[\bar{x}], E[\bar{y}])$ approaches a normal distribution with mean 0.

[5] The only time the mean exists is if $\bar{y} = c\bar{x}$ with certainty, for some constant c.

Example 7.1. Model-Based Variance of the Dual System Estimator (DSE). A hypergeometric model for dual system estimation based on a Post Enumeration Survey (PES) was discussed in Section 6, Chapter 2. The model treated the number of enumerations in the census, n_1, and the number of enumerations in the PES, n_2, as fixed, and the number in both, m, as random. As discussed in Exercise 11 of Chapter 2, a variance estimate of the DSE can be estimated obtained using the delta method as $(n_1 n_2)^2 m^{-3}(n_1 - m)(1 - m/n_2)$ (Chandra Sekar and Deming 1949; Bishop, Fienberg, and Holland 1975, 233). Wolter (1986) presents some data from the U.S. Census Bureau's 1980 Post-Enumeration Program showing, for black males, (weighted) counts $n_1 = 11{,}306{,}493$, $n_2 = 11{,}233{,}060$, $m = 9{,}803{,}540$. The weights are needed because the PES was based on a sample of areas (blocks), so a DSE based on unweighted counts would only estimate the population size of the sample of areas. If we divide the counts by the average sampling weight, say w, we can estimate the population for the sampled area as $(n_1/w)(n_2/w)/(m/w) = n_1 n_2/(mw)$. Multiplying this by w to estimate the total population, we have the usual form of the DSE but based on the weighted counts, or 12,955,169. The estimated standard error according to the hypergeometric model is 1,809,549. ◊

7.3. Estimating Equations[6]

We review here some principles of statistical inference in a sampling context. Consider again a population of size N. Many of the quantities we estimate from sample surveys can be defined in a roundabout way as solutions to equations. Denote the population characteristic of interest by θ and note that the population mean is the solution to

$$\sum_{i=1}^{N}(y_i - \theta) = 0, \tag{7.3}$$

the population ratio is the solution to

$$\sum_{i=1}^{N}(y_i - \theta x_i) = 0, \tag{7.4}$$

and the population cumulative distribution function at a point y is the solution to

$$\sum_{i=1}^{N}(I_{(-\infty,y]}(y_i) - \theta) = 0. \tag{7.5}$$

(Exercise 12). These equations are all of the form $\psi_T(\theta) = 0$, with

$$\psi_T(\theta) = \sum_{i=1}^{N} \psi(y_i, x_i, \theta). \tag{7.6}$$

[6] The section is somewhat theoretical and may be skipped without loss of continuity.

As a sum over population values, $\psi_T(\theta)$ can be thought of as a population total. For any θ, $\psi_T(\theta)$ can be unbiasedly estimated by the sample-weighted total

$$\psi_s(\theta) = \sum_{i=1}^{n} \psi(y_i, x_i, \theta)/\pi_i. \tag{7.7}$$

A sample estimate, say $\hat{\theta}$, can be obtained as a solution to the *estimating equation* $\psi_s(\theta) = 0$.

Vector-valued estimating equations are useful for estimating vectors of characteristics. For example, consider how the least-squares estimates for the linear regression model satisfy a vector-valued estimating equation. Let y_i denote the variable y for element i and let $\mathbf{x}_i$ be a $q \times 1$ vector of covariates for element i in the population. Consider a sample of n observations and write the sample values as $\mathbf{X} = (\mathbf{x}_1, \ldots, \mathbf{x}_n)^T$ and $\mathbf{y} = (y_1, \ldots, y_n)^T$. The classical multiple regression model asserts that y_i is random with conditional mean (given $\mathbf{x}_i$) equal to $\mathbf{x}_i^T \boldsymbol{\theta}$ for a $q \times 1$ coefficient vector $\boldsymbol{\theta}$. The least-squares estimate of $\boldsymbol{\theta}$ minimizes the sum of $(y_i - \mathbf{x}_i^T \boldsymbol{\theta})^2$ and can be shown to satisfy the "normal equations" $\mathbf{X}^T \mathbf{y} = \mathbf{X}^T \mathbf{X} \boldsymbol{\theta}$. Alternatively, without resorting to assumptions about a model, we can define $\boldsymbol{\theta}$ as the solution to the normal equations when they are based on the N sets of population values. Specifically, define $\boldsymbol{\psi}(\mathrm{y}_i, \mathbf{x}_i, \boldsymbol{\theta}) = \mathbf{x}_i(y_i - \mathbf{x}_i^T \boldsymbol{\theta})$ and note that the solution to $\boldsymbol{\psi}_T(\boldsymbol{\theta}) = \mathbf{0}$, where

$$\boldsymbol{\psi}_T(\boldsymbol{\theta}) = \sum_{i=1}^{N} \boldsymbol{\psi}(y_i, \mathbf{x}_i, \boldsymbol{\theta}), \tag{7.8}$$

satisfies the normal equations based on the whole population. The sample-weighted estimate of $\boldsymbol{\theta}$, say $\hat{\boldsymbol{\theta}}$, is a solution of $\boldsymbol{\psi}_s(\boldsymbol{\theta}) = \mathbf{0}$, where

$$\boldsymbol{\psi}_s(\boldsymbol{\theta}) = \sum_{i=1}^{n} \boldsymbol{\psi}(y_i, \mathbf{x}_i, \boldsymbol{\theta})/\pi_i, \tag{7.9}$$

or $\mathbf{X}^T \mathbf{D}_w \mathbf{y} = \mathbf{X}^T \mathbf{D}_w \mathbf{X} \hat{\theta}$, where $\mathbf{D}_w$ is a diagonal matrix with elements $1/\pi_i$.

Suppose the function $\boldsymbol{\psi}$(y, $\mathbf{x}$, .) is continuously differentiable for all y and $\mathbf{x}$, and write $\mathbf{H}(\boldsymbol{\theta}) = \partial \boldsymbol{\psi}_s(\boldsymbol{\theta})/\partial \boldsymbol{\theta}^T$ for the matrix of partial derivatives. Then, we may expand $\boldsymbol{\psi}_s(\boldsymbol{\theta})$ in a Taylor series about $\boldsymbol{\psi}_s(\hat{\boldsymbol{\theta}})$ to yield (Complement 14)

$$\hat{\boldsymbol{\theta}} - \boldsymbol{\theta} \approx -\mathbf{H}(\boldsymbol{\theta})^{-1} \boldsymbol{\psi}_s(\boldsymbol{\theta}) \approx -E[\mathbf{H}(\boldsymbol{\theta})]^{-1} \boldsymbol{\psi}_s(\theta). \tag{7.10}$$

The elements of $\boldsymbol{\psi}_s(\boldsymbol{\theta})$ are weighted sample totals and for large samples $\boldsymbol{\psi}_s(\boldsymbol{\theta})$ typically is distributed approximately as multivariate normal. The covariance matrix of the asymptotic normal distribution (cf., Section 3 of Chapter 1) is

$$E[\mathbf{H}(\boldsymbol{\theta})]^{-1} \mathrm{Cov}(\boldsymbol{\psi}_s(\boldsymbol{\theta})) E[\mathbf{H}(\boldsymbol{\theta})^T]^{-1}. \tag{7.11}$$

The elements of $\mathrm{Cov}(\boldsymbol{\psi}_s(\boldsymbol{\theta}))$ for any fixed $\boldsymbol{\theta}$ can be estimated in the usual manner (e.g., (5.2) or (5.9), or using replication methods of Section 8). Evaluating the estimate at $\boldsymbol{\theta} = \hat{\boldsymbol{\theta}}$ leads to an estimate of the actual covariance under the population value of $\boldsymbol{\theta}$, say $\hat{\text{Cov}}_{\psi_s}$. A consistent estimator of the asymptotic covariance matrix of $\hat{\boldsymbol{\theta}}$ given by (7.11) is

$$\hat{\text{Cov}}(\hat{\boldsymbol{\theta}}) = \mathbf{H}(\hat{\boldsymbol{\theta}})^{-1}\hat{\text{Cov}}_{\psi_s}\left(\mathbf{H}(\hat{\boldsymbol{\theta}})^T\right)^{-1}. \tag{7.12}$$

We have that, approximately,

$$\frac{\hat{\theta}_i - \theta_i}{\hat{\sigma}_{\hat{\theta}_i}} \sim N(0, 1), \tag{7.13}$$

where $\hat{\sigma}^2_{\hat{\theta}_i}$ denotes the ith diagonal element of $\hat{\text{Cov}}(\hat{\boldsymbol{\theta}})$.

Much of statistics starts from assumptions about the population. In fact, in many studies with causal aims there may not exist any finite population of which the observations are a sample. Instead, the population values are assumed to be drawn by nature from a density $f(y, \mathbf{x}|\boldsymbol{\theta})$ that belongs to a parametric family indexed by $\boldsymbol{\theta}$. Then we set $\psi(y, \mathbf{x}, \boldsymbol{\theta}) = \partial/\partial\boldsymbol{\theta} \log(f(y, \mathbf{x}|\boldsymbol{\theta}))$. In this case we can take $n = N$ so $\pi_i = 1$. Even if $n < N$, so an actual sample is selected, but each component of ψ is uncorrelated with the selection probabilities then (recall Section 4.1) we do not need to use unequal sampling weights in $\psi_s(\boldsymbol{\theta})$ and we say the sampling is *non-informative* or *ignorable* (Valliant, Dorfman, and Royal 2000, 36–39). In that case we may also replace $1/\pi_i$ in (7.9) by 1. (Exercise 15). In these cases the root of $\psi_s(\boldsymbol{\theta}) = \mathbf{0}$ yields a maximum likelihood estimator, as introduced in Chapter 1. Recall the definition of the Fisher information as $\mathcal{I}(\boldsymbol{\theta}) = -E[\mathbf{H}(\boldsymbol{\theta})]$. Then, we have that $E[n^{-1}\psi_s(y, \mathbf{x}, \boldsymbol{\theta})] = N^{-1}\psi_T(\boldsymbol{\theta}) \approx \mathbf{0}$ and $n^{-1}\,\text{Cov}(\psi_s(\boldsymbol{\theta})) \approx \mathcal{I}(\boldsymbol{\theta})$, so (7.11) may be replaced by $\mathcal{I}(\boldsymbol{\theta})^{-1}$ (Exercise 15). Instead of (7.12), the covariance of $\hat{\boldsymbol{\theta}}$ may be estimated by $\mathcal{I}(\hat{\boldsymbol{\theta}})^{-1}$. The latter is an example of a *model-based estimator of covariance*, as compared to a *design-based estimator of covariance* such as (7.12).

Even if the aims of a study are causal and the real target population transcends the sampling frame, it can be useful to calculate the covariance estimates both ways to see if there is evidence of possible model mis-specification or informative sampling (Horowitz 1994). Or, one can include characteristics of the sample design (such as indicators for clusters or strata) in the model to see if they have explanatory power. If they do, then the specification of the presumed causal model may be incomplete in some respect. Furthermore, if estimates of the parameters of interest change after the inclusion of variables related to the sampling design, then a revision of the causal assumptions, collection of better data that allows one to address possible confounding, or both may be called for. For further discussion, see Binder and Roberts (2003), Korn and Graubard (1999), Chambers and Skinner (2003), Skinner, Holt and Smith (1989), and Valliant, Dorfman, and Royall (2000).

Example 7.2. Design-Based Variance of the Dual System Estimator (DSE). In Example 7.1 we considered a model-based estimate of variance of the DSE based on a hypergeometric model. Such a model is unrealistic, in part because the enumeration rates vary by subgroups, and the hypergeometric model assumes equal enumeration probabilities. Separate DSEs can be constructed for different post-strata and summed, but the variance of the sum is not equal to the sum of the variances because selections in different poststrata are not independent due to the cluster sampling used in the PES (Example 3.3). In the PES, a stratified sample of clusters was selected with unequal probabilities. Let n_1 denote the weighted number of census enumerations in the sample clusters, n_2 the weighted number of enumerations in the second, sample enumeration (the "P sample"), and m the weighted number in both, where the weights are reciprocals of design-based selection probabilities. For simplicity, ignore erroneous enumerations. Consider a single poststratum. The simple DSE for the poststratum is $n_1 n_2$/m. We can improve on this using the known total number of census enumerations, N_1, to yield $\tilde{N} = N_1 n_2/m$. The ratio $\hat{R}_c = n_2/m$ is called an "adjustment factor", because the DSE is equal to the adjustment factor times the census count, N_1. The variance of $\tilde{N}$ for a given poststratum can be estimated by N_1^2 times the quantity after the first summation sign in (5.12), with a_h the number of clusters in stratum h, $\breve{e}_{h\alpha} = \breve{y}_{h\alpha} - \hat{R}_c \breve{x}_{h\alpha}$, $\breve{y}_{h\alpha} = n_2$ for cluster $h\alpha$, and $\breve{x}_{h\alpha} = m$ for cluster $h\alpha$. To find the covariance between estimates for the poststratum and another poststratum, which we will indicate with a $'$, simply replace $(\breve{e}_{h\alpha} - \bar{\breve{e}}_h)^2/\hat{T}^2_{x,st}$ in (5.12) by $(\breve{e}_{h\alpha} - \bar{\breve{e}}_h)(\breve{e}'_{h\alpha} - \bar{\breve{e}}'_h)/(\hat{T}^2_{x,st}\, \hat{T}^2_{x,st})$. Applying this to the estimated number of black males from the 1980 Post-Enumeration Program (Example 7.1) under a simplification of the actual sample design (Wolter 1986, 343–344) yielded an estimated standard error of 51,000, which is more than twice the model-based standard error. The differences are due partly to weighting but also to clustering. The clustering will inflate the variance if the enumeration probabilities have a positive intraclass correlation, which means that the enumeration probabilities are variable and give rise to a clustering of census misses (Hengartner and Speed 1993). It is possible that some of what appears as intraclass correlation is due to interviewer effects or other operational effects in the PES that were similar within clusters. In a careful analysis they would be estimated and taken into account where feasible. By themselves, clusters cannot serve to define poststrata, so although there is some geographic heterogeneity in the enumeration probabilities, how to revise the estimation method to account for the heterogeneity is not obvious. ◇

Although models do not have to be correct to be useful, as John Tukey has noted, it is important to appreciate that the advantages of using assumptions about the population distribution do depend on the validity of the assumptions. Note that $\boldsymbol{\theta}$ is implicitly defined by (7.6) and a consistent estimate of $\boldsymbol{\theta}$ can be obtained whether or not the density is correctly specified. Similarly, (7.12) provides automatically a correct covariance estimator for the implicitly defined parameter even under a wrong model. However, the usefulness of the estimates depends on the degree of mis-specification.

Example 7.3. Parameter Interpretation Under An Erroneous Model. Suppose we assume erroneously that $Y_i \sim N(0, \theta)$, $i = 1, \ldots, n$ are independent and take $\psi(y_i, x_i, \theta) = y_i^2 - \theta$, but in reality $Y_i \sim N(\mu, \sigma^2)$. A consistent estimate of θ is obtained by setting (7.7) to zero, so $\hat{\theta} = (y_1^2 + \cdots + y_n^2)/n$, but in this case $\theta = E[Y_i^2] = \mu^2 + \sigma^2$. Any attempt at calculating one-sided prediction intervals for a future value is likely to fail if μ^2 is not small compared to σ^2. Although two-sided prediction intervals with nominal coverage levels between 68% and 99% will have approximate probability α of covering the future value even for $|\mu/\sigma|$ as large as 0.6 (Cochran 1977, 15), the non-coverage is asymmetric. For example, when $\mu = 0.6\sigma$ the probability that the future value falls above a nominal 95% interval is 0.0459 and the probability it falls below the interval is 0.0020. Furthermore, consider variance estimation via (7.12). In this case $H(\theta) = -n$, so (7.11) equals exactly $\text{Var}(\hat{\theta})$. (Using the properties of the normal distribution one can show that $E[Y_i^4] = 3\sigma^4 + 6\sigma^2\mu^2 + \mu^4$, so in this case (7.11) equals $(2\sigma^4 + 4\sigma^2\mu^2)/n$.) Thus, (7.12) leads to asymptotically correct inferences about mean squared error θ. The problem is that the user of the mis-specified model believes that the inferences are about a variance. ◊

Interval estimates can be developed in several ways. Let θ_0 denote the root of $\psi_T(\theta) = 0$. One way to produce a two-sided $100(1-\alpha)\%$ confidence interval for θ_0 is to use (7.13) to obtain the interval $\hat{\theta} \pm z_{1-\alpha/2}\hat{\sigma}_{\hat{\theta}}$, with z_p the pth fractile of the $N(0, 1)$ distribution for $0 < p < 1$. A second way, often but not always applicable, is to use the approximate normality of $\psi_s(\theta)$ so that, approximately,

$$\frac{\psi_s(\theta) - \psi_T(\theta)}{\hat{\sigma}_{\psi(\theta)}} \sim N(0, 1). \tag{7.14}$$

Consider testing the null hypothesis $H_0 : \psi_T(\theta) = 0$ versus the two-sided alternative, $H_A : \psi_T(\theta) \neq 0$. A $100(1-\alpha)\%$ confidence interval for θ_0 is the set of θ values for which H_0 is not rejected, i.e., the set of θ such that

$$\psi_s(\theta)^2/\hat{\sigma}^2_{\psi(\theta)} \leq z^2_{1-\alpha/2}. \tag{7.15}$$

Note that $z^2_{1-\alpha/2}$ is also the $1-\alpha$ fractile of the χ^2 distribution with one degree of freedom. This approach leads to alternative confidence limits for the ratio, as developed by Fieller (1932).

Example 7.4. Fieller Intervals for a Ratio Estimator. Define $\hat{T}_{Y,HT}$ by (4.5) and define $\hat{T}_{X,HT}$ analogously. The ratio estimator $\hat{\theta} = \hat{T}_{Y,HT}/\hat{T}_{X,HT}$ from an unequal probability sample is the solution to (7.7) with $\psi(y_i, x_i, \theta) = y_i - \theta x_i$. To find the endpoints of the interval for θ such that (7.15) holds, we solve the quadratic equation obtained by setting $\psi_s(\theta)^2 = \hat{\sigma}^2_{\psi(\theta)} z^2_{1-\alpha/2}$. Note that $\psi_s(\theta) = \hat{T}_{Y,HT} - \theta\hat{T}_{X,HT}$ and $\hat{\sigma}^2_{\psi(\theta)} = \text{Var}(\hat{T}_{Y,HT}) - 2\theta\text{Cov}(\hat{T}_{Y,HT}, \hat{T}_{X,HT}) + \theta^2\text{Var}(\hat{T}_{X,HT})$. After some algebra, we find that the roots and hence the endpoints of the interval are

$$\hat{\theta}\frac{1 - z^2_{1-\alpha/2}c_{xy} \pm z_{1-\alpha/2}\sqrt{c_{yy} + c_{xx} - 2c_{xy} - z^2_{1-\alpha/2}(c_{yy}c_{xx} - c^2_{xy})}}{1 - z^2_{1-\alpha/2}c_{xx}} \tag{7.16}$$

where the relative variances and relative covariances are $c_{yy} = \hat{\text{V}}\text{ar}(\hat{T}_{Y,HT})/\hat{T}^2_{Y,HT}$, $c_{xx} = \hat{\text{V}}\text{ar}(\hat{T}_{X,HT})/\hat{T}^2_{X,HT}$, and $c_{xy} = \hat{\text{C}}\text{ov}(\hat{T}_{X,HT}, \hat{T}_{Y,HT})/(\hat{T}_{X,HT}\hat{T}_{Y,HT})$. The roots in (7.16) are imaginary for any sample if we take α small enough, and in this case the interval is the whole real line. However, for commonly used significance levels, this is rare if c_{xx} and $c_{yy} < 0.09$ (Cochran 1977, 156). For comparison, the confidence interval obtained from (7.13) is $\hat{\theta}(1 \pm z_{1-\alpha/2}\sqrt{c_{yy} + c_{xx} - 2c_{xy}})$. ◊

8. Replication Estimates of Variance

Although the delta method can often be used to derive approximations to the variances of complex nonlinear statistics, its practical application can be hard. It can be tedious to determine analytically the partial derivatives needed, and errors of programming can occur, as the process must be repeated afresh for each new statistic. The error of approximation is often difficult to assess. The so-called resampling methods circumvent these problems via brute force computation that is implemented formally the same way, no matter what the statistic of interest. We will discuss two such methods, and comment on a shortcut that is sometimes available.

8.1. Jackknife Estimates

Consider a with-replacement sample of n units such that unit i is chosen with probability $z_i > 0$ (as in Section 5.2) and let $\hat{\theta}$ denote an estimator that is a smooth function of sample means or totals, e.g., a mean, a ratio, a regression coefficient, etc. Denote by $\hat{\theta}_{(i)}$ the estimate when the ith unit is omitted from the calculation. A *jackknife estimate of variance* is defined as

$$\hat{\mathrm{V}}\mathrm{ar}_{\mathrm{jack}}(\hat{\theta}) = \frac{n-1}{n}\sum_{i=1}^{n}\left(\hat{\theta}_{(i)} - \hat{\theta}_{(\cdot)}\right)^2, \qquad \hat{\theta}_{(\cdot)} = \frac{1}{n}\sum_{j=1}^{n}\hat{\theta}_{(j)}. \tag{8.1}$$

This is sometimes called a "delete-1" jackknife. $\hat{\mathrm{V}}\mathrm{ar}_{\mathrm{jack}}(\hat{\theta})$ reduces to the usual unbiased one when $\hat{\theta}$ is a linear function of the data, such as $\hat{T}_{HH}$ in (5.1). For example, if the selections are made with equal probabilities, then $\hat{\mathrm{V}}\mathrm{ar}_{\mathrm{jack}}(\bar{y}) = s^2/n$ (Exercise 19). Therefore, the concept is primarily useful when the statistic of interest is a nonlinear function of the data.

In multi-stage sampling, if the n sample units are PSUs, we delete all sample selections within the PSU (i.e., we delete the whole ultimate cluster) when we obtain $\hat{\theta}_{(i)}$. If simple random sampling without replacement is used, $\hat{\mathrm{V}}\mathrm{ar}_{\mathrm{jack}}(\hat{\theta})$ may be multiplied by the finite population correction factor $1 - f$. An alternative form of the jackknife uses $\hat{\theta}$ in place of $\hat{\theta}_{(\cdot)}$ in $\hat{\mathrm{V}}\mathrm{ar}_{\mathrm{jack}}(\hat{\theta})$. If n is large, we may reduce computations by randomly sorting the sample into groups and deleting a group at a time.

If we want to apply the jackknife method to an estimate from a stratified simple random sample, we may use

$$\sum_{h=1}^{H}(1 - \lambda_h f_h)\frac{n_h - 1}{n_h}\sum_{i=1}^{n_h}\left(\hat{\theta}_{(hi)} - \hat{\theta}_{(h)}\right)^2, \quad \text{with} \quad \hat{\theta}_{(h)} = \frac{1}{n_h}\sum_{j=1}^{n_h}\hat{\theta}_{(hj)}, \tag{8.2}$$

where $\hat{\theta}_{(hi)}$ is the estimate calculated without observation i in stratum h; $\lambda_h = 1$ if the sampling is without replacement and $= 0$ if with replacement; $f_h = n_h/N_h$ is the sampling fraction, and n_h is the number of groups in stratum h. A variety of alternative jackknife estimators can be obtained by replacing $\hat{\theta}_{(h)}$ in (8.2) by $\hat{\theta}$, by the unweighted average across strata of $\hat{\theta}_{(h)}$'s, or by the unweighted average of all of the $\hat{\theta}_{(hi)}$'s (Rao and Wu 1985).

To accommodate without replacement sampling in multi-stages or when selections are made with unequal probabilities, we need to use modifications of these methods or special versions of the bootstrap, as in Sitter (1992) and Rao and Wu (1988). For application of the jackknife (or bootstrap or similar replication methods) for variance estimation in multiple frame surveys such as the Survey of Consumer Finance discussed in Example 4.3, see Lohr and Rao (1997).

Many computer programs use the standard deviation of variance estimates from (8.2) in computing t statistics with degrees of freedom equal to $n_1 + \cdots + n_H - H$, but that may be optimistic if the sample allocation is very disproportionate, the strata have unequal variances, or one is analyzing a subgroup that may be absent in the sample from numerous PSUs (Cochran 1977, Korn and Graubard 1999, 193*ff*.).

8.2. Bootstrap Estimates

Again, we begin by considering a with-replacement sample of n units such that unit i is chosen with probability $z_i > 0$, and let $\hat{\theta}$ denote a smooth estimator (e.g., Shao and Tu 1995, 86ff). Keeping the sampled values fixed, draw a simple random with-replacement subsample of size m from the original sample and compute $\hat{\theta}$ for the subsample; repeat this independently B times and denote the estimates by $\hat{\theta}^{*1}, \hat{\theta}^{*2}, \ldots, \hat{\theta}^{*B}$. A bootstrap estimate of the variance of $\hat{\theta}$ is

$$\hat{\mathrm{V}}\mathrm{ar}_{\mathrm{boot}}(\hat{\theta}) = \sum_{b=1}^{B} (\hat{\theta}^{*b} - \hat{\theta}^{*\cdot})^2/(B-1), \quad \hat{\theta}^{*\cdot} = \frac{1}{B}\sum_{b=1}^{B} \hat{\theta}^{*b}. \tag{8.3}$$

Notice that when the original sample is viewed as fixed, for $B < \infty$ the bootstrap estimator (8.3) is still random as its value depends on the subsamples chosen. Efron and Tibshirani (1993, 50–53) rely on theory and experience to suggest that B between 50 and 200 usually suffices for estimating variance. The additional variability from having B at 200, say, rather than ∞ is dwarfed by the variability from the original sample. The expected value of $\hat{\mathrm{V}}\mathrm{ar}_{\mathrm{boot}}(\hat{\theta})$ with respect to the subsampling and conditional on the original sample will be denoted by $E_*[\hat{\mathrm{V}}\mathrm{ar}_{\mathrm{boot}}(\hat{\theta})]$. When $\hat{\theta}$ is a linear statistic, $E_*[\hat{\mathrm{V}}\mathrm{ar}_{\mathrm{boot}}(\hat{\theta})]$ is equal to $(n-1)/m$ times the usual unbiased estimator of variance (Exercise 22). For example, if the selection probabilities are equal, then we have that $E_*[\hat{\mathrm{V}}\mathrm{ar}_{\mathrm{boot}}(\bar{y})] = (n-1)m^{-1}s^2/n$. Although many applications of the bootstrap choose subsamples of size n, as in the original sample, the resulting variance estimates for linear statistics will be downward biased by the factor $(1 - 1/n)$. Choosing $m = n - 1$ eliminates that bias.

To account for without-replacement simple random sampling, one can multiply $\hat{\mathrm{V}}\mathrm{ar}_{\mathrm{boot}}(\hat{\theta})$ by the finite population correction factor $1 - n/N$. More generally, however, the bootstrap can be modified to directly account for unequal probability sampling without replacement by with-replacement subsampling from the $n(n-1)$

pairs of sample units with unequal probabilities that reflect the original joint selection probabilities (Rao and Wu 1988, 237–239).

To account for multi-stage sampling, one can use the ultimate cluster method (with the simplifications that entails) and subsample whole ultimate clusters (i.e., all sampled elements in the PSU) and use (8.3). That method parallels the jackknife treatment in Section 8.1. One can also, however, choose the subsamples with multi-stage sampling; Sitter (1992, 761–764) and Rao and Wu (1988, 239) provide details for two-stage sampling.

The simplest way to get a bootstrap estimate of sampling variance in stratified simple random sampling is to draw a simple random with-replacement subsample of size m_h from stratum $h = 1, \ldots, H$ in the original sample, then compute the bootstrap estimates $\hat{\theta}^{*b}$, $b = 1, \ldots, B$ for independent subsamples, calculate $\hat{\mathrm{V}}\mathrm{ar}_{\mathrm{boot}}(\hat{\theta})$ as in (8.3), and sum across strata. If $m_h = n_h - 1$ then $E_*[\hat{\mathrm{V}}\mathrm{ar}_{\mathrm{boot}}(\bar{y}_w)]$ is equal to (3.2) but without the finite population correction factors $1 - f_h$. If the sampling fractions are negligible, this is fine, or if the sampling fractions are equal, the bootstrap variance estimate may be multiplied by $1 - f$. To estimate sampling variance under stratified multi-stage sampling using the ultimate cluster method, subsample m_h ultimate clusters from the n_h in the sample from stratum $h = 1, \ldots, H$ and apply (8.3).

Should one prefer to use the bootstrap or the jackknife for variance estimation? The bootstrap is better able to accommodate sampling without replacement than the jackknife, although at the cost of some complexity. The bootstrap can also be used to obtain one-sided and other asymmetric confidence intervals; see Efron and Tibshirani (1993). The jackknife can involve less computing, however, and simulations suggest that in some cases its variance estimates have somewhat smaller mean square error than those from the bootstrap (Shao and Tu 1995, 251–258). In terms of the accuracy of the variance estimates, if the ultimate cluster method is acceptable and the estimator is a smooth function of sample means or totals, either the jackknife or bootstrap may be used, with the choice based on convenience. For very small sample sizes, as may occur in highly stratified samples, the jackknife appears to be preferable to the bootstrap.

8.3. *Replication Weights*

Replication weights provide a simple method for computing variances for secondary analysis of data. When preparing a public use data file, some statistical agencies include with each case a set of r replicate weights. Calculating an estimate using any one of the r replicate weights yields an estimate of the form $\hat{\theta}^{*b}$ (if the bootstrap is used) or $\hat{\theta}_{(i)}$ (if the delete-1 jackknife is used) or something similar (if other replication methods are used for the variance estimation). The variance of a statistic can be estimated by a constant c times the sum of squared deviations of the weighted estimates about their mean or about the full-sample estimate, $\hat{\theta}$. The constant c depends on the replication method being used, and guidance is provided along with documentation for the public use data file. If available, replicate weights are quite useful. They may be derived from more efficient replication methods than the delete-1 jackknife or the bootstrap, such as balanced repeated replication, which allow r to be fairly moderate. The creation of the replicate weights may also take into account weighting adjustments for poststratification and other calibration and nonresponse.

Exercises and Complements (*)

1. (a) Derive (1.2). (Hint: Notice that the y_k's are constants, and find the expected value of $\bar{y}$ by substituting $E[I_k]$ for I_k in (4.1).) Show that for simple random sampling, $E[I_k] = n/N$, and hence $E[\bar{y}] = \bar{Y}$. (b) Use the properties of the variance of a linear combination to show that the variance of $\bar{y}$ is

$$\operatorname{Var}(\bar{y}) = \frac{1}{n^2}\left[\sum_{k=1}^{N} \operatorname{Var}(I_k)y_k^2 + \sum_{k=1}^{N}\sum_{l \neq k}^{N} \operatorname{Cov}(I_k, I_l)y_k y_l\right].$$

(c) Show that for simple random sampling, $\operatorname{Var}(I_k) = (n/N)(1 - n/N)$ and $\operatorname{Cov}(I_k, I_l) = -(n/N)(1 - n/N)/(N - 1)$. Substitute and simplify the algebra to obtain (1.3). (d) Finally, write $s^2 = [n/(n-1)]\left[\sum_1^n y_i^2/n - \bar{y}^2\right]$. Show that the expected value of the first term in the square brackets is $\sum_1^N Y_i^2/N$ and note that $E[\bar{y}^2] = \bar{Y}^2 + \operatorname{Var}(\bar{y})$. Substitute and simplify to obtain $E[s^2] = S^2$.

2. In with-replacement simple random sampling, elements are selected in n independent draws with equal probabilities at each draw. Define $\sigma^2 = (N - 1)S^2/N$ and show that

$$E[\bar{y}] = \bar{Y}, \quad \operatorname{Var}(\bar{y}) = \sigma^2/n, \quad E[s^2] = \sigma^2, \quad E[s^2/n] = \operatorname{Var}(\bar{y}).$$

*3. The ratio of the absolute bias of $\hat{R}$ to the standard error of $\hat{R}$ is less than or equal to the CV of $\bar{x}$. The accuracy of the approximation in (2.2) depends on $\bar{x}$ being close to $\bar{X}$. In practice, the approximation should be adequate for typical purposes if the CV of $\bar{x}$ is less than 0.1 (Cochran 1977). In that case the bias may be neglected in relation to the standard error. The estimate of variance (2.3) tends to be biased downward, particularly for $n \leq 12$, unless the CV of $\bar{x}$ is less than 0.1.

*4. The ratio estimator provides an alternative to estimating $\bar{Y}$ by the sample mean, provided that the population mean of X is known. The ratio-estimate of the mean is $\hat{R}\bar{X}$ and the ratio estimate of the total is $N\bar{X}\hat{R}$. The variances may be estimated by multiplying (2.3) by $\bar{X}^2$ or $N^2\bar{X}^2$ respectively. If the population scatterplot of y_i against x_i lies close enough to a straight line through the origin, the ratio-estimate of the mean (or total) will be superior to that based on the sample mean. A practical guide is to choose $\hat{R}\bar{X}$ over $\bar{y}$ only if its estimated variance is appreciably smaller than that of $\bar{y}$.

*5. The square root of the design effect is abbreviated as *Deft*. There is some inconsistency in practice concerning finite population corrections. Some authors define Deft as the ratio of (i) the actual standard error of the statistic, under the given design with sample size n, to (ii) $S/\sqrt{n}$ – without the finite population correction; e.g. Kish (1995, 56).

6. Show that the analysis of variance identity holds in a stratified population,

$$(N-1)S^2 = \sum_{h=1}^{H}(N_h - 1)S_h^2 + \sum_{h=1}^{H} N_h(\bar{Y}_h - \bar{Y})^2.$$

Show that if proportional allocation is used in stratified sampling, then for large N_h's

$$\text{Deff}\,(\bar{y}) = \frac{\sum_{h=1}^{H} \frac{N_h}{N} S_h^2}{S^2} \approx 1 - \frac{\sum_{h=1}^{H} \frac{N_h}{N}(\bar{Y}_h - \bar{Y})^2}{S^2} \leq 1.$$

This shows that to a good approximation, proportional allocation helps efficiency (the ratio of sampling variances) if the strata are chosen propitiously and does not hurt it if the strata are chosen unwisely.

7. Prove that the Horvitz-Thompson estimator (4.5) is unbiased for the population total. (Hint: Extend (4.1) to include weights and omit the factor $1/n$.)
8. To obtain the variance of (5.1), denote by m_α the number of times unit α is selected in the sample. The joint distribution of the m_α's is given by the *multinomial distribution.* Mult($n; z_1, \ldots, z_A$). The probability of observing $(m_1, \ldots, m_A)$ is $n!(m_1! \ldots m_A!)^{-1} z_1^{m_1} \ldots z_A^{m_A}$ and we have $E[m_\alpha] = nz_\alpha$, $\text{Var}(m_\alpha) = nz_\alpha(1 - z_\alpha)$, and $\text{Cov}(m_\alpha, m_{\alpha'}) = -nz_\alpha z_{\alpha'}$. Write $\hat{T}_{HH} = (m_1 y_1/z_1 + \cdots + m_A y_A/z_A)/a$ and use the moments of m_α's to derive

$$\text{Var}(\hat{T}_{HH}) = a^{-1} \sum_{\alpha=1}^{A} z_\alpha (y_\alpha / z_\alpha - N\bar{Y})^2$$

and show that this equals (5.2). Show that (5.1) and (5.2) are unbiased. (Cf., Cochran 1977, 253–254.)

*9. Justification of ultimate cluster method of estimating variances. The variance of the Horvitz-Thompson estimator (5.3) in one-stage cluster sampling may be expressed as (see e.g., Cochran 1977, 260–261 for the complex details)

$$\text{Var}_1(\hat{T}_{HT}) = \sum_{\alpha=1}^{A} \sum_{\alpha'>\alpha}^{A} (\pi_\alpha \pi_{\alpha'} - \pi_{\alpha\alpha'})(\breve{y}_\alpha - \breve{y}_{\alpha'})^2.$$

Several variance estimators have been derived, including the ("Sen-Yates-Grundy") estimator

$$\hat{\text{V}}\text{ar}_1(\hat{T}_{HT}) = \sum_{\alpha=1}^{a} \sum_{\alpha'>\alpha}^{a} (\pi_\alpha \pi_{\alpha'} - \pi_{\alpha\alpha'})\pi_{\alpha\alpha'}^{-1}(\breve{y}_\alpha - \breve{y}_{\alpha'})^2,$$

but they are unbiased only if $\pi_{\alpha\alpha'} > 0$ for all (not just sampled) pairs of PSUs. Furthermore, depending on the design used, the unbiased estimators may take negative values for some samples. In two-stage sampling, let $\breve{y}_\alpha$ denote the

Horvitz-Thompson estimate of the total for PSU α and let $V(\breve{y}_\alpha)$ denote its variance. The variance of (5.3) under two-stage cluster sampling is

$$\mathrm{Var}_2(\hat{T}_{HT}) = \mathrm{Var}_1(\hat{T}_{HT}) + \sum_{\alpha=1}^{A} \mathrm{Var}(\breve{y}_\alpha)/\pi_\alpha.$$

The estimator $\hat{\mathrm{V}}\mathrm{ar}_1(\hat{T}_{HT})$ in fact accounts for a good portion of the variance due to subsampling, and under two-stage sampling its expected value is

$$\mathrm{Var}_2(\hat{T}_{HT}) - \sum_{\alpha=1}^{A} \mathrm{Var}(\breve{y}_\alpha).$$

An unbiased estimator of variance is provided by

$$\hat{\mathrm{V}}\mathrm{ar}_2(\hat{T}_{HT}) = \hat{\mathrm{V}}\mathrm{ar}_1(\hat{T}_{HT}) + \sum_{\alpha=1}^{a} \hat{\mathrm{V}}\mathrm{ar}(\breve{y}_\alpha)/\pi_\alpha.$$

which typically is only slightly larger than $\hat{\mathrm{V}}\mathrm{ar}_1(\hat{T}_{HT})$. (This discussion is based on Särndal, Swensson and Wretman (1992), 135–141; see their pp. 141–150 for three and higher-stage sampling.)

*10. The *separate ratio estimator* of the total is

$$\sum_{h=1}^{H} \hat{R}_h T_{Xh}$$

with T_{Xh} the known population total for x in stratum h and

$$\hat{R}_h = \sum_{\alpha=1}^{a_h} \breve{y}_{h\alpha} \Big/ \sum_{\alpha=1}^{a_h} \breve{x}_{h\alpha}.$$

The variance of the separate ratio estimator may be estimated by

$$\sum_{h=1}^{H} \hat{\mathrm{V}}\mathrm{ar}(\hat{R}_h) T_{Xh}^2$$

with (from (5.5))

$$\hat{\mathrm{V}}\mathrm{ar}(\hat{R}_h) = \frac{a_h}{a_h - 1} \sum_{\alpha=1}^{a_h} \breve{e}_{h\alpha}^2 \Big/ \left(\sum_{\alpha=1}^{a_h} \breve{x}_{h\alpha} \right)^2$$

and $\breve{e}_{h\alpha} = \breve{y}_{h\alpha} - \hat{R}_h \breve{x}_{h\alpha}$. A possible drawback of the separate ratio estimator is bias, if the coefficients of variation of the denominators of $\hat{R}_h$ are not all small; in that case the variance estimator may well underestimate, leading to overconfidence in the accuracy of the estimate.

*11. An alternative to the separate ratio estimator is the *combined ratio estimator* of the total, $\hat{R}_c T_X$, with $\hat{R}_c$ defined by (5.10). The variance of the combined ratio-estimator of the total may be estimated by the numerator of (5.12).

12. Show that the estimating equation method estimates the population mean by a weighted sample mean with weights as in (4.6) and that it estimates the ratio with an estimator of the form (5.4).
*13. If additional information on the population is available, modifications to the ψ functions may be put in place. Consider, for example, the case of population mean that has $\psi(y_i, x_i, \theta) = \psi_0(y_i - \theta)$. If the population values were known to be symmetrically distributed about the mean, we could identify the population mean by setting (7.6) to zero when ψ_0 is any odd function about zero. For example, take $\psi_0(z) = z$ for $z \in [-k, k]$, $\psi_0(z) = k$ for $z > k$, and $\psi_0(z) = -k$ for $z < -k$, for some $k > 0$. This leads via (7.7) to a *Winsorized estimate* of the mean, insensitive to outliers (Lehmann 1983, 376*ff.*).
14. We consider a linear approximation to the solution from an estimating equation. Let $\hat{\boldsymbol{\theta}}$ denote the estimate and $\boldsymbol{\theta}$ the population value. Under regularity conditions the estimator is consistent. This justifies using a linear approximation to ψ_s as $-\psi_s(\boldsymbol{\theta}) = \psi_s(\hat{\boldsymbol{\theta}}) - \psi_s(\boldsymbol{\theta}) \approx \mathbf{H}(\boldsymbol{\theta})(\hat{\boldsymbol{\theta}} - \boldsymbol{\theta})$. Assuming the inverse exists, we may solve this to yield the first part of (7.10). Similarly, under regularity conditions, for large samples $\mathbf{H}(\boldsymbol{\theta})$ is close to its mean, so $\mathbf{H}(\boldsymbol{\theta})(\hat{\boldsymbol{\theta}} - \boldsymbol{\theta}) \approx E[\mathbf{H}(\boldsymbol{\theta})](\hat{\boldsymbol{\theta}} - \boldsymbol{\theta})$. This yields the second part. Binder (1983) and Thompson (1997, 104 *ff.*) discuss conditions under which these approximations are valid.
15. Suppose nature selects the N population values from a density $f(y, \mathbf{x}|\boldsymbol{\theta})$ that belongs to a parametric family indexed by $\boldsymbol{\theta}$. For simplicity ignore $\mathbf{x}$. Set $\psi(y, \boldsymbol{\theta}) = \partial/\partial\boldsymbol{\theta} \log(f(y|\boldsymbol{\theta}))$. Use a law of large numbers to show that $N^{-1}\psi_T(\boldsymbol{\theta})$ approaches $E[\partial/\partial\boldsymbol{\theta} \log(f(y|\boldsymbol{\theta}))]$ as N gets large. In the classical formulation, one considers an infinite population with $\int \psi_T(\boldsymbol{\theta})dy = E[\partial/\partial\boldsymbol{\theta} \log(f(y|\boldsymbol{\theta}))]$. Recall the discussion of scores in Section 3 of Chapter 1 and show that if the order of differentiation and integration can be switched, $E[\partial/\partial\boldsymbol{\theta} \log(f(y|\boldsymbol{\theta}))] = \mathbf{0}$. Next, suppose that non-informative sampling is used to select a sample of size n. Recall that in the finite population setting (Section 7.1) both n and N get large, and in the classical formulation the population is infinite. Consider $\psi_s(\boldsymbol{\theta})$ with weights $1/\pi_i$ in (7.12) replaced by 1. Observe that $n^{-1}\text{Cov}(\psi_s(\boldsymbol{\theta})) \approx n^{-1}E[\psi_s(\boldsymbol{\theta})^T\psi_s(\boldsymbol{\theta})]$, which tends to a matrix whose (i, j) element is $E[(\partial/\partial\theta_i \log f(y|\boldsymbol{\theta}))(\partial/\partial\theta_j \log f(y|\boldsymbol{\theta}))]$. As discussed in Section 3 of Chapter 1, conclude that $n^{-1}\text{Cov}(\psi_s(\boldsymbol{\theta})) \approx \mathcal{I}(\boldsymbol{\theta})$.
16. Show that the population cumulative distribution function at a point y, say $F(y)$, is the root of (7.5). Let $u_i = I_{(-\infty,y]}(y_i)$ and $w_i = 1/\pi_i$ and show that

$$\hat{F}(y) = \sum_{i=1}^{n} w_i u_i \Big/ \sum_{i=1}^{n} w_i$$

is the solution, when one sets (7.7) to zero and $\psi(y_i, x_i, \theta) = u_i - \theta$. If the denominator in $\hat{F}(y)$ were replaced by its expected value, would the resulting estimator take all its values on [0, 1]? Note that if the w_i's vary other than across strata, then $\hat{F}(y)$ is a ratio of sample totals and its variance may be estimated as described in (2.3), (5.6), or (5.12), or as in Section 8. Denote the variance estimate by $\hat{\sigma}^2_{\hat{F}(y)}$ and use the delta method to show that an approximate $100(1-\alpha)\%$ confidence interval is given by $\hat{F}(y) \pm z_{1-\alpha/2}\hat{\sigma}_{\hat{F}(y)}$.

*17. Population quantiles. We would like to define the pth population quantile, or the $100p^{\text{th}}$ percentile, say θ_p, as the solution to $F(\theta_p) = p$, with F the population c.d.f.. An exact solution may not exist, however, if F is not continuous, and the solution may not be unique if F is not strictly increasing. Even if F is continuous, however, $\hat{F}$ is discrete. Lohr (1999, 311–313) and especially Korn and Graubard (1999, 68–74) discuss problems and solutions for discrete distributions, including various interpolation methods to define $\hat{F}^{-1}$. One way (Woodruff 1952) to develop an approximate $100(1-\alpha)\%$ confidence interval for θ_p is to transform the endpoints of the interval from Exercise 16 using $\hat{F}^{-1}$. This leads us to take $(\hat{F}^{-1}(p - z_{1-\alpha/2}\hat{\sigma}_{\hat{F}(\hat{\theta}(p))}), \hat{F}^{-1}(p + z_{1-\alpha/2}\hat{\sigma}_{\hat{F}(\hat{\theta}(p))}))$ as the interval, with $\hat{\sigma}_{\hat{F}(\hat{\theta}(p))}$equal to $\hat{\sigma}_{\hat{F}(y)}$ evaluated at $y = \hat{\theta}(p)$.

*18. Alternative confidence sets for population quantiles. The quantile θ_p is approximately a zero of (7.6) with $\psi(y_i, x_i, \theta) = I_{(-\infty,\theta)}(y_i) - p$. Using (7.16) we may develop alternative confidence intervals for θ_p as (Francisco and Fuller 1991)

$$\left\{\theta \,|\, \hat{F}(\theta) - z_{1-\alpha/2}\hat{\sigma}_{\hat{F}(\theta)} < p < \hat{F}(\theta) + z_{1-\alpha/2}\hat{\sigma}_{\hat{F}(\theta)}\right\}.$$

19. Verify that $\hat{\text{V}}\text{ar}_{\text{jack}}(\hat{T}_{HH})$ gives the estimator (5.2) and that if the selections are made with equal probabilities, $\hat{\text{V}}\text{ar}_{\text{jack}}(\bar{y}) = s^2/n$.

*20. Grouped jackknife. Given a with-replacement sample of n units, we may randomly assign the sampled units to form groups of (equal or nearly equal) size $d = n/r$, and let $\hat{\theta}_{(g)}$ denote the value of the statistic $\hat{\theta}$ when the gth group is omitted. A *grouped jackknife* estimate of the variance or of the mean square error of $\hat{t}$ is

$$\frac{r-1}{r}\sum_{g=1}^{r}\left(\hat{t}_{(g)} - t_{(\cdot)}\right)^2.$$

with $\hat{t}_{(\cdot)}$ the average of the $\hat{t}_{(g)}$'s or alternatively

*21. In the grouped jackknife, we form the sample into groups at random one time, and then delete d observations at a time. Let N_d denote the number of without-replacement subsamples of size $n-d$, and $\hat{\theta}_{(g)}$ denote the value of the statistic based on the gth subsample, $g = 1, \ldots, N_d$. A *delete-d jackknife* estimate of the variance of $\hat{t}$ is

$$\frac{n-d}{N_d}\sum_{g=1}^{N_d}\left(\hat{\theta}_{(g)} - \hat{\theta}_{(\cdot)}\right)^2,$$

with $\hat{\theta}_{(\cdot)}$ the average of the $\hat{\theta}_{(g)}$'s. A consistent estimate of variance of the sample median is obtained if $d > n^{1/2}$ and $n - d \to \infty$. Generally, in cases where the delete-1 jackknife does not give consistent variance estimates but the delete-d jackknife does, it is necessary that both d and $n - d \to \infty$. Typically N_d is too large for manageable computing, and a random subsample (either with or without replacement) of the N_d subsamples may be used to estimate the variance. (Shao and Tu 1995, 49–55).

*22. The delete-1 jackknife applies to many statistics, including linear statistics, ratios and regression coefficients in linear and generalized linear models, and statistics that are smooth functions of the data (see Shao and Tu 1995, chapter 2, for further information). As described in Complement 21, the delete-1 jackknife does not give good estimates of the variance of the sample median. The performance of jackknife estimates of variance in stratified and stratified multi-stage sampling has been studied for statistics that are smooth functions (having continuous second derivatives) of vectors of population means and such that the function evaluated at the vector of means is proportional to the function evaluated at the vector of totals – such statistics include linear statistics, ratios, and regression coefficients in linear and generalized linear models. The sampling designs use with-replacement sampling of PSUs and it is assumed that as n increases, $\max_h(N_h/N)/(n_h/n)$ remains bounded (this allows for increasing number of strata or for constant number of strata), and that as N increases the W_h-weighted averages of within-stratum covariances are bounded.

*23. If $n_h = 2$ for each stratum, a convenient way to form a jackknife estimator of variance is to pick one unit from each stratum, say unit $h1$ from stratum h, and only delete it. The estimator is then

$$\hat{\text{V}}\text{ar}_{\text{jack}}(\hat{\theta}) = \sum_{h=1}^{H} \left(\hat{\theta}_{(h1)} - \hat{\theta}\right)^2.$$

Balanced repeated replication (BRR) is an alternate method of variance estimation that can be used with $n_h = 2$ (and other stratum sizes too but less easily), in which half of the units are omitted from the calculation of each replicate, with the half chosen according to a systematic design.

24. Show that $E_*[\hat{\text{V}}\text{ar}_{\text{boot}}(\hat{T}_{HH})]$ is equal to $(n-1)/m$ times the estimator (5.2). What is it equal to if the selection probabilities are all equal?

25. To use the bootstrap to estimate variance from a stratified without-replacement simple random sample, denote the original sample values by $y_{hi}, i = 1, \ldots, n_h$, denote the stratum means by $\bar{y}_h$, and denote the values in any subsample by $y^*_{hi}, i = 1, \ldots, m_h$, all for $h = 1, \ldots, H$. Calculate the estimate $\hat{\theta}^{*b}$ not from the y^*_{hi}, but rather from scaled values $\tilde{y}_{hi}$ defined as $\tilde{y}_{hi} = \bar{y}_h + m_h^{1/2}(n_h - 1)^{1/2}(y^*_{hi} - \bar{y}_h)$, and then estimate the variance with (8.3). A simple choice for m_h is $n_h - 1$, in which case $\tilde{y}_{hi} = y^*_{hi}$. (Rao and Wu (1988); see Sitter (1992) for methods based on without-replacement subsampling.)

*26. *Successive sampling* is a method of drawing a sample of size n with unequal probabilities and without replacement from a population of size N. Let $z_1, \ldots, z_N$ be positive numbers summing to 1. At each draw, choose unit i if not selected at a previous draw with probability proportional to z_i. For example, at the first draw unit i has probability z_i of being selected. If unit

j was selected at the first draw, the probability that unit $i(\neq j)$ is selected at the second draw is $z_i/(1 - z_j)$. Hájek (1981) analyzes this method in detail.

*27. The degrees of freedom for variance estimates from complex sample designs is a complicated question. The degrees of freedom, say d, may be chosen so the asymptotic second moment of the variance estimator agrees with the second moment of a chi-squared random variable on d degrees of freedom. Cochran (1977, 96) presents a formula for d stratified simple random sampling with n_h observations from stratum $h = 1, \ldots, H$, and shows d lies between $\min\{n_h - 1)$ and n, with $n = n_1 + \cdots + n_H$. The result assumes the underlying observations are normally distributed, and if their actual distribution has heavier tails, the formula will overstate the degrees of freedom. The approach may be extended to multi-stage samples, in which case sample sizes refer to numbers of PSUs. As a practical rule, d should not exceed $n - H$, which is optimistic but utilized in some software packages.

When one is analyzing data from a sparse subgroup, instead of all H strata and all n PSUs, it is better to consider only those containing at least one sample member from the subgroup. Also, when using a replication method to estimate variance, it is commonly recommended that d should not exceed the number of replicates minus 1. Rust and Rao (1996) present a clear discussion.

4 Waiting Times and Their Statistical Estimation

We will first describe the simplest model for survival data, the exponential distribution. Its demographic significance often goes unnoticed, because it assumes a constant hazard rate. This is unfortunate, because many of the key issues of demographic estimation can already be discussed in this simple case. We continue in Section 2 by treating the classical model for a general waiting time. The emphasis is on the probability of survival function and its estimation based on individual level or grouped data. Section 3 discusses the estimation and use of survival probabilities in forecasting. A probabilistic handling of fertility measures is given in Section 4. In particular, we will give an introduction to Poisson processes in this setting. In Section 5 we consider the magnitude of random variability in demographic rates and the commonly used Poisson assumption. Section 6 discusses the simulation of waiting times and counts. For a classical presentation, see Pressat (1972).

1. Exponential Distribution

Consider a *waiting time* until a specified event. The event can be death, so for a newborn the waiting time is the length of life. The waiting time can also be the time of appearance of the first cancer, the time between the first and second births, the time of first marriage, duration of marriage etc. In this section we develop a simple exponential model for a waiting time. Although the model is a crude one, it provides a direct way to introduce statistical concepts that are central to more realistic models. We also obtain optimality results that provide a foundation for the age-specific estimation of general waiting times.

We let a nonnegative random variable $X \geq 0$ represent the waiting time. As described in Chapter 1, the distribution function of X is $F(x) = P(X \leq x)$. Suppose $F(.)$ is differentiable, so $F'(.) = f(.)$ is the density function of X. Then, the expectation of X is

$$E[X] = \int_0^\infty x f(x)\, dx. \tag{1.1}$$

In demography, $E[X]$ may correspond to *life expectancy*, for example.

The variable X has an *exponential distribution* with parameter $\mu > 0$, or $X \sim \text{Exp}(\mu)$, if its *survival function* $p(x;\mu) = P(X > x)$ is equal to $\exp(-\mu x)$ for $x \geq 0$. For reasons to be explained in Section 2, μ is called a *hazard rate*[1]. In this case $F(x;\mu) = 1 - \exp(-\mu x)$ and $f(x;\mu) = \mu\exp(-\mu x)$. When viewed as a function of μ, $f(x;\mu)$ is the likelihood function of the observation. Integrating by parts gives us the result $E[X] = 1/\mu$. In Section 2 we show a simpler way to calculate the integral.

Example 1.1. Memorylessness of Exponential Waiting Time. The exponential distribution has the so-called *memorylessness property*: $p(x+t)/p(x) = p(t)$ for all $x > 0$. In words, this means that the probability of surviving an additional time t, given survival beyond time x, does not depend on x. It follows that $E[X|X > x] = x + 1/\mu$, for example. Starting from the equation $p(x+t) = p(x)p(t)$ one can prove that no other distribution has the memorylessness property (Feller 1968, 459–460). ◊

Example 1.2. Independent Causes of Death. Suppose $X_1, \ldots, X_k$ are independent, exponentially distributed waiting times with parameters $\mu_1, \ldots, \mu_k$, respectively. Define $X = \min\{X_1, \ldots, X_k\}$. Then (Exercise 1), we have that $P(X > x) = \exp(-(\mu_1 + \cdots + \mu_k)x)$ or, in other words, the minimum has also an exponential distribution with the parameter $\mu_1 + \cdots + \mu_k$. In demography, $X_1, \ldots, X_k$ might represent waiting times to death from k independent causes of death and X would be the actual duration of life. ◊

The method of moments provides a way to estimate μ. (Complement 3.) Suppose $X_i \sim \text{Exp}(\mu)$, $i = 1, \ldots, n$, are *independent and identically distributed (i.i.d.)*. Define $\bar{X} = (X_1 + \cdots + X_n)/n$, so $E[\bar{X}] = 1/\mu$. The method of moments sets $\bar{X} = 1/\hat{\mu}$, giving us $\hat{\mu} = 1/\bar{X}$ as the estimator of μ. As we discuss next, $\hat{\mu}$ is also a MLE of μ.

Maximum likelihood estimation can accommodate censoring, which may occur if individuals exit the population for reasons other than death. For simplicity of language let us think of the X_i's as representing the independent lengths of life of n individuals. In practice, we may not observe an individual's full lifetime: if $X_i \leq c_i$ we will observe X_i but if $X_i > c_i$, then we only know that i died after c_i, or X_i was *censored* at time c_i. Suppose there are fixed numbers $c_i > 0$ such that each i is followed only until the censoring time c_i. Let m denote the number of deaths that were not censored, and assume (with no loss of generality) that they were the ones with the first m indices. The likelihood function of the observed times of deaths $X_i = x_i$ can then be written as

$$L(\mu) = \prod_{i=1}^{m} \mu \exp(-\mu x_i) \prod_{i=m+1}^{n} \exp(-\mu c_i). \tag{1.2}$$

[1] The word hazard comes from Arabic *al zahr* meaning dice.

Define the loglikelihood function as $\ell(\mu) = \log L(\mu)$. We leave it as an exercise for the reader to prove that by differentiating $\ell(\mu)$ and setting the derivative to zero, one obtains the solution,

$$\hat{\mu} = \frac{m}{K + K'}, \tag{1.3}$$

where K is the number of *person years* lived by those whose deaths were observed, and K' is the number of person years lived by those who were censored, or

$$K = \sum_{i=1}^{m} x_i, \quad K' = \sum_{i=m+1}^{n} c_i\,. \tag{1.4}$$

We see that the MLE is of the form: "observed cases divided by person years lived". It is customary to call it an *occurrence-exposure rate.* We will be talking about *o/e rates* for short.[2] By taking each $c_i = \infty$, we get that $m = n$, and the result that the moment estimator is the MLE when there is no censoring. Thus, in the absence of censoring the estimator $\hat{\mu} = 1/\bar{X}$ is actually an *o/e* rate!

Above we have assumed that the censoring variables are fixed numbers. We will see below that this is an extremely common situation in the age-specific estimation of waiting times of demography. However, suppose now that the c_i's are values of random variables C_i that are independent of the X_i's, and have distributions that do not depend on μ. Let $p_{C_1,\ldots,C_m|C_{m+1},\ldots,C_n}(x_1, \ldots, x_m|c_{m+1}, \ldots, c_n)$ denote the conditional probability that the first m censoring times equal or exceed the corresponding x values, given the values of $C_{m+1}, \ldots, C_n$, and let $f_{C_{m+1},\ldots,C_n}(c_{m+1}, \ldots, c_n)$ denote the joint density of $C_{m+1}, \ldots, C_n$. Define $L_C = f_{C_{m+1},\ldots,C_n}(c_{m+1}, \ldots, c_n) \times p_{C_1,\ldots,C_m|C_{m+1},\ldots,C_n}(x_1, \ldots, x_m|c_{m+1}, \ldots, c_n)$. Then, the full likelihood is $L(\mu) \times L_C$. Since L_C does not depend on μ, it does not affect the maximum likelihood estimation, and $\hat{\mu}$ is also the MLE under general independent censoring. (For more details about likelihood construction under various censoring mechanisms, see Klein and Moeschberger 1997, 66–67.) This result is important in demographic applications, because censoring by migration, or by death, is often independent of the risk being estimated.

Similarly, if an individual i enters the follow-up after the beginning of the observation period, say at time $d_i > 0$, his or her survival experience is *left censored* (as opposed to *right censoring* considered above). Due to the memorylessness property of the exponential distribution the late arrivals can be accommodated by adjusting their entry times to zero, and by defining their time of death as $X_i - d_i$, and their time of censoring as $c_i - d_i$. This shows that in the case of exponential distribution the *o/e* rate is the MLE under both right and left censoring.

Note that this corresponds precisely to the observational scheme in which the data are collected from the rectangles of a Lexis diagram (e.g., ABCD in Figure 1 of Chapter 2). Individuals spend varying times in any given rectangle based on the

[2] In epidemiology an *o/e* rate is often called "incidence" or "incidence rate" (e.g., Rothman 1986).

time of year they were born. This leads to fixed left and right censoring. Other mechanisms of censoring can often be assumed independent of the waiting time being studied. Hence, if constant hazard can be assumed to hold in each rectangle, then the exponential model provides a full estimation theory for parameter estimation, rectangle by rectangle.

Here, we digress to comment on the calculation of person years when the population being studied is open. In large populations that are open to migration, person years lived during a year are typically approximated by the average of the population sizes in the beginning and at the end of the year. So, if $V(t)$ is the size of the population of interest at exact time t, the person years lived during $[t, t+1)$ are approximated as $K(t) \approx (V(t) + V(t+1))/2$. Consider two cases. (i) Let the population of interest be *those in age x at exact time t* (meaning those whose exact age is in the interval $[x, x+1)$ at exact time t). Referring to Figure 1 of Chapter 2 again, let V_{AD} be the number of life lines crossing AD, and let V_{CE} be the number of life lines crossing *CE*. Suppose the number of deaths in the parallelogram *ACED* is D_{ACDE}. Then the *o/e* rate is approximately $D_{ACDE}/(V_{AD} + V_{CE})/2$. (ii) Let the population of interest be *those in age x during t.* In obvious notation, the approximate *o/e* rate is $D_{ABCD}/(V_{AD} + V_{BC})/2$. Note that it is not easy to express the latter notion in words, in an unequivocal manner. The difficulty comes up when individual level data are available, and one wants to use a computer to compute the person years exactly. The algorithms are surprisingly tricky (e.g., Breslow and Day 1987, 362), especially if the population is open.[3]

Returning to inference, we note that classical results of maximum likelihood estimation can be used to draw inferences concerning μ. Subject to regularity conditions on censoring, as a MLE the *o/e* rate, $\hat{\mu}$, is a consistent, asymptotically normal estimator of μ as the number of cases gets large (e.g., Rao 1973, 365; also Chapter 1, Section 3). The asymptotic variance of the *o/e* rate is $\text{Var}(\hat{\mu}) = -1/\ell''(\mu)$. Since $\ell(\mu) = m\log(\mu) - \mu(K + K')$, we have that $\ell''(\mu) = -m/\mu^2$, and the asymptotic variance is μ^2/m. Hence, in large samples (say, when the expected count m is > 30) we can test, for example, the hypothesis H_0: $\mu = \mu_0$ by noting that the distribution of the standardized variable $Z = m^{1/2}(\hat{\mu} - \mu_0)/\mu_0$ is approximately normal $N(0, 1)$ when H_0 is true. We leave it as an exercise to show that confidence intervals can similarly be constructed for μ, and for its monotone functions such as the survival probability $e^{-\mu t}, t > 0$.

As an aside, we note a partial justification of the Poisson model for demographic events. There is a relation between the estimate of variance of the *o/e* rate under the exponential model, and under a Poisson model. Under the exponential model, we estimate the variance of the MLE $\hat{\mu}$ by $\hat{\mu}^2/m$. On the other hand, suppose we condition on the person years lived, K and K', and consider m to have a Poisson distribution with mean $\mu(K + K')$, where $K + K'$ is assumed to be a known constant. Then, the MLE of μ is formally given by (1.3) and its variance $\mu/(K + K')$ is estimated as $\hat{\mu}^2/m$. The equality of the estimates under the exponential and

[3] Software capable of computing person years is increasingly becoming available, e.g., Stata, $S+$, R, and SAS have such modules.

Poisson models is of interest, because under the exponential model the count m does not have an exact Poisson distribution. In fact, when there is no censoring, $m = n$ with probability one, or m is fixed. The above derivation can be used as a justification of a Poisson *assumption* in many demographic settings in which other arguments cannot be used (cf., Section 5).

In all its simplicity the exponential model may serve as a building block for more complex models, when population heterogeneity is introduced in one way or another.

Example 1.3. Cross-Sectional Heterogeneity of Constant Hazard Rates. Suppose the lifetimes of those born at $t > 0$ have a constant hazard rate $\mu e^{-\alpha t}$, where $\mu > 0$ and $\alpha > 0$. Those who are in age x at t, and thus were born at $t - x$, have hazard $\mu e^{-\alpha(t-x)} = \mu e^{-\alpha t} e^{\alpha x}$. Notice that the survivors at t are a heterogeneous population with the hazard increasing exponentially with age $x > 0$. If the quality of industrial production improves over time, such a pattern of hazard rates might be observed in a cross sectional sample of products. It is not unthinkable that human cohorts adopt increasingly healthier life styles and benefit from public health improvements. If so, one would expect a similar patterns in human period mortality. $\Diamond$

Example 1.4. Gamma Distribution for Frailty. Again consider that an individual has a constant hazard, μ, but suppose that μ is heterogeneous in the population. One convenient model is that μ has probability density function $g(\mu; \alpha, \beta) = \beta^\alpha \mu^{\alpha-1} e^{-\beta\mu} / \Gamma(\alpha)$ for $\mu > 0$, with $\alpha > 0$, $\beta > 0$, and $\Gamma(\alpha) = \int_0^\infty x^{\alpha-1} e^{-x} dx$. This distribution is known as the *gamma distribution* with shape parameter α and scale parameter β, and it has mean α/β (e.g., DeGroot 1987, 286–290). The *gamma function* $\Gamma(\alpha)$ is a generalization of the factorial, and satisfies $\Gamma(n) = (n-1)!$ for positive integer n and $\Gamma(x+1) = x\Gamma(x)$ more generally. Suppose we pick an individual at random. Then, the probability that he is alive in age $x > 0$ is $(\beta^\alpha/\Gamma(\alpha)) \int_0^\infty e^{-\mu x} \mu^{\alpha-1} e^{-\beta\mu}\, d\mu = (\beta/(x+\beta))^\alpha \int_0^\infty g(\mu; \alpha, x+\beta)\, d\mu = (\beta/(x+\beta))^\alpha$. (You can check the first equality by substituting in the definition of $g(\mu; \alpha, x+\beta)$.) Although we do not exploit the fact here, we note also that the gamma distribution itself serves as a model of lifetimes and includes the exponential distribution as a special case ($\alpha = 1$). $\Diamond$

The gamma distribution describes the heterogeneity of the population in this example. The bigger μ is, the higher the hazard is. Therefore, it is called a *frailty distribution*. Notice that if we would use the average hazard α/β to assess the probability of surviving to $x > 0$, the result would be $\exp(-(\alpha/\beta)x)$. Because the probability of survival $e^{-\mu x}$ is a convex function of the hazard μ, it follows from Jensen's inequality (Complement 8) that the probability of surviving to x, at average hazard, is *smaller* than the average of the probabilities of survival, $(\beta/(x+\beta))^\alpha$. Since Jensen's inequality does not depend on the particular form of the distribution of hazards, the result actually holds for *any* frailty distribution with a finite expectation. We will see below that the result can be extended into a much more general form still.

2. General Waiting Time

Section 2.1, below, introduces the concept of a hazard function and relates it to probability of survival function. In Section 2.2, we discuss how to calculate the expectation of life, given the survival function. We also define life table populations and stable populations, consider the effect of heterogeneity and change of mortality on survival, and apply the concepts to pension funding. Section 2.3 discusses estimation of the survival function and cumulative hazard function from individual level data. Section 2.4 considers aggregated data.

2.1. Hazards and Survival Probabilities

We derive now a basic identity between hazard rates and survival probabilities. Many, but not all, of the details of this development will carry over to the analysis of multistate demographic systems in Chapter 6.

Let X be a nonnegative random variable representing a waiting time. Again, to simplify language, we will be talking about a length of life. Recall the definition, $p(x) = P(X > x)$.[4] Let us assume that $p(0) = 1$. Assume also that there is a piece-wise right-continuous function $\mu(.) \geq 0$ on $[0, \infty)$ such that $P(x < X \leq x + h | X > x) = \mu(x)h + o(h)$, where $o(h)/h \to 0$ when $h \to 0$. This a mathematical way of saying that the conditional probability of dying at or before age $x + h$, given survival beyond age x, is approximately proportional to h, with the constant of proportionality depending on x. The function $\mu(.)$ will be called a *hazard*.[5] In mortality analysis it has traditionally been called *force of mortality*.

In terms of the survival function $p(.)$ the condition can be written as

$$\frac{p(x) - p(x + h)}{p(x)} = \mu(x)h + o(h). \tag{2.1}$$

Dividing both sides by h and letting $h \to 0$, we obtain a differential equation

$$\frac{p'(x)}{p(x)} = -\mu(x). \tag{2.2}$$

Since the left hand side equals the derivative $d/dx \log p(x)$, we have

$$p(x) = \exp\left(-\int_0^x \mu(t)\,dt + C\right). \tag{2.3}$$

[4] In demography, survival is traditionally described via a function $\ell(x)$ defined as $100{,}000 \times p(x)$. The idea is that we follow a cohort of 100,000 individuals, and $\ell(x)$ gives the expected number alive at age x.

[5] Terms *hazard rate*, *incidence*, *incidence density*, *incidence rate*, *intensity*, or *instantaneous probability* are also sometimes used for $\mu(.)$.

The constant C must satisfy the boundary condition $p(0) = 1$, so we must have $C = 0$. In summary, we have the representation

$$p(x) = \exp(-\Lambda(x)), \tag{2.4}$$

where

$$\Lambda(x) = \int_0^x \mu(t)\, dt \tag{2.5}$$

is the so-called *cumulative hazard*. Formula (2.4) shows that the distribution of a general waiting time can be obtained from the exponential distribution with parameter $\mu = 1$ by transforming the time axis: $p(x)$ at time $x > 0$ is the same as the survival probability under the exponential model at time $\Lambda(x)$. The estimation of $\mu(.)$, $\Lambda(.)$, and $p(.)$ from individual-level data will be discussed in Section 2.3, and estimation from grouped data in Section 2.4. In Section 3, we discuss a numerical procedure for estimating $p(.)$ given estimates of $\mu(x)$ for integer ages x.

Example 2.1. Weibull Distribution. If $\mu(x) = (\beta/\alpha)(x/\alpha)^{\beta-1}$ for some $\alpha > 0$ and $\beta > 0$, then we have a so-called *Weibull distribution* with $\Lambda(x) = (x/\alpha)^{\beta}$. We see from the formula that α influences the scale of the distribution, whereas β determines its shape. For $\beta > 1$ the hazard is increasing, for $\beta < 1$ it is decreasing. Taking $\beta = 1$ we get, as a special case, the exponential distribution Exp$(1/\alpha)$. ◊

Example. 2.2. Linear Survival Functions. Consider the ages $t \in [x, x+1)$, and assume that $\mu(t) = b_x/(1 - b_x(t - x))$ for some $b_x < 1$. Then, $p(t)/p(x) = 1 - b_x(t - x)$. In other words, the function $p(.)$ is *linear* on interval $[x, x+1)$. On the other hand, if $p(t)/p(x) = 1 - b_x(t - x)$ on $[x, x+1)$, then $\mu(t) = -d/dt \log p(t) = b_x/(1 - b_x(t - x))$, or it is of the form given. The linearity of the survival function means that the deaths are expected to be *uniformly distributed* over the interval $[x, x+1)$. This is in contrast to the exponential model, in which a constant hazard leads to an exponential decline in the numbers of deaths, as the population at risk is depleted. We see from Figure 2 that this model is more realistic than the exponential model in ages, say, $x > 30$. We will show in Example 2.9 how this model leads to the so-called actuarial estimator of survival. ◊

Example 2.3. Balducci Model for Survival Function. G. Balducci proposed the following model in 1920. Let $t \in [x, x+1)$, and assume that $\mu(t) = a_x/(1 + a_x(t - x))$ for some $a_x > 0$. Then, $p(t)/p(x) = 1/(1 + a_x(t - x))$. In this case the declining hazard leads to an even faster decline in the numbers of deaths during the interval $[x, x+1)$ than the exponential model. We see from Figure 2 that this model is more realistic than the other two for the youngest ages such as $x < 15$. ◊

Example 2.4. Competing Risks. Adding demographic realism to Example 1.2, suppose there are k causes of death with hazards $\mu_1(x), \ldots, \mu_k(x)$ in age x. Then, the overall hazard of death can be taken as $\mu(x) = \mu_1(x) + \cdots + \mu_k(x)$. This is the classical model of *competing risks of death*. Forecasts of future mortality are sometimes formulated in terms of cause-specific death rates. For example, the

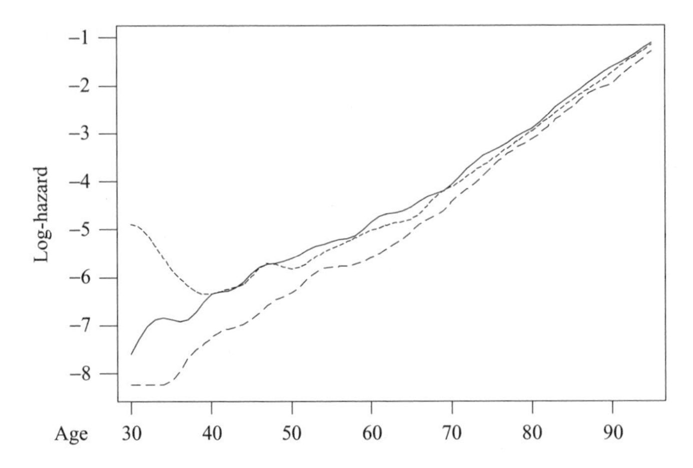

FIGURE 1. Log of Mortality Hazard for the Married (Dashed Line), Widowed (Dotted Line), and Single and Divorced (Solid Line) Women in Finland, in 1998.

U.S. Office of the Actuary (1987) has used the following classification: (1) heart disease, (2) cancer, (3) vascular diseases, (3) violence, (4) respiratory diseases, (5) diseases of the infancy, (6) digestive diseases, (7) diabetes mellitus, (8) cirrhosis of the liver, and (9) other diseases. ◊

Mortality can vary by many characteristics of the individual, sometimes in an unexpected manner.

Example 2.5. Mortality and Marital Status in Finland. Figure 1 shows estimates of the logarithms of age-specific mortality rates for females in Finland in 1998 by marital status. The rates were calculated from single year of age data provided by Statistics Finland. The estimates have been smoothed using a robust smoother (RSMOOTH of Minitab, which applies a carefully selected sequence of moving averages and running medians to the data). We see that the mortality of those who are married is the lowest, and the mortality of the singles and the divorced is the highest. The mortality of the widows is in between, except in young ages. We will come back to the latter issue in Example 3.2 of Chapter 5. There appears not to be agreement as to whether marriage lowers mortality hazards by providing a less risky life style, or whether there is a selection mechanism in operation such that those who are more "fit" are also more likely to find a spouse (e.g., Gove 1973; Hu and Goldman 1990; Lillard and Panis 1996). We will consider this problem in Section 1.5 of Chapter 6, and show that both points of view may have a certain justification. ◊

Note that the approximate linearity of the log-hazard as function of age $x(> 55)$ in Figure 1 is not compatible with a Weibull distribution. However, it is compatible with the *Gompertz model* $\mu(x) = \alpha c^x$, with $\alpha, c > 0$, that was introduced by B. Gompertz in 1825. Note also that the hazards of the three marital statuses are roughly parallel in the log-scale in higher ages. This implies that their hazards are equal, up to a multiplicative constant. That is, we have approximately a *proportional hazards* situation in the higher ages.

2.2. Life Expectancies and Stable Populations

Instead of relying on parametric models, demographers have traditionally described mortality nonparametrically. Starting from *o/e* rates of the type (1.3) and, e.g., the linearity hypothesis of Example 2.2, one obtains estimates of $p(x)$ for $x = 0, 1, 2, \ldots$ The resulting estimates are then presented (usually as multiplied by 100,000) in a tabular form, together with some related quantities.[6] This is the *life table*. Shryock and Siegel (1976), Chiang (1968, 1984), and Smith (1992) provide details of the many variants that are in use. With the development of user-friendly computer programs, tabular representations of the relevant quantities are gradually becoming obsolete. Nevertheless, life table is a central concept in demographic theory.

2.2.1. Life Expectancy

The expectation of the general waiting time can be calculated using (1.1). However, the following result is often simpler. Define $I(t)$ to be the *indicator process* of a waiting time X, or $I(t) = 1$ if $X > t$, and $I(t) = 0$ otherwise. It follows that we can represent X in a roundabout way, as follows:

$$X = \int_0^\infty I(t)\, dt. \tag{2.6}$$

We may call this an *integral representation of a waiting time X*. Note that the probability that $X > t$ equals $p(t) = E[I(t)]$. Take the expectation of both sides in (2.6), and change the order of expectation and integration (which is permissible here because $I(t) \geq 0$; Chung 1974, 59) to get the formula

$$E[X] = \int_0^\infty p(t)\, dt. \tag{2.7}$$

Alternative methods of proof that rely on calculus are given in exercises (see also Çinlar 1975, 24–25).

Proving the result $E[X] = 1/\mu$ for the exponential distribution is a one-step integration using (2.7).

In demography, special notation is used for life expectancies. The additional life expectancy, given survival to age x, is denoted by e_x. (Sometimes e_x is used for the discrete time version, and e_x° for continuous time. We will not make the distinction.) Using our notation this is $e_x = E[X - x | X > x]$. Since the conditional

[6] Thus, instead of speaking of a "nonparametric" representation, one could equally well say that a very high-dimensional parametric model is used!

probability of surviving to age $x+t$ given survival to age x, is $p(x+t)/p(x) = \exp(-(\Lambda(x+t) - \Lambda(x)))$, we can also write

$$e_x = \int_0^\infty p(x+z)/p(x)\,dz. \\ = \int_0^\infty \exp\left(-\int_0^z \mu(x+s)ds\right) dz. \tag{2.8}$$

Since only weak assumptions are typically made concerning the hazard rate $\mu(.)$, the estimation of $p(.)$, $\Lambda(.)$, or $\mu(.)$ itself, is difficult. A relatively crude approach is as follows. If one approximates $\mu(.)$ by a piecewise constant function, then the theory of Section 1 can be used to derive the MLEs of the constant hazards. For example, if we assume that $\mu(t) = \mu_x$ for $t \in [x, x+h)$ and we know the total number of deaths and the total number of person years lived in the population during age $[x, x+h)$, then $\hat{\mu}_x$ is simply the *o/e* rate (1.3). Similarly, if we define the increment of the hazard as,

$$\Lambda_{x,h} = \Lambda(x+h) - \Lambda(x), \tag{2.9}$$

then we can estimate $\hat{\Lambda}_{x,h}$ by $h\hat{\mu}_x$. If $h = 1$ and x takes integer values, for example, the estimate of $p(x)$ would be $\hat{p}(x) = \exp(-\hat{\mu}_0 - \cdots - \hat{\mu}_{x-1})$. Under a piecewise constant hazard model, we can estimate $\text{Var}(\hat{\mu}_x) \approx \hat{\mu}_x{}^2/m_x$, where m_x is the number of deaths in age x. Relying on a normal approximation, a 95% confidence interval for $p(x)$ can be given approximately as $\hat{p}(x)\exp(\pm 1.96 \times (\hat{\mu}_0{}^2/m_0 + \cdots + \hat{\mu}_{x-1}{}^2/m_{x-1})^{1/2})$, for example. Chiang (1968, 1984) and Smith (1992) provide extensive variance formulas under several alternative models.

Life expectancy is one of the most widely used summary measures of mortality. The suggestive terminology may lead some non-demographers to think that life expectancy at birth, or e_0, is a forecast made at the given time for how long a particular birth cohort might live. However, life expectancy is almost universally calculated from age-specific data of a given period. Thus it typically refers to a synthetic cohort rather than an actual cohort. An alternative concept of synthetic cohort is considered by Coleman (1997) in the context of diffusion of HIV infection in a social network.

Apart from a limited number of analytical models, numerical integration must be used to calculate the life expectancies e_x in (2.8). Suppose $p(x)$ has been specified for a set of ages x, say $x = 0, 1, 2, \ldots$ The most common approximation assumes the linearity of $p(t)$ in each interval $[x, x+1)$. This is equivalent to the so-called *trapezoidal method* of numerical integration. It leads to the approximate formula

$$e_x \approx \frac{1}{2} + \sum_{t=1}^{\infty} p(x+t)/p(x). \tag{2.10}$$

The formula can be used independently of the way $p(x)$ has been estimated. In particular, it follows from Example 2.2 that (2.10) is compatible with hazards of the form $\mu(t) = b_x/(1 - b_x(t - x))$.

2.2.2. *Life Table Populations and Stable Populations*

Life expectancies and survival probabilities have a peculiar interpretation in demography that appears not to be generally known among statisticians. Suppose individuals are born into a population at a constant rate of 1 person per unit of time, and the survival probability of a person aged x is $p(x)$, unchanging over time. Then, at any given time we expect there to be $p(x)dx$ individuals in the narrow age interval $[x, x + dx]$. The expected total size of this population is given by the right hand side of (2.7) (draw a Lexis diagram!). The function $p(.)$ is then the density of the expected population. (Note that it integrates to $E[X]$, not to 1.) The expected population is called the *life table population* determined by $p(.)$. Assume that $E[X]$ is finite. It follows that in the life table population the expected person years per new born are specified by the right hand side of (2.7). Thus, $1/E[X]$ can be interpreted as an *o/e* rate. However, as the population size does not change over time, there must also be one death per year, so the *o/e* rate $1/E[X]$ can also be interpreted as the (crude) *life table mortality rate*, calculated as number of deaths divided by total population size.

As part of classical mathematical demography, the theory of life table populations is deterministic. It typically assumes a continuous population density and does not require the size of the total population to be an integer. As shown by Keiding and Hoem (1976), the theory can be reconciled with statistical models of the type we discuss here. Instead of pursuing those details, we will use the traditional language when discussing life table populations, stable populations (below), and later in discussing population renewal.

The population interpretation of life expectancies can be carried further. Suppose that individuals are born at rate $Be^{\rho t}$ where ρ is some constant. Consider the number of people in age x at time t. They were born at time $t - x$, so their number is $Be^{\rho(t-x)}p(x)$. Let $V(t)$ be the size of the population at time t, or

$$V(t) = Be^{\rho t} \int_0^\infty e^{-\rho x} p(x)\, dx. \tag{2.11}$$

We see that the population grows (or declines) exponentially at rate ρ, and its age distribution is proportional to $e^{-\rho x}p(x)$. Note the effect of growth on age distribution. If ρ is increased, the age distribution becomes younger, if ρ is decreased, the age distribution becomes older. Exponentially growing populations with unchanging age distribution are called *stable* (e.g. Coale 1972). If $\rho = 0$ we have a life table population. Since it does not grow, it is called *stationary*.

Although the assumption underlying stable populations (exponential births, unchanging mortality schedule, no migration) are highly restrictive, the model can

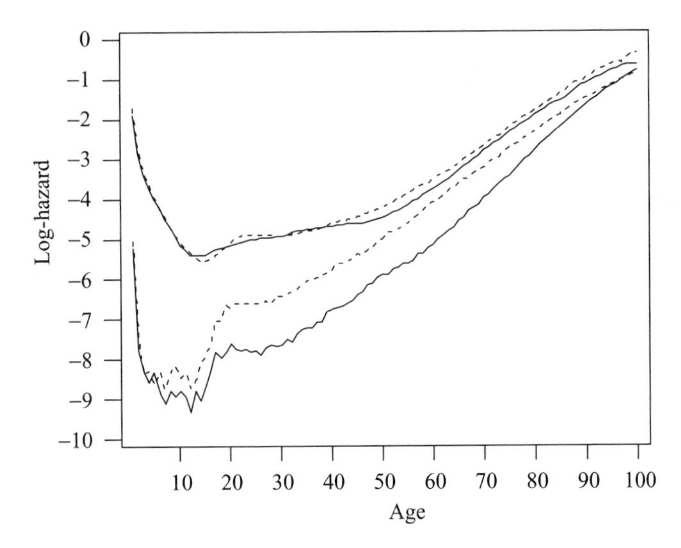

FIGURE 2. Log of the Hazard Increment of Mortality in Finland in 1881-1890 (Upper Curves) and 1986–1990 (Lower Curves), for Females (Solid line) and Males (Dashed Line).

be valuable in situations in which the data are poor. For example, since the growth rate, life table population, and age distribution are functionally related, knowing two of them allows us to guess the third. For a list of relations one can use, see Keyfitz (1977, 174–185).

2.2.3. *Changing Mortality*

What happens to life expectancy when mortality changes over time? We consider first some historical data and then an analytical example.

Figure 2 shows empirical estimates of the logarithm of the hazard increments (2.9) with $h = 1$ for $x = 0, \ldots, 99$, based on Finnish data from 1881-1890, and from 1986–1990. We have calculated the estimates as $\log(\hat{\Lambda}_{x,1}) = \log(-\log p(x+1)/p(x))$ based on Tables 4A and 4B of Kannisto and Nieminen (1996) that give the probabilities of death $1 - p(x+1)/p(x)$.

The figure shows first that mortality in ages 0 to 45 has decreased dramatically during the hundred year period. In higher ages the decrease has been much less pronounced. To appreciate the difference, note that around age 13 the hazard declined from about $e^{-5.3} \approx 0.005$ to $e^{-8.7} \approx 0.00017$, whereas in age 70 the decline was from about $e^{-3.2} \approx 0.041$ to $e^{-4.2} \approx 0.015$. In other words, in the younger ages the earlier hazard was about 30-fold as compared to the rate a century later, whereas in the older ages is was merely 3-fold. Second, in relative terms, female life expectancies have remained steadily higher than male life expectancies. During 1881–1890 we had e_0 of 41.3 for males and 44.1 for females, or the female figure was 7% higher than the male figure. In 1986–1990 we had e_0's of 70.7 and 78.8, respectively, or the female figure was nearly 11% higher. In older ages the change was even more pronounced. We had e_{50}'s of 19.4 and 21.1 during

1881–1890, and e_{50}'s of 24.6 and 30.7 during 1986–1990 for males and females, respectively. Or the female advantage had grown from 9% to 25%.

Past mortality schedules form the basis on which forecasts of future mortality must be based, in one way or another. To set the reader thinking about the problem, let us consider two simple (even simplistic!) approaches. Suppose we assume that life expectancy increases linearly. Since the improvement for males was 29.4, and for females 34.7 years, during 1890-1990, the linearity assumption would imply a forecast of 100.1 for males and 113.5 for females, in 2090. On the other hand, let $\Lambda_{x,1}(t)$ be the hazard increment of year t, and define $y(x,t) = \log \Lambda_{x,1}(t)$. From the data of Figure 2 we get estimates of $y(x, 1890)$ and $y(x, 1990)$. Consider a year $t > 1990$. A linear trend extrapolation (in the log-scale) would assume that $\hat{y}(x,t) = y(x, 1990) + [y(x, 1990) - y(x, 1890)](t - 1990)/100$. Taking $t = 2090$, we get the schedule $\hat{y}(x, 2090)$, and the corresponding survival probabilities $\hat{p}(x,t) = \exp[-\exp\{\hat{y}(0, 2090)\} - \cdots - \exp\{\hat{y}(x - 1, 2090)\}]$. The implied life expectancy would be $\hat{e}_0(2090) = 78.7$ for males, and $\hat{e}_0(2090) = 87.2$ for females. These forecasts are over twenty years less than those based on the linearity of the life expectancy itself. The methods that start from the mortality rates but put more weight on the most recent rates of decline lead to intermediate values. For example, a recent Finnish forecast puts the median of the predictive distribution (Section 2 of Chapter 9) of e_0 for the males as 83.8 in 2065, and as 88.2 for the females. In either case the loglinear model leads to an eventual deceleration in the increase of life expectancy. During the period we are considering Finnish life expectancy appears to be a slightly *concave* function of time.

In general, there are infinitely many mortality schedules that correspond to a given life expectancy. A connection can be established, if mortality is parametrized in some way.

Example 2.6. Effect of Changes in Hazards on Life Expectancy. Suppose the hazard of mortality in age x at time $t \geq 0$ is of the form $\mu(x,t) = \mu(x) - g(t)\delta(x)$, where $g(0) = 0$, $\delta(x) \geq 0$ for $x \geq 0$, and let the corresponding life expectancy at birth be $e_0(t)$. How does $e_0(t)$ change over time? One way to investigate that is to calculate the derivative with respect to t. Recall (2.5) and define

$$\Delta(x) = \int_0^x \delta(s)\, ds. \tag{2.12}$$

Differentiating under the integral sign yields

$$\frac{d}{dt} e_0(t) = g'(t) \int_0^\infty p(x,t)\Delta(x)\, dx. \tag{2.13}$$

For example, if $g(t) = t$, then $g'(t) = 1$ and $p(x,t) = p(x,0)e^{\Delta(x)t}$. In this case, as t increases, the derivative of $e_0(t)$ increases. Therefore, the graph of $e_0(t)$ is *convex* if the decline in mortality rates is linear in each age. Of course, linear decline cannot continue forever. ◊

2.2.4. *Basics of Pension Funding*

Suppose a person starts working at age $\alpha > 0$ and retires at age $\beta > \alpha$. During work the person pays continuously an amount c per year to a fund that earns an interest r. This entitles the worker to a unit pension (or *annuity*) per year that is paid continuously until death. How large should c be? To determine c we discount both the contributions and the pension payments to time of birth. The discounted value of all contributions is

$$C = c \int_{\alpha}^{\beta} e^{-rt} \, I(t) \, dt, \tag{2.14}$$

where $I(t)$ is the indicator process of time at death, as in (2.6). Suppose the highest age is ω, so $p(\omega) = 0$. The discounted value of pensions is

$$A = \int_{\beta}^{\omega} e^{-rt} \, I(t) \, dt, \tag{2.15}$$

Setting $E[C] = E[A]$ yields an equation from which c can be solved as

$$c = \int_{\beta}^{\omega} e^{-rt} \, p(t) \, dt \Bigg/ \int_{\alpha}^{\beta} e^{-rt} \, p(t) \, dt. \tag{2.16}$$

In an infinite population the laws of large numbers would guarantee that this value of c would exactly balance the contributions and payments. In practice, a pension institution would have to take into account that the number of participants in the scheme is finite.

Suppose we have n participants. Let C_i be the contribution and A_i the pension of person $i = 1, \ldots, n$, and define $D_i = C_i - A_i$. Let us determine c so that with probability 0.999 the fund is sufficient to cover the pensions. Defining

$$D = \sum_{i=1}^{n} D_i, \tag{2.17}$$

the task is to determine c so that $P(D \geq 0) \geq 0.999$. An approximate way of doing this is to appeal to the central limit theorem (CLT). Suppose the D_i's are independent with common mean $E[D_i] = \mu$ and variance $\mathrm{Var}(D_i) = \sigma^2$, $i = 1, \ldots, n$. It follows from the CLT that $Z = (D - n\mu)/(n^{1/2}\sigma) \sim N(0, 1)$ asymptotically, as $n \to \infty$. Note that the event $\{D \geq 0\}$ is the same as the event $\{Z \geq -\mu n^{1/2}/\sigma\}$. Thus the condition is $\mu n^{1/2}/\sigma = 3.09$, the 0.999 fractile of the N(0, 1) distribution. Here μ and σ depend on c. We indicate in Exercises 18 and 19 how the solution can be found.

The system considered thus far is *funded* meaning that contributions are collected into a fund from which annuities are later paid. Most current pension systems are not funded, however. Instead, they are *Pay-As-You-Go (PAYG)*, which means that current workers pay the pensions of current pensioners. In a *defined benefit* system pension rules determine how much each pensioner is entitled to get and contribution rates are set so that the needs are met, each year. In a *defined contribution* system the contribution rate is fixed and the level of pensions may fluctuate.

Consider, for example, a defined benefit PAYG system under a stable population (2.11) that grows at the rate ρ. As above, we simplify and assume that contributions are made at the constant rate c. What value of c produces a unit annuity for each pensioner? A moment's reflection shows that we must have

$$c = \int_{\beta}^{\omega} e^{-\rho x} p(x)\, dx \Big/ \int_{\alpha}^{\beta} e^{-\rho x} p(x)\, dx. \tag{2.18}$$

The expression is formally the same as (2.16) but population growth rate replaces the interest rate. Note that c is a declining function of ρ: the smaller the growth rate, the higher the contribution rate. Although the stable population model is based on highly restrictive assumptions, (2.18) indicates correctly the root cause of the problems that have become acute in many countries at the turn of the millennium. Populations of many industrialized countries are expected turn into a decline, so the PAYG principle is becoming unsustainable.

2.2.5. *Effect of Heterogeneity*

Returning to the problem of heterogeneity (cf., Example 1.4 and the discussion thereafter), suppose $\xi > 0$ is a measure of a person's frailty, such that the person's hazard is $\mu(x, \xi) = \mu(x)\xi$. The probability of surviving to age $x > 0$, $p(x, \xi) = \exp(-\Lambda(x)\xi)$, is a convex function of the frailty ξ. Therefore, by Jensen's inequality the probability of survival for a person with average frailty $E[\xi]$, or $\exp(-\Lambda(x)E[\xi])$, is smaller than the average probability of survival $E[\exp(-\Lambda(x)\xi)]$. Define life expectancy at frailty ξ as $e_0(\xi) = \int p(x, \xi)\, dx$. By changing the order of integration we have that $E[e_0(\xi)] = \int E[p(x, \xi)]\, dx \geq \int p(x, E[\xi])\, dx$. Therefore, *the life expectancy of a person with average frailty is smaller than the average life expectancy of a population*, whenever frailty influences the hazard of mortality multiplicatively.

We caution the reader not to misinterpret the above result. For example, a person with median frailty *does* have a median life expectancy, because under the assumed model, life expectancy is a decreasing function of ξ.

2.3. *Kaplan-Meier and Nelson-Aalen Estimators*

Although our primary interest will be with grouped data, as noted in Section 5 of Chapter 2, individual level data are increasingly becoming available from population registries, epidemiologic databases, and reconstructed historical records. Kaplan and Meier (1958) discussed an estimator of $p(.)$ using such data, under censoring.

Consider a cohort of size n. Let X_i be the time until death, and let c_i be the censoring time, for individual $i = 1, \ldots, n$. Define the observable *withdrawal times* $T_i = \min\{X_i, c_i\}$ and order them: $0 \leq T_{(1)} < T_{(2)} < \cdots < T_{(n)}$. Define the *indicators of not being censored*: $\delta_{(i)} = 1$ if $T_{(i)}$ corresponds to a death, and $\delta_{(i)} = 0$ if it corresponds to a censoring. Then, we may estimate $p(t)$ for any $t \geq 0$ by

$$\hat{p}(t) = \prod_{T_{(i)} \leq t} \left\{ \frac{n-i}{n-i+1} \right\}^{\delta_{(i)}}. \tag{2.19}$$

This is the celebrated *Kaplan-Meier* or *product limit* estimator. To understand its rationale, suppose $n = 4$ and the withdrawal times are 1.0, 1.5, 2.5, and 4.0. Consider $p(t)$ for $1.5 \leq t < 2.5$, so two withdrawals have occurred by t. If neither was a censoring, the estimate is (3/4)(2/3) = 2/4, or it is the fraction remaining in the cohort. If the second withdrawal was a censoring, then we have seen one death out of four, and the estimate is 3/4. If the first withdrawal was a censoring and the second was not, then we have seen one death out of three, and the estimate is 2/3. In general, a death decreases the estimate by the fraction it represents out of those remaining in the cohort.

Example 2.7. Life Expectancy Calculation from Kaplan-Meier Estimates. Expected waiting times (such as a life expectancies) can be calculated based on Kaplan-Meier estimates. Take $x = 0$ in (2.10), and suppose that in the example above we have no censoring. Then, we have $\hat{p}(1) = 3/4$, $\hat{p}(2) = 1/2$, $\hat{p}(3) = 1/4$, and $\hat{p}(4) = 0$. Therefore, $\hat{e}_0 = 1/2 + 3/4 + 1/2 + 1/4 + 0 = 2$. Since the Kaplan-Meier estimator is a step function, the integral (2.8) can be evaluated directly as $1.0 \times 1.0 + 0.5 \times 0.75 + 1.0 \times 0.5 + 1.5 \times 0.25 = 2.25$. This is the correct value of the integral that avoids the approximation involved in the trapezoidal method. In order not to forget first principles, recall that the latter figure must agree with the simple average of the survival times, when there is no censoring. And it does: $(1.0 + 1.5 + 2.5 + 4.0)/4 = 2.25$! ◊

The same principle applies if there are tied waiting times: if r persons are at risk and d die simultaneously at time t', then from t' on a factor $(r - d)/r$ is included in the product (2.19). The only difficulty arises if d deaths and c censorings occur simultaneously among r who are at risk at t'. Typically such an event would be an artifact due to imprecise data collection. If we place the censorings first, then the term $(r - c - d)/(r - c)$ is included in (2.19) from t' on. If we place the deaths first, then the term $(r - d)/r$ is included. The latter is always bigger. In this way we can bracket the value of the estimator we would get if the exact withdrawal times were known.

Example 2.8. Survival Probabilities for Habsburgs. Figure 3 has a graph of Kaplan-Meier estimates of survival probabilities for the males and females of the Habsburgs family of Austria. The data relate to 175 members of the main line of the family through which the throne was passed from generation to the next. The birth years range from 1218 to 1895. The survival curves are for females and males separately. Sex was not known for 10 of the members, so those have been left out. These individuals have typically died very young, so leaving them out exaggerates survival. We see that after the first year or so, the survival curves are surprisingly linear. From the right triangle that has height 0.85 at age 1, and the length of the

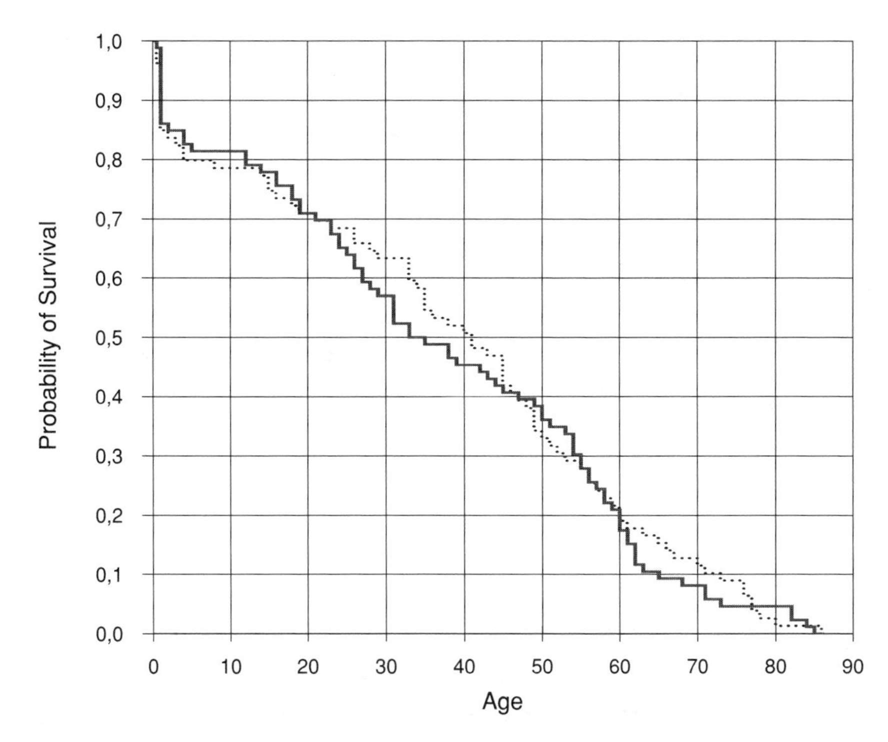

FIGURE 3. Survival Probabilities for Females (Solid) and Males (Dashed) Among the Members of the Main Line of the Family of Habsburgs.

base of 85 years, we can estimate that the life expectancy is approximately 36 years. The correct arithmetic result (that includes those whose sex is not available) is 35 years. More details about the data will be given in Chapter 5, starting from Example 2.1. ◊

The estimation of the cumulative hazard could be based on the Kaplan-Meier estimator, by taking $\hat{\Lambda}(t) = -\log \hat{p}(t)$. However, an alternative that generalizes more easily to regression settings is as follows. Suppose the interval $[0, t]$ is divided into short subintervals of length h. If there are n individuals in the population in the beginning of the interval $[x, x + h)$ and the probability of two or more deaths is negligible, the probability of exactly one death during the interval is approximately $n\mu(x)h$. If there is a death, then a moment estimator for the hazard increment is $\hat{\Lambda}_{x,h} = 1/n$. If there is no death, the moment estimator is $= 0$. Combining the estimates from the subintervals we obtain the so-called *Nelson-Aalen estimator*

$$\hat{\Lambda}(t) = \sum_{T_{(i)} \leq t} \frac{\delta_{(i)}}{n - i + 1}. \tag{2.20}$$

This estimator was independently introduced by Nelson (1969) and Aalen (1976). A comprehensive discussion of the Kaplan-Meier and Nelson-Aalen estimators is given in Andersen et al. (1993).

In survival theory literature it has become customary to write the sum in (2.20) as a stochastic *Stieltjes integral* (e.g., Klein and Moeschberger 1997, 70–79). Suppose we follow a cohort of size n. Let $Y(t)$ be the size of the cohort at time t, and let $N(t)$ be the number of deaths that have occurred during time $[0, t]$. Then, we have that

$$\hat{\Lambda}(t) = \int_0^t \frac{dN(s)}{Y(s)}, \tag{2.21}$$

if $Y(t) > 0$. The denominator $Y(s)$ keeps track of the size of the population that has neither died nor become censored by s.

2.4. Estimation Based on Occurrence-Exposure Rates

We showed in Section 1 that the *o*/*e* rate is the MLE of the hazard rate if the true hazard is constant. The actuarial method and Balducci hypothesis provide estimators that are based on more realistic models for various ages. Over the years, demographers have devised ever more refined methods that attempt to minimize biases due to an erroneous parametric model. Their motivation is the fact that because the populations being studied usually are large, random variability in the counts is small (compared to the expected values of the counts) and hence, unless models are pushed to extremes, biases from incorrect models can be more detrimental than random error in estimation of parameters. Also balancing this tendency away from parametric models is the fact that the data typically are grouped by year.

A second desideratum involves the intended use of the estimates. Life tables contain various summaries (such as e_x) based on an estimated version of the survival function $p(.)$. For those purposes, all one needs, roughly speaking, is to be able to estimate the one-year survival probabilities $p(x+1)/p(x) = \exp(-\Lambda_{x,1})$ for $x = 0, 1, 2, \ldots$

We continue to use mortality as our paradigm case. Consider a one-year age-interval $[x, x+1)$, and suppose first that data are available from the rectangles of the Lexis diagram (e.g., ABCD in Figure 1 of Chapter 2). Let $k(t)$ be the density of population in age $t \geq 0$ at a fixed time. Define

$$K_x = \int_x^{x+1} k(t)\,dt. \tag{2.22}$$

If the density of the population remains the same during the year in which the observations are made, then K_x is the number of person years lived by the x-year olds during the year. Let us assume this. Suppose the observed *o*/*e* rate is M_x and that we observe M_x's and K_x's. How then to estimate $\mu(.)$? Using the method of moments we equate the observed rate with the expected average hazard of the population in age x,

$$M_x = \int_x^{x+1} \mu(t)k(t)\,dt/K_x. \tag{2.23}$$

Note that $k(t)/K_x$ defines a probability density on $[x, x+1)$ that integrates to one and $\mu(.)$ is assumed to be continuous. By the mean value theorem of integral calculus there is some point $\xi_x \in [x, x+1)$ such that $M_x = \mu(\xi_x)$. In other words, the *o/e* rate estimates $\mu(.)$ at some age between x and $x+1$, but without additional assumptions we don't quite know which.

Keyfitz (1977, 19–21) suggested the following *local linearity* approximation. Suppose that the true rate is linear in interval $[x, x+1)$, say $\mu(t) = \mu_{0,x} + \mu_{1,x}(t - x - 1/2)$ for some constants $\mu_{0,x}$ and $\mu_{1,x}$. Similarly, assume that the population density is piecewise linear, $k(t) = k_{0,x} + k_{1,x}(t - x - 1/2)$ for $t \in [x, x+1)$. It follows that $K_x = k_{0,x}$ and $\Lambda_{x,1} = \mu_{0,x}$. By a direct calculation one can show that the right hand side of (2.23) is equal to $\mu_{0,x} + \mu_{1,x}k_{1,x}/(12k_{0,x})$. Thus, if we have estimates of the slopes $\mu_{1,x}$ and $k_{1,x}$, we have from (2.23) the estimate

$$\hat{\Lambda}_{x,1} = M_x - \frac{\hat{\mu}_{1,x}\hat{k}_{1,x}}{12k_{0,x}}. \tag{2.24}$$

Keyfitz suggested that we estimate the slopes by

$$\hat{\mu}_{1,x} = (M_{x+1} - M_{x-1})/2, \quad \hat{k}_{1,x} = (K_{x+1} - K_{x-1})/2. \tag{2.25}$$

These estimates are available for $x = 1, 2, \ldots, \omega - 1$, where ω refers to the open ended age-group $[\omega, \infty)$. One could thus obtain the estimates $\hat{\mu}(t) = \hat{\Lambda}_{x,1} + \hat{\mu}_{1,x}(t - x - 1/2)$ for $t \in [x, x+1)$.

Keyfitz's approach is a reasonable one. It takes care of the first order deviation from constancy both in $\mu(.)$ and $k(.)$. It also has the merit of being non-iterative. Although the estimates $\hat{\mu}(t)$ typically are not continuous, a continuous estimate of the whole curve $\mu(.)$ can be obtained using Keyfitz's method. Under the assumption of piecewise linearity for $\mu(.)$ and $k(.)$, it follows that $\mu_{0,x} = \mu(x + 1/2)$. Therefore, the right hand side of (2.24) can also be interpreted as an estimate of the mid-interval mortality $\mu(x + 1/2)$. Having these estimates available we can use any interpolation method (e.g., splines) to get continuous estimates of the intermediate values of $\mu(.)$. Some bias will inevitably be introduced.

Example 2.9. Actuarial Estimator. The so-called *actuarial estimator* of survival is of the form $p(x+1)/p(x) = (2 - M_x)/(2 + M_x)$, where M_x is the age-specific mortality rate of age x. It is probably the most widely used estimator of survival due to its simplicity. As discussed in Exercise 9, it is based on the linearity assumption of Example 2.2. ◊

No matter how the intermediate ages are handled, the highest age must be handled separately. It is typically an open-ended age-group such as 100+. Let the

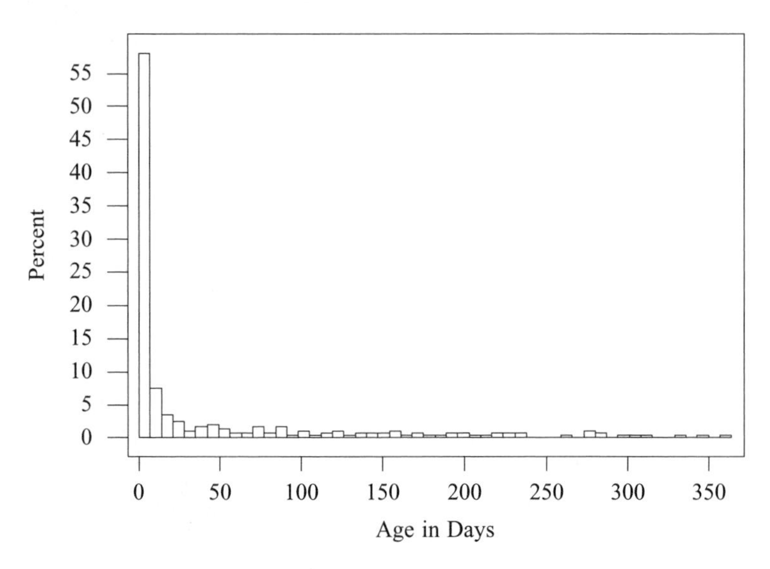

FIGURE 4. The Distribution of Life Times of Those Born in 1994, Who Died in Age Zero, in Finland.

lower end point of the highest age be ω and denote the crude mortality rate in this age as M_ω. Under a constant hazard assumption, the corresponding probability of surviving for one year would be $\exp(-M_\omega)$, and under a more realistic "uniform distribution of death" hypothesis of Example 2.2 the probability would be $(2 - M_\omega)/(2 + M_\omega)$. The numerical effect of the approximation errors can be reduced simply by continuing the calculations to sufficiently high ages so that the populations involved are small. For the purpose of completing a life table, we can equate observed mortality rate with the life table mortality rate, and solve e_ω from the identity $M_\omega = 1/e_\omega$.

For later use we also need estimates of the distribution of life times among those who die during their first year of life.

Example 2.10. Distribution of Death During First Year. Figure 4 has a histogram of the death times of those who died before reaching their first birthday. The data are for the cohort of 1994, in Finland. The columns correspond to weeks. A total of 58% died during the first week, with 23% dying during the first day. A total of 71% died during the first four weeks. The total number of deaths on which these estimates are based is 291. The total number of live births in 1994 was 65,231, so the proportion dying during the first year of life was 0.45% for both sexes combined. For males the proportion dying during the first year of life was 0.5% and for females it was 0.4%. The average number of days lived by those who died before reaching their first birthday was 43, corresponding to 0.12 years. ◊

Example 2.11. Proportion of Deaths During First Days. Later on, we will need estimates of hazards $\mu(0)$ and $\mu(28/365)$, for example. They can be based on parametric models or direct empirical estimates. Consider the data of Example 2.10. The proportion of births dying during the first year of life was 0.0045. Given the low level of mortality, we can also interpret this as an *o*/*e* rate. The proportion of deaths during the first day of life (out of all deaths before first birthday) was 23.0%. Therefore, on an annual basis the rate of death is $0.23 \times 365 = 84$ times the age-specific rate of age $[0, 1)$. Therefore we can estimate $\mu(0) = 84 \times 0.0045 = 0.378$. For the two-week period of days 22–35 the proportion of deaths was 3.8%, so on an annual basis we can estimate $\mu(28/365)$ as $0.038 \times (365/14) = 0.99$ times the age-specific rate of age $x = 0$. In this case it would be 0.0045. ◊

The concept of hazard that leads to survival probabilities and life tables appears so self-evident that it is hard to detect the conventional aspects of its adoption. Although slightly philosophical, we ask the reader to consider the following case of "randomness or predestination".

Suppose a waiting time X can take three values 1, 2, 3. Consider two models. (a) Suppose we toss a die once. If we get 1 or 2, then $X = 1$; if we get 3 or 4, then $X = 2$; and if we get 5 or 6, then $X = 3$. (b) Suppose we toss a die once. If we get 1 or 2, then $X = 1$. Otherwise, we toss again. If we get 1, 2 or 3, then $X = 2$. Otherwise $X = 3$. We interpret (a) as an extreme form of frailty that completely determines survival – your time of death is set at birth – and (b) as a pure hazard model with no frailty. Under both models $P(X = j) = 1/3$, for $j = 1, 2, 3$, so an outside observer could not tell which of the two models is valid, even based on a large number of independent observations. The models are incompatible but the classical deterministic life table theory does not distinguish between them. If we could have repeated observations on the "same X" after the first toss, then we could, in principle, distinguish between the models. A realistic point of view may be that there are elements of both (a) and (b) in the world we live in. As in (a), some individuals are better programmed to live long than others, yet as in (b), we all face outside risks that are unpredictable. A challenge of life table theory is not to lose sight of either model. We will come back to this topic in Section 8 of Chapter 5 and in Section 1.3.4 of Chapter 6.

3. Estimating Survival Proportions

In population forecasts one needs estimates of the proportions of survivors from age x to age $x + 1$, where "age x" refers to the interval $[x, x + 1)$. Here we take as a starting point estimates of survival probabilities as derived in Section 2.4. Let $k(s, t)$ denote the density of the actual population aged exactly s at time t. In the absence of migration, the proportion in question may be written as

$$\int_{x+1}^{x+2} k(s, t + 1)\, ds \Big/ \int_{x}^{x+1} k(s, t)\, ds. \tag{3.1}$$

Letting ${}_1p(s)$ denote the probability that an individual aged s at time t survives 1 year, and defining the weight function $v(s) = k(s,t)/\int_x^{x+1} k(y,t)\,dy$, we may rewrite (3.1) as a weighted average of the one-year survival rates,

$$\int_x^{x+1} v(s)\,{}_1p(s)\,ds. \tag{3.2}$$

The usual way of estimating (3.1) is use the *life table survival proportion* L_{x+1}/L_x, where

$$L_x = \int_x^{x+1} p(t)\,dt. \tag{3.3}$$

(Traditionally, the right hand side of (3.3) is multiplied by 10,000 or by 100,000. We will *not* follow this practice.) These integrals are usually evaluated using the linearity assumption, so

$$\frac{L_{x+1}}{L_x} = \frac{(p(x+2)+p(x+1))/2}{(p(x+1)+p(x))/2}. \tag{3.4}$$

Rewrite (3.4) as a weighted average of the one-year survival probabilities,

$$\frac{L_{x+1}}{L_x} = {}_1p(x+1)\frac{p(x+1)}{p(x)+p(x+1)} + {}_1p(x)\frac{p(x)}{p(x)+p(x+1)}, \tag{3.5}$$

where ${}_1p(x) = p(x+1)/p(x)$. We note two things. First, if the true population density $k(.,t)$ is not proportional to the density of the life table population whose age distribution is determined by $p(.)$, then the weights in (3.5) may be incorrect. Second, a correct survival proportion from age x to age $x+1$ can, in principle (by the mean value theorem of calculus), always be obtained as a weighted average of the one-year survival probabilities ${}_1p(x)$ and ${}_1p(x+1)$, if the one-year survival probabilities ${}_1p(x+t)$ are monotone for $t \in [0,1)$. Alternative, and potentially more accurate, methods can be devised.

For example, suppose the density of population is piecewise linear, $k(t) = k_{0,x} + k_{1,x}(t - x - 1/2)$ for $t \in [x, x+1)$. Suppose also that the one-year survival probabilities ${}_1p(x) = p(x+1)/p(x)$ are piecewise linear, ${}_1p(t) = {}_1p_{0,x} + {}_1p_{1,x}(t - x - 1/2)$, $t \in [x, x+1)$. (Note that this linearity assumption involves one-year survival probabilities rather than probabilities $p(x)$.) Then, instead of (3.5) the average survival probability is given by

$${}_1\overline{p}(x) = {}_1p(x)\left\{\frac{2k(x)+k(x+1)}{3(k(x)+k(x+1))}\right\} + {}_1p(x+1)\left\{\frac{2k(x+1)+k(x)}{3(k(x)+k(x+1))}\right\}. \tag{3.6}$$

For the unknown densities $k(x)$ and $k(x+1)$ we can use the estimates $\hat{k}(x) = (K_{x-1} + K_x)/2$, for example. We would expect to see differences between (3.6) and (3.5), if (a) one-year survival probabilities ${}_1p(x)$ change rapidly as a function of x, and (b) fertility was rapidly changing approximately x years ago.

Surviving births must be handled separately. Consider the Lexis diagram of Figure 1 of Chapter 2. Suppose $x = 0$. The life lines of the births during year t start in AB, and we are interested in the proportion that cross BC. Suppose that only data from rectangles are available. Consider all deaths that occur in ABCD, and let f denote the fraction that occur in the triangle ACD and thus represent deaths to persons born during year $t - 1$. This fraction f is called a *separation factor*, and it gives more weight to deaths at ages closer to $x + 1$ than to x. Values of f have historically been in the range 0.15 to 0.3 (Keyfitz 1977, 11). However, in the Finnish data of Example 2.10, the fraction was 0.08. This is a reflection of the low level of infant mortality in Finland. In any case, the probability of surviving from birth to the end of the year (i.e., survival in triangle ABC) is approximately $L_0 \approx \exp(-(1 - f)M_0) \approx 1 - (1 - f)M_0$.

Note also that if we want to consider cohort survival during the first year of life (i.e., survival in ABFC), then separation factors can be used to get estimates.

The difficulties encountered in the handling of the surviving births stem from data collection when information is available only for the rectangles of the Lexis diagram. However, when triple classified data (by age, year, and cohort) are available, the most obvious choice is to estimate the proportion of deaths in triangle ABC out of the births in AB (when $x = 0$). This gives directly an average probability of survival to the end of the year (provided that net migration is not large).

A similar remark can be made for the one-year survival probabilities L_{x+1}/L_x for $x = 0, 1, \ldots, \omega - 1$. Referring again to the Lexis diagram of Figure 1 of Chapter 2, we could assume that the mortality rate M_x has been calculated on a birth cohort basis from the *parallelogram* ACED. Then, a natural estimate of the one-year ahead survival is $\exp(-M_x)$ for ages in which mortality does not change much. The actuarial estimator $(2 - M_x)/(2 + M_x)$ discussed in Example 2.9 and Exercise 9 would be appropriate for ages with increasing mortality hazards (such as $x > 30$). Finally, an estimator could also be based on the Balducci model (cf., Example 2.3). It might be appropriate for ages with declining hazards (such as $x < 10$).

4. Childbearing as a Repeatable Event

4.1. Poisson Process Model of Childbearing

A statistical model for a repeatable event can be given in terms of counting processes. We call a set of random variables $\{N(t)|t \geq 0\}$ a *counting process* (or an *arrival process* or a *point process* – the terms will be used interchangeably), if $N(0) = 0$ and $N(t)$ increases by jumps of size one only. Then, $N(t)$ counts the number of events of interest (or "arrivals") by time t. In the case of childbearing, each woman starts childless and a counting process can keep track of her pregnancies that result in one or more live births (e.g., Keiding and Hoem 1976, Mode 1985). Since a single pregnancy can result in multiple births, we can attach a "mark" to each arrival indicating how many live births $(1, 2, \ldots)$ occurred. In this case one speaks of a *marked counting process*.

A particularly simple arrival process is obtained if we assume that the *interarrival times* are independent and exponentially distributed with some parameter $\lambda > 0$. This defines the so-called *Poisson process with intensity parameter* λ, because in this case $N(t) \sim \text{Po}(\lambda t)$, or $P(N(t) = k) = e^{-\lambda t}(\lambda t)^k/k!$ for $k = 0, 1, 2, \ldots$ (cf., Çinlar 1975, Chapter 4). We give a direct proof of the distributional result using the properties of the exponential distribution.

Proof of the Poisson distribution property. Let $T_1 \leq T_2 \leq \cdots$ be the arrival times such that $T_1, T_2 - T_1, T_3 - T_2, \ldots$ are independent with exponential distributions with parameter λ. For the following argument, let $p_k(t)$ denote $P(T_k > t)$ for $k = 1, 2, \ldots$ We show first by induction that

$$p_k(t) = \sum_{i=0}^{k-1} e^{-\lambda t}(\lambda t)^i / i!. \tag{4.1}$$

This is the survival function of the so-called *Erlang-k distribution*. Since $p_1(t) = e^{-\lambda t}$, the equality (4.1) holds for $k = 1$. Now make the induction assumption that the result holds for $k = j$, and consider $k = j + 1$. A moment's reflection shows that the event $\{T_{j+1} > t\}$ occurs if and only if one of two mutually exclusive events occur, either $\{T_j > t\}$ or $\{T_{j+1} > t \geq T_j\}$. Recall that the density of T_j is the negative of the first derivative of $p_j(t)$, i.e., $-p'_j(t)$. Therefore, we have the equality

$$p_{j+1}(t) = p_j(t) + \int_0^t -p'_j(s)e^{-\lambda(t-s)}\,ds. \tag{4.2}$$

Integrate by parts and observe that the integral on the right hand side can be written as the sum of $-p_j(t)$ and the right hand side of (4.1) for $k = j + 1$. This completes the induction proof of (4.1).

Having proved (4.1), we conclude by noting that $\{N(t) = k\}$ is equivalent to $\{T_k \leq t < T_{k+1}\}$, and we know from the proof of (4.2) that the probability of this event is $P(N(t) = k) = p_{k+1}(t) - p_k(t) = e^{-\lambda t}(\lambda t)^k/k!$. $\diamond$

We note that the Erlang-k distribution defined by (4.1) has many applications in telecommunications, where it is used in the analysis of incoming phone calls to a switching board, for example. It could still be of some demographic interest on its own right, because it can be used to gain intuition on waiting times until the kth child the kth unemployment spell, the kth relapse of a disease etc.

The Poisson process model is useful in statistical demography because it leads directly to a MLE of λ. Suppose we observe n independent Poisson processes $N_i(t)$ with the same parameter λ. Assume that the observation time of the ith process is $t_i > 0$, and define $K = t_1 + \cdots + t_n$. Now the total count is $N = N_1(t_1) + \cdots + N_n(t_n)$. It has the distribution $\text{Po}(\lambda K)$, where K is known. The MLE of λ is $\hat{\lambda} = N/K$, with an estimated variance of $\hat{\lambda}/K$. We see that this is an *o/e* rate of the same type we considered in the analysis of mortality. A different argument was, nevertheless, needed to motivate it in the case of a repeatable phenomenon, such as births.

Since the birth rate varies considerably by a woman's age, estimation is typically carried out by assuming constancy over a one-year or a five-year age interval. The *childbearing ages* are often operationally defined to be the ages 15–44, or 15–49, because outside these ages fertility is low. Fertility rates have also been quite erratic, and, hence, hard to forecast, during the past century.

Example 4.1. Age-Specific Fertility Rates for Italy and the U.S. The following table has *o/e* rate estimates (multiplied by 1,000) of age-specific fertility by 5-year age-groups in the United States in 1940–1970, and in Italy in 1975–1985.

Age-Specific Fertility Rates in the United States and Italy

	United States				Italy	
Age	1940	1950	1960	1970	1975	1985
15–19	45.3	70.0	79.4	57.4	32.5	12.1
20–24	131.4	165.1	252.8	163.4	129.8	72.5
25–29	123.6	165.1	194.9	145.9	140.2	101.8
30–34	83.4	102.6	109.6	71.9	84.1	65.7
35–39	45.3	51.4	54.0	30.0	40.7	25.2
40–44	15.0	14.5	14.7	7.5	12.6	5.0
45–49	1.6	1.0	0.8	0.7	0.9	0.3
Total	2.23	2.98	3.53	2.39	2.21	1.41

"Total" refers to the total fertility rate that is discussed in more detail below, but here it is defined simply as 5 $\times$ (sum of the five-year age-specific rates)/1,000. In the U.S. data we see the famous *baby-boom* of the post-war times. Within a decade, fertility went up by 1/3, stayed at a high level for a decade, and then dropped by 1/3. Neither the increase nor the decline was anticipated by population forecasters in the United States. In the mid-1940's it was believed that total fertility would decline to 2.06 by 1960 (Whelpton, Eldridge, and Siegel 1947). Ten years later, in the forecast for 1960–1980 (U.S. Census Bureau 1958) the highest of the four forecast variants for white total fertility in 1970 was 3.90 and the lowest 2.54. Later, in Italy, fertility also declined by 1/3 in a decade. This too was not anticipated by forecasters. In an official Italian forecast published in 1969 ("Tendenze evolutive della popolazione delle regioni italiana fino al 1981") the low scenario for the total fertility rate in 1979 was 2.6 and the high scenario was 2.8. By 1985 the forecasters had changed their minds, and forecasted a future total fertility of about 1.3. Similar decreases were observed in other Mediterranean countries. ◊

In causal analyses, birth rates are needed for sub-populations defined by education, region etc. A curious problem arises when birth order (first birth, second birth, etc.) is taken into account. By *parity* we refer to the number of children previously borne. Women who have had no children are said to be of parity zero, for example. Let $B_{x,i}$ be the number of births of order $i = 1, 2, \ldots$ to women in age x, and let K_x be the person years lived by women in age x, during a given year. The ith order-specific (or parity-specific) fertility rate is usually defined as

$B_{x,i}/K_x$ (cf., Shryock and Siegel 1976, 280). We caution that this is *not* an *o/e* rate, however, since the denominator is not restricted to women of parity $i-1$. The calculation of the measure in this manner can be motivated, however, if the proper exposure data are not available.[7]

Alternatively, we may consider parity from the perspective of the interarrival times of births for a woman. The so-called *parity progression ratios*, i.e. the ratio of women in parity i that reach parity $i+1$, can be illuminating as a tool to understand changes in childbearing behavior (e.g., Mode 1985, 119–120; Smith 1992, 235–237).[8] The meaning of such ratios is rather subtle, however, and multistate techniques (Chapter 6) appear to be required for a proper treatment of parity progression. We will illustrate the problems in Section 4.3.3.

4.2. Summary Measures of Fertility and Reproduction

As seen in Example 4.1, fertility varies considerably within the childbearing ages. We will apply the Poisson process model to define the most important summary measures of fertility. A *nonstationary Poisson process* can be obtained from a stationary process defined in Section 4.1 by a change of the time scale. Consider an *intensity function* $\lambda(.) \geq 0$ for $t \geq 0$. In analogy with (2.5) we define a *cumulative intensity*

$$\Lambda(x) = \int_0^x \lambda(t)\,dt. \tag{4.3}$$

Define an arrival process $N(x)$ such that $P(N(x)=k) = e^{-\Lambda(x)}\Lambda(x)^k/k!$. In other words, the number of arrivals by time x equals the number of arrivals of a stationary Poisson process with intensity 1, by time $\Lambda(x)$.

We use these concepts in the following way to describe childbearing. Suppose $N(x)$ counts the number of children a woman has by age x. Then, we call $\lambda(x)$ the *age-specific fertility rate* at exact age x. In human population we would typically have bounds $0 < \alpha < \beta$ such that $\lambda(x) = 0$ for $x < \alpha$ and $x > \beta$. Then, the interval $[\alpha, \beta]$ is said to consist of the *childbearing ages*. In Example 4.1 we displayed estimated age-specific fertility rates (for five-year age groups) with $\alpha = 15$ and $\beta = 50$.

The most important summary measure of fertility is $\Lambda(\beta)$, which is called the *total fertility rate*. Notice that $E[N(x)] = \Lambda(\beta)$ for all $x \geq \beta$. Thus, the *total fertility rate can be interpreted as the expected number of children a woman will have during her lifetime, provided that she survives to the end of the childbearing ages and the rates do not change with time.*

[7] In demography, such measures are sometimes called *rates of the second kind* (e.g., International Encyclopedia of the Social and Behavioral Sciences (2001)), 3482–3483.

[8] In particular, the so-called "children ever born" methods enjoy wide use in countries with deficient data (e.g., United Nations 1983, Chapter II).

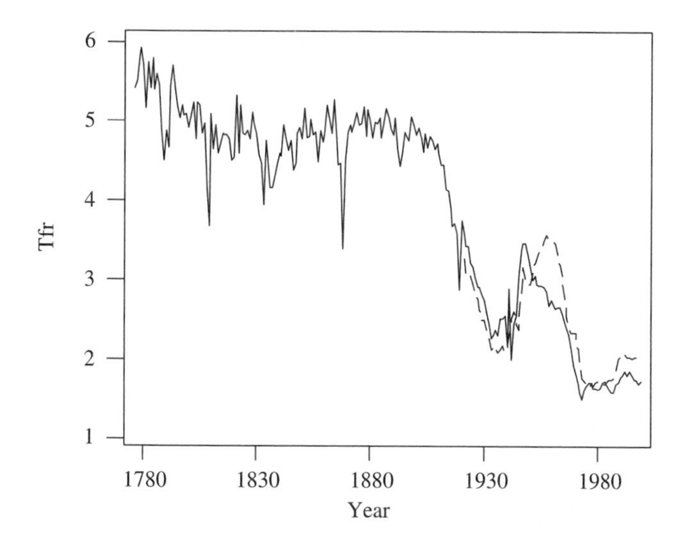

FIGURE 5. Total Fertility Rate in Finland in 1776–1999, and in the United States (Dashed line) in 1920–1999.

Example 4.2. Finnish Fertility, 1776-1999. Figure 5 has a plot of the Finnish total fertility rate during 1776–1999. We see that fertility remained high to the beginning of the 20th century. It then declined until the early 1930's. The peak of the Finnish baby-boom was in 1947. Figure 5 has also a plot of the U.S. total fertility rate in 1921–1998, with a peak in 1957. It is sometimes thought that the baby-booms were caused by postponement of births during war time and subsequent recovery. Figure 5 suggests that this cannot be the case, since fertility rose already before and during the war. We will come back to this issue in Section 4.3.1. ◊

The usual procedure for estimating age-specific fertility treats the intensity $\lambda(.)$ as constant over one-year or five-year age-intervals. As in Example 4.1, a total fertility rate is then obtained by approximating the integrand of (4.3) by the piece-wise constant estimate of age-specific fertility. If single year data are used, then an estimate of the total fertility rate is simply the sum of the age-specific *o/e* rates. Under a Poisson model for births, an estimate of *variance for the estimated total fertility rate* is obtained by summing the variances of the *o/e* rates.

The reproduction of the population is traditionally measured by the extent to which the female population reproduces itself. Define κ as the *sex-ratio at birth*, i.e., it is the ratio of male births to female births. It follows that the fraction of female births is $1/(1+\kappa)$. The so-called *gross reproduction rate* is the total fertility rate when only female births are considered, or it is defined as $\Lambda(\beta)/(1+\kappa)$.

The value of κ varies from one culture to another. The value $\kappa = 1.05$ is fairly typical in industrialized countries, but values in the range $1.01 - 1.08$ seem to occur in populations in which technologies for detecting the sex of a fetus (at an age when an abortion has been a medically safe option) have not been available (e.g., Shryock and Siegel 1976, 109). Statisticians might be interested to know that in 1710 John

Arbuthnot conducted what may have been one of the earliest applications of the so-called sign-test by calculating the probability that male births would exceed female births for eighty two consecutive years (1629–1710) in London, provided that $\kappa = 1$. He found this probability to be exceedingly small, thus proving the operation of Divine Providence (cf., Stigler 1986, 225–226). Karlin and Lessard (1986) consider the optimality of the sex ratio.

In addition to regional variation, κ may vary by age of mother. We note that if such variation is numerically important, then gross reproduction rate can be defined as $\int_0^\beta \lambda(t)/(1+\kappa(t))\,dt$, with the $\kappa(x)$ the sex-ratio for births to a mother of age x.

Example 4.3. Time Trends in Sex Ratios in Finland. Sex ratio at birth may also vary in unexpected ways over time. Figure 6 has a plot of the Finnish ratio from 1751–2000. The actual ratios vary quite a bit around the smoothed curve that was obtained by running the RSMOOTH procedure of Minitab twice. The variation is due to random fluctuations in Bernoulli trials. The interesting thing, however, is the trend of the time series. We will see later that the series is nonstationary by usual measures, indicating that there have been real changes in the ratio. The causes of changes have been investigated, but no obvious demographic factor such as paternal age, maternal age, age difference of parents or birth order can explain the nonstationarity (Vartiainen, Kartovaara and Tuomisto 1999). $\Diamond$

Let T be the waiting time until a woman's death and define $p(x) = P(T > x)$. Then, $N(T)$ is the total number of children she has over her lifetime. (Note how death may cause censoring here via T.) The expected number of girls she will have

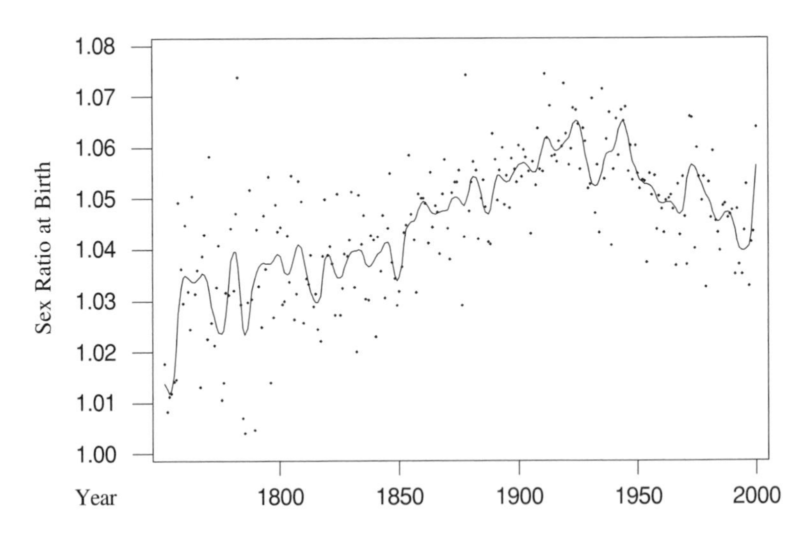

FIGURE 6. Sex Ratio at Birth (Actual and Smoothed) in Finland in 1751–2000.

is $E[N(T)]/(1+\kappa)$. This is called the *net reproduction rate*.[9] To evaluate it, note that conditionally on $T = t$, a woman is expected to have $\Lambda(t)$ children. Recall that $-p'(t)$ is the density of T and integrate by parts to show that $E[N(T)]/(1+\kappa)$ equals

$$\frac{1}{(1+\kappa)} \int_0^\infty \Lambda(t)(-p'(t))\,dt = \frac{1}{(1+\kappa)} \int_\alpha^\beta \lambda(t)p(t)\,dt, \tag{4.4}$$

because $\lambda(.)$ vanishes outside $[\alpha, \beta]$. The right hand side of (4.4) is the usual definition given for the net reproduction rate. It can be interpreted as the *expected number of girls a new born baby girl will have over her life time (provided that fertility and mortality schedules do not change over time)*. The gross reproduction rate is the expected number of girls a new born baby girl will have if she survives to age β. A stationary life table population is obtained if the net reproduction rate is $= 1$. As discussed in Section 2.2 of Chapter 6, a growing or declining *stable population* is obtained if it is > 1 or < 1, respectively. The integrand $\lambda(.)p(.)$ is called the *net maternity function*.

To determine the growth rate ρ of the stable population corresponding to $\lambda(.)$ and $p(.)$, suppose the female births at time t are $Be^{\rho t}$ and the female population density at time t is $Be^{\rho(t-x)}p(x)$. From the equality $Be^{\rho t} = \int Be^{\rho(t-x)}p(x)\lambda(x)\,dx/(1+\kappa)$ we get the equation

$$1 = \int_0^\infty e^{-\rho x}\,\lambda(x)p(x)\,dx/(1+\kappa). \tag{4.5}$$

By computing the derivative of the right hand side with respect to ρ we note that the right hand side is monotone function that declines from $+\infty$ to 0. Therefore, (4.5) has a unique real root in ρ.

If we would have $p(x) = 1$ for $x < \beta$, then a value $1+\kappa \approx 2.05$ of the total fertility rate would guarantee the reproduction of the population. Due to mortality in ages $< \beta$ a somewhat higher value, such as 2.1 is often mentioned as the threshold value. In countries with a low level of mortality an intermediate value such as 2.07 may be more accurate.

A possible definition for the *length of generation* is the number of years until the annual births become multiplied by the net reproduction rate (4.4). If we denote the generation length by G and the net reproduction rate by N, we then have for the stable population the equation $N = e^{\rho G}$ or $G = \log(N)/\rho$. Being determined by the life table age-distribution and period fertility, ρ is also called an *intrinsic growth rate*.

[9] The invention of the net reproduction rate is often attributed to Robert R. Kuczynski (1876–1947) although several authors entertained similar ideas in the 1920's and 1930's (DeGans 1999, 65).

One measure of the timing of the births is the *mean age at childbearing*. In statistical terms, this is an expected value of the age of the mother. There are at least four logical densities with respect to which the expectation might be taken. (i) Suppose $b(x)$ is the density of births by mother's age, $\alpha \leq x \leq \beta$. We get the actual mean age if we use $b(x)$. (ii) If we use a density proportional to the age-specific fertility rate $\lambda(x)$, we get a hypothetical mean age that would occur if there were constant past births and no mortality by age β. (iii) If we use a density proportional to $\lambda(x)p(x)$, we get a hypothetical mean age that assumes constant past births but takes into account mortality. (iv) If we use a density proportional to $e^{-\rho x}\lambda(x)p(x)$, we get a hypothetical mean age that takes into account intrinsic growth. As shown by Keyfitz (1977, 126), this mean age is close to the length of generation, as defined above. Usually, mean age is calculated assuming (ii) (Shryock and Siegel 1976, 279). To develop a sense of the practical meaning of the various measures, consider the data from Finland in 2000.

Example 4.4. Alternative Measures of Mean Age at Childbearing, Finland 2000. The total fertility rate was 1.73 and the sex ratio at birth was 1.06. Therefore, the gross reproduction rate was $1.73/2.06 = 0.84$. The net reproduction rate was $N = 0.83$, so the effect of mortality during childbearing ages on reproduction was negligible. The mean age at childbearing was approximately 29.9 using definition (i), and 29.5 using (ii). The reason the actual mean age is higher than that determined by the age-specific rates is that the cohorts in the youngest childbearing ages are smaller than those in the older childbearing ages. The other two definitions lead to slightly lower values lower than 29.5. If the length of the generation would be $G \approx 29$ years, then the corresponding population growth rate would be $\rho \approx \log(0.83)/29 = -0.006$. ◊

In the Finnish example, the current fertility and mortality rates would imply, in the absence of migration, a decline at the rate of about 0.6% per year. The low level of natural reproduction has not been a topic of interest in public debate because the baby-boom generations have produced large numbers of births during the past decades. The situation will change when the small generations born after 1970 form the bulk of the child-bearing population, and may be changing already as underfunding of pensions is increasingly a topic in the news.

As childbearing is largely voluntary activity, but subject to social norms, it is of interest to consider to what extent the sex distribution of their children can be controlled by the parents, by means other than genetic testing or X-ray determination of the sex of a fetus and abortion.

Suppose a couple can potentially have some finite number of children. They may elect to cease childbearing earlier. Let $X_i = 1$ if the ith potential child is a boy, and $X_i = 0$ otherwise. Assume that the X_i are independent and identically distributed Bernoulli random variables with parameter p, $X_i \sim \text{Ber}(p)$. The number of boys among the first n potential births, say $S_n = X_1 + \cdots + X_n$, has mean np. Define $Y_n = S_n - np$, with $Y_0 = 0$. Suppose the couple has elected to have n births, and they are deciding whether to have one more. There are two possibilities. (1) If the nth birth was the last feasible birth, or if the couple decides not to have further births, then the final Y value is Y_n. (2) If additional births are available and the couple decides to continue, then $E[Y_{n+1}|Y_n] = Y_n + E[X_{n+1}] - p = Y_n$. In both cases the expected Y value at the next step is the current value, no matter what the circumstances. The same argument applies to the previous step, so no matter what strategy the couple is following, their expected final Y value was then Y_{n-1}. Continuing in this way we see that their expected final Y value must have been $Y_0 = 0$. A similar argument can be made for the girls, so that *the ratio of the expected number of boys to the expected number of girls* is always $p : 1 - p$, no matter what decision rule the couple follows. This is an elementary example of the celebrated *optional sampling theorem of Doob* (cf., Chung 1974, 324–327): "No strategy in a fair game improves your chances." One implication of this finding is that in large populations the overall sex ratio does not depend on the strategies couples use. Although the result, as we have presented it, is straight forward, we point out additional subtleties in exercises.

4.3. *Period and Cohort Fertility*

4.3.1. *Cohort Fertility is Smoother*

The total fertility rate is usually interpreted in terms of a hypothetical (synthetic) cohort whose evolution is determined by the vital rates of year t. If the population is stable, the period total fertility rate may also correspond to the experience of actual cohorts. However, as amply demonstrated by Example 4.1 and Figure 5, fertility rates have been highly variable in the past. One possibility is that period fluctuations might be due to changes in the *timing* of fertility in different cohorts (cf., Ryder 1956). We will discuss this issue in the context of the baby-boom in Finland. As discussed in Example 4.2, it is unlikely that it could be explained simply as a recovery of births postponed during World War II.

In fact, consider the total numbers of (live) births in Finland, in consecutive five-year periods, during 1925–1954:

years	births
1925–1929	384,300
1930–1934	349,200
1935–1939	366,000
1940–1944	372,600
1945–1949	521,300
1950–1954	466,200

We see that the number of births reached a low during the years following the economic depression of the 1930's. After that there was a recovery, and during the five-year period that was most influenced by the war, the recovery continued: the total number of births was *higher* during 1940–1944 than during the previous five-year period of peace. A more plausible explanation can potentially be given in terms of a longer term postponement caused by both the depression and the war. This can be investigated by studying completed cohort fertility. The difficulty with

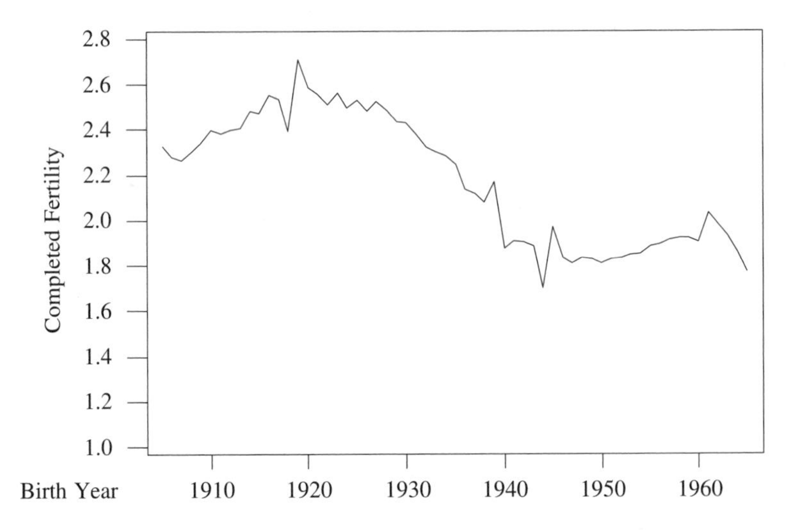

FIGURE 7. Approximate Completed Fertility for Birth Cohorts Born in Finland in 1905–1965.

cohort analysis is that it takes 30–35 years to observe the whole completed fertility of a cohort. Instead, Figure 7 presents the sum of age-specific fertility rates in ages 15–40 for the birth cohorts born in 1905–1965.[10] Before analyzing the data, two technical remarks are in order.

First, the estimates are based on the rectangles of the Lexis diagram rather than the genuine cohort parallelograms. This can have a notable numerical effect for some birth cohorts that were born at a time when fertility was rapidly changing from month to month due to war. The years 1918–1919, 1939–1940, and 1944–1946 are examples of this (Fougstedt 1977, 19).

Second, for the last five cohorts the values have been forecasted by adding 0.16 to the cumulative fertility of ages 15–35. This is the difference observed for the last available cohort born in 1960. Given that the fertility in ages 40–49 has been approximately 0.05 during the 1940's and 0.01 recently, the (forecasted) cumulative sum for the ages 15–40 approximates cohort total fertility rate well.

Turning to Figure 7, completed fertility presents a much smoother picture of the evolution of fertility than period fertility of Figure 5. This is to be expected, since fertility is heavily influenced by period factors that tend to compensate for each other over time for actual cohorts. Nevertheless, completed fertility has changed during the period we are investigating. It started at level 2.3 for the cohort of 1905 and rose to a high of 2.7 for the cohort of 1919. As argued by Fougstedt (1977, 18), the method of estimation has slightly exaggerated this value and decreased the low value of the previous year. Perhaps, 2.6 is closer to the actual maximum. From there, a decline to about 1.8 takes place. In other words, the increase during

[10] The authors are grateful to Timo Nikander of Statistics Finland for providing these data.

the early part of the period is about 0.3 children, and the subsequent decline is about 0.8 children, or 31%. Therefore, the baby-boom still appears as a reversal of a declining trend that started in the late 1800's and continued after the 1950's. Although timing has certainly contributed to the creation of the baby-boom in Finland, it cannot be explained *merely* by timing. Major changes in completed cohort fertility also occurred.

In thinking about the possible reasons for a reversal of a long-term decline, it seems useful to look at other countries, as well. Sweden did not participate in the war, but had a baby-boom that peaked in 1945, and a smaller peak in 1964. Great Britain and Belgium had lesser peaks in 1947–1948 and a bigger one in 1964. France and the Netherlands had higher peaks in 1946–1947 and a lesser one in 1964. The United States and Canada had major peaks in 1957 and 1960, respectively. (I.N.E.D. 1976, 46–54)

In summary, all the countries appear to have experienced a temporary reversal of a long time declining trend. Looking for an analogy in physical systems theory, we may observe that this corresponds to an *underdamped* system (Box and Jenkins 1976, 344). That is, when such a system is perturbed, its equilibrium state may change, but this value is only reached after a sequence of oscillations.[11]

4.3.2. Adjusting for Timing

Although timing cannot explain all of the fluctuations in childbearing we observe, it can certainly play a role. Therefore, if one has reason to believe that childbearing is currently being postponed (or that it occurs earlier than before) it is of interest to see how its effect might be assessed.

Let $\lambda(x, t)$ be the age-specific fertility rate in exact age x at exact time t and define the *period total fertility rate* as

$$\Lambda(t) = \int_0^\infty \lambda(x, t)\, dx. \tag{4.6}$$

Correspondingly, define the *cohort total fertility rate* of those born at s as

$$C(s) = \int_0^\infty \lambda(x, s + x)\, dx. \tag{4.7}$$

Assume that $\lambda(x, t) = g(x)$ for $t \leq 0$ and write $\Lambda = \Lambda(0)$, for short. Let us assume that $g(x) = 0$ for $x < \alpha$ and $x > \beta$.

Suppose that during $t > 0$ two things happen. First, all age-specific rates are multiplied by $(1 - r)$, where $|r| < 1$. Second, the schedule $g(x)$ is shifted at a rate of r per year towards older ages ($r > 0$) or towards younger ages ($r < 0$). In other words, assume that $\lambda(x, t) = (1 - r)g(x - rt)$ for $t > 0$. As a result $\Lambda(t) = (1 - r)\Lambda$, so the period total fertility rate is multiplied by $(1 - r)$.

---

[11] Readers who have ever hit a pothole driving in a car with worn-out shock absorbers have experienced underdamped systems.

To see what happens with cohort total fertility, note first that the lowest and highest ages of childbearing at t are $\alpha(t) = \alpha + rt$, and $\beta(t) = \beta + rt$, respectively. Consider a cohort born at $s \geq -\alpha$. Its lifeline in the Lexis diagram is $L(t) = t - s$. Therefore, it enters the childbearing ages when $L(t) = \alpha(t)$, or at time $t = (\alpha + s)/(1 - r)$, when its members have age $(\alpha + rs)/(1 - r)$. Similarly, the cohort ends childbearing at $t = (\beta + s)/(1 - r)$ in age $(\beta + rs)/(1 - r)$. We have that

$$C(s) = \int_{(\alpha+rs)/(1-r)}^{(\beta+rs)/(1-r)} (1 - r)g(x - r(s + x))\,dx. \tag{4.8}$$

After a variable change $y = (1 - r)x - rs$ we see directly that $C(s) = \Lambda$. In other words, the completed total fertility of the cohort born at $s \geq -\alpha$ equals that of period $t = 0$ despite the transformation of the age-specific schedules. Moreover, a similar argument shows that $C(s) = \Lambda$ for all s. The interpretation is that we can have a level change in period fertility and no change in completed cohort fertility, if the period level change is suitably matched by a translation type delay in fertility.

If a translation at speed r occurs, then the mean age at childbearing changes by r each year (if we define mean age with respect to a population whose age distribution is proportional to $\lambda(x, t)$, as in definition (ii) preceding Example 4.4). *Conversely*, if the mean age at childbearing changes by r per year, then we would expect period fertility to be multiplied by $(1 - r)$ *if* no change in completed cohort fertility occurs and fertility schedules are simply being translated. *Conditionally on this hypothesis*, $\Lambda(t)/(1 - r)$ would be a possible measure of fertility for year t that would "adjust" for the timing effect observed during t (see Bongaarts and Feeney 1998; and for extensions Van Imhoff and Keilman 2000, Kohler and Philipov 2001). Of course, the hypothesis may be false. For an alternative statistical formulation, see Example 3.1 of Chapter 5.

4.3.3. *Effect of Parity on Pure Period Measures*

Could the argument be pushed further to birth order specific fertility rates discussed in Example 4.1? Suppose a woman can give up to I births. We can then write

$$\lambda(x, t) = \varphi_1(x, t) + \cdots + \varphi_I(x, t), \tag{4.9}$$

where $\varphi_i(x, t), i = 1, \ldots I$, is the parity-specific fertility rate (defined just after Example 4.1) or the *component of fertility* that is due to births of order i. Let $\lambda_i(x, t)$ be the age-specific rate for order i births and let $w_i(x, t)$ be the fraction of women in age x who are at parity $i - 1$ at t. Then, the components can be written as $\varphi_i(x, t) = w_i(x, t)\lambda_i(x, t)$. Suppose we repeat the argument given above for the component of total fertility that is due to order i,

$$T_i(t) = \int_0^\infty \varphi_i(x, t)\,dx. \tag{4.10}$$

An adjusted measure would then be $T_i(t)/(1 - r_i)$, where r_i is the speed at which the components $\varphi_i(x, t)$ have been translated. The sum of the order-specific adjusted measures would be the adjusted total fertility rate. The reasoning is problematic, however, since changes in $\varphi_i(x, t)$ can be due to changes in $\lambda_i(x, t)$, $w_i(x, t)$, or both. Moreover, if $\lambda_i(x, t)$ changes it necessarily affects $w_{i'}(x', t')$ for $x' \geq x, t' \geq t$, and $i' \leq I$ (cf., Van Imhoff 2001, and references therein).

From a methodological point of view, a more serious problem is revealed by the consideration of parity. By a *pure period measure* one might refer to summary measures that depend on the transition intensities of the current period only. This seemingly simple definition depends on the setting, in a surprising way. For example, given this definition, the components $\varphi_i(x, t)$ are not pure period measures, because the weights $\mathrm{w}_i(x, t)$ depend on the fertility by birth order before time t. It follows that the actual *age-specific rates* $\lambda(x, t)$ *are not pure period measures either*, because they are sums of components that depend on earlier events. Hence, the measures $T_i(t)$ are not pure period measures, nor is their sum, the "period" total fertility rate $\Lambda(t)$! A multistate analysis (cf., Chapter 6) can produce a period measure that takes parity into account, but it is clear that if further disaggregation, e.g., by economic or social status, were entertained then the same problem would reappear.

On the other hand, suppose we stick with parity as the only criterion of disaggregation beyond age. Although transition intensities from one parity to the next can depend on any aspect of the past event history of the person, we will here formulate an example in which only the time of the previous birth has an effect (cf., Mode 1985, 144). It also serves as an example of a *multiple decrement model*: each parity can be left via two routes: death and having an additional birth.

Example 4.5. Parity Progression Ratios. Consider a new-born baby girl with mortality hazard $\mu(x)$ in age x. Suppose childbearing ends in age $\beta > 0$. Set $T_0 = 0$, and let $0 < T_1 < T_2 < \cdots$ be the times of birth of her children. The woman is at parity i in age x, provided that she is alive in age x, and $T_i \leq x < T_{i+1}$. Suppose that the hazard of a new birth is of the form $P(x < T_{i+1} \leq x + h \mid \text{woman is alive in age } x, T_{i+1} > x \geq T_i = u) = \nu_i(x, u)h + o(h)$. Write

$$\Lambda_i(x, u) = \int_u^x (\nu_i(s, u) + \mu(s))\, ds, \tag{4.11}$$

for short. Let $g_i(x)$ be the density of the entry time to parity $i + 1$, in age x. Using (2.4), we get $g_0(x) = \exp(-\Lambda_0(x, 0))\nu_0(x, 0)$ for $i = 0$. For $i = 1, 2, \ldots$ we have the recursion

$$g_i(x) = \int_0^x g_{i-1}(u) \exp(-\Lambda_i(x, u))\, \nu_i(x, u)\, du. \tag{4.12}$$

The probability of ever entering parity $i = 1, 2, \ldots$ is

$$G_i = \int_0^\beta g_{i-1}(x)\,dx, \tag{4.14}$$

so the probability of remaining childless is $1 - G_1$, for example. The *parity progression ratio* is G_{i+1}/G_i, or it is the conditional probability of entering parity $i + 1$, given entry to i. This can be estimated from period data based on estimates of $\mu(x)$ and $\nu_i(x, u)$. The interpretation of the ratio is more complex than one might think, because it depends on the hazards of entering earlier parities $j \leq i$ via (4.12). ◊

4.4. Multiple Births and Effect of Pregnancy on Exposure Time

Apart from the repeatable/nonrepeatable distinction, fertility rates differ from mortality rates because of the possibility of simultaneous multiple births. In addition, even though a pregnancy is a precondition of a later birth, after fertilization a woman is essentially incapable of giving birth for nine months or so. This is a form of censoring from the perspective of the Poisson model. We will show that neither factor typically has an effect that would invalidate the Poisson process approximation.

Historical statistics from Finland since the year 1900 show that the fraction of multiple births increases until age 35–39, but appears to decrease thereafter. The number of live births resulting in twins has been in the range 1.0–1.5% out of the total number of live births. The number of live births resulting in triplets has been approximately 0.01–0.02%, or one tenth of the twins. The fraction of live births resulting in quadruplets used to be approximately 0.0005%, but since the 1980's the fraction has increased to about 0.002, or to one tenth of the triplets. The increase may have been caused by the introduction of fertility-enhancing drugs that tend to produce multiple births. In summary, the total number of live born babies is, therefore, 1–2% higher than the number pregnancies resulting in live births.

Multiple births can be handled via *marked counting processes*. For example, for each woman i we can *superpose* independent Poisson processes $N_{ij}(t)$ for the arrival of each type of pregnancies ($j = 1$ corresponds to a single live birth, $j = 2$ corresponds to twins etc.; cf., Çinlar 1975, Section 4.4). The total number of children born to woman i by age t, is then a (finite) sum of the form $L_i(t) = N_{i1}(t) + 2 \times N_{i2}(t) + 3 \times N_{i3}(t) + \cdots$. Due to the independence of the arrival processes the probabilistic characteristics of the process $L_i(t)$ are easily derived. For example, let us ignore the effect of triplets, quadruplets etc. Suppose the expected number of live births per person year is $\lambda > 0$. Then, we get approximately that $N_{i1}(t) \sim \mathrm{Po}(0.97 \times \lambda)$ and $N_{i2}(t) \approx \mathrm{Po}(0.015 \times \lambda)$, because $0.97 + 2 \times 0.015 = 1$. It follows that $\mathrm{Var}(L_i(t)) \approx \lambda t(0.97 + 0.015 \times 2^2) = 1.03 \times \lambda t$. In other words, by ignoring multiple births we would *underestimate* the variance of the births by about 3%. Although this is a topic of considerable interest in *micro demography* i.e., the branch of demography dealing with small groups, families, or individuals (e.g., Sheps and Menken 1973), it has no practical effect in the analysis of aggregate fertility data usually considered in demography, where the dominant source of variation is in the expected values λ rather than the Poisson variance conditional on λ.

The second problem has to do with the fact that the usual duration of pregnancy is nine months, or 3/4 years. It follows that women who give birth during the period of observation, or have given birth during the latter 3/4 of the preceding year, do not actually contribute a whole year of exposure to risk of birth, only a part. This is in contrast with mortality: everybody is exposed to death while living! The usual method of calculating person years currently exaggerates *the number of person years of the population exposed to births* by 3/4 of the fraction giving birth. We saw in Example 4.1 that during the baby-boom, 20–25% of women in ages 20–30 gave birth each year. Subsequently, the fraction has declined to 5–15%.

Again, the problem is of interest in micro demography, but in aggregate studies the calculation of person years is rarely corrected. There are at least two reasons for this. First, *infecundity*[12] (i.e., physiological inability of a woman in a childbearing age to conceive or carry a pregnancy to a term) also occurs for reasons unrelated to births (infections, blocking of Fallopian tubes etc.). Even if a woman is fecund, she may not be at risk of pregnancy because she is not sexually active, by choice or by external constraints. Lack of exposure to pregnancy of these types would remain uncorrected. Second, when fertility statistics are used at an aggregate level, a possible correction would often *cancel out* in applications. For example, in forecasting one would apply a corrected fertility estimate to a risk population that is smaller than the total population in the age-group of interest.

5. Poisson Character of Demographic Events

For many kinds of demographic events, the distribution of the number of occurrences is well approximated by the Poisson distribution. For example, in Section 1 we saw that in the case of censored exponential waiting times, the number of events can be taken to have a Poisson distribution for inferential purposes. A classical result for proportions of events says that the distribution of the number of successes in trials, with a small probability of success but a large number of trials, is approximately Poisson. Specifically, suppose there are n independent trials, such that the outcome of trial i is "success" with probability $p_{i,n}$ and "failure" with probability $1 - p_{i,n}$. Now consider a sequence of such trials as $n \to \infty$ such that $p_{1,n} + \cdots + p_{n,n} \to \lambda > 0$ and $\max\{p_{i,n} | i = 1, \ldots, n\} \to 0$. Then the distribution of the number of successes is approximately Poisson Po(λ).

---

[12] In English "fertility" refers to actual realized fertility and "fecundity" refers to physiological ability to have children. In French it is the other way round!

Proof of asymptotic Poisson distribution property. The following proof is taken from Feller (1968, 282). Suppose $P(Y_{i,n} = 1) = p_{i,n}$ and $P(Y_{i,n} = 0) = 1 - p_{i,n}$ for independent Bernoulli variables $Y_{i,n}$, and define $S_n = Y_{1,n} + \cdots + Y_{n,n}$. The probability generating function (of argument s) of $Y_{i,n}$ is $E[s^{Y_{i,n}}] = (1 - p_{i,n} + p_{i,n}s)$, so the probabilities generating function of S_n is the product $E[s^{S_n}] = (1 - p_{1,n} + p_{1,n}s) \cdots (1 - p_{n,n} + p_{n,n}s)$. Taking the logarithm we get that

$$\log(E[s^{S_n}]) = \sum_{i=1}^{n} \log(1 - p_{i,n}(1 - s)). \tag{5.1}$$

The first order Taylor series approximation to the logarithm is $\log(1 - x) = -x$ with the (Lagrange form) remainder term $-x^2/[2(1 - \xi)^2]$, where ξ is a point between x and 0. Taking $x = (1 - s)p_{i,n}$ in the ith summand, it follows that as $n \to \infty$, we have that

$$\begin{aligned}\log(E[s^{S_n}] &= -(1 - s) \sum_{i=1}^{n} p_{i,n} - (1 - s)^2 \sum_{i=1}^{n} p_{i,n}^2 / 2(1 - \xi_{i,n})^2 \\ &\to -\lambda(1 - s). \end{aligned} \tag{5.2}$$

This proves that as $n \to \infty$, $E[s^{S_n}] \to \exp(-\lambda(1 - s))$, which is the probability generating function of the Poisson distribution $\text{Po}(\lambda)$. The convergence of the generating functions implies the convergence of the corresponding distributions (Feller 1968, 264, 280). $\diamond$

Feller's proof shows that we may have both population heterogeneity and different censoring times in a population and still get a Poisson limiting distribution for a count, provided that the event in question is rare. Section 4.1, on the other hand, says that if we are dealing with a repeatable event, then a Poisson model may be appropriate irrespective of the relative frequency of the event, provided that the interarrival times are exponential. One can show the latter result to agree with Feller's result by dividing the time interval into short subintervals. Then, the rarity assumption can be invoked within each subinterval, and we have an approximate Poisson distribution within each subinterval. The counts within subintervals will be independent because of the memorylessness property of the exponential distribution and the fact that no one is removed from exposure by the event of interest. For additional discussions, see Breslow and Day (1987, 131–135, and references therein).

When intensities of events are compared across small regions, for example, it is useful to note that the Poisson model assumes more variability than the binomial model. In addition, a sum of heterogeneous Bernoulli variables has a smaller variance than a sum of homogeneous Bernoulli variables. Therefore, the Poisson model leads to a more conservative inference.

Is Poisson variation important? The coefficient of variation of the Poisson distribution is $\lambda^{-1/2}$. In the early days of stochastic population modeling considerable interest centered on the so-called branching processes (Galton-Watson processes, in particular) as models of population growth. This theory is very interesting on its own right. However, it is not an adequate descriptor of the actual variability of the observed vital rates in human populations. Consider a simple example. Annual changes of several percent are common in age-specific mortality and fertility rates. However, for a Poisson model the coefficient of variation remains under 0.05 as soon as the expected count is greater than 400, and it remains below 0.01 when the expected count is over 10,000. It follows that from the point of view of population forecasting the Poisson variability and, *a fortiori* the binomial or Bernoulli variability, are negligible, unless we are dealing with small populations with expected counts that are in the hundreds or less (Pollard 1968, Goodman 1968).

6. Simulation of Waiting Times and Counts

Consider a waiting time $X \geq 0$ with survival probability $p(x) = P(X > x)$. Suppose first that $p(.)$ is strictly decreasing, so the inverse $p^{-1}(.)$ exists. Let U be a random variable with a uniform distribution on [0, 1] and define $T = p^{-1}(U)$. Then, we have that $P(T > x) = P(U < p(x)) = p(x)$. In other words, T has the same distribution as X. Several methods are available for the generation of uniformly distributed pseudo random numbers (e.g., Ripley 1987). Therefore, this method can be used to generate observations from any strictly decreasing survival function: simply generate U and set $X = p^{-1}(U)$. The method is equivalent to using the inverse of the distribution function.

Example 6.1. Simulation of Weibull Random Variates. Consider the Weibull distribution of Example 2.1. that has survival probabilities of the form $p(x) = \exp(-(x/\alpha)^{\beta})$, so $p^{-1}(u) = \alpha(-\log(u))^{1/\beta}$. If we randomly generate U uniform on (0, 1], $X = p^{-1}(U)$ will be Weibull with the desired parameters. In the case of the exponential distribution, or $\beta = 1$, we have simply $p^{-1}(u) = -\alpha \log(u)$. ◊

More generally, we have $p(x) = \exp(-\Lambda(x))$, so if $\Lambda^{-1}(.)$ exists, $p^{-1}(u) = \Lambda^{-1}(-\log(u))$. Provided that it is easy to compute the values of the inverse function, a straightforward way to simulate waiting times is thus available.

Consider counts now. Suppose X has the binomial distribution $\mathrm{Bin}(n, p)$. In that case X is the sum of n independent Bernoulli distributed random variables, $X = Y_1 + \cdots + Y_n$, where $P(Y_i = 1) = p$ and $P(Y_i = 0) = 1 - p$. If U_i's are n independent random variables that are uniformly distributed on [0, 1], then we can define $Y_i = 1$, if $U_i < p$, and define $Y_i = 0$ otherwise. Now X has the desired distribution. More complex methods are available for large n (cf., Ripley 1987, 92).

One way to simulate observations from a Poisson distribution is to resort to Poisson processes. Suppose $X \sim \mathrm{Po}(\lambda)$. Then, X equals the number of arrivals in a Poisson process with intensity 1 during time $\lambda > 0$. Hence, all we need to do is to generate waiting times from the survival function $p(x) = \exp(-x)$, until their sum exceeds λ. If the nth waiting time brings the sum over the value λ, then we take $X = n - 1$. Again, other methods may be faster when λ is large (Ripley 1987, 92). The same methods can be applied to other processes related to the Poisson process. For example, as in Section 4.2 we may consider the total number of births per woman as a sum of Poisson processes bringing her single births, twins, triplets etc. We stop the processes at the simulated time of death of the woman, or at the end of childbearing ages, whichever comes first.

Exercises and Complements (*)

1. Show that if $X_1, \ldots, X_k$ are independent and exponentially distributed waiting times with parameters $\mu_1, \ldots, \mu_k$, respectively, and $X = \min\{X_1, \ldots, X_k\}$ then $P(X > x) = \exp(-(\mu_1 + \cdots + \mu_k)x)$, or the minimum has also an exponential distribution with the parameter $\mu_1 + \cdots + \mu_k$. (Hint: the minimum exceeds x if and only if all of the waiting times exceed x.)
2. Consider two cohorts of N (statistically independent) individuals. Suppose the lifetimes within each cohort have exponential distributions with parameters $\mu_j > 0$, $j = 1, 2$. How many individuals do you expect to be alive in age $x > 0$ in each cohort? Show that the average force of mortality in the population formed by the two cohorts is $(\mu_1 \exp(-\mu_1 x) + \mu_2 \exp(-\mu_2 x))/(\exp(-\mu_1 x) + \exp(-\mu_2 x))$, in age x. How does the force of mortality change over time if the cohorts are heterogeneous with $\mu_1 > \mu_2$? For more discussion about population heterogeneity, see Keyfitz (1985), Chapter 14, or Vaupel and Yashin (1985).

*3. *Method of Moments. Suppose* $X_1, \ldots, X_n$ are i.i.d. from some distribution with a k dimensional parameter $\boldsymbol{\theta}$. The method of moments estimates $\mu^{(j)} = E[X_i^j] < \infty$ with $m^{(j)} = (x_i^j + \cdots + x_n^j)/n$. It is an application of estimating functions (Chapter 3, Section 7.3): it uses functions $\psi(x_i, \boldsymbol{\theta})$, whose j^{th} component is $\psi_j(x_i, \boldsymbol{\theta}) = y_i{}^j - \mu^{(j)}(\boldsymbol{\theta})$, $j = 1, \ldots k$.

4. Derive formula (1.3).
5. Consider exponentially distributed waiting times with m units observed and with $\hat{\mu}$ the *ole* rate. Since $Z = m^{1/2}(\hat{\mu} - \mu)/\mu$ has an asymptotic standard normal distribution, when μ is the true hazard rate, we have that asymptotically Z^2 has a χ_1^2 distribution. Let k_α be the $(1 - \alpha)$ fractile of the χ_1^2 distribution. It follows that an approximate $(1 - \alpha)$ level confidence interval for μ consists of all those values of μ that satisfy the inequality $Z^2 \leq k_\alpha$. Solve this quadratic equation for μ to get the end points of the confidence interval.
6. Continuation. Construct a $(1 - \alpha)$ level confidence interval for $e^{-\mu}$.
7. Consider the setting of Example 1.4. Assume $\alpha/\beta = 0.02$. Study numerically the probability of survival to age $0 < x < 80$, comparing an individual with the average hazard to the average probability of survival, for $\beta = 400$, 100, 25.

*8. *Jensen's Inequality*. If g is a convex function and $E[X]$ is finite, $E[g(X)] \geq g(E[X])$. The result is geometrically obvious once we note that for a convex function, $g(X) \geq g(E[X]) + s(X - E[X])$, where s is the slope of the tangent of $g(.)$ at $E[X]$.

9. In reference to Example 2.2, assume that $p(t) = 1 - bt$ for $t \in [0, 1]$ or equivalently that $\mu(t) = b/(1 - bt)$, where we take $0 < b < 1$. Then we have that $-p'(t) = b$. Note that if there are m deaths in a cohort of n individuals, then the likelihood of the data is $L(b) = b^m(1 - b)^{n-m}$, and the MLE of b is simply $\hat{b} = m/n$. This is quite reasonable, since b can be interpreted as the probability of death during the interval. From the latter perspective $\hat{b}$ can also be seen to be a moment estimator of b. Note also that the expected number of person years in the cohort is $n(1 - b/2)$ and the expected number of deaths is nb. Therefore, in large samples we expect the *o/e* estimator to be $\mu = b/(1 - b/2)$. One can solve for b from this to derive the *actuarial estimator* for the probability of death $b = 2\mu/(2 + \mu)$, and for the probability of survival $1 - b = (2 - \mu)/(2 + \mu)$. Neither formula seemed particularly intuitive to us without the derivation! We see that the actuarial estimator is reasonable when the force of mortality is well approximated by the formula $\mu(t) = b/(1 - bt)$.
10. Under the Balducci model of Example 2.3 one assumes that $\mu(t) = a/(1 + at)$ for $t \in [0, 1]$, where $a > 0$, so $p(t) = 1/(1 + at)$ (cf., Keyfitz and Beekman 1984, 34). In a cohort of n individuals the expected number of deaths is $na/(1 + a)$ and the expected person years are $n\log(1 + a)/a$. Therefore, in a large cohort we would expect the *o/e* estimator to be $\mu = a^2/[(1 + a)\log(1 + a)]$. This is a nonlinear equation that can be solved numerically for a.
11. (a) An alternative proof of (2.7) can be based on double integrals starting from

$$\int_0^\infty x(-p'(x))\,dx = \int_0^\infty \int_0^x -p'(x)\,dt\,dx.$$

(Hint: Change the order of integration.)

(b) Prove (2.7) by partial integration (i.e., integrating by parts), starting from

$$E[X] = \int_0^\infty x(-p'(x))\,dx.$$

12. (a) As in 11(b), show by partial integration that

$$E[X^2] = 2\int_0^\infty tp(t)\,dt.$$

(b) Prove the result starting from $P(X^2 > u) = p(u^{1/2})$, and making a change of variable $u = t^2$.
13. Show that cause-specific hazards are additive under an independent competing risks model (cf., Examples 1.2 and 2.4) by determining first the cumulative hazard of $X = \min\{X_1, \ldots, X_k\}$, and then differentiating.
14. Consider a model of independent competing risks of death with $\mu(x) = A + Re^{\alpha x}$, where $A, R, \alpha \geq 0$. This is the so-called *Gompertz-Makeham* family of

hazards. Gavrilov and Gavrilova (1991) present evidence that in many human populations changes in mortality over time can be described by varying the term A only. How can this be interpreted? If this were the only way mortality can be lowered, what would it imply concerning the further reduction of mortality?

15. (a) Show that the Gompertz model $\mu(x) = \alpha c^x$, with $\alpha, c > 0$, satisfies $\mu(x+1)/\mu(x) = c$. (b) Show that a Gompertz-Makeham model of Exercise 14 satisfies $\log\{(\mu(x+1) - \mu(x))/(\mu(x) - \mu(x-1))\} = \alpha$.
16. Derive the approximation (2.10) starting from (2.8).
17. Calculate the expectation of the general Weibull distribution in terms of the gamma function.
18. Suppose $c(t)$ is an integrable function, let $I(t)$ be the indicator process defined in Section 2.2.1, and define the random variables

$$X_1 = \int_{\alpha}^{\beta} c(t)I(t)\,dt, \quad X_2 = \int_{\beta}^{\omega} c(t)I(t)\,dt.$$

(a) The expectations of the variables are obtained by changing the order of integration and expectation, as in (2.6) and (2.7). (b) To calculate the second moments, note first that

$$X_1 X_2 = \int_{\alpha}^{\beta} c(t)\,dt \int_{\beta}^{\omega} c(t)I(t)\,dt,$$

because $X_1 X_2 = 0$ unless $I(\beta) = 1$. Now take the expectation under the integral sign to get $E[X_1 X_2]$. (c) To calculate $E[X_1{}^2]$ note first that $X_1{}^2$ can be written as

$$\int_{\alpha}^{\beta}\int_{\alpha}^{\beta} c(s)I(s)c(t)I(t)\,ds\,dt = 2\int_{\alpha}^{\beta}\int_{\alpha}^{t} c(s)\,ds\,c(t)I(t)\,dt.$$

Now take expectation under the integral sign.

19. Apply the results given above to derive expressions for the moments of D and solve for c, in Section 2.2.4.
20. Consider a cohort of size N with withdrawal times 1.1, 1.5, 2.0, and 2.2. Draw a graph of the Kaplan-Meier estimator for these data if (a) $N = 4$, and all events are deaths, (b) $N = 4$, and third withdrawal was a censoring, (c) $N = 4$, and last withdrawal was a censoring (how does the estimator defined by (2.19) behave for large t? Is this realistic?), (d) $N = 5$, and there are two tied deaths at the third withdrawal time, (e) $N = 5$, and there is a tied death and censoring at the third withdrawal time (present an upper and lower estimate in this case).
21. Continuation. Draw a graph of the Nelson-Aalen estimator in each case.

*22. An estimate of the variance of the Kaplan-Meier estimator is given by the formula introduced by *Greenwood* in 1926,

$$\text{Var}(\hat{p}(t)) = \hat{p}(t)^2 \sum_{T_{(i)} \leq t} \frac{\delta_{(i)}}{(n+1-i)(n-i)}.$$

Suppose that there is no censoring, and let the number of cases by time t be $c(t)$. Note that we then have $\hat{p}(t) = (n - c(t))/n$. Using this, show that Greenwood's formula reduces to $\hat{p}(t)(1 - \hat{p}(t))/n$ (cf., Andersen et al. 1993, 258). For a version applicable to grouped (or tied) data, see Woodward (1999, 203–204). If the Kaplan-Meier estimate is applied to data from a complex sample, sample-weighted numbers may be used for n and i in (2.19) and alternative variance estimates may be appropriate, as discussed in the next complement.

*23. (a) Show that the Nelson-Aalen estimator of the cumulative hazard is equal to the first order Taylor expansion of the estimator $\log \hat{p}(t)$. (Hint: a Taylor expansion yields $\log((n-i)/(n-i+1)) \approx -1/(n-i+1)$.) (b) Recall that if $Y \sim \text{Bin}(N, p)$, then $\text{Var}(\hat{p}) = p(1-p)/N$. Suppose we have $N = n - i + 1$ individuals at risk just before the ith death and assume that one dies in a short time interval around the time of death. Given one death, we would estimate the probability of survival in the interval as $\hat{p}_i = (n-i)/(n-i+1)$. A Taylor series expansion yields the approximation $\text{Var}(\log \hat{p}_i) \approx \hat{p}_i{}^{-2}\text{Var}(\hat{p}_i)$. Assume that the "trials" consisting of death times are independent, to arrive at a variance for the Nelson-Aalen estimator (2.20) as

$$\text{Var}(\hat{\Lambda}(t)) = \sum_{T_{(i)} \leq t} \frac{\delta_{(i)}}{(n-i+1)(n-i)}.$$

(c) Derive Greenwood's formula using the delta method approximation $\text{Var}(\hat{p}(t)) = \text{Var}(\exp(\log \hat{p}(t))) \approx \hat{p}(t)^2\text{Var}(\log \hat{p}(t))$. For a rigorous discussion, see Andersen et al. (1993). If the data come from a survey, the sampling variance of the estimate can be obtained using replication methods (Chapter 3, Section 8).

24. Derive formula (2.24).

25. Derive a formula for K_x defined by (2.22), when $k(t) = Be^{\beta t}$. Suppose the number of deaths is $D_x = K_x M_x$. Using (2.20), derive a formula for D_x, when hazard is of the Gompertz-Makeham form $\mu(t) = A + Re^{\alpha t}$, with $A = 0.00376$, $R = 0.0000274$, and $\alpha = 0.104$. (These values correspond to Swedish male data from 1926–1930; cf., Gavrilov and Gavrilova 1991, 75–76). Similarly, using (2.9) and (2.5) derive a formula for $\Lambda_{x,1}$. Let $B = 10{,}000$, and $\beta = -0.01$. Verify that you get the following table (the number of deaths is not an integer but this won't matter),

x	K_x	D_x	M_x	$\Lambda_{x,1}$
70	9851.49	449.692	0.0456471	0.0456580
71	9560.33	480.292	0.0502380	0.0502501
72	9277.78	513.358	0.0553320	0.0553454
73	9003.58	549.077	0.0609843	0.0609992

26. Apply Keyfitz's method to the table of Exercise 25. For the first age, 70, use slope estimates $\hat{\mu}_{1,70} = M_{71} - M_{70}$ and $\hat{k}_{1,70} = K_{71} - K_{70}$. Similarly for the last age. Verify that you get the following estimates (2.24): 0.0456584, 0.0502501, 0.0553454, 0.0609986. Calculate the exact values of the hazard increments based on the Gompertz-Makeham model, and show that for the two central ages these agree with the values given here.
27. Derive the weights in (3.6).
28. Derive a formula for the expectation of the Erlang-k distribution (a) by integrating $pk(t)$, (b) by using the difinition directly.
29. Consider a couple that continues to have children until they get the first boy, and then they stop. Suppose the probability of a boy is $0 < p < 1$, and let X denote the number of children the family will have, so $1/X$ is the fraction of boys. Under our model the family size has the *geometric distribution* $P(X = k) = p(1 - p)^{k-1}, k = 1, 2, \ldots$ Use it to show that under this strategy $E[1/X] = -\log(1 - p)$. In the case $p = 1/2$ the expected proportion is ≈ 0.693. For more discussion, see Yamaguchi (1989), or Keyfitz (1985, 335–344).
30. We have shown in Section 4.2 that a couple cannot influence the ratio of the expected number of boys they will have to the expected number of girls they will have. However, Exercise 29 shows that they can influence the expected fraction of boys in their own family. How can the two facts be reconciled? (a) Use the geometric distribution to show that in the setting of Exercise 29 the expected number of girls in the family is $(1 - p)/p$. Since the couple is certain to have exactly one boy, the expected number of children is $E[X] = 1/p$. (b) By Jensen's inequality, $E[1/X] > 1/E[X] = p$. Thus, the discrepancy is due to nonlinearity (or "ratio bias"). Intuitively, the fraction of boys is larger (smaller) than expected in small (large) families.
31. Suppose a couple can have at most two children, but they stop at one if they have a boy. Let the probability of a boy be $0 < p < 1$. Let X be the total number of children they will have. (a) Show that $E[X] = 2 - p$. (b) Show that the expected number of boys is $p(2 - p)$ and the expected number of girls is $(1 - p)(2 - p)$, so their ratio is $p/(1 - p)$. (c) Show that $E[1/X] = p(3 - p)/2$. (d) Conclude that $E[1/X] > p$. This shows that the conclusion of Exercise 30 was not due to the unrealistic assumption of being able have an unlimited number of children.
32. Consider an individual exposed to a carcinogenic agent at dose level $s > 0$. A *one-hit model* for carcinogenicity assumes that cells are bombarded by molecules or by radiation and cancer occurs if there is even a single hit. Assume that hits arrive as a Poisson process with intensity λs. Show that

during a period of length L, the probability of at least one hit is $1 - e^{-\alpha s}$, where $\alpha = \lambda L$. This probability is $\approx \alpha s$ for small α and s. Therefore, one also speaks of a *linear dose-response* model.

33. Derive formula (4.4).
34. Suppose the age-specific fertility rate of year t is of the form $\lambda(x, t) = \lambda_0(x) \exp(\gamma(x - M)t)$, for $x = \alpha, \ldots, \beta$, where M is the mean age of childbearing of the form $M = \Sigma_x x\lambda_0(x)/\Sigma_x \lambda_0(x)$. Suppose that at $t = T$ the mean age at childbearing is M'. Set up a calculation using Newton's method to find a value of γ such that $M' = \Sigma_x x\lambda(x, T)/\Sigma_x \lambda(x, T)$. This is an example of loglinear models to be discussed in Chapter 5.
35. The OECD publishes comparative statistics on the "probability" of ever starting studies in institutions of higher learning (universities, polytechnic institutions etc.). For year t, the measure is $c(t) = c(\alpha, t)w(\alpha, t) + \cdots + c(\beta, t)w(\beta, t)$, where $c(x, t)$ is the probability that a person of age $x = \alpha, \alpha + 1, \ldots, \beta$, who has not started such studies earlier, will do so during year t, and $w(x, t)$ is the share of those who have not started such studies earlier out of the total population in age x (in the beginning of year t). Think of α as the lowest age in which the studies could be started, and β as some (conventionally chosen) upper age: $\alpha = 16$ and $\beta = 44$, for example. Show that this is the life table probability of starting such studies by age β, if $c(x, t)$ *does not depend* on t. If this assumption fails, the measure is influenced by earlier events, and we may even have $c(t) > 1$!
36. Consider Example 4.5. Define $p_i(x)$ = probability that the woman is at parity i in age x. (a) Show that $p_0(x) = \exp(-\Lambda_0(x, 0))$ for $i = 0$, and for $i = 1, 2, \ldots$
$$p_i(x) = \int_0^x g_{i-1}(u) \exp(-\Lambda_i(x, u))\, du.$$
(b) Note that *if there were no mortality* until age β, then we would have $G_i = p_i(\beta) + p_{i+1}(\beta) + \cdots$.
37. Use simulation to estimate the variance of the Weibull distribution, when $\alpha = 1$ and $\beta = 2$.
38. Consider exposed and unexposed cohorts of size n, with risks of death p_j, $j = 1, 2$. Suppose the relative risk $\rho = p_1/p_2$ is estimated from binomial data $X_j \sim \text{Bin}(n, p_j)$, $j = 1, 2$, with $\hat{\rho} = \hat{p}_1/\hat{p}_2$, where $\hat{p}_j = X_j/n$. Use simulation to study the skewness of the distribution of $\hat{\rho}$ for $n = 10, 20, 30, 50, 100$, when $p_1 = 0.3$ and $p_2 = 0.15$ by drawing the histogram of the results. (Note in programming that $\hat{\rho}$ is not defined for all data sets.)
39. A non-obvious consequence of the duration of pregnancy is that it creates a negative autocorrelation into annual data. To evaluate the magnitude of the negative autocorrelation in births caused by 9-month pregnancy, consider a population of fixed size N and a constant birth rate f. Assume there are B_t births during year $[t, t + 1)$. Show that a randomly chosen woman who gave birth during year t spends an expected time 9/32 of the year $t + 1$

in a state of not being able to give birth. The expected loss of due to this is $9fB_t/32$ births. Using the result $\text{Cov}(B_{t+1}, B_t) \approx \text{Cov}(-9fB_t/32, B_t) = -9f\text{Var}(B_t)/32$ show that the autocorrelation must be $-9f/32$. For $f = 0.1$, we get the approximate numerical value $-.03$, for example.

40. Consider a Gompertz distribution with $\mu(x) = \alpha c^x$, $x > 0$, $c > 1$. Show that we can simulate its values by taking $U \sim U(0, 1]$ and computing $T = \log(1 - \log(c) \times \log(U)/\alpha)/\log(c)$.

5
Regression Models for Counts and Survival

Populations studied in demography are often large. There has been relatively little need to introduce parsimonious parametric models that are common in other fields of applied statistics, such as epidemiology. For example, the classical life table uses one parameter to describe each age. Therefore, it is not unusual that a hundred or more parameters are estimated from the data. Similarly, age-specific fertility and mortality rates can be viewed as estimators of age-specific parameters, one for each age-group. When demographers have used parametric models, the uses have been to induce smooth changes in the estimates from one age to the next (e.g., Gompertz-Makeham models for mortality; Lotka, Wicksell and Hadwiger have introduced analogous graduation models for fertility; cf., Keyfitz 1977). In contrast, epidemiologists studying the occurrence of diseases often have to resort to small data sets. The biases that might arise from imperfect parametric models have been outweighed by the increased precision the models provide. Optimality of statistical estimation procedures and statistical significance testing have become an important aspect of epidemiologic inference.

In this chapter we will provide a brief introduction to the most commonly used statistical models for relative risk, namely logistic regression, Poisson regression, and Cox regression. It turns out that the estimation theory of all these models can be viewed from a unified point of view. The likelihoods they lead to are examples of the so-called generalized linear models. Therefore, we will start by describing some general features of the theory in Section 1. Then, we proceed to discuss logistic regression in Section 2, and Poisson regression in Section 3. Standardization and loglinear models are specifically noted. In Section 4 we discuss ways of incorporating random effects into these models. Heterogeneity in capture-recapture data will be considered in Section 5. In Section 6 we consider bilinear models that have been used both in forecasting and data analysis. In Section 7 we consider proportional hazards models for survival type data. In Section 8 we discuss selection by survival. Section 9 discusses some aspects of spatial point patterns. We conclude in Section 10 by discussing methods for simulating regression data.

1. Generalized Linear Models

1.1. Exponential Family

The exponential family of statistical distributions is a family of parametric distributions that includes the binomial, Poisson, exponential, normal, beta, gamma, inverse Gaussian, and other distributions. The exponential family is characterized by the fact that parametric inferences can be based on a limited set of summary statistics no matter how large the sample. This leads to an elegant statistical theory that applies *verbatim* to most distributions of the family. We will discuss only a subset of the exponential family below, so as to be able to introduce logistic, Poisson, and Cox regression in as direct a way as possible later. The methods provide tools for analyzing relative risks in slightly varying settings. More details about exponential families and generalized linear models can be found in Andersen (1980) and McCullagh and Nelder (1989), for example.

Suppose a random variable Y takes values y and has a density function (or probability function in the discrete case; we will speak of densities, for short) of the form

$$f(y,\theta) = \exp(y\theta - b(\theta) + c(y)), \tag{1.1}$$

where θ is the so-called *canonical parameter* of the distribution, and $b(.)$ and $c(.)$ are known functions. Densities of the form (1.1) belong to the (1-parameter) *exponential family*.

Example 1.1. Exponential Distribution. Suppose $Y \sim \text{Exp}(\mu)$ with density $f(y;\mu) = \mu e^{-\mu y}$, where $y > 0$ and $\mu > 0$. This can be written in the form $f(y;\mu) = \exp(-\mu y + \log(\mu))$, so by taking $\theta = -\mu$, $b(\theta) = -\log(-\theta)$ for $\theta < 0$, and $c(y) = 0$, we see that the exponential distribution is of the form (1.1). In this case $b'(\theta) = -1/\theta$. As noted below (2.7) of Chapter 4, $E[Y] = 1/\mu$, so $E[Y] = b'(\theta)$. ◊

Example 1.2. Bernoulli Distribution. Suppose Y $\sim$ Ber(p) with $f(y;p) = p^y(1-p)^{1-y}$, where $0 < p < 1$ and $y \in \{0, 1\}$. In this case we can write $f(y;p) = \exp(y\log(p/(1-p)) + \log(1-p))$. By taking $\theta = \log(p/(1-p))$, $b(\theta) = \log(1+\exp(\theta))$, and $c(y) = 0$, we see that the Bernoulli distribution belongs to the 1-parameter exponential family (1.1). In this case $b'(\theta) = \exp(\theta)/(1+\exp(\theta))$, so again $E[Y] = b'(\theta)$. We will see below that this is generally true. ◊

Since our interest will primarily be in the modeling of counts, in the following we will assume that Y takes integer values. Similar arguments go through in the continuous case, when sums are replaced by integrals. Since $f(.;\theta)$ defines a probability distribution, we must have

$$\sum_y f(y;\theta) = 1 \tag{1.2}$$

for all values of θ. Let us differentiate both sides of (1.2) with respect to θ. The left hand side can be differentiated termwise provided that the resulting series converges. Since $d/d\theta f(y,\theta) = (y - b'(\theta))f(y,\theta)$, we get the result,

$$E[Y] = b'(\theta). \tag{1.3}$$

In other words, whenever $E[Y]$ exists, it is given by $b'(\theta)$. Furthermore, differentiating (1.2) the second time yields $d^2/d\theta^2 f(y,\theta) = -b''(\theta)f(y,\theta) + (y - b'(\theta))^2 f(y,\theta)$, so that

$$\text{Var}(Y) = b''(\theta). \tag{1.4}$$

Returning to Example 1.1, we note that in that case $\text{Var}(Y) = 1/\theta^2$. In Example 1.2, we get $\text{Var}(Y) = b''(\theta) = \exp(\theta)/(1+\exp(\theta))^2 = p(1-p)$.

1.2. Use of Explanatory Variables

Suppose now that we have independent variables Y_i, each with a density of type (1.1), but with individually varying parameters $\theta_i, i = 1, \ldots, n$. The key idea in the formulation of *generalized linear models* is that a linear model is assumed for some function of θ_i. In the simplest case, suppose there is a vector of explanatory variables $\mathbf{X}_i = (X_{i1}, \ldots, X_{ik})^T$ and a vector of unknown parameters $\boldsymbol{\beta} = (\beta_1, \ldots, \beta_k)^T$, such that

$$\theta_i = \mathbf{X}_\mathbf{i}^\mathbf{T}\boldsymbol{\beta}. \tag{1.5}$$

In practice, we usually take $X_{i1} \equiv 1$, i.e., the model has a constant term. This is not required for the theory to be presented below, however.

McCullagh and Nelder (1989) discuss more complicated mappings between the canonical parameter θ_i, and the *linear predictor* $\mathbf{X}_i^T\boldsymbol{\beta}$. In fact, the usual formulation is in terms of *link functions* between the mean $b'(\theta)$ and the linear predictor. Our formulation corresponds to the special case of a *canonical link function* that leads to a linear mapping between the canonical parameter and the explanatory variables. The generalized linear models were introduced by Nelder and Wedderburn (1972).

1.3. Maximum Likelihood Estimation

The likelihood function of the observed data is

$$L(\boldsymbol{\beta}) = \exp(\mathbf{U}^\mathbf{T}\boldsymbol{\beta} - B(\boldsymbol{\beta}) + C(\mathbf{Y})), \tag{1.6}$$

where $\mathbf{Y} = (Y_1, \ldots, Y_n)^T$, and

$$\mathbf{U} = \sum_{i=1}^{n} Y_i\mathbf{X}_i; \quad B(\boldsymbol{\beta}) = \sum_{i=1}^{n} b(\mathbf{X}_\mathbf{i}^\mathbf{T}\boldsymbol{\beta}); \quad C(\mathbf{Y}) = \sum_{i=1}^{n} c(Y_i). \tag{1.7}$$

Note that $L(\boldsymbol{\beta})$ is the product of two factors, $\exp(\mathbf{U}^T\boldsymbol{\beta} - B(\boldsymbol{\beta}))$ and $\exp(C(\mathbf{Y}))$. Treating the explanatory variables $\mathbf{X}_i$ as known constants, the former involves the random data only through the summary statistic $\mathbf{U}$, and the latter does not involve the parameter $\boldsymbol{\beta}$. The Neyman factorization theorem (e.g., Lehmann 1986, 54–55) implies that $\mathbf{U}$ is *sufficient* for $\boldsymbol{\beta}$. For inferential purposes, we only need to pay attention to $\mathbf{U}$. Furthermore, $\boldsymbol{\beta}$ has k components and the likelihood (1.6) corresponds to a *k-parameter exponential family*. As in (1.3), one can show that $E[U_j] = \partial/\partial\beta_j B(\boldsymbol{\beta})$ or, in vector form, $E[\mathbf{U}] = \partial/\partial\boldsymbol{\beta} B(\boldsymbol{\beta})$.

To estimate $\boldsymbol{\beta}$, we use maximum likelihood. Define $\ell(\boldsymbol{\beta}) = \log L(\boldsymbol{\beta})$, and differentiate with respect to $\boldsymbol{\beta}$. Setting the derivative to $\mathbf{0}$, we get that $\mathbf{U} = \partial/\partial\boldsymbol{\beta}$ $\mathrm{B}(\boldsymbol{\beta})$. Hence we have the elegant equation

$$\mathbf{U} = E[\mathbf{U}]. \tag{1.8}$$

Defining the *design matrix* $\mathbf{X} = [\mathbf{X}_1, \ldots, \mathbf{X}_n]^T$ we may write $\mathbf{U} = \mathbf{X}^T\mathbf{Y}$. Therefore, (1.8) is equivalent to $\mathbf{X}^T\mathbf{Y} = \mathbf{X}^T\mathrm{E}[\mathbf{Y}]$.

As opposed to ordinary linear regression, (1.8) may be a nonlinear equation in the parameters $\boldsymbol{\beta}$ that doesn't admit an explicit, let alone linear, solution. Instead, the solution has to be found using numerical methods, and it is typically a nonlinear function of the observations. Instead of exact normality and unbiasedness that we obtain in normal theory ordinary regression, we get *asymptotic normality* and *asymptotic unbiasedness* (and *consistency*), when the number of observations n is large.

1.4. Numerical Solution

Newton's method is frequently used to solve (1.8). Define the Hessian, or the $k \times k$ matrix of second partial derivatives of the loglikelihood function, as

$$\mathbf{H} = \partial^2/\partial\boldsymbol{\beta}\partial\boldsymbol{\beta}^T \ell(\boldsymbol{\beta}). \tag{1.9}$$

From (1.6) we see that $-\mathbf{H} = \partial^2/\partial\boldsymbol{\beta}\partial\boldsymbol{\beta}^T B(\boldsymbol{\beta})$, and as in (1.4), one can show that $-\mathbf{H} = \mathrm{Cov}(\mathbf{U})$. Let $E_{(i)}[.]$ and $\mathbf{H}_{(i)}$ refer to the expectation and covariance as estimated based on the ith iterated value of $\boldsymbol{\beta}$, or $\boldsymbol{\beta}_{(i)}$, and note that Newton's method provides the recursion,

$$\boldsymbol{\beta}_{(i+1)} = \boldsymbol{\beta}_{(i)} - \mathbf{H}_{(i)}^{-1}(\mathbf{U} - E_{(i)}[\mathbf{U}]), \quad i = 0, 1, 2, \ldots, \tag{1.10}$$

that must be started from some initial value $\boldsymbol{\beta}_{(0)}$ and repeated until convergence.

Although the numerical calculations are carried out using a computer, a closer look of how Newton's method works gives us some insight as to the nature of the solution. Note that $-\mathbf{H} = \mathrm{Cov}(\mathbf{X}^T\mathbf{Y}) = \mathbf{X}^T\mathbf{W}\mathbf{X}$, where $\mathbf{W} = \mathrm{Cov}(\mathbf{Y})$, a diagonal matrix with $\mathrm{Var}(Y_i)$ as the ith diagonal element. Equation (1.4) provides a general formula for computing $\mathbf{W}$, but, e.g., in the binomial and Poisson cases the variances are known from introductory statistics courses. As noted by Finney (1952) already, (1.10) can be written as

$$\boldsymbol{\beta}_{(i+1)} = (\mathbf{X}^T\mathbf{W}_{(i)}\mathbf{X})^{-1}\mathbf{X}^T\mathbf{W}_{(i)}\mathbf{h}_{(i)}, \quad i = 0, 1, 2, \ldots, \tag{1.11}$$

where

$$\mathbf{h}_{(i)} = \mathbf{X}\boldsymbol{\beta}_{(i)} + \mathbf{W}_{(i)}^{-1}(\mathbf{Y} - E_{(i)}[\mathbf{Y}]) \tag{1.12}$$

is the so-called *working variate*. The right hand side of (1.11) is a *generalized least squares* (GLS) estimator when $\mathbf{X}$ is the design matrix, (1.12) is the vector of observations, and $\mathbf{W}_{(i)}$ is the diagonal matrix of weights. This shows that maximum likelihood estimation for the generalized linear models (of the form described here) can be carried out by a repeated use of *weighted least squares* (WLS) (e.g., Thisted 1988, 215*ff*.).

1.5. Inferences

When the MLE $\hat{\beta}$ has been obtained, its variance-covariance matrix can be estimated as

$$\hat{\text{Cov}}\,(\hat{\boldsymbol{\beta}}) = (\mathbf{X}^T \hat{\mathbf{W}} \mathbf{X})^{-1}, \tag{1.13}$$

where $\hat{\mathbf{W}}$ is the MLE of $\mathbf{W}$. To compute this in practice, we simply plug the MLE of $\boldsymbol{\beta}$ into (1.5), and use the result in (1.4). A heuristic derivation for (1.13) can be obtained from (1.11) and (1.12). The MLE is (subject to regularity conditions that typically obtain) consistent, so the essential part of the randomness in (1.12) comes from $\mathbf{Y}$. Ignoring all other sources we get that the covariance matrix of $\mathbf{h}_{(i)}$ in (1.12) is approximately $\hat{\mathbf{W}}^{-1}$, because $\text{Cov}(\mathbf{Y}) = \mathbf{W}$. Therefore, the approximate covariance of (1.11) should be (1.13). (See also Section 3 of Chapter 1 and the discussion related to (7.11) of Chapter 3.)

Often, inferences concerning the parameters utilize Wald tests (Section 3 of Chapter 1) in which we compare the estimates of the parameters (or their linear combinations) with their estimated standard errors, as calculated from (1.13). When the number of observations is large enough and the number of parameters is moderate, the asymptotic normality of $\hat{\boldsymbol{\beta}}$ can be assumed. For example, let $\boldsymbol{\lambda}^T \boldsymbol{\beta}$ be a linear combination of interest and consider the hypothesis $H_0 : \boldsymbol{\lambda}^T \boldsymbol{\beta} = \boldsymbol{\lambda}^T \boldsymbol{\beta}_0$. Based on (1.13), the estimated standard error of $\boldsymbol{\lambda}^T \hat{\boldsymbol{\beta}}$ is $(\boldsymbol{\lambda}^{\mathbf{T}}(\mathbf{X}^{\mathbf{T}}\hat{\mathbf{W}}\mathbf{X})^{-1}\boldsymbol{\lambda})^{1/2}$, and the test statistic $T = \boldsymbol{\lambda}^{\mathbf{T}}(\boldsymbol{\beta} - \boldsymbol{\beta}_0)/(\boldsymbol{\lambda}^{\mathbf{T}}(\mathbf{X}^{\mathbf{T}}\hat{\mathbf{W}}\mathbf{X})^{-1}\boldsymbol{\lambda})^{1/2}$ is distributed approximately as $N(0, 1)$ when H_0 is true. A 95% confidence interval for $\boldsymbol{\lambda}^T \boldsymbol{\beta}$ is correspondingly $\boldsymbol{\lambda}^{\mathbf{T}}(\hat{\boldsymbol{\beta}}) \pm 1.96 \times (\boldsymbol{\lambda}^{\mathbf{T}}(\mathbf{X}^{\mathbf{T}}\hat{\mathbf{W}}\mathbf{X})^{-1}\boldsymbol{\lambda})^{1/2}$.

If $f(\boldsymbol{\beta})$ is a (smooth) nonlinear transformation of the parameters, then a confidence interval for it can be based on the delta method (Section 7.2. of Chapter 3). In this case, the approximate 95% interval is $f(\hat{\boldsymbol{\beta}}) \pm 1.96 \times (\boldsymbol{\lambda}^{\mathbf{T}}(\mathbf{X}^{\mathbf{T}}\hat{\mathbf{W}}\mathbf{X})^{-1}\boldsymbol{\lambda})^{1/2}$, where $\boldsymbol{\lambda} = \partial f/\partial \boldsymbol{\beta}$.

Both score and likelihood ratio testing can be used as an alternative to Wald tests in generalized linear models. In the case of likelihood ratio tests it has become customary to carry out these calculations via a related measure called *deviance*. Define a *saturated model* (or a *full model*) as a model that has as many parameters as there are data points. It can fit the data perfectly. The deviance of a regression model is defined by

$$2(\ell^* - \hat{\ell}), \tag{1.14}$$

where ℓ^* is the loglikelihood of the saturated model and $\hat{\ell}$ is the maximum loglikelihood of the regression model being entertained. The deviance does not, in general, have a known distribution, although in special cases approximations are available. However, the difference in deviance between two nested models yields the usual *likelihood ratio test statistic* $2(\hat{\ell}_1 - \hat{\ell}_0)$ for testing the larger model, which has loglikelihood $\hat{\ell}_1$, against the smaller one with loglikelihood $\hat{\ell}_0$ (cf., Section 3 of Chapter 1).

Specifically, consider a generalized linear model with canonical parameter $\boldsymbol{\theta} = (\theta_1, \ldots, \theta_k)^T \in \Theta$, an interval in $\mathbb{R}^k$. Define two subspaces of the form $\Theta_i = \{\boldsymbol{\theta} \in \Theta \,|\, g_1(\theta) = \cdots = g_{m_i}(\boldsymbol{\theta}) = 0\}$, $i = 0, 1$, where $m_0 > m_1$, and consider two hypotheses, $H_0 : \boldsymbol{\theta} \in \Theta_0$ and $H_1 : \boldsymbol{\theta} \in \Theta_1$. In this case $\Theta_0 \subset \Theta_1$, and we say that H_0 is *nested* in H_1. Suppose the "restrictions" g_j are subject to mild regularity conditions (e.g., continuous first partial derivatives and no redundancy, so that one cannot derive one restriction from the others; e.g., Rao 1973, 416*ff*.). In this case, $2(\hat{\ell}_1 - \hat{\ell}_0)$ has an asymptotic χ^2 distribution with $m_0 - m_1$ degrees of freedom when H_0 is true.

Among other things, these results provide a method for constructing confidence intervals for the parameters, or their linear combinations. In the simplest case, take $m_0 = 1, m_1 = 0$, and $g_1(\boldsymbol{\theta}) = \theta_k - c$, for some c. Denote the maximum of the log-likelihood, conditionally on $\theta_k = c$, by $\hat{\ell}_0(c)$. This is the so-called *profile likelihood*. Then, an approximate 95% confidence interval for θ_k is $\{c | 2(\hat{\ell}_1 - \hat{\ell}_0(c)) < 3.841\}$, for example. Both analytical considerations (e.g., Jennings 1986; Cox and Hinkley 1974) and simulations suggest that the likelihood ratio approach may be preferable to Wald testing in small samples. An illustration is given in Exercise 17.

1.6. Diagnostic Checks

In ordinary linear regression the predicted values are given by $\hat{\mathbf{Y}} = \mathbf{X}(\mathbf{X}^T\mathbf{X})^{-1}\mathbf{X}^T\mathbf{Y}$, where $\mathbf{X}$ is as above. The matrix $\mathbf{X}(\mathbf{X}^T\mathbf{X})^{-1}\mathbf{X}^T$, which converts $\mathbf{Y}$ to $\hat{\mathbf{Y}}$ ("Y hat"), is called the *hat matrix*. In ordinary least squares (OLS) regression, the ith diagonal element of the hat matrix gives the so-called *leverage* of the ith observation (cf., Exercise 10). Note that leverage depends on the design matrix $\mathbf{X}$ but not on $\mathbf{Y}$. Analogously, in generalized linear models leverage is sometimes measured by the diagonal elements of the matrix $\mathbf{W}^{1/2}\mathbf{X}(\mathbf{X}^T\mathbf{W}\mathbf{X})^{-1}\mathbf{X}^T\mathbf{W}^{1/2}$ based on (1.11) (cf., Pregibon 1981). Some care is needed when interpreting the leverages, since the variances in $\mathbf{W}$ typically depend on the mean (Hosmer and Lemeshow 2000, 153).

Example 1.3. Leverage in Simple Generalized Linear Model. Consider simple linear regression, $Y_i = \beta_1 + \beta_2 X_i + \varepsilon_i$, where $\varepsilon_i \sim N(0, \sigma^2)$ are independent, $i = 1, \ldots, n$. In this case $k = 2$, $X_{i1} = 1$, and we have written $X_{i2} = X_i$, for short. One can show, by a direct calculation, that the ith diagonal element of the hat matrix equals $1/n + (X_i - \bar{X})^2 / \Sigma_j (X_j - \bar{X})^2$. In other words, the further the value of the explanatory variable is from the mean, the larger the leverage of the ith observation. Consider now a simple generalized linear model with $\theta_i = \beta_1 + \beta_2 X_i$ and $\text{Var}(Y_i) = W_i$, $i = 1, \ldots, n$. Define $V = \Sigma_j W_j$, $\tilde{X} = \Sigma_j W_j X_j / V$, and $S = \Sigma_j W_j (X_j - \tilde{X})^2 / V$. The details are somewhat tedious, but one can then show that the leverage of the ith observation is $W_i(1 + (X_i - \tilde{X})^2/S)/V$. This is harder to interpret, because X_i can also affect W_i. $\Diamond$

The *influence* of data points refers to how much the estimates would change if the data points were omitted. In ordinary regression the most widely used measure of the influence of the ith observation is the so-called *Cook's distance* $(\hat{\beta} - \hat{\beta}_{(i)})^T(\mathbf{X}^T\mathbf{X})(\hat{\beta} - \hat{\beta}_{(i)})/k\hat{\sigma}^2$, where $\hat{\beta}_{(i)}$ is the MLE that has been computed without the ith observation (Weisberg 1985, 119). Defining $\hat{\mathbf{Y}}_{(i)} = \mathbf{X}\hat{\beta}_{(i)}$ as the vector of predictions when observation i is not used in the estimation of β, notice that the numerator of Cook's distance equals $(\hat{\mathbf{Y}} - \hat{\mathbf{Y}}_{(i)})^T(\hat{\mathbf{Y}} - \hat{\mathbf{Y}}_{(i)})$. The rationale of the particular weighting (denominator) used in the definition of Cook's distance derives from the sampling distribution of $\hat{\beta}$ (cf., Exercise 12). An analogous measure in generalized linear models is $(\hat{\beta} - \hat{\beta}_{(i)})^T(\mathbf{X}^T\mathbf{W}\mathbf{X})(\hat{\beta} - \hat{\beta}_{(i)})$ (cf., Pregibon 1981).

If the data are obtained with random sampling, one can compare estimated means and variances from the model with estimates derived using sampling weights (cf., Chapter 3). Then, (1.8) would be replaced by a weighted version that incorporates the inverses of selection probabilities, as in (7.9) of Chapter 3. Similarly **H** of (1.9) would be replaced by a version including the weights (cf., Chapter 3, Section 7.3; Hosmer and Lemeshow 2000, 211–221). This is sometimes called a "pseudo maximum likelihood" approach.

2. Binary Regression

2.1. Interpretation of Parameters and Goodness of Fit

Consider a binomial random variable $Y \sim \text{Bin}(n, p)$. As in Example 1.2, we write $\theta = \log(p/(1-p))$, or $p = \exp(\theta)/(1+\exp(\theta))$. Thus, the canonical parameter θ equals the *log-odds* of the individual trials. Often, the notation logit(p) $= \log(p/(1-p))$ is used. Therefore, these models are also referred to as *logit models*. Assuming the model (1.5) for θ leads to *logistic regression*. A detailed introduction to these models is given in Hosmer and Lemeshow (2000), for example. Here we will first discuss the interpretation of the parameters of the models using a simple example relating to the probability of death. We then discuss statistical inference for these models. In Section 2.2 we discuss a series of examples.

Suppose $q(x, t)$ is the probability that an individual in exact age x dies within one year, if the mortality level of calendar year t applies. Consider two logistic models,

$$q(x, t) = \exp(\alpha_0 + \alpha_1 x + \beta t)/(1 + \exp(\alpha_0 + \alpha_1 x + \beta t)), \tag{2.1}$$

and

$$q(x, t) = \exp(\alpha_x + \beta t)/(1 + \exp(\alpha_x + \beta t)). \tag{2.2}$$

It is easy to see that under both models

$$\frac{q(x, t+1)}{1 - q(x, t+1)} \Bigg/ \frac{q(x, t)}{1 - q(x, t)} = \exp(\beta), \tag{2.3}$$

or the *odds-ratio* (OR) of death during year $t+1$ versus year t equals $\exp(\beta)$, irrespective of age x. Equivalently, β can be interpreted as a log-odds-ratio. A similar interpretation can be given to α_1 in (2.1), but under (2.2) logit $q(x, t+1) -$ logit $q(x, t) = \alpha_{x+1} - \alpha_x$. Therefore, model (2.1) is a special case of the *analysis of covariance* model (2.2).

Under (2.1) the odds-ratio for those in age $x+1$ at $t+1$ divided by that for those in age x at t is $\exp(\alpha_1 + \beta) = \exp(\alpha_1)\exp(\beta)$. Under (2.2) the ratio is $\exp(\alpha_{x+1} - \alpha_x)\exp(\beta)$. Therefore, time and age affect the odds-ratio *multiplicatively*.

When the probability of death is small, the left hand side of (2.3) is close to the relative risk $q(x, t+1)/q(x, t)$, and it is customary to say that the parameters of logistic regression models measure relative risk. However, if the probability of death is large, then this interpretation is not valid, so it is the safest to refer to odds-ratios at all times. Of course, once a model has been fitted, we can estimate relative risk $q(x, t+1)/q(x, t)$ (or, say, risk differences $q(x, t+1) - q(x, t)$) by simply plugging in the estimates of the model parameters. As discussed in 1.5, a standard error for the measure can be based on the delta method.

One can test model (2.2) against (2.1) using likelihood ratio tests as discussed in Section 1.5. If both models are applied to ages $x = 1, \ldots, m$, then the test statistic (1.14) will have an approximate χ^2 distribution with $m - 2$ degrees of freedom, when (2.1) holds.

Measuring the goodness of fit is possibly the most important difference between binary regression and ordinary (normal distribution theory based) regression.[1] In the latter a single residual may give important clues as to the possible lack of fit. In the former, especially in the Bernoulli case ($n = 1$), we have to group or smooth the data in some way to see if the group means differ locally more from the predicted than one would expect under the correct model (e.g., Landwehr, Pregibon, and Shoemaker 1984; Fowlkes 1987). Hosmer and Lemeshow (2000, 140–145) have derived approximate critical values for one such test, in which the groups are formed based on the deciles (or other percentiles) of the predicted probability of success. Their simulations suggest that if J groups are used one can get approximate critical values from a χ^2 distribution with $J - 2$ degrees of freedom. Of course, if the data are initially binomial, Y - Bin(n, p) with np moderately large, then one can study the lack of fit for each binomial separately using the standard normal approximation to the *Pearson residuals* $(Y - n\hat{p})/(n\hat{p}(1 - \hat{p}))^{1/2}$.

2.2. Examples of Logistic Regression

Logistic regression can be used in a multitude of ways in demographic contexts. We will here introduce a historical data set, discuss confounding, and analyze attitudes.

Example 2.1. Sex Ratios of the Habsburgs. We consider a data set collected from *Encyclopædia Britannica* concerning the Habsburgs of Austria.[2] A section of

---

[1] More subtle differences exist. Gail (1986) shows that omitting a covariance that has the same distribution among the exposed and unexposed biases logistic regression, but not ordinary regression, for example.

[2] The authors would like to thank Prof. Weyss of I.I.A.S.A., who had tables of the Habsburg family that were in some respects more accurate and complete than those in the *Britannica*. Visitors to Vienna may want to visit *Kaisergruft* in the basement of *Kapuzziner Kirche* that houses the graves of many in our data set.

the family tree begins with Guntram the Rich who lived around 950. Only male descendants were recorded in the earliest times, so our data set starts from Rudolf I (1218–1291) who was a German king. He forms our generation 0, his children are the generation 1 etc. We follow the throne to generation 20 consisting of Charles I (1887–1922) and Maximilian Eugene (1895–1952). Only the part of the family tree is included through which the throne went. For example, all of Maria Theresa's (1717–1780) sixteen children are included, but out of their descendants only those of Leopold II (1747–1792) are included, since Leopold's son Francis I (1768–1835) inherited the throne. We have already used this data set for Figure 3 of Chapter 4, and we will analyze several aspects of the data later. However, here we would like to inspect the reliability of the data using regression techniques.

Maria Theresa was the only woman to hold the throne and pass it on to her children. All other were men. We therefore expect that both the actual and reported sex-ratio at birth would be tilted in favor of the males among the 20 families. This is not the case, however. There are a total of 175 individuals in the data set. Sex is given for all but 10 individuals who have died young. Among the remaining 165 persons, there were 79 males. If all births can be considered to be i.i.d. with respect to sex, then we have a model $Y \sim \text{Bin}(n, p)$ with $n = 165$ and $Y = 79$. The MLE of the probability of a male is $\hat{p} = 79/165 = 0.479$. The common method of calculating a 95% confidence interval for the proportion of males is $\hat{p} \pm 1.96(\hat{p}(1-\hat{p})/n)^{1/2} = 0.479 \pm 0.076$. Or, we get the interval [0.403, 0.555] that easily includes the value $105/205 = 0.512$ that we might expect. Overall, we see no indication of the omission of females from the data set.

As a second step we might wonder whether the fraction of the males has remained constant over time. We consider the model $Y_i \sim \text{Ber}(p_i)$, $\text{logit}(p_i) = \beta_0 + \beta_1 X_i$, where $Y_i = 1$ if i is a male and $Y_i = 0$ otherwise, and X_i is the birth year of individual $i = 1, \ldots, 165$. The MLE is $\hat{\beta}_1 = -0.001$ with an estimated standard error of 0.00085. This finding is consonant with the notion that the fraction of females has increased over the years due to more accurate reporting. However, the P-value is only 0.244, so the evidence is weak at best. ◊

Example 2.2. Child Mortality Among the Habsburgs. As a second check of the quality of the Habsburgs data we consider deaths in early age among the children who did not pass on the crown. We consider the model $Y_i \sim \text{Bin}(n_i, p_i)$, $\text{logit}(p_i) = \beta_0 + \beta_1 X_i$, where n_i is the number of children in generation i excluding the one whose descendants formed generation $i + 1$, Y_i is the number of them that died in age < 2, and X_i is the birth year of the individual founding the generation $i = 1, \ldots, 20$. The P-value under the hypothesis of zero slope was 0.936, which does not suggest any systematic change in the fraction of those who have died young. Therefore, child mortality appears not to have improved in a gradual manner (although we certainly know from other sources that it has improved in the 20th century), or if it has, then infant deaths may have been omitted from the data set in earlier times. ◊

Example 2.3. Testing Effects of Exposure on Illness. Consider an epidemiologic study of the effect of exposure on the risk of illness. Suppose the following (artificial) data have been obtained during a follow-up period:

	Ill	Not	Total
Exposed	36	64	100
Non-Exposed	24	76	100
Total	60	140	200

Let us assume binomial models for the data: Y_1 is the number of illnesses among the exposed with $Y_1 \sim \text{Bin}(100, p_1)$, Y_0 is the number of illnesses among the non-exposed with $Y_0 \sim \text{Bin}(100, p_0)$, and Y_1 and Y_0 are independent. Relative risk can be measured directly as $\text{RR} = (36/100)/(24/100) = 1.5$, or via the odds ratio $\text{OR} = (36 \times 76)/(64 \times 24) = 1.781$. The data can be analyzed in different ways. For example, we may condition on the number of illnesses (= 60), non-illnesses (= 140), and the total number of exposed (= 100). Under the null hypothesis that $p_1 = p_0$ the number of those who are ill among the exposed has a hypergeometric distribution and we can calculate the probability of obtaining 36 or more such cases as $P(36; 60, 140, 100) + \cdots + P(60; 60, 140, 100) = 0.0446$, where $P(x; \alpha, \beta, \gamma)$ is as defined in (6.1) of Chapter 2. This probability may be interpreted as a P-value for the one-sided alternative hypothesis that illness is more likely among the exposed than the non-exposed, or $p_1 > p_0$. This is *Fisher's exact test.*

There is no unique method for calculating a P-value corresponding to the two-sided alternative hypothesis $p_1 \neq p_0$. Often it is calculated simply by doubling (the smaller of the two tail probabilities), in this case $2(0.0446) = 0.0892$.[3] The results would indicate that there may well be an association. However, we may also pursue the analysis based on the assumption of two binomial models. Defining $\beta_0 = \log(p_0/(1 - p_0))$ and $\beta_1 = \log([p_1/(1 - p_1)]/[p_0/(1 - p_0)])$, we can write $p_0 = \exp(\beta_0)/(1 + \exp(\beta_0))$ and $p_1 = \exp(\beta_0 + \beta_1)/(1 + \exp(\beta_0 + \beta_1))$. Defining $X_1 = 1$ for the exposed group and $X_0 = 0$ for the non-exposed group, we can write $p_i = \exp(\beta_0 + \beta_1 X_i)/(1 + \exp(\beta_0 + \beta_1 X_i))$. Now we have a logistic regression model that can be fitted with any number of statistical packages, but it is simple enough that we can solve it by hand. The MLE of p_0 is 0.24 and the MLE of p_1 is 0.36, and so the MLEs are $\hat{\beta}_0 = \log(0.24/0.76) = -1.1528$ and $\hat{\beta}_1 = \log([0.36/0.64]/[0.24/0.76]) = 0.5773$. Taking $\mathbf{Y} = (Y_1, Y_0)^T$ the matrix (1.13) is evaluated as

$$\widehat{\text{Cov}}\begin{pmatrix} \hat{\beta}_0 \\ \hat{\beta}_1 \end{pmatrix} = \left(\begin{pmatrix} 1 & 1 \\ 1 & 0 \end{pmatrix} \begin{pmatrix} 23.04 & 0 \\ 0 & 18.24 \end{pmatrix} \begin{pmatrix} 1 & 1 \\ 1 & 0 \end{pmatrix} \right)^{-1} = \begin{pmatrix} 0.05482 & -0.05482 \\ -0.05482 & 0.098227 \end{pmatrix}.$$

[3] These values are based on the exact hypergeometric distribution. They are easily obtained from the program StatXact, for example. If a χ_1^2 distribution is used as an approximation, we get the one-sided P-value of 0.0324 and the two-sided P-value of 0.0649. The StatXact manual has additional discussion on the various definitions of the two-sided P-values. SAS sums the probabilities of the possible tables whose probabilities are not greater than the probability of the observed table (Cox and Hinkley 1974, 106), and Haberman (1978, 107) sums the probabilities of the possible tables whose cell value deviates from its expectation by as much or more than the observed table, which yields the exact significance level for the Pearson chi-square test.

The estimated standard error obtained from the diagonal of the matrix (1.13) is $0.3134 = 0.098227^{1/2}$, so a Wald test statistic for H_0: $\beta_1 = 0$ gets the value $0.5773/0.3134 = 1.842$. Referring this to the standard normal distribution leads to the same P-value as the χ_1^2 approximation to Fisher's exact test. ◊

Logistic regression is well suited to the study of joint effects of several variables. In particular, it can be used to assess confounding by factors that have been measured in the study (cf., Section 5.4 of Chapter 2). Let us continue in the setting of the previous example.

Example 2.4. Detecting Confounding. Suppose there was a dichotomous third variable Z such that the 2×2 table of Example 2.3 is actually a sum of two 2×2 tables as follows:

	Overall			$Z = 1$			$Z = 0$		
	Ill	Not	Total	Ill	Not	Total	Ill	Not	Total
Exposed	36	64	100	32	48	80	4	16	20
Non-Exposed	24	76	100	8	12	20	16	64	80
Total	60	140	200	40	60	100	20	80	100

Whereas the previous analysis seemed to suggest that exposure increased the risk of illness, we now see the relative risk of illness is = 1.0 for those with $Z = 1$ and for those with $Z = 0$! Clearly, exposure does not have any effect, but Z may. In this (artificially constructed) example it is easy to detect the source of confounding. In practice, there can be many potential confounders and they may be measured in continuous scales. Then a tabular analysis becomes very cumbersome. In contrast, using logistic regression it is easy to study complex patterns of confounding by simply adding and subtracting explanatory variables from regression. For the case at hand we might define $X_{ij} = 1$ for $j = 1$ and $X_{ij} = 0$ for $j = 0$; $Z_{ij} = 1$ for $i = 1$ and $Z_{ij} = 0$ for $i = 0$; and then assume four independent binomial models the number of those ill, $Y_{ij} \sim \text{Bin}(n_{ij}, p_{ij})$, where $\text{logit}(p_{ij}) = \beta_0 + \beta_1 X_{ij} + \beta_2 Z_{ij}$ and $n_{00} = n_{11} = 80$, $n_{01} = n_{10} = 20$. ◊

Logistic regression is also suitable for the study of attitudes. The following example shows that sometimes attitudes may depend on birth cohort. Some practical aspects of model choice are also illustrated.

Example 2.5. Choosing the Sword. The University of Joensuu has arranged Doctoral Promotions once or twice a decade. This is a festive event in which a Doctor's hat and a sword are given to those who have completed their doctorate since the previous Promotion. Participation is voluntary and some do not. One reason is that the promotees must pay themselves for the hat, sword, fancy dinner, formal clothing etc. In 1999, a controversy arose. Some promotees wanted to omit the sword from the ceremony, because they felt it is a militaristic symbol, and expensive to the bargain. Others said that this would undermine tradition. A compromise was reached, and the choice was left to the promotees. A total of $n = 104$ promotees participated with 70 taking the sword. Can we explain why some did but others did not?

We know, for each promotee $i = 1, \ldots, 104$, their SEX (= 1, if i is female, otherwise 0), AGE (in years), and SCHOOL (Education, Forestry, Humanities, Natural Sciences, Social Sciences), and if they took a sword ($Y_i = 1$) or not ($Y_i = 0$). Define $P(Y_i = 1) = p_i$, as before. Beforehand we thought that possibly men are more likely to take the sword than women, and so might those in natural sciences be more likely than those in education, humanities, or social sciences.

Treating SCHOOL as a factor (i.e., dummy variables were created for four of the five categories), and including it as an explanatory variable together with AGE and SEX, showed that the probability of taking the sword did not depend on SCHOOL at all: the smallest P-value of the four indicators was 0.48. Omitting SCHOOL we fitted the equation $\text{logit}(p_i) = 4.57 - 0.70 \times \text{SEX}_i - 0.086 \times \text{AGE}_i$. The estimated standard error of the coefficient of SEX was 0.45 corresponding to a P-value of 0.12 and the estimated standard error of the coefficient of AGE was 0.029 corresponding to a P-value of 0.03. Hence, there was some evidence that the women were less likely to take the sword, but there was clear evidence that the older you were the *less* likely you were to take the sword. The youngest promotee was 26 years old, and the oldest 64 years old, a difference of 38 years, so the odds-ratio comparing the youngest and oldest (holding SEX constant) would be $\exp(0.086 \times 38) = 26.3$. The 95% confidence interval for that odds ratio is $\exp((0.086 \pm 1.96 \times 0.029) \times 38) = (3.0, 228)$, which does not include 0. Hence the age effect was not only statistically significant (i.e., too large to plausibly be due to random error), but implied a large difference in preferences.

As older people would be expected to be more respectful of tradition than younger ones, the finding appeared puzzling. To examine the relationship between age and the probability of taking the sword more closely, a factor variable AGE2 was defined corresponding to 10-year age-groups 26–34, ..., 55–64. Using the youngest age as a comparison or reference group, the dummy variables of the three older ages had negative coefficients, but only that of age-group 45-54 was significant[4]. Defining just a single dummy A for this age-group and entering it to the equation with SEX, produced the equation $\text{logit}(p_i) = 1.35 - 0.82 \times \text{SEX}_i - 0.97 \times A_i$. The P-values for the two explanatory models are now 0.044 and 0.049 respectively. However, the model does fit the data slightly less well than the original model using SEX and AGE.

We conclude that women have been less likely than men to choose the sword. The older promotees have similarly been less likely to take the sword than the younger ones. In addition, there is some evidence that especially those in ages 45–54 at the time of the Promotion were reluctant to take the sword. We note that they were born during 1945–1954 and so most of them belong to the baby-boom cohorts in Finland. They carried out their university studies 20–30 years later, roughly during the 1970's, when student radicalism was fashionable. We speculate that this may have influenced their preferences. ◊

---

[4] We say that a statistic is "significant" if it is significantly different than zero at some significance level, which usually is 0.05 unless specifically stated.

2.3. Applicability in Case-Control Studies

Logistic regression can be applied in a cohort study to explain, in terms of background characteristics, why an event of interest occurs during follow-up to some but not to others. It is less obvious that it could be applied in a case-control setting, because of the outcome selective method of data collection. However, we show now that the method is valid under certain conditions.

Consider an individual with vector of characteristics $\mathbf{X}$. Define $Y = 1$ if the individual is ill, and $Y = 0$ otherwise. Define $S = 1$, if the individual is selected into the study, and $S = 0$ otherwise. Assume that the logistic model $P(Y = 1) = \exp(\alpha + \mathbf{X}^T\beta)/(1 + \exp(\alpha + \mathbf{X}^T\beta))$ holds, where we have displayed the constant term separately. The probability that an individual is selected into the study depends on Y, and we denote the selection probabilities by $\tau_j = P(S = 1|Y = j)$, $j = 0, 1$. We would like to determine the probability of being ill, given that the individual is selected into the study. Following Breslow and Day (1980, 203), we can use Bayes' formula and write

$$P(Y = 1|S = 1) = \frac{P(S = 1|Y = 1)P(Y = 1)}{P(S = 1|Y = 1)P(Y = 1) + P(S = 1|Y = 0)P(Y = 0)}. \tag{2.4}$$

Substituting in the logistic probabilities, and simplifying, yields the result

$$P(Y = 1|S = 1) = \frac{\exp(\alpha^* + \mathbf{X}^{\mathrm{T}}\beta)}{1 + \exp(\alpha^* + \mathbf{X}^{\mathrm{T}}\beta)}, \tag{2.5}$$

where $\alpha^* = \alpha + \log(\tau_1/\tau_0)$. Thus, the same logistic model is valid for the study of relative risk in both cohort and case-control studies, but unless $\tau_1/\tau_0 = 1$ the constant term from a case-control study α^* cannot be interpreted as representing the risk of those with $\mathbf{X} = \mathbf{0}$.[5]

Suppose now that $\tau_j = \tau_j(\mathbf{X})$, but in such a way that $\tau_1(\mathbf{X}) = c\tau_0(\mathbf{X})$. We see from (2.5) that the logistic model is still valid, as long as both selection probabilities depend in a similar way on $\mathbf{X}$.

However, if the relative risk of selection depends on $\mathbf{X}$ and is of the form $\tau_1(\mathbf{X})/\tau_0(\mathbf{X}) = \exp(\alpha' + \mathbf{X}^T\gamma)$, we have

$$P(Y = 1|S = 1) = \frac{\exp(\alpha'' + \mathbf{X}^{\mathrm{T}}(\beta + \gamma))}{1 + \exp(\alpha'' + \mathbf{X}^{\mathrm{T}}(\beta + \gamma))}, \tag{2.6}$$

where $\alpha'' = \alpha + \alpha'$. We note that the coefficients become biased. This conclusion is of practical importance in studies such as the Doll and Hill study (Example 5.2 of Chapter 2). Suppose all available cases are taken into the study ($\tau_1(\mathbf{X}) = 1$), and controls are selected from among patients who have come to a hospital for

[5] If prior information about baseline risk (when $\mathbf{X} = \mathbf{0}$) is available, absolute risks can still be estimated (Neutra and Drolette 1978; King and Zeng 2002 review several of the alternative formulations).

reasons other than the disease under study. If similar exposures increase the risk of both types of disease, then the bias represented by γ in (2.6) is likely to be present.

There are several variants of the cohort and case-control designs in which the use of logistic regression may be valid. Keinänen (2002) investigated factors influencing the recruitment of workers into information technology (IT) branch in Finland, during 1999. The data source was the employee database of Statistics Finland (cf., Statistics Finland 2002), which has detailed data on employment histories of everyone employed in Finland. Three random samples were first selected from among those who were either outside the labor force, in the labor force but unemployed, and in the labor force but outside the IT sector, in the beginning of the year. Since recruitment into the IT sector is a rare event, massive samples would have been necessary to get reliable estimates using this approach alone. However, a fourth sample was selected from among those who had moved into the IT sector during 1999.

The use of logistic regression in this setting can be justified much the same way as above. For example, restrict attention to those who are unemployed in the beginning of the year. Consider an individual with characteristics $\mathbf{X}$ in the beginning of the year. Let $Y = 1$ if the individual is employed in IT sector at the end of the year and let $Y = 0$ otherwise. Define $S = 1$ if the individual was selected into the study and $S = 0$ otherwise. Assume that $P(Y = 1) = \exp(\alpha + \mathbf{X}^T\beta)/(1 + \exp(\alpha + \mathbf{X}^T\beta))$. Let τ_0 be the probability of being selected into the study in the beginning (i.e., the first three samples). Let τ_2 be the probability of selecting a case into the study, provided that he or she was not already selected in the beginning, and denote the marginal selection probability $P(S = 1)$ by $\tau_1 = \tau_0 + (1 - \tau_0)\tau_2$. It follows that $P(S = 1, Y = 0) = \tau_0/(1 + \exp(\alpha + \mathbf{X}^T\beta))$, and $P(S = 1, Y = 1) = \tau_1 \exp(\alpha + \mathbf{X}^T\beta)/(1 + \exp(\alpha + \mathbf{X}^T\beta))$. With these conventions the conditional probability that the individual becomes employed in the IT sector, given that the individual selected into the study, is given exactly by (2.5). As this was a register based study, the selections into the samples could be made independently of $\mathbf{X}$. As noted in Chapter 2, studies of this type are sometimes called case-cohort or case-base studies.

Both case-control and case-cohort studies may include matching as part of data collection. We will indicate in Example 7.5 how this changes the likelihood.

3. Poisson Regression

3.1. Interpretation of Parameters

Suppose $Y \sim \text{Po}(\lambda)$. By taking $\theta = \log(\lambda)$, $b(\theta) = \lambda = \exp(\theta)$, and $c(y) = -\log(y!)$, we see that the Poisson distribution belongs to the 1-parameter exponential family (1.1). In this case the canonical parameter is the log of the expectation. The Poisson regression model is *loglinear*, because the expectation is related to the linear predictor (1.5) in the log-scale. Let $K_{x,t}$ be the number of *person years* lived by those in age x during year t in a population, and let $Y_{x,t}$ be the corresponding number of deaths. Suppose $\lambda_{x,t} K_{x,t}$ is the expected number of deaths[6], so $\lambda_{x,t}$ is the hazard (cf., Chapter 4). Then, a model corresponding to (2.1) would be

$$\lambda_{xt} = \exp(\alpha_0 + \alpha_1 x + \beta t). \tag{3.1}$$

It is easy to see that $\lambda_{x,t+1}/\lambda_{x,t} = \exp(\beta)$ irrespective of x. Using hazards as risk measures, we note that the parameters of the Poisson regression model have an exact interpretation in terms of the log of *relative risk*. The same way logistic regression assumed multiplicativity for the odds-ratios, Poisson regression assumes multiplicativity for the relative risk. Using terminology introduced in Chapter 4, we note that (3.1) is actually a proportional hazards model.

Once the parameters have been estimated, other measures that can be estimated including hazard differences (e.g., $\lambda_{x,t+1} - \lambda_{x,t}$) and expected values $\lambda_{x,t} K_{x,t}$. Confidence intervals for them can be derived using the delta method (Section 1.5).

If (3.1) holds, the Poisson expectation is of the form

$$\lambda_{xt} K_{xt} = \exp(\alpha_0 + \alpha_1 x + \beta t + \log(K_{xt})). \tag{3.2}$$

We see that the person years can be accommodated by incorporating an additional regression term $\log(K_{x,t})$ with a fixed coefficient $= 1$ to the regression model. Many computer programs such as GLIM, EGRET, R, S+, SAS and Stata allow such *offset* regressors.

Inference concerning Poisson regression can be carried out the same way as for logistic regression. The goodness of fit of the Poisson models is easier to study, however, since the deviance is known to have an asymptotic χ^2 distribution when the expectations of the Poisson counts are sufficiently large (cf., Conover 1980, 191). In addition, several more refined tools for diagnostic checking have been developed (e.g., Bishop et al. 1975, 136–137; Haberman 1978, 77–79). In Section 4 we will also note that count data often display more variability than one would expect under a strict Poisson assumption. Alternative models are provided for this situation.

3.2. *Examples of Poisson Regression*

Poisson regression is a standard tool of demographic analysis. Here we give a few simple illustrations, and others will appear later in several places.

Example 3.1. Poisson Models for Births. Estimates of age-specific fertility in Example 4.1 of Chapter 4 are based on a saturated model, where the number of births in age $x = \alpha, \ldots, \beta$ during year $t = 1, \ldots, T$, is $Y_{xt} \sim \text{Po}(\lambda_{xt} K_{xt})$. More parsimoniously, consider models of the form $\log(\lambda_{xt}) = \delta_x + \eta_t + \gamma(x - M)t + \zeta(x - M)^2 t$, where M is the mean age at childbearing at $t = 0$ (for the various

[6] Although $K_{x,t}$ depends on $Y_{x,t}$, this dependence can be ignored at least as long as the expected count is small relative to the person years. In a data set on old-age mortality (Alho and Nyblom 1997) alternative estimates of relative risk could be calculated using a binomial model. In this case, the estimates were essentially the same as those obtained from a Poisson model even though Y_{xt} represented a *large* proportion of K_{xt}.

definitions, see Example 4.4 of Chapter 4). For identifiability, assume that $\Sigma_x \delta_x = 0$. If $\gamma = \zeta = 0$, we have a main effects (or a "2-way analysis of variance" model) in which the δ_x's determine the shape of the age-specific fertility schedule and the η_t's determine the level of total fertility. If $\zeta = 0$, then (as discussed in Exercise 34 of Chapter 4) the model incorporates a systematic change in the mean age at childbearing: for $\gamma > 0$ the mean age increases and for $\gamma < 0$ it decreases over time. Finally, if we also have $\zeta \neq 0$, it is possible to capture a systematic change in the spread of fertility around the mean age: for $\zeta > 0$ the spread increases over time, for $\zeta < 0$ it decreases over time. The role of M is to center the x values, so a better interpretation for the parameters γ and ζ is obtained. ◊

Example 3.2. Mortality of Young Widows. A notable feature in Figure 1 of Chapter 4 is the high mortality of widows in young ages. Is the effect significant? Consider ages 26–34. The number of deaths among married were $Y_0 = 35$, and the number of person years were $K_0 = 145,651$. For the widowed the deaths were $Y_1 = 3$, and person years were $K_1 = 663$. Assume that $Y_i \sim \text{Po}(\lambda_i K_i)$, $i = 0, 1$, are independent, and consider the model $\log(\lambda_i) = \mu + \alpha_i$, with $\alpha_0 = 0$. We obtain the estimate $\hat{\alpha}_1 = 2.9355$, so an estimate of relative risk is $\exp(2.9355) = 18.83$ with a 95% confidence interval [5.79, 61.2]. Thus, the excess risk appears to be real. The finding agrees with those of Hu and Goldman (1990, 241) from several countries. The authors suggest that the circumstances leading to the spouse's death may also increase the hazard of the remaining partner. ◊

Example 3.3. Age-Period-Cohort Problem. Model (3.1) treats both age and period effects linearly (in the log-scale). In many demographic applications it is also of interest to consider cohort effects. For example, harsh conditions in childhood may adversely effect later survival. Note, however, that if a term $\beta_3(t - x)$ is added to the linear predictor, then the model is *not identifiable*: to any value for β_3 there corresponds a model containing age and period effects only that provides the same fit. The root cause for the problem is that the three effects are perfectly collinear in this case. This is the famous *age-period-cohort problem.* If there is a basis for deciding which two of the effects are the most important, then the effect of the third can be determined conditionally on the estimates of the first two. For a review, see Clayton and Schiffers (1987a,b), and for an example of a potential resolution in a non-parametric setting, see Ogata et al. (2000). ◊

Example 3.4. Number of the Habsburg Offspring. Continuing in the setting of Example 2.2, consider the sizes of the generations $i = 1, \ldots, 20$. Let Y_i be the number of children in generation i minus one (i.e., excluding the one who passed on the throne). A possible model assumes that $Y_i \sim \text{Po}(\lambda_i)$, $i = 1, \ldots, 20$ are independent. To investigate time trends, let us assume the model $\log(\lambda_i) = \alpha + \beta X_i$, where X_i is the birth year of the person generating the generation i. We obtain the MLE $\hat{\beta} = 0.000114$ and an estimated standard error of 0.000415. We conclude that there appears to be no overall trend in family size over the observation period. ◊

Example 3.5. Regression Models for Rates of Small Areas. Summary measures such as life expectancy or total fertility rate are sometimes desired for *small areas.*

In Finland, for example the median size of a municipality is 5,000, and the annual number of births and deaths is of the order of 50. In the U.S., there are more than 40,000 places, municipalities, and minor civil divisions, and the median size is around 1,000. Even though data by municipality are available, the numbers are so small that Poisson variation makes the results unreliable. Poisson regression provides a way to stabilize the estimates by "borrowing strength" in estimation from neighboring areas. Suppose $Y_{xm} \sim \mathrm{Po}(\lambda_{xm} K_{xm})$ is the number of events in age x in municipality m. Fit a main effects model $\log(\lambda_{xm}) = \alpha_x + \beta_m$ to data from several municipalities. This yields the MLEs $\hat{\lambda}_{xm}$. Suppose the counts are births. We can then estimate the age-specific fertility rates for each municipality m by $\hat{\lambda}_{xm}$'s. Similarly, if the counts are deaths, we can estimate age-specific mortality rates by $\hat{\lambda}_{xm}$'s. In an analysis of a few small municipalities we may want to use *external baseline rates* in estimation. If the α_x's are known, this can be effected by offsetting $\alpha_x + \log(K_{xm})$, instead of just $\log(K_{xm})$, in estimation. ◊

3.3. Standardization

Poisson regression has a close connection to *standardization*, a topic that is central to classical demography (e.g., Breslow and Day 1987, 128; Hoem 1987). For concreteness, we consider mortality, but the concepts and results of this section apply generally. Denote the number of deaths in age x at time t by Y_{xt} and the corresponding person years of exposure by K_{xt} for $x = 0, \ldots, \omega$ and $t = 1, \ldots, T$. A dot (.) in place of a subscript will denote summation over the subscript,

$$Y_{x\cdot} = \sum_{t=1}^{T} Y_{xt}, \quad Y_{\cdot t} = \sum_{x=0}^{\omega} Y_{xt}, \quad Y_{\cdot\cdot} = \sum_{t=1}^{T} Y_{\cdot t}$$

$$K_{x\cdot} = \sum_{t=1}^{T} K_{xt}, \quad K_{\cdot t} = \sum_{x=0}^{\omega} K_{xt}, \quad K_{\cdot\cdot} = \sum_{t=1}^{T} K_{\cdot t}. \tag{3.3}$$

Often we are interested in comparing $Y_{\cdot t}$ across years, but we want to eliminate the effect of age distributions (K_{xt}) varying with t. Denote the age-specific mortality rates by $m_{xt} \equiv Y_{xt}/K_{xt}$, and note that the crude mortality rate of year t can be written as a weighted average of the age-specific rates,

$$\frac{Y_{\cdot t}}{K_{\cdot t}} = \sum_{x=0}^{\omega} \left(\frac{K_{xt}}{K_{\cdot t}} \right) m_{xt}. \tag{3.4}$$

The fact that the weights depend on t is problematic – do differences in crude rates reflect different risks or different weights?

Direct standardization solves the problem by the use of *standard weights* $w_x > 0$ with $w_0 + \cdots + w_\omega = 1$. The *directly standardized mortality rate* is defined simply as

$$\sum_{x=0}^{\omega} w_x m_{xt}. \tag{3.5}$$

Since (3.5) depends on the chosen weights, standardized rates can generally be used for comparative purposes only. A common choice is $w_x = K_{x\cdot}/K_{\cdot\cdot}$. For the purpose of standardizing time series, external standard weights are used (cf., Anderson and Rosenberg 1998).

Calculation of the directly standardized rate requires knowledge of the individual m_{xt}'s. If only the crude rate is known for time t, an alternative, indirect standardization may be used. Taking the reference group to be the aggregate over t, with $w_x = K_{x\cdot}/K_{\cdot\cdot}$ and $m_x = Y_{x\cdot}/K_{x\cdot}$, notice that the ratio of the direct standardized rate to the crude rate in the reference group is $\Sigma_x w_x m_{xt}/\Sigma_x w_x m_x$. If we replace the standard weights w_x by $K_{xt}/K_{\cdot t}$, that ratio transforms to the *standardized mortality ratio* (SMR),

$$\frac{Y_{\cdot t}/K_{\cdot t}}{\sum_{x=0}^{\omega}(K_{xt}/K_{\cdot t})m_x} = \frac{Y_{\cdot t}}{\sum_{x=0}^{\omega} K_{xt}m_x}. \tag{3.6}$$

Note that (3.6) can be interpreted as an *observed/expected ratio*. If we multiply the SMR by the crude rate for the reference group, we obtain the *indirectly standardized mortality rate*,

$$\left(\frac{Y_{\cdot t}}{\sum_{x=0}^{\omega} K_{xt}m_x}\right)\frac{Y_{\cdot\cdot}}{K_{\cdot\cdot}}. \tag{3.7}$$

For additional insight into indirect standardization, suppose that Y_{xt} are mutually independent and distributed as Po($\lambda_{xt}K_{xt}$), and consider a main-effects analysis of variance model as,

$$\lambda_{xt} = \exp(\alpha_x + \beta_t). \tag{3.8}$$

If we write out the likelihood and apply the factorization criterion, we see that the vector $\mathbf{U} = (Y_{0\cdot}, \ldots, Y_{\omega\cdot}, Y_{\cdot 1}, \ldots, Y_{\cdot T})^T$ is sufficient for $(\alpha_0, \ldots, \alpha_\omega, \beta_1, \ldots, \beta_T)^T$. Recalling (1.8), we note that the MLEs are the solution to $\mathbf{U} = \mathrm{E}[\mathbf{U}]$. Equating first $Y_{x\cdot} = E[Y_{x\cdot}]$ and setting $\beta_t = 0$ leads to the estimates

$$\exp(\hat{\alpha}_x) = Y_{x\cdot}/K_{x\cdot}. \tag{3.9}$$

In other words, the initial estimates for the α_x's are the logs of the age-specific rates when the data have been aggregated across years. If we insert these estimates into the equations $Y_{\cdot t} = E[Y_{\cdot t}]$, we get

$$\exp(\hat{\beta}_t) = Y_{\cdot t}\Big/\sum_{x=0}^{\omega}\exp(\hat{\alpha}_x)K_{xt}, \tag{3.10}$$

which is equal to the standardized mortality ratio (3.6). Multiplying $\exp(\hat{\beta}_t)$ by the crude mortality rate across age and years, we obtain the indirectly standardized mortality rate (3.7). Upon further iteration the estimates may change, but (3.10) shows that the "main effects" model (3.8) can be viewed as a way of carrying out indirect standardization (Hoem 1987).[7]

The variance of the directly standardized rate (3.5) is usually calculated under the assumption that $Y_{xt} \sim \text{Po}(\lambda_{xt} K_{xt})$ are independent. Hence, the estimated variance (3.5) is

$$\sum_{x=0}^{\omega} w_x^2 Y_{xt} / K_{xt}^2. \tag{3.11}$$

Statistical inference can then be based on a normal approximation to the distribution of (3.5).

Example 3.6. Relative Risk of Mortality for Unemployed. To illustrate standardization, let us consider the relative risk of mortality among the unemployed as compared to the employed in Finland, in 1998. Whereas previously t had referred to year, now we let $t = 1, 2$ distinguish employed from unemployed. The deaths D_{xt}, the person years in thousands K_{xt}, and the mortality rates (per thousand) m_{xt}, for $x = 0, 1, \ldots, 5$, were the following.

	Employed ($t = 1$)			Unemployed ($t = 2$)			SDPOP	SDRATE
Age (x)	Y_{x1}	K_{x1}	m_{x1}	Y_{x2}	K_{x2}	m_{x2}	$K_{x\cdot}/K_{\cdot\cdot}$	m_x
(0) 15–19	11	16.7	0.659	24	10.3	2.33	0.021	1.30
(1) 20–29	89	177.7	0.501	113	57.4	1.97	0.185	0.86
(2) 30–39	259	296.1	0.874	246	55.0	4.47	0.277	1.44
(3) 40–49	565	313.8	1.80	526	59.1	8.90	0.294	2.93
(4) 50–59	759	199.2	3.81	555	54.5	10.18	0.200	5.18
(5) 60–69	176	24.3	7.24	51	4.29	11.86	0.023	7.94
Total	1859	1027.8	1.81	1515	240.6	6.3	1.000	

The crude mortality rates are $Y_{\cdot 1}/K_{\cdot 1} = 1859/1028 = 1.81$ and $Y_{\cdot 2}/K_{\cdot 2} = 1515/240.6 = 6.3$, so the relative risk appears to be $6.3/1.81 = 3.48$, indicating that mortality among the unemployed is three to four times as high as among those employed. Can this be due to a difference in age-distribution?

The column SDPOP contains the age-distribution of the whole population, $K_{x\cdot}/K_{\cdot\cdot}$. Multiplying the age-specific rates m_{xt} by the population shares SDPOP yields the directly standardized rates 6.57 for the unemployed and 1.80 for the employed. These yield a relative risk of $6.58/1.81 = 3.64$. An indirectly standardized relative risk estimate can be obtained by first calculating the standardized mortality ratios for both groups. As an observed/expected ratio the standardized mortality ratio (3.6) equals $1859/2743.0 = 0.678$ for the employed and $1515/631.0 = 2.400$ for the unemployed. Hence, the relative standardized mortality ratio is

[7] The functional iteration we have used to solve the likelihood equations is not identical to Newton's method. The latter does not yield the same insight provided by (3.9) and (3.10).

2.400/0.678 = 3.54. Fitting the main effects model (3.6) $\log(\lambda_{xt}) = \alpha_x + \beta_t$, with $\beta_1 = 0$ for identifiability, yields the estimate $\hat{\beta}_2 = 1.2795$. The standard error of the estimate is 0.0348. Therefore, the relative risk is exp(1.2794) = 3.59 with a 95% confidence interval of [3.36, 3.85]. In this case the estimates of relative risk are nearly the same if one uses crude rates, directly standardized rates, indirectly standardized rates, or Poisson regression estimates. An advantage of the latter is the easy access to confidence intervals, although they can be calculated for the other estimates fairly easily.

However, the real power of the regression approach comes from the facility of elaboration. In this case, many of the age-effects were within sampling error of the mean age effect. By entering age as a continuous explanatory variable, $\log(\lambda_{xt}) = \mu + \alpha x + \beta_t$, one obtains a smaller model with a significant age effect. The deviance of model (3.8) is 99.47 and the deviance of the model with continuous ages is 125.14. Comparing the difference 125.14 −99.47 = 25.67 to χ^2 distribution with 4 degrees of freedom, we find a P-value < 0.0001, so the smaller model is not adequate. However, there appears to be *interaction between age and employment status*. Extending the main effects model to a form $\log(\lambda_{xt}) = \alpha_x + \beta_t + \gamma \text{AGE2}(x)$, where $\text{AGE2}(x) = x$ for the unemployed and $\text{AGE2}(x) = 0$ for the employed, we get the deviance 42.14. This is a major improvement on the main effects model, because comparing 99.47 − 42.14 = 57.33 to χ^2 distribution with 1 degree of freedom, we find a P-value much below 0.0001. In this model, the we have $\hat{\beta}_2 = 2.1222$, and the coefficient of the interaction term is $\hat{\gamma} = -0.2608$. All age effects, except that of age group 1 are significantly different from the age-group 0. Thus, our estimate of the relative risk of the unemployed as compared to employed, in age group x, is $\exp(2.1222 - 0.2608x)$ for $x = 0, 1, \ldots, 5$, which ranges from 8.3 to 2.3. Due to the interaction, the main effects model that underlies indirect standardization is not valid, and even direct standardization is somewhat crude. The more refined analysis reveals that for the young unemployment is a greater risk factor than suggested by standardization techniques, whereas for the old the relative risk is less than suggested by the standardization techniques. A possible explanation for the change in relative risk can be given in terms of the notion of multiple decrements: those in ill health are selected out of the labor force before death. ◊

3.4. Loglinear Models for Capture-Recapture Data

There is a large literature on the application of loglinear models to contingency tables (e.g., Bishop, Fienberg and Holland 1975, Haberman 1978,1979). These models are of interest to demographers, since demographic data are often collected as classified by variables such as age, sex, race, or region. Here, we will briefly show how they can be used to analyze capture-recapture data.

By taking $K_{xt} = 1$ in the model of Section 3.3 and generalizing from deaths to counts more generally, we get formally a contingency table of counts $Y_{xt} \sim \text{Po}(\lambda_{xt})$. The model (3.8) is called a *main effects* model, because it has parameters α_x relating to the $\omega + 1$ rows and parameters β_t relating to the T columns. A (saturated) model including *interactions* between rows and columns would be of the form $\log(\lambda_{xt}) = \alpha_x + \beta_t + \gamma_{xt}$.

Suppose now that a census and a subsequent survey have been conducted for the same population. Let $Y_{ij} \sim \text{Po}(\lambda_{ij})$ be the population counts: Y_{11} = the number of those counted on both occasions; Y_{10} = the number of those counted the first time but not the second time; Y_{01} = the number of those counted the second time but not the first time; Y_{00} = the number of those not counted at all. The total population is then $N = Y_{11} + Y_{10} + Y_{01} + Y_{00}$, where Y_{00} is unknown. Suppose we have a main effects model $\lambda_{ij} = \exp(\alpha_i + \beta_j)$, where we set $\beta_0 = 0$ to attain identifiability. Setting the three observed values equal to their expectation one gets the estimates $\hat{\alpha}_1 = \log(Y_{10})$, $\hat{\beta}_1 = \log(Y_{11}/Y_{10})$, and $\hat{\alpha}_0 = \log(Y_{10}Y_{01}/Y_{11})$. The MLE of the expectation of the unknown count is $\hat{\lambda}_{00} = Y_{10}Y_{01}/Y_{11}$. By a direct calculation one can show that $\hat{N} = Y_{11} + Y_{10} + Y_{01} + \hat{\lambda}_{00}$ agrees with the classical dual systems estimator, or $\hat{N} = (Y_{11} + Y_{10})(Y_{11} + Y_{01})/Y_{11}$.

There are several variants of the derivation of the classical estimator. In particular, one may bypass the Poisson assumption of the counts and resort to multinomial distribution of the observed counts (Y_{11}, Y_{10}, Y_{01}) (cf., Bishop, Fienberg and Holland 1975; we will apply a similar argument in Section 5). The MLEs are similar, however, since the multinomial model is obtained from the Poisson model by conditioning on the observed total $Y_{11} + Y_{10} + Y_{01}$. Moreover, if one conditions further on the marginals $Y_{1.} = Y_{11} + Y_{10}$ and $Y_{.1} = Y_{11} + Y_{01}$, one obtains the hypergeometric model mentioned in Chapter 2 in which Y_{11} is the only free variable. All models lead to the same MLEs albeit that their (model-based) variances need not be the same.

The interest in applying loglinear models in capture-recapture data is not that it provides yet another derivation of the classical results. However, suppose the two captures are positively (negatively) dependent, in the sense that having been captured on the first occasion changes the person in such a way that his or her probability of capture during the second occasion is higher (lower) than the probability of capture of those who were not captured during the first occasion. Conditioning on the marginals $Y_{1.}$ and $Y_{.1}$, one then expects a larger (smaller) number of those captured twice, Y_{11}, than under a model of independence. Thus, the classical estimator is expected to underestimate (overestimate) the true population. Such behavioral response to the capture event is essentially impossible to assess based on two captures, but if three or more captures are available, loglinear models can help.

Suppose $Y_{ijk} \sim \text{Po}(\lambda_{ijk})$ are the population counts: Y_{111} = the number of those counted on all occasions, Y_{110} = the number of those counted the first two times but not the last time, etc. In this case Y_{000} = the number of those not counted at all, and the total population size to be estimated is $N = Y_{111} + Y_{110} + Y_{101} + Y_{100} + Y_{011} + Y_{010} + Y_{001} + Y_{000}$. A main effects loglinear model would be $\lambda_{ijk} = \exp(\alpha_i + \beta_j + \gamma_k)$, where $\beta_0 = \gamma_0 = 0$ for identifiability. However, this is not the only possibility. A model allowing an interaction between the first two captures, but keeping the third capture independent of the first two, assumes that $\lambda_{ijk} = \exp(\alpha_i + \beta_j + \gamma_k + \delta_{ij})$. Details of the analysis of these models are given in Bishop, Fienberg and Holland (1975, Chapter 6).

Examples of the application of triple-systems estimation in the context of the 1990 U.S. census data are given by Zaslavsky and Wolfgang (1993) and Darroch et al. (1993). In this case the three captures are formed by the census, the post-enumeration survey, and pre-census administrative records from Employment Security, driver's license administration, Internal Revenue Service, Selective Service, and Veteran's Administration. There seems to be some evidence that the capture by administrative records was only weakly, if at all related to capture by the census or the survey.

We conclude by expanding on Example 6.2 of Chapter 2 on drug use in Finland.

Example 3.7. Triple Systems Estimates of Numbers of Drug Users. In addition to the Hospital Discharge Register ($i = 0, 1$) and the Criminal Report Register ($j = 0, 1$), there is a Register for Driving Under the Influence of Alcohol and other Drugs ($k = 0, 1$) that contain information about drug users. The following capture data that we analyze under the model $Y_{ijk} \sim \text{Po}(\lambda_{ijk})$, were obtained in year 2000:

i	1	0	1	0	1	0	1
j	1	1	0	0	1	1	0
k	1	1	1	1	0	0	0
Captures	3	77	9	87	50	695	384

The total number of captures is 1,305. The model $\log(\lambda_{ijk}) = \alpha_i + \beta_j + \gamma_k$ has deviance 85.81 (residual d.f. = 3); the model $\log(\lambda_{ijk}) = \alpha_i + \beta_j + \gamma_k + \delta_{ij}$ has deviance 27.16 (d.f. = 2); the model $\log(\lambda_{ijk}) = \alpha_i + \beta_j + \gamma_k + \pi_{ik}$ has deviance 81.68 (d.f. = 2); and the model $\log(\lambda_{ijk}) = \alpha_i + \beta_j + \gamma_k + \xi_{jk}$ has deviance 2.30 (d.f. = 2). Thus, the last mentioned model is the best among the ones considered. In the Poisson case deviance has approximately a χ^2 distribution with 2 degrees of freedom, so we find that it is acceptable based on goodness-of-fit. The estimate for the expectation of the missing cell is $\hat{\lambda}_{000} = \exp(8.5793) = 5{,}320$. Adding this to the total number of captures yields the estimate $5{,}320 + 1{,}305 = 6{,}625$. This is about 5% less than the estimate of 6,942 obtained from two registers in Example 6.2 of Chapter 2. A 95% prediction interval for the count of the missing cell is [4,035; 7,015]. This translates into an interval [5,340; 8,320] for the total population. ◊

4. Overdispersion and Random Effects

Consider the model (1.5). As noted in Chapter 4, Section 5, often demographic data show more variability than can be accounted by the binomial or Poisson model we may be using. The excess variability is called *overdispersion*. In Section 4.1 we will first describe a simple extension of model (1.1) that can be used as a diagnostic tool to investigate the presence of overdispersion. Then, in Section 4.2 we discuss two classical marginal models for handling the overdispersion in these settings. Section 4.3 presents alternative random effect models that are intended for more general forms of overdispersion.

4.1. Direct Estimation of Overdispersion

The classical formulation of Nelder and Wedderburn (1972) includes a scale factor that corresponds to the variance in the case of a normal distribution, for example. However, we can also use an estimate of the scale as a diagnostic tool to investigate the possible presence of overdispersion or *underdispersion* (i.e., the case in which observed variability is smaller than expected under the chosen model). Suppose we have independent counts Y_i that correspond to person years K_i, $i = 1, \ldots, n$, such that $E[Y_i] = \exp(\mathbf{X}_i^T\beta)K_i$, where $\mathbf{X}_i$ is a vector of characteristics of observation i. Suppose $\hat{\beta}$ is a solution to (1.8) under a Poisson assumption for the data. By the law of large numbers, (1.8) provides a consistent solution for β provided that the Y_i's, K_i's and $\mathbf{X}_i$'s are sufficiently well-behaved, even if the Poisson assumption does not hold (cf., Rao 1973, 112-114, theorems (i) and (iii)). Consider another estimating equation (cf., Section 7.3 of Chapter 3) for a parameter ϕ, of the form

$$\sum_{i=1}^{n}\left\{\frac{\left(Y_i - \exp\left(\mathbf{X}_i^T\hat{\beta}\right)K_i\right)^2}{\exp\left(\mathbf{X}_i^T\hat{\beta}\right)K_i} - \phi\right\} = 0. \tag{4.1}$$

Under a Poisson assumption, the laws of large numbers imply that $\phi = 1$ asymptotically, but if we have overdispersion, or $\mathrm{Var}(Y_i) > E[Y_i]$ for all i, then (under regularity conditions) the solution to (4.1) is asymptotically $\hat{\phi} > 1$. Similarly, for underdispersion we get $\hat{\phi} < 1$. Thus, (4.1) provides us with a diagnostic tool to check for possible overdispersion under fairly general conditions (McCullagh and Nelder 1989). More definite results can be obtained in specific settings.

4.2. Marginal Models for Overdispersion

Suppose $Y_i \sim \mathrm{Bin}(n_i, p_i), i = 1, \ldots, n$, are conditionally independent given $p_1, \ldots, p_n$, but that each p_i has been sampled independently from a beta distribution $\mathrm{Be}(\alpha_i, \beta_i)$ with mean $\mu_i = \alpha_i/(\alpha_i + \beta_i)$ and variance $\sigma_i^2 = \alpha_i\beta_i/[(\alpha_i + \beta_i)^2 (\alpha_i + \beta_i + 1)]$ (cf., DeGroot 1987, 294–296). It follows that $E[Y_i] = E[E[Y_i| p_i]] = E[n_i p_i] = n_i\mu_i$. Similarly, using the fact that $\mathrm{Var}(Y_i) = \mathrm{Var}(E[Y_i|p_i]) + \mathrm{E}[\mathrm{Var}(Y_i|p_i)]$, one can show that $\mathrm{Var}(Y_i) = n_i\mu_i(1 - \mu_i) + n_i(n_i - 1)\sigma_i^2$. Here we have binomial variance + an overdispersion term determined by σ_i^2. It is convenient to model the overdispersion as being proportional to the binomial variance. Thus, given $0 < \mu_i < 1$ and a *single* variance parameter σ^2, we can reparametrize each beta distribution by choosing $\alpha_i = \mu_i(\sigma^{-2} - 1)$ and $\beta_i = (1 - \mu_i)(\sigma^{-2} - 1)$, which yields $E[Y_i] = n_i\mu_i$ and $\mathrm{Var}(Y_i) = n_i\mu_i(1 - \mu_i)[1 + (n_i - 1)\sigma^2]$. In this parametrization a multiplicative increase in variance due to overdispersion is assumed. For modeling, we can assume that $\mathrm{logit}(\mu_i) = \mathbf{X}_i^T\beta$, if there is a vector of explanatory variables $\mathbf{X}_i$ available for unit $i = 1, \ldots n$. Maximum likelihood can then be used to estimate both the regression parameters β and the dispersion parameter σ^2. This is the so-called *beta-binomial model* (cf., Williams 1982). It has been implemented in the program EGRET, for example. To examine whether the overdispersion specification is appropriate, denote the fitted value of Y_i by $\hat{Y}_i = n_i\hat{\mu}_i = n_i \exp(\mathbf{X}_i^T\hat{\beta})/(1 + \exp(\mathbf{X}_i^T\hat{\beta}))$ and plot scaled residuals

$(Y_i - \hat{Y}_i)/\sqrt{n_i\hat{\mu}_i(1-\hat{\mu}_i)}$ versus n_i; the model implies that the variance of the residuals should increase approximately as a linear function of n_i (McCullagh and Nelder 1989, 126).

Suppose that $Y_i \sim \text{Po}(\lambda_i)$ are independent, $i = 1, \ldots, n$ and that each λ_i has been sampled independently from a gamma distribution with parameters α_i and β_i (cf., Example 1.4 of Chapter 4) that has mean $\mu_i = \alpha_i/\beta_i$ and variance $\sigma_i^2 = \alpha_i/\beta_i^2$ (cf., DeGroot 1987, 258–261). It follows that marginally the Y_i's have a *negative binomial distribution* with expectation $E[Y_i] = \mu_i$ and variance $\text{Var}(Y_i) = \mu_i + \sigma_i^2$ (cf., Johnson and Kotz 1969, 124–125; these formulas provide the connection to the parametrization given in Exercise 1). As in the case of beta-binomial distribution, we can reparametrize the negative binomial distribution in terms of the μ_i's and a single variance parameter $\sigma^2 \geq 1$ that provides a multiplicative increase in the variance. Choosing $\alpha_i = \mu_i/(\sigma^2 - 1)$ and $\beta_i = 1/(\sigma^2 - 1)$ leads to $E[Y_i] = \mu_i$ and $\text{Var}(Y_i) = \mu_i\sigma^2$. A loglinear model $\log(\mu_i) = \mathbf{X}_i^T\beta$ can be used if there is a vector of explanatory variables $\mathbf{X}_i$ available for unit $i = 1, \ldots, n$. Maximum likelihood can be used to estimate the parameters. Such models can be fitted using the program STATA, for example. As in the beta-binomial situation, to examine whether the Poisson-gamma overdispersion specification is appropriate, denote the fitted value of Y_i by $\hat{Y}_i = \hat{\mu}_i = \exp(\mathbf{X}_i^T\hat{\beta})$ and plot scaled residuals $(Y_i - \hat{Y}_i)/\sqrt{\hat{\mu}_i}$ versus $\hat{\mu}_i$; the model implies that the variance of the residuals should be approximately homoscedastic.

4.3. Random Effect Models

The formulations for the binomial and Poisson case lead to nice, closed form probability models, for which maximum likelihood is a feasible estimation strategy. Note, however, that the choice of the beta and gamma distributions is based on mathematical convenience (they form so-called *conjugate families* with the binomial and Poisson distribution, respectively) rather than substantive reasoning. Unfortunately, no attempt to handle more general cases that we have seen is entirely free from theoretical complications. There are a number of promising frequentist methods (e.g., Lee and Nelder 1996, 2001; Durbin and Koopman 2000) and corresponding Bayesian methods (e.g., Zeger and Karim 1991, West, Harrison, and Migon 1985). We will briefly discuss the philosophy of the latter approach and then present two examples that have been implemented with generally available software.

In the Bayesian paradigm *all* unknown parameters are treated as being random, not just the random effects. Randomness may then interpreted in various ways, including in subjective terms: *a priori* we may have a more or less vague idea of the values of the unknown parameters, and those beliefs are represented by a *prior distribution* for the unknown parameters.[8] *A posteriori* – after we have seen the

[8] Alternative, non-subjective interpretations include frequency distributions for prior data and "normative and objective representations of what it is rational to believe about a parameter, usually in a situation of ignorance" (Cox and Hinkley 1974, 375); see also Berger (1980).

data – a more definite, but still not exact, view of their values arises. The conditional distribution of the parameters, given the data, is called the *posterior distribution*. The updating of the views is carried out using the famous Bayes formula (e.g., DeGroot 1987, 66; a particular case was used in Section 2.3), which says that the posterior distribution for the parameters given the data is proportional to the product of conditional distribution of the data given the parameters (i.e., the likelihood) and the prior distribution. Until the 1990's the numerical implementation of the Bayes formula was considered a major obstacle in the Bayesian analysis. However, the phenomenal increase in computing speed together with some theoretical innovations has largely removed these problems. For example, *Gibbs sampling* (cf., Gelman et al. 1995, 326–327)[9] is a simulation technique that produces a Markov chain whose invariant distribution (see Exercise 23 of Chapter 6) coincides with the posterior distribution of the parameters (whence the term *Markov Chain Monte Carlo* or *MCMC*; we will illustrate the method in Chapter 9). This approach is logically consistent, and produces results to the desired degree of accuracy. The price one has to pay for the advantages is the increased complexity of the model. In particular, a joint prior distribution has to be formed for all parameters. There are routine ways of doing this. For example, one can use priors that are nearly "non-informative" (Kass and Wasserman 1996). However, if the sample size is not large, the particular choice may have unintended effects on the results that are hard to detect. Moreover, in complex situations priors that are thought to be non-informative may actually put strong constraints on some parts of the model that are similarly hard to detect.

Experience with Bayesian methods is rapidly increasing, but still limited, in part because they are not yet routinely available in most statistical packages. In the past, there has been much debate in statistics about the relative merits of the Bayesian and frequentist methods. We remain agnostic is this respect: while a simple analysis is usually preferable to a more complex one, in some cases the essence of the matter may be lost if too much is simplified.[10] The methods must match the problem. We will now briefly review both frequentist and Bayesian models that are readily available for the demographic user.

First, Goldstein (2003) reviews the so-called *multilevel models* that are widely used in education and other social sciences. Suppose we are modeling mortality as a function of age x and time t, either via logistic or Poisson regression. In either case we might model the canonical parameter as $\theta_{xt} = \mu + \alpha_x + \beta t$, for example. Under this model there would be a systematic linear time trend and otherwise a constant age pattern. Due to extra-binomial or extra-Poisson variability, the model might not fit the data of each year well. A possible extension would be a *1-level model* $\theta_{xt} = \mu + \alpha_x + \beta t + \varepsilon_{xt}$, where the random effects $\varepsilon_{xt} \sim N(0, \sigma_1^2)$ are independent. However, there might be years during which the linear trend

[9] J. Willard Gibbs (1839-1903) developed models in statistical physics. A probability distribution for a random number of interacting particles in different energy states bears his name.

[10] "Things should be made as simple as possible – but no simpler." A. Einstein.

would be too high for all ages, and other years for which it would be too low. This could be represented by a *2-level model* $\theta_{xt} = \mu + \alpha_x + \beta t + \varepsilon_{xt} + \eta_t$, where the annual random effects $\eta_t \sim N(0, \sigma_2^2)$ are independent. Such models can be fitted using the software program MLwiN, for example. The fitting algorithm is based on an approximation to the likelihood function. The resulting estimates are sometimes called *quasi-likelihood estimates*.

Second, Gilks, Richardson, and Spiegelhalter (1995) present several examples of the so-called *hierarchical Bayesian models*. As an example, consider the 1-level model of the previous example. The random effect $\varepsilon_{xt} \sim N(0, \sigma_1^2)$ would further be described by treating the unknown σ_1^2 as random, with some prior distribution. A common choice is to assume the inverse of the variance, or *precision* $1/\sigma_1^2$, to have a gamma distribution with a large variance. In addition, one would assume that $\mu \sim N(0, \sigma_\mu^2)$, $\alpha_x \sim N(0, \sigma_\alpha^2)$ are i.i.d., and $\beta \sim N(0, \sigma_\beta^2)$, all with large variances. One would then use numerical simulation techniques to determine the joint posterior distribution of the parameters μ, α_x, β, and σ_1^2 given the observed data. The 2-level model can similarly be generalized. For practical calculations, WinBUGS software can be used (cf., Thomas, Speigelhalter, and Gilks 1992).

Example 4.1. Overdispersion in Habsburg Cohort Sizes. Returning to the Habsburgs of Example 2.1, consider the possible time trends in the number of children per generation $i = 1, \ldots, 20$. Since all families include the child who later became emperor/empress, define Y_i = (number of children in generation i) - 1 as the outcome variable. As explanatory variable we use X_i = birth year of parent i whose children are being considered. The values ranged from 1218 to 1865. The outcome variable had the mean = 7.75 and standard deviation 4.85. Since the variance is much larger than the mean, and no major time trends are apparent, extra-Poisson variability is a possibility.

The data were analyzed under three models: (i) negative binomial model; (ii) a 1-level Poisson model; and (iii) Bayesian hierarchical model with weakly informative priors. The basic model was $Y_i \sim \mathrm{Po}(\lambda_i)$, where $\lambda_i = \exp(\mu_i + \varepsilon_i)$, and the linear predictor μ_i depends on X_i. The following estimates were obtained (standard errors in parenthesis): (i) $\hat{\mu}_i = 1.85 + 0.00013(0.00078) \times X_i$ and $\hat{\sigma}^2 = 0.28(0.13)$; (ii) $\hat{\mu}_i = 1.87 - 0.00007(0.00083) \times X_i$ and $\hat{\sigma}^2 = 0.38(0.16))$; (iii) $\hat{\mu}_i = 1.85 + 0.00006(0.00084) \times X_i$ and $\hat{\sigma}_1^2 = 0.39(0.21)$. In the Bayesian case, the means of the posterior distributions were used as point estimates, and standard deviations of the posterior distributions as standard errors. None of the models suggest that there would be a time trend. All models suggest that there is extra-Poisson variability. ◊

Modelers are sometimes confused about whether random or fixed effects should be used to represent a particular factor. Econometricians (cf., Hausman 1978) have even devised ingenious tests to solve the problem. We prefer the advice of Searle (1971, 376-380) who argues that the choice be made on substantive grounds. If we are interested in making inferences about only those factors being analyzed, the corresponding parameters should be viewed as fixed effects. If we are viewing the factors as being sampled from a larger population, and we are interested in generalizing to that population, we want to consider random effects. For example, in analyses of mortality, dependence on age is almost always of interest, and the age effects usually would be treated as fixed effects. The rate of decline in mortality is also typically of interest, but variation around the declining trend need not be. If we are not specifically interested in those variations, we might consider the yearly deviations from the trend as random.

Usually, the inclusion of a factor as a random effect tends to increase the standard errors of the fixed effects. This decreases the risk of overfitting in regression, and leads to a more conservative statistical analysis. In some cases, inclusion of a factor as a random effect is necessitated by technical considerations concerning number of parameters and the number of data points. For example, if we are analyzing data on individuals and want to include a fixed effect for each individual, the number of parameters will grow with the sample size and the MLEs may be substantially biased even in large samples; in such a case we would consider the individual effects to be sampled from some distributions.

5. Observable Heterogeneity in Capture-Recapture Studies

As discussed in Section 3.4, if capture events are behaviorally correlated on an individual level, the classical population estimator can be biased. Alternatively, population heterogeneity may create a population level correlation and cause a capture-recapture estimator of population size to become biased. We will now briefly indicate how heterogeneity may be handled statistically, when there are two capture occasions.

Consider a closed population of unknown size N. For each individual $i = 1, \ldots, N$, define indicator variables u_{ji} and m_i such that $u_{ji} = 1$ if and only if i is captured on occasion j only, $j = 1, 2$; and $m_i = 1$ if and only if i is captured twice. Otherwise, $u_{ji} = m_i = 0$. Define $n_{ji} = u_{ji} + m_i$ as the indicator of capture on the jth occasion. Let $M_i = u_{1i} + u_{2i} + m_i$ indicate capture at least once. Define the individual capture probabilities as $p_{ji} = E[n_{ji}]$, $j = 1, 2$; and $p_{12i} = E[m_i]$. We assume that the first and second captures are *independent* for each i, so that $p_{12i} = p_{1i}p_{2i}$. We now have for each individual $M_i \sim \text{Ber}(\varphi_i)$, with $\varphi_i = p_{1i} + p_{2i} - p_{1i}p_{2i}$. For those with $M_i = 1$ (i.e., for those that have been captured at least once), we have the multinomial model

$$(u_{1i}, u_{2i}, m_i) \sim \text{Mult}\left(1; p_{1i}(1 - p_{2i})/\varphi_i,\ (1 - p_{1i})p_{2i}/\varphi_i,\ p_{1i}p_{2i}/\varphi_i\right). \quad (5.1)$$

The classical dual systems estimator is $\hat{N} = n_1 n_2/m$, where $n_j = n_{j1} + \cdots + n_{jN}$, $j = 1, 2$, and $m = m_1 + \cdots + m_N$. Define $\bar{p}_{jN} = (p_{j1} + \cdots + p_{jN})/N$ and define $\bar{p}_{12N} = (p_{11}p_{21} + \cdots + p_{1N}p_{2N})/N$. Consider asymptotics, in which the limits $\bar{p}_{jN} \to \bar{p}_j$, and $\bar{p}_{12N} \to \bar{p}_{12}$, exist when $N \to \infty$. By the law of large numbers we have that

$$\hat{N}/N \to \bar{p}_1 \bar{p}_2/\bar{p}_{12}, \quad \text{as} \quad N \to \infty. \quad (5.2)$$

For any N, let us formally define the covariance between the probabilities p_{ji} as $C_N = \bar{p}_{12N} - \bar{p}_{1N}\bar{p}_{2N}$. Under the assumptions we have made, there is a limit $C_N \to C$. It follows that

$$\hat{N}/N \to 1 - C/\bar{p}_{12}. \tag{5.3}$$

We see that the classical estimator is not consistent, unless $C = 0$. This asymptotic bias is called *correlation bias*.[11]

Can correlation bias matter? Unfortunately it can. Using a linear Taylor-series approximation, one can show (e.g., Alho 1994) that the variance of $\hat{N}/N$ is approximately

$$\text{Var}(\hat{N}/N) = N^{-1}(1 - p_1)(1 - p_2)/(p_1 p_2). \tag{5.4}$$

Comparing (5.3) and (5.4), we see that the ratio of the bias to the standard error is of order $\sqrt{N}$. It follows that even a small correlation bias dominates the standard error in large populations.

In demographic applications, factors that cause a person to be missed in the first count (e.g., life style, attitude towards authorities, peer pressure etc.) often cause him or her to missed in the second count. In such cases $C > 0$, so population *underestimation* is the typical direction of bias. To the extent that such explanatory factors can be measured, they can be accounted for by a statistical analysis.

Suppose now that there are characteristics $\mathbf{X}_i$ that explain the probability that individual $i = 1, \ldots, N$ is captured on occasion $j = 1, 2$ via logistic regression models

$$\text{logit}(p_{ji}) = \mathbf{X_i^T}\boldsymbol{\beta}_\mathbf{j}. \tag{5.5}$$

By a direct calculation one can show that the probabilities appearing in (5.1) are as follows, $p_{1i}(1 - p_{2i})/\varphi_i = \exp(\mathbf{X}_i^T\boldsymbol{\beta}_1)/K_i$; $(1 - p_{1i})p_{2i}/\varphi_i = \exp(\mathbf{X}_i^T\boldsymbol{\beta}_2)/K_i$; and $p_{1i}p_{2i}/\varphi_i = \exp(\mathbf{X}_i^T\boldsymbol{\beta}_1 + \mathbf{X}_i^T\boldsymbol{\beta}_2)/K_i$, where

$$K_i = \exp\left(\mathbf{X}_i^T\boldsymbol{\beta}_1\right) + \exp\left(\mathbf{X}_i^T\boldsymbol{\beta}_2\right) + \exp\left(\mathbf{X}_i^T\boldsymbol{\beta}_1 + \mathbf{X}_i^T\boldsymbol{\beta}_2\right). \tag{5.6}$$

We see that model (5.1) belongs to an exponential family. It is also a generalized linear model, so its parameters can be estimated using the methods of Section 1. Details of the ML-estimation of $\boldsymbol{\beta}_j$'s are given in Alho (1990b).

Once the MLE's of $\boldsymbol{\beta}_j$'s have been obtained, we get MLE's of φ_i's. Using these we can define a Horvitz-Thompson type estimator for N,

$$\hat{N} = \sum_{i=1}^{N} M_i/\hat{\varphi}_i. \tag{5.7}$$

The rationale for (5.7) is that $E[M_i] = \varphi_i$, and if the error in $\hat{\varphi}_i$ is negligible, (5.7) is nearly unbiased. We emphasize that only those individuals contribute to the sum

[11] In Section 4.1 of Chapter 3 we discussed a similar bias arising from the correlation of sampling probabilities and the variable of interest. In Section 5.6 of Chapter 10 we will consider the estimation of correlation bias in a post enumeration survey.

that have $M_i = 1$, and covariates $\mathbf{X}_i$ are needed only for them. It is shown in Alho (1990b) that (5.7) reduces to the classical estimator given in Section 6 of Chapter 2, if the population is homogeneous.

Example 5.1. Heterogeneity in Reporting of Occupational Disease. In Example 6.1 of Chapter 2 we pointed out that under reporting of occupational diseases depended heavily on diagnosis in Finland in 1980. The methods outlined above were used to study whether the probability of reporting depended on other characteristics, such as age (Alho 1990b). A significant effect was found for insurance companies' reporting of noise-induced hearing loss: the older the patient the more likely the case was reported. Presumably the cases for older workers were more severe. Interestingly, age did not have an influence on the reports through the other information channel, so there was no correlation bias (a constant is uncorrelated with everything!) and the estimate for the total number of cases did not change. $\diamond$

Example 5.2. Heterogeneity in Census Enumeration Probabilities. In an analysis of the 1990 U.S. census data Alho et al. (1993) applied the conditional regression techniques to the minority, central city post-strata in various parts of the country. (A post-stratum is defined as a set of enumerations with specified values of the covariates $\mathbf{X}_i$; see Chapter 10, Section 5.2.) Comparison of the characteristics of those hard-to-enumerate (i.e., those individuals with estimated enumeration probability $< 75\%$) to the rest of the post-stratum showed that the hard-to-enumerate typically were young, black, unmarried renters, who lived among similar neighbors in an area of high vacancy and multi-unit housing rates. In many cases the information concerning them had been reported by an unrelated person. $\diamond$

An alternative and somewhat simpler approach can also be considered. The local independence assumption $p_{12i} = p_{1i} p_{2i}$ means that $p_{1i} = P(m_i = 1 | n_{2i} = 1)$, and hence we can use ordinary logistic regression to estimate p_{1i} from data on those individuals who were captured in the second survey ($n_{2i} = 1$). Instead of the estimator (5.7), we can then use

$$\hat{N} = \sum_{i=1}^{N} n_{1i} / \hat{p}_{1i}. \tag{5.8}$$

The estimator (5.8) will be less efficient than (5.7). In certain contexts, such as the first capture being enumeration in the census and the second capture enumeration in a far smaller survey, the loss in efficiency may be unimportant compared to the gain from simplicity.

Estimators (5.7) and (5.8) may be used to provide estimates for subgroups (or domains or small areas), say, G. The idea is to restrict the summation in (5.7) or (5.8) to $i \in G$. In census applications (5.8) is especially useful, because p_{1i} is estimated from a sample, but the estimation of the size of G can be based on the more precise census count via (5.8).

A methodological issue one has to consider in the application of (5.7) or (5.8) is that in practice the population being studied may not be closed. Individuals may enter or exit between the two captures. As discussed by Alho et al. (1993), it may still be possible to carry out estimation based on (5.1) and (5.2), using the following principles: (i) define N as the population of the, say, first capture; (ii) exclude from the second capture all those who were not present in the area during the first capture; (iii) define the second capture probability as referring to both being captured *and* being in the area. If the logistic model (5.5) still applies, the estimators given by (5.7) or (5.8) will still be approximately unbiased, although variance may be increased. The degree to which (5.5) holds for $j = 2$ now depends on how well the logistic regression explains not only capture but non-movement.

Above we have assumed that the data are without other errors, besides the enumeration errors being discussed. As discussed in Chapter 10, this can be far from reality!

6. Bilinear Models

All models considered thus far have been linear (in the chosen scale). The simplest nonlinear extension is based on conditional linearity in a sense to be explained below. The models are closely related to factor analysis.

Consider a two-way table consisting of I rows and J columns with counts Y_{ij} in the ith row and j^{th} column; this is called a (two-dimensional) contingency table. As discussed in Section 3.4 such data can arise from a Poisson model for the counts; from a multinomial model, if we condition on the total $Y_{..} = \Sigma_{ij} Y_{ij}$; and it can arise from a (multivariate) hypergeometric model, if we condition on the row totals $Y_{i.} = \Sigma_j Y_{ij}, i = 1, \ldots, I$, and the column totals $Y_{.j} = \Sigma_i Y_{ij}, j = 1, \ldots, J$. In fact, it can also arise from I independent multinomials, if we condition on the row totals only, or from J independent multinomials if we condition on the column totals.

In any case, define $E[Y_{ij}] = \lambda_{ij}$ and consider loglinear models for the expectations. Under the main effects model we can write $\log(\lambda_{ij}) = \mu + \alpha_i + \beta_j$. In this case we have that $\lambda_{ij} = \exp(\mu)\exp(\alpha_i)\exp(\beta_j)$, so the row and column effects multiply. For identifiability, we may apply suitable "analysis of variance type" identifiability conditions $\Sigma_i \alpha_i = 0 = \Sigma_j \beta_j$. Conditioning on $Y_{..}$ and considering the $Y_{..}$ realizations to be mutually independent, we can consider the probability of the observation falling into cell (i, j). The probability is $\lambda_{ij}/\lambda_{..} = \exp(\alpha_i)\exp(\beta_j)/\Sigma_{ij} \exp(\alpha_i + \beta_j)$, so the row and column effects are independent under the main effects model. In fact, the probability of falling into row i is $\lambda_{i.}/\lambda_{..} = \exp(\alpha_i)/\Sigma_i \exp(\alpha_i)$ and the probability of falling into column j is $\lambda_{.j}/\lambda_{..} = \exp(\beta_j)/\Sigma_j \exp(\beta_j)$, under the main effects model.

As noted earlier, including all interaction terms we would have $\log(\lambda_{ij}) = \mu + \alpha_i + \beta_j + \gamma_{ij}$, where $\Sigma_j \gamma_{ij} = 0$ for each $i = 1, \ldots, I$, and $\Sigma_i \gamma_{ij} = 0$ for each $j = 1, \ldots, J$. This permits arbitrary patterns of interdependence between rows and columns. Unfortunately, the model would be saturated and would not really add to our understanding of the possible dependencies. In case there is a natural ordering in the categories (as in the case when i is age and j is time), then models of the type $\log(\lambda_{ij}) = \mu + \alpha_i + \beta_j + \gamma \times ij$, where γ is a scalar parameter to be estimated, and i and j are treated as integers, might be valuable in the study of the possible association of the row and column factors. However, there are many interesting categorical variables for which no such ordering exists. For example, marital status (never married, married, divorced, widowed), race, or region cannot be easily thought of in such terms.

A possible intermediate formulation is the so-called *association model* of Goodman (1991),

$$\log(\lambda_{ij}) = \mu + \alpha_i + \beta_j + \varphi \nu_i \eta_j, \tag{6.1}$$

where $\varphi > 0$, and the *row scores* satisfy the conditions $\Sigma_i \nu_i = 0$ and $\Sigma_i \nu_i^2 = 1$ and *column scores* satisfy the conditions $\Sigma_j \eta_j = 0$ and $\Sigma_j \eta_j^2 = 1$. This is a *log-bilinear model*, because given the parameters that depend on i, the model is linear in the parameters that depend on j; and given the parameters that depend on j, it is linear in the parameters that depend on I. We will call the model *bilinear*, for short. The model adds $1 + (I - 2) + (J - 2) = I + J - 3$ new parameters after the main effects. The model with full interactions adds $(I - 1)(J - 1)$, or the number of degrees of freedom of the usual χ^2-statistic for testing the independence of the columns and the rows. The model with known integer scores adds only 1 degree of freedom. Therefore, the bilinear association model can be a useful compromise.

The reason the parameters ν_i (and η_j) may be called "scores" (not to be confused with the scores of Section 3 of Chapter 1!) is that they can be used to quantify the distance between the otherwise categorical rows (columns) of the contingency table. If two rows have similar values of υ_i, their dependence on the columns is similar. In this manner, the rows can be ordered on a line, and presented graphically (cf., Goodman 1991). The distance between rows i and i' is $|\nu_i - \nu_{i'}|$, and we order the rows based on their estimated ν values.

The association model can similarly be formulated for the general Poisson regression. Suppose that $Y_{ij} \sim \text{Po}(\lambda_{ij} K_{ij})$ is the number of deaths in age i during year j, where K_{ij} is the number of person years lived in age i during year j, and λ_{ij} is the age-specific death rate. Then, (6.1) defines an association model for the mortality counts.

Example 6.1. Lee-Carter Model for Mortality. If we set $\beta_j \equiv 0$ and fix $\mu + \alpha_i$ to equal the average of the log-mortality rates during $j = 1, \ldots, J$, (6.1) essentially becomes the model proposed by Lee and Carter (1992) for the forecasting of the U.S. age-specific mortality. Eklund (1995) investigated the approach of Lee and Carter with Finnish male and female mortality data for ages 65, 66, ..., 99 for the years 1972-1989. The data show quite a bit of random variability in the highest ages due to the small number of deaths. One consequence of this is that the estimated model produces non-monotone period mortality patterns in ages over 90. This suggests that in some circumstances either smoothing, or some further constraint on the model parameters, may be desirable. Girosi and King (2003) have come to a similar conclusion using a much larger data set. ◊

The model (6.1) can be generalized further. For example, we can have two sets of scores so that

$$\log(\lambda_{ij}) = \mu + \alpha_i + \beta_j + \varphi_1 \nu_{i1} \eta_{j1} + \varphi_2 \nu_{i2} \eta_{j2}, \tag{6.2}$$

where both scores are similarly normalized as in (6.1), and furthermore $\Sigma_i \nu_{1i}\nu_{2i} = 0$ and $\Sigma_j \eta_{1j}\eta_{2j} = 0$. Therefore, the number of new parameters introduced is $I + J - 5$. Extension to higher order scores is immediate.

In the case of the higher order methods the parameters $\varphi_1 > \varphi_2 > \cdots > 0$ measure the importance of the scores in explaining the deviations of from independence of the rows and the columns. As in ordinary factor analysis, a choice has to be made, in practice, as to how many terms are included in the model. Methods for making such a choice on statistical grounds are given in Goodman (1991) for the contingency table case. In general, it is also useful to consider the interpretation of the resulting scores. If no sensible interpretation can be given, one may be overfitting the data.

Models of this general type appear to have been introduced in demography by Ledermann and Breas (1959) and further developed by Bozik and Bell (1989) and Bell (1992). The approach of Lee and Carter is particularly elegant, because after subtracting the mean of the series it uses just a one-dimensional approximation to describe differences from the mean.

We discuss two approaches to the numerical solution of bilinear models. Suppose first, for definiteness, that we have observed mortality rates $m_{x,t}$ for ages $x = 0, 1, \ldots, \omega$ and years $t = 1, \ldots, T$. Define an $(\omega + 1) \times T$ matrix $\mathbf{L}$ with the (x, t) element equal to $\log(m_{x,t})$. We can make the so-called *singular value decomposition* (cf., Rao 1973, 42–43) $\mathbf{L} = \mathbf{U}\mathbf{\Gamma}\mathbf{V}^T$, where $\mathbf{\Gamma}$ is a diagonal matrix of dimension $\min\{\omega + 1, T\}$ that has the nonnegative values γ_i in decreasing order and $\mathbf{V}^T\mathbf{V} = \mathbf{U}^T\mathbf{U} = \mathbf{I}$, where $\mathbf{I}$ is an identity matrix of dimension $\min\{\omega + 1, T\}$. Let r denote the rank of $\mathbf{L}$. The first r diagonal elements of Γ are called the singular values of $\mathbf{L}$ and are the square roots of the eigenvalues of $\mathbf{L}\mathbf{L}^T$. (Eigenvalues are discussed in more detail in Chapter 6, Section 2.2.) The ith column vectors of $\mathbf{U}$ and $\mathbf{V}$, $\mathbf{U}_i$ and $\mathbf{V}_i$, are called the right and left singular vectors corresponding to γ_i. We have a one dimensional approximation $\mathbf{L} \approx \gamma_1 \mathbf{U}_1 \mathbf{V}_1^T$. Here $\mathbf{U}_1$ represents the average relative level of mortality by age. Then, the vector $\gamma_1 \mathbf{V}_1^T$ tells us the approximate level of log-mortality during years $t = 1, \ldots, T$. A two-dimensional approximation is of the form $\mathbf{L} \approx \gamma_1 \mathbf{U}_1 \mathbf{V}_1^T + \gamma_2 \mathbf{U}_2 \mathbf{V}_2^T$. One can prove that the approximations mentioned above are *the best one and two dimensional approximations* to the log-mortality rates, *under the least squares criterion* (e.g., Greenacre 1984, 343-344). Unfortunately, the assumption of homogeneous variances underlying OLS is not satisfied in the Poisson setting.

The second approach relies on maximum likelihood. Many bilinear association models for exponential family observations can be fitted with standard software, such as GLIM, by starting out from the main effects model and, e.g., the assumption that the column scores are proportional to j. Fixing the β_j's, all parameters that depend on i can be re-estimated, and normalized (for simplicity, one can absorb φ into ν_i's and not require that their squares sum to 1). Then, one can fix the parameters that depend on i, re-estimate those that depend on j, and normalize the estimates to satisfy the constraints. However, specialized software for handling some of these models have also been written. For example, LEM (cf., Vermunt 1997a, 1997b) can handle a wide class under a Poisson assumption. In that program bilinear models are called "log-multiplicative".

Independently of how the likelihood equations are solved it is useful to note that unlike the SVD based approach, these calculations do not require that we have observations for all ages for all years of observation. Similarly, the standard properties of the MLE's carry over to this case under regularity conditions (e.g., that the φ's are non-zero and separated).

Example 6.2. Mortality Among Elderly. To illustrate models (6.1) and (6.2), let $Y_{xts} \sim \text{Po}(\lambda_{xts} K_{xts})$ be the number of deaths in age $x = 81, 82, \ldots, 101$ during year $t = 1991, \ldots, 1994$ for sex $s = \text{M, F}$, in Finland. Although separate models could be fitted for the two sexes, a potentially more reliable estimate of time trends is obtained if the age-effects α_{xs} depend on s but the year-effects β_t do not. In the same vein, we assumed that the association model has the same effects for males and females. The log-likelihood of the larger model (6.2) was -1153039.4 and that of the smaller model (6.1) was -1153064.1. Therefore, the likelihood ratio test statistic was $2(-115039.4 + 1153064.1) = 49.4$. The larger model has $20 + 14 - 5 = 29$ additional free parameters. Based on the χ^2 distribution with 29 degrees of freedom, we find the P-value 0.01. $\diamond$

As in the one-dimensional case, under (6.2) one can use graphical displays to characterize the locations of the rows with respect to each other. A two-dimensional plot of the points $(\varphi_1 \upsilon_{i1}, \varphi_2 \upsilon_{i2})$, $i = 1, \ldots, I$, can characterize the way different rows depend on the columns. The plot shows how close the rows are in the space spanned by the vectors $(\eta_{11}, \ldots, \eta_{J1})$ and $(\eta_{12}, \ldots, \eta_{J2})$. Note that the two vectors form an orthonormal basis of a 2-dimensional subspace of $\mathbb{R}^J$, the space in which the rows lie. The plot of the points $(\upsilon_{i1}, \upsilon_{i2})$, $i = 1, \ldots, I$, gives similar comparative information, but does not take into account the relative importance of the two sets of scores (cf., Goodman 1991).

In many applications neither the row categories nor the column categories are of a dominant interest. In this case, plots of $(\varphi_1 \eta_{j1}, \varphi_2 \eta_{j2})$, $j = 1, \ldots, J$, can also be made to compare, how columns differ in their association with rows, in the space spanned by the orthonormal vectors $(\nu_{11}, \ldots, \nu_{I1})$ and $(\nu_{12}, \ldots, \nu_{I2})$.

A final, and slightly controversial question relating to plotting (cf., the discussion of the paper Goodman 1991), concerns the simultaneous description of rows and columns. Define the points $\boldsymbol{\nu}_i = (\nu_{i1}, \nu_{i2})^T$, $i = 1, \ldots, I$, $\boldsymbol{\eta}_j = (\eta_{j1}, \eta_{j2})^T$, $j = 1, \ldots, J$, and the matrix $\boldsymbol{\varphi} = \text{diag}(\varphi_1, \varphi_2)$. We see from (6.2) that if $\boldsymbol{\nu}_i^T \boldsymbol{\varphi} \boldsymbol{\eta}_j$ is large, in absolute value, then row i and column j produce a large deviation from independence in the table. This is an inner product, but weighted with $\boldsymbol{\varphi}$. A seemingly reasonable way the represent such data would be to plot the points $\boldsymbol{\varphi}^{1/2} \boldsymbol{\nu}_i$, $i = 1, \ldots, I$, and the points $\boldsymbol{\varphi}^{1/2} \boldsymbol{\eta}_j$, $j = 1, \ldots, J$ into the same plot. Such plots are examples of the so-called *biplots* (cf., Gower and Hand 1996). Note, in particular, that if one simply plots the points $\boldsymbol{\nu}_i$ and $\boldsymbol{\eta}_j$, then the angle between the points is not necessarily related to the inner product of interest. We will illustrate the scores in connection with migration modeling, in Chapter 6.

The discussion we have given is closely related to *correspondence analysis* (e.g., Greenacre 1984). The starting point there is a contingency table with counts Y_{ij}. It is first transformed into empirical probabilities $p_{ij} = Y_{ij}/Y_{\cdot\cdot}$, and they are normalized to deviations of the form $d_{ij} = (p_{ij} - p_{i\cdot} p_{\cdot j})/(p_{i\cdot} p_{\cdot j})^{1/2}$. Note that the sum of the squared normalized deviations d_{ij}^2 is then the usual χ^2-statistic divided by $Y_{\cdot\cdot}$. Therefore, the deviations also characterize how the assumption of independence between rows and columns might not hold. A singular value decomposition is carried out for the matrix of the deviations $\mathbf{D} = (d_{ij})$. If one retains the first two singular values, one gets formally a bilinear representation of the form $(p_{ij} - p_{i\cdot}p_{\cdot j})/(p_{i\cdot}p_{\cdot j})^{1/2} \approx \mathbf{v}_i^T \varphi \eta_j$, so similar plots as those described above can be made. A practical advantage of the correspondence analysis formulation is that software for simple correspondence analysis are available in several general purpose statistical packages, such as Minitab.

7. Proportional Hazards Models for Survival

Poisson regression provides a basic tool for the analysis of aggregated demographic data. However, when individual event histories are available, the information can be handled more efficiently by concentrating on individual waiting times, and their determinants. We will call the smaller of waiting time and censoring time a *withdrawal time*.

Cox (1972) introduced a semiparametric regression model for the hazard function. Suppose the survival function of an individual is given by (2.4)–(2.5) of Chapter 4 with hazard of the form

$$\mu(t, \mathbf{X}) = \mu_0(t) g(\mathbf{X}^T \boldsymbol{\beta}), \tag{7.1}$$

where $g(.) > 0$ is an increasing function with $g(0) = 1$, $\mathbf{X}$ is a vector of covariates, and $\boldsymbol{\beta}$ is a vector of regression parameters to be estimated. Since $\mu(t, \mathbf{0}) = \mu_0(t)$, the function $\mu_0(.)$ can be viewed as a *baseline hazard*. The equation (7.1) defines a *proportional hazards model*, because time t and covariates $\mathbf{X}$ act multiplicatively on the hazard. In the so-called *Cox regression* we take $g(.) = \exp(.)$. The model is *semiparametric*, because no parametric assumptions are made about the baseline hazard, but relative risk is represented parametrically.

Example 7.1. A Simple Example of Cox Regression. Consider an epidemiologic study of the survival of two internally homogeneous groups, those who are exposed $(X = 1)$ and those who are not exposed $(X = 0)$. Assume a Cox regression model, so for the exposed we have $g(X\beta) = \exp(\beta)$ and for the non-exposed we have $g(X\beta) = 1$. Then the relative risk is simply $\exp(\beta)$. ◊

Although many aspects of ordinary linear regression, logistic regression, and Poisson regression carry over to (7.1) as such, there are some special aspects that need to be observed when modeling survival times via (7.1). Suppose $T_{(1)} < \cdots < T_{(n)}$ are ordered withdrawal times of a cohort of n individuals and let $X_{(i)}$ denote the covariate vector of the individual who was the ith withdrawal. Let $R_{(i)}$ be the set of those who were at risk just prior to the ith withdrawal. Hence, $R_{(1)} = \{1, \ldots, n\}$, and if $i = 2$ is the first to withdraw, then $R_{(2)} = \{1, 3, \ldots, n\}$, for example. Suppose the ith withdrawal is a death. Consider the probability that the individual to die then is exactly the one who did, given that we know who were at risk just prior to $T_{(i)}$ and that exactly one individual died during $[T_{(i)}, T_{(i)} + h)$. Recall the definition of hazard in Section 2.1 of Chapter 4. Using those notations we can write the probability as

$$\frac{(\mu(T_{(i)}, \mathbf{X}_{(i)})h + o(h)) \prod_{j \in R_{(i+1)}} (1 - \mu(T_{(i)}, \mathbf{X}_j)h - o(h))}{\sum_{k \in R_{(i)}} (\mu(T_{(i)}, \mathbf{X}_k)h + o(h)) \prod_{j \in R_{(i)} \setminus \{k\}} (1 - \mu(T_{(i)}, \mathbf{X}_j)h - o(h))}, \tag{7.2}$$

where if $i = n$ the product in the numerator equals 1. In the denominator $R_{(i)} \setminus \{k\}$ is the set of those at risk just before $T_{(i)}$ but excluding k. Although (7.2) looks complicated, let us divide both the numerator and the denominator by h and then let $h \downarrow 0$. This gives us the limit

$$\frac{\mu(T_{(i)}, \mathbf{X}_{(i)})}{\sum_{k \in R_{(i)}} \mu(T_{(i)}, \mathbf{X}_k)}. \tag{7.3}$$

Under the proportional hazards model (7.1), we can go one step further and simplify (7.3) by canceling the baseline risks for the ith death,

$$L_{(i)}(\boldsymbol{\beta}) = \frac{g(\mathbf{X}_{(i)}^T \boldsymbol{\beta})}{\sum_{k \in R_{(i)}} g(\mathbf{X}_k^T \boldsymbol{\beta})}. \tag{7.4}$$

A similar probability can formally be written for the censored individuals but we want to exclude those terms from estimation. Define $\delta_{(i)} = 0$ if the ith withdrawal was a censoring and $\delta_{(i)} = 1$ otherwise. The part of the likelihood involving only non-censored individuals and not their exact times of withdrawal is

$$L(\boldsymbol{\beta}) = \prod_{i=1}^{n} L_{(i)}(\boldsymbol{\beta})^{\delta_{(i)}}. \tag{7.5}$$

Example 7.2. A Simple Example of Cox Regression with Censoring. Continuing Example 7.1, let us suppose that just prior to the ith withdrawal there were n_{1i} exposed individuals and n_{0i} non-exposed individuals present in the cohort. Then, the loglikelihood corresponding to (7.5) is of the form

$$\ell(\beta) = \sum_{i=1}^{n} \delta_{(i)} [X_{(i)}\beta - \log(n_{1i} \exp(\beta) + n_{0i})], \tag{7.6}$$

where $X_{(i)} = 1$ if the ith withdrawn person was exposed, and $X_{(i)} = 0$ otherwise. $\diamond$

A number of remarks about Cox regression are in order.

(1) The likelihood (7.5) belongs to an exponential family, so the theory of Section 1 applies. However, the numerical implementation requires additional considerations (McCullagh and Nelder 1989, 429).
(2) Since the baseline terms cancel, only relative risks can be studied via (7.5).
(3) Mechanisms related to censoring have been stripped away from (7.5). Therefore, this likelihood is called a *partial likelihood* (cf., Cox 1975).

(4) Since the baseline hazard cancels, the exact times of the withdrawals are not relevant in estimation, only their order is.

(5) On the other hand, no aspect of the above derivation would change, if we would let the covariate vectors be functions of time, or $\mathbf{X}_{(k)} = \mathbf{X}_{(k)}(t)$. The covariates are evaluated at the times of withdrawals. In this case, as in (3), a description of the processes that produced changes in the covariates is not included in (7.5). This is an additional reason for calling it a partial likelihood. Apart from technicalities, an important thing in the extension is the choice of covariates in the model. For example, if A influences both B and the hazard, but B has no influence on survival, then including only B in regression may lead to an erroneous conclusion that it does. For another example, suppose that A influences B and B influences the hazard. A may or may not have a direct influence. Then including both A and B into the model may mask (the possibly more fundamental role) of A in the process (e.g., Andersen 1986). Example 7.4, below, provides further discussion.

(6) If the covariates $\mathbf{X}$ are fixed in (7.1), then the reasoning behind (2.4) in Chapter 4 implies that the survival function $p(t, \mathbf{X}) \equiv P$(lifetime for individual with covariate $\mathbf{X}$ is $> t$) satisfies the equation $-\log p(t, \mathbf{X}) = \log(p(t, \mathbf{0}))g(\mathbf{X}^T\beta)$. Therefore, we have that $\log(-\log p(t, \mathbf{X})) = \log(-\log p(t, \mathbf{0})) + \log g(\mathbf{X}^T\beta)$. In other words, the curves $t \to \log(-\log p(t, \mathbf{X}))$ are equidistant for different $\mathbf{X}$. This provides a possible way to check the appropriateness of the proportional hazards assumption, if some estimates of the survival curves (e.g., Kaplan-Meier) are available for the functions $p(., \mathbf{X})$. We caution that there are many applications in which the assumption of proportionality is not valid (e.g., Example 1.4 of Chapter 6). Fully nonparametric models (e.g., Section 1.4 of Chapter 6) may then be used to estimate the hazards.

(7) Although the baseline risk disappeared from (7.5), it is possible to estimate the baseline risk, once the regression parameters have been estimated. Breslow (1974) proposed a procedure based on the cumulative hazard (2.5) of Chapter 4. Recall the definition of the hazard in terms of probabilities in (2.1) of Chapter 4. In analogy with the derivation of the Nelson-Aalen estimator (2.20) in Chapter 4, we can equate the expected number of deaths with the observed number in the interval $[T_{(i)}, T_{(i)} + h)$ to get the equation

$$1 = \int_{T_{(i)}}^{T_{(i)}+h} \mu_0(t)dt \sum_{k \in R_{(i)}} g(\mathbf{X}_\mathbf{k}^\mathbf{T}\hat{\beta}), \tag{7.7}$$

where we have taken the sum outside the integral sign. We can solve (7.7) for the integral on the right hand side. A similar equation can be written for intervals of length h that contain no deaths. For such intervals the left hand side would be zero, and the resulting estimate of the integral would be zero, as well. Putting together such estimates for a fine enough partition of the interval $[0, x]$ yields the following estimator,

$$\int_0^x \mu_0(t)dt \approx \sum_{T_{(i)} \leq x} \frac{\delta_{(i)}}{\sum_{k \in R_{(i)}} g(\mathbf{X}_\mathbf{k}^\mathbf{T} \hat{\beta})}. \tag{7.8}$$

(8) Finally, tied survival times are possible. This complicates both the argument and the result corresponding to (7.4). In practice, approximations are used to replace the resulting complicated likelihood by a simpler one (Cox and Oakes, 1984), although methods for efficiently computing the exact likelihood are becoming available. In formula (7.8) tied observations lead to replacing the 1's (i.e., $\delta_{(i)} = 1$) in the numerator by the numbers of deaths.

Example 7.3. Changes in Mortality of the Habsburgs. A question of interest in connection with the Habsburgs' data is the possible change in the longevity of the members of the privileged family. Did mortality change over the centuries and did gender matter? Since the study population follows the throne, it is selective. One expects better than average survival among the members. On the other hand, excluding the person who passed on the crown to his/her children might bias the sample the other way. In situations like this it is frequently the best to carry out the analyses both ways to see, if the results change. In addition, the age at death is not accurately recorded for all children who have "died young". We consider the effect of excluding those who did not survive to age 2. The data set contained the life times of 175 individuals, and for 165 sex was known. The latter form our basic data set.

It is not clear how – if at all – mortality might have changed over the years, so time-period indicators for the birth centuries 13th through 19th were used in regression as explanatory variables. In addition, an indicator variable for sex was used. The coefficient for being male (standard error in parenthesis) was for (a) the complete data −0.02 (0.16), (b) the data omitting progenitors 0.19 (0.18), (c) among those who survived to age 2, −0.10 (0.18), and (d) among non-progenitors who survived to age 2, 0.065 (0.21). Although the sex effect is not significant in any of the cases, we see that including progenitors probably biases the sample by exaggerating chances of male survival.

Under data set (c) none of the period indicators are significant. However, under data sets (a), (b) and (d) the indicator of the 19th century is, indicating a lower hazard during that period. Defining an indicator for the 19th century alone we get the following estimates for its coefficient using the four data sets, (a) −0.76 (0.29), (b) −0.87 (0.34), (c) −0.75 (0.30), (d) −0.89 (0.36). All results are significant. We conclude that mortality did appear to decline during the 19th century, but *no progress appears to have been made during the previous six centuries*. Results on the effect of sex do not materially change. Given that we have found no evidence of under reporting of females in the data, the conclusion is that the *difference between the mortality of males and females has been too small to be detectable* in the available data. For additional discussion, see McKeown (1976). ◊

Example 7.4. Time-Varying Covariates. Consider the effect of smoking on cancer risk. In a follow-up study one might want to construct a time-varying covariate $X(t)$ to quantify the amount of smoking. A possible representation is

$$X(t) = \int_0^t w_t(s)A(s)\,ds, \tag{7.9}$$

where $A(s)$ is, say, the number of cigarettes per day at time s, and $w_t(s)$ is some weight function. Taking $w_t(s) \equiv 1$ implies that the total ever smoked is the relevant risk measure; taking $w_t(s) = e^{-\alpha(t-s)}$, $\alpha > 0$, says the most recent smoking is the most relevant; taking $w_t(s) = 1_{[0,t-a]}(s)$ implies there is a *latency period* of length $a > 0$, so that the most recent smoking should not be counted etc. Summarizing the risk history is quite demanding in practice (cf., Hoel 1985). The problem also arises in controlled experiments such as the long-term rodent experiments on carcinogenicity (e.g., Crouch and Wilson 1981, 108). ◊

Example 7.5. Likelihood for Matched Studies. Somewhat surprisingly, the likelihood used in matched studies is *formally* equivalent to (7.4). Suppose the probability of person k falling ill is of a logistic form $\exp(\mathbf{X}_k^T\beta)/(1+\exp(\mathbf{X}_k^T\beta))$. Suppose one case i is matched to a set of controls. Together they form a set of individuals that we denote $R_{(i)}$. Thus, the controls form the set $R_{(i)}\setminus\{i\}$. Then, the conditional probability that the person to have fallen ill among those in $R_{(i)}$ is the one that did, is given by (7.4), when $g(.) = \exp(.)$. A similar result holds for matched cohort studies, as well. This is the so-called *conditional logistic regression* model. Epidemiologic data sets, such as the lung cancer study described in Example 5.2 of Chapter 2, would nowadays be analyzed using such methods. ◊

8. Heterogeneity and Selection by Survival

Consider a simple random sample from a homogeneous cohort. We expect that, within sampling variation, the sample will display similar features as the original cohort. This intuition may fail in some demographic contexts if the sampling mechanism has something to do with the measure being studied. We will discuss two examples in which the sampling mechanism is simply survival and the measure of interest is life expectancy or the hazard.

Suppose a sample is drawn by picking all those members of the cohort who survive to age $t > 0$. At birth all members of the cohort have a life expectancy $E[X]$ defined by formula (2.7) of Chapter 4. The life expectancy of the sampled individuals is $E[X|X \geq t]$. It is a simple matter to prove that

$$E[X|X \geq t] \geq E[X]. \tag{8.1}$$

In other words, the sampled individuals always have a higher life expectancy than those of the original cohort. Recall that in Example 1.1 of Chapter 4 we have shown that in the case of the exponential distribution the left hand side of (8.1) is $t + E[X]$, for example.

In actual populations consisting of individuals with differing probabilities of survival the method of *selection by survival* would not produce a simple random sample. Those with higher probabilities of survival would have a higher probabilities of being included than those with lower probabilities of survival. Therefore, the inequality (8.1) would hold with even greater force. However, it is important to understand that if we observe (8.1) to hold empirically for some cohort, then we cannot conclude that the individuals who have survived to age $t > 0$ are necessarily "hardier" or "more fit" than those who do not. They may simply have been lucky!

In some situations the effect of selection by survival can be more subtle. The introduction to the book by Bienen and Van de Walle (1991, 9) on leadership duration ($= X$) describes a theoretical model and empirical findings. The theoretical model is that

> "leaders take a "random walk" through history. A hypothesis that leaders face constant risks of falling from power could be put forward. Perhaps leaders stand at the edge of a precipice, which is loss of power. They must initially take a step to the right or the left. The step could be expressed as policy or personnel choices. If they go the wrong way, they topple. But if, by chance, their moves take them three steps away from the edge of the cliff, then they can survive an exogenous shock, say falling commodity prices, which pushes them only one step back towards the cliff. Leaders are eliminated randomly over time, but a few survive for long periods through no particular merit of their own. This is not a completely implausible theory of leadership survival. It will be shown, however, that the risks of falling from power are not constant but they decline as leaders remain longer in power."

An important empirical finding is that the risk of losing power peaks in the first years in power and decreases thereafter. This leads us to back to the "randomness or predestination" discussion of Section 2.4 of Chapter 4: is the finding a result of different initial characteristics of the leaders, so that the frail ones fall from power early and leave the stronger to stay longer, or does staying in power increase a leader's power and make longer duration more likely, or both?

It is shown in Spencer (1997a) that the random walk model is actually consistent with the empirical findings (at least as they are simplistically summarized here). This shows that the findings can be supported by the hypothesis that differences in innate characteristics of leaders do not matter. While it is quite plausible that differences in innate characteristics *do* matter, such a hypothesis is not necessary to explain the empirical results if one believes the random walk model is a useful characterization of leadership duration.

Intuitively the result can be understood as follows. Suppose a leader starts from point 0, and advances one step up or down at each epoch depending on the success of his/her actions. Positive rewards can be accumulated without limit, so the leader may advance upwards without limit. However, suppose there is some lower limit $r < 0$, such that if the random walk reaches r, the leader falls from power. The hazard of falling from power during any epoch n is defined as $\mu(n) = P(\text{falls from power during epoch } n | \text{ has not fallen from power before } n)$. This is the discrete time version of (2.1) of Chapter 4 with $x = n, h = 1$, and $o(h) = 0$. Now, the leader has zero probability of falling during the first $r - 1$ epochs. After that the probability of falling becomes positive and it may increase for a while. However, among the leaders who have survived for a long time only a few are close to r, the less so the larger n. Therefore, the hazard will eventually decrease.[12] The details of the calculations are given in Spencer (1997a) and Carvalho and Spencer (2001).[13]

In the leadership example, many of those who have managed to survive have been lucky many times. Although each leader has initially the same chance to survive to any epoch n, the ones who actually do have become heterogeneous with respect to their probability of falling from power. Under this model luck may accumulate!

9. Estimation of Population Density

Up to this point we have thought of events as indexed by age or time. Logically, they can also be indexed by place of occurrence. All difficulties one encounters in time domain appear in this case. However, new problems are created by the fact that, unlike time, points in space do not have a unique natural ordering.

Variations in population density or in population characteristics across geographic locations belong to the domain of geographers. A specialized statistical literature addressing such issues has developed (e.g., Griffith 1988). Especially since the introduction of GIS (geographic information systems), one can expect that micro demographers will increasingly become interested in spatial aspects of population data. A sophisticated statistical theory involving spatially mapped data is being developed (e.g., Ripley 1981, Diggle 1983, Cressie 1993, Ghosh and Rao 1994, Wackernagel 1998) that cannot be done justice here. We will briefly consider population density.

From a spatial perspective, a population of size N can be viewed as a collection of points $\mathbf{x}_i = (x_{1i}, x_{2i}) \in \mathbb{R}^2$, $i = 1, \ldots, N$, on a plane. A set[14] $A \subset \mathbb{R}^2$ can then be characterized by the number of points $n(A)$ it contains. For example, suppose a country is partitioned into municipalities $A_j = 1, \ldots, J$. Then, $n(A_j)$ would be the population size of the municipality. Suppose $d(A_j)$ is the *area* of the set A_j. Then the average population density of A_j is the ratio $n(A_j)/d(A_j)$.

More generally, let us think of a changing population that is depleted by deaths, increased by births, and subject to migration. Then the population size

---

[12] Strictly speaking, in the most elementary random walk model we have described here, the leader can topple only during every other epoch. For example at epoch $n = r$ the smallest possible values of the process are $r + 2$ and r, so a survivor who is at $r + 2$ when $n = r$ cannot topple at $n = r + 1$. This artificial aspect can be eliminated by permitting the process not to move during an epoch, or by considering jumps with continuous distributions, for example.

[13] Although parametric distributions including inverse Gaussian distributions exhibit non-monotonic hazard functions, generalized linear models based on such distributions did not give a better fit to the data than Bienen and Van de Walle obtained with Cox regression.

[14] More precisely, a *Borel set*, i.e., a set that can be obtained from rectangles by countable unions and intersections.

at any given time can be viewed as a realized value of a random process. In fact, one may often assume that for any partition into disjoint subsets, the counts $n(A_j) \sim \text{Po}(\lambda(A_j)d(A_j))$, where $\lambda(A_j)$ is the expected density of area A_j, are independent. In this case, one speaks of a *spatial Poisson process*,[15] and the MLE of the average population density is $\hat{\lambda}(A_j) = n(A_j)/d(A_j)$. Such estimates may have high sampling variability, so smoother estimates may be desired. Suppose the center of A_j is at $\mathbf{z}_j = (z_{1j}, z_{2j})$. We might then have a 1st degree polynomial model for the density, $\log \lambda(A_j) = \beta_0 + \beta_1 z_{1j} + \beta_2 z_{2j}$. A 2nd degree polynomial surface would be of the form $\log \lambda(A_j) = \beta_0 + \beta_1 z_{1j} + \beta_2 z_{2j} + \beta_3 z_{1j}^2 + \beta_4 z_{2j}^2 + \beta_5 z_{1j} z_{2j}$, etc. The parameters of the models can be estimated using Poisson regression, as described in Section 3.

A potential defect of the regression formulation is that the density may not change in as regular a manner as the simple polynomial, or other parametric, models assume. If individual level data are available, nonparametric methods provide feasible alternatives. Suppose the expected population of A is given by an intensity function $\lambda(\mathbf{x}) \geq 0$ for $\mathbf{x} \in \mathbb{R}^2$. Then, the expected count is of the form

$$E[n(A)] = \int_A \lambda(\mathbf{x})\, d\mathbf{x}, \tag{9.1}$$

for a set $A \subset \mathbb{R}^2$. Suppose the points $\mathbf{x}_i$ come from a region B with $d(\text{B})$ finite. In *kernel estimation* one chooses a symmetric kernel function $\kappa_h(.) \geq 0$ that integrates to 1 and has a smoothing parameter $h > 0$. For any point $\mathbf{x} \in \mathbb{R}^2$, one estimates (cf., Cressie 1993, 600)

$$\tilde{\lambda}(\mathbf{x}) = \sum_{i=1}^{N} \kappa_h(\mathbf{x} - \mathbf{x}_i)/d(B). \tag{9.2}$$

For any $\mathbf{x}$, one or more of the N kernels may spread mass outside B. Apart from these "edge effects", the integral of (9.2) over $\mathbf{x} \in \text{B}$, would equal $N/d(\text{B})$, as it should. By far the most popular choice for a kernel function is the *Gaussian kernel* $\kappa_h(y_1, y_2) = \exp(-(y_1^2 + y_2^2)/2h^2)/2\pi h$. We see that for small values of h the points nearest to $\mathbf{x}$ are primarily relevant in estimation. If h is increased, the points further away make increasingly a contribution. A data dependent choice of the smoothing parameter h can be made using cross-validation (cf., Wahba and Wold 1975, Härdle 1990, Green and Silverman 1994). We will illustrate the method in Section 1.4 of Chapter 6.

Note that the right hand side of (9.1) is a spatial analogue of the cumulative intensity ((4.3) of Chapter 4) of a birth process that depends on a two-dimensional location $\mathbf{x}$ rather than a one-dimensional age x. This shows that a kernel estimator similar to (9.2) is also available to the nonparametric estimation of age-specific fertility. In fact, most demographic rates can be similarly handled.

---

[15] The term *Poisson random measure* is also used, since $n(A)$ is a measure of the size of A, and it takes a random value for each set A.

The spatial Poisson process is a model of spatial randomness in the sense that if $n(B) = N$ is given, then the points $\mathbf{x}_i, i = 1, \ldots, N$, can be viewed as a random sample from a distribution with density $\lambda(\mathbf{x}) / \int_B \lambda(\mathbf{x})\, d\mathbf{x}$ on B. In the case of constant intensity $\lambda(\mathbf{x}) \equiv \lambda$, the density is uniform, and one speaks of *complete spatial randomness* (e.g., Diggle 1983, 32). Complex patterns of deviations from randomness may occur in a spatial setting. In the so-called *Cox processes* the intensity $\lambda(\mathbf{x})$ is a realization of a random process much like the random effects in Section 4.3. They can serve as models for disease outbreaks, for example. The so-called *Neyman-Scott process* is generated by a mechanism that first samples "mother points" from a Poisson process and then distributes points around them according to some probability density. This might correspond to housing patterns in some societies. Spatial interaction processes may display *inhibition* in which a point may outright exclude other points in its neighborhood, or at least make them improbable (cf., Diggle 1983, Chapter 4). Explanatory variables may be included into the density of such a process, in addition to the distance between the points. Such processes may well have applications in enterprise demography for example.

A natural way to understand spatial interaction processes, is in terms of the conditional distribution of the location of a single point, given the locations of all other points. Moreover, in regression analyses of other population characteristics that can be mapped, such as income of families, or crime rates of cities, the so-called *conditional autoregressive models* (Whittle 1954) are often used. These models are also formulated conditionally, by specifying the conditional distribution of the characteristic at one location given the values of the same characteristic in all other locations. Such conditional distributions are the foundation of Gibbs sampling mentioned in Section 4.

10. Simulation of the Regression Models

The basic principles of simulating counts were discussed in Chapter 4. Only minor additional considerations are needed to apply those techniques to the regression settings. Consider logistic regression first with $Y_{xt} \sim \text{Bin}(n_{xt}, q(x,t))$. Knowing how to simulate a single binomial count as a sum of n_{xt} independent Bernoulli variables with probability of success $q(x,t)$ is all we need. If $q(x,t)$ is defined by (2.1), for example, then the only additional programming task is to recalculate $q(x,t)$ for each x and t. Poisson regression can be handled exactly the same way. For large expected counts we may want to resort to special methods not discussed in Chapter 4.

The random effects model requires one additional layer of computation. Suppose the random effects ε are independent for different values of t, with $\varepsilon(t) \sim N(0, \sigma^2)$. Then, we would first generate an effect from $N(0, \sigma^2)$ for each t, add them to the fixed (nonrandom) part of the canonical parameter, and generate the Poisson count after that. Possibly the most widely used method of generating normal random variables is the so-called Box-Muller method and its various refinements (Ripley 1987, 54; Press et al. 1992, 289). In its classical form the method generates a pair of independent standard normal variables via the following steps:

(1) Generate two independent uniformly distributed variables U_1 and U_2;
(2) Set angle $\Theta = 2\pi U_1$ and an independent radius $R = (-2\log(U_2))^{1/2}$;
(3) Get two independent standard normals $X_1 = R\cos(\Theta)$ and $X_2 = R\sin(\Theta)$.

The formal proof that this actually produces the desired standard normals is a somewhat tedious exercise in multivariate calculus. However, note that conditionally on R the pair (X_1, X_2) is uniformly distributed on a circle with radius R. Therefore, X_1 and X_2 are uncorrelated, and their distance from the origin is the square root of an exponential variable with expectation 2. This exponential distribution is the same as a χ^2 distribution with two degrees of freedom. This no proof, but note that *if* X_1 and X_2 are independent standard normals, then they will have exactly those properties!

Observations from a spatial Poisson process with a constant intensity can be easily simulated. Suppose the region of interest is B with the expected count C. One can then generate a Poisson variable with expectation C, denote the realized value as $n(B)$. One can enclose B into a rectangle, and generate uniformly distributed points inside the rectangle, as long as $n(B)$ of them fall into B.

Exercises and Complements (*)

1. Consider an infinite sequence of trials with probability $0 < p < 1$ of success. Let Y be the number of failures before the rth success. Then,

$$P(Y = y; r, p) = \binom{r + y - 1}{y} p^r(1 - p)^y, \; y = 0, 1, 2, \ldots$$

is the *negative binomial distribution*. The definition can be generalized to non-integer $r > 0$ by the same formula (cf., DeGroot 1987, 259). It has expectation $E[Y] = r(1 - p)/p$ and variance $\text{Var}(Y) = r(1 - p)/p^2$. If r is known, show that this belongs to an exponential family.

2. Consider the likelihood (1.6). Show that the Hessian (1.9) does not depend on random data, so $E[\mathbf{H}] = \mathbf{H}$. This simplifies the theory of exponential families.

*3. Suppose $\mathbf{Y}$ has density $f(\mathbf{y}; \boldsymbol{\theta})$. A statistic $\mathbf{U}(\mathbf{Y})$ is a *sufficient* for $\boldsymbol{\theta} \in \Theta$ if the conditional distribution of $\mathbf{Y}$ given $\mathbf{U} = \mathbf{u}$ does not depend on $\boldsymbol{\theta}$. *Neyman's factorization criterion* shows that $\mathbf{U}$ is sufficient if and only if we can write $f(\mathbf{y}; \boldsymbol{\theta}) = g(\mathbf{y})h(\mathbf{U}(\mathbf{y}), \boldsymbol{\theta})$. A sufficient statistic $\mathbf{U}(\mathbf{Y})$ is *minimal sufficient* if $\mathbf{U}$ is a function of any other sufficient statistic. Intuitively this means that the set of values taken by a minimal sufficient statistic is more "coarse" than that of any other sufficient statistic. Consider, for example, $u(x) = x$ and $v(x) = x^2$ for $x \in \mathbb{R}$. Is $u(.)$ a function of $v(.)$ or $v(.)$ a function of $u(.)$?

*4. A random variable Y belongs to the *exponential family* of distributions parameterized by $\boldsymbol{\theta} = (\theta_1, \ldots, \theta_k)^T$ if its density $f(y; \boldsymbol{\theta})$ (or probability function) may be expressed as

$$\exp\left[\sum_{j=1}^{k} u_j(y)\theta_j - b(\boldsymbol{\theta}) + c(y)\right].$$

When might this expression be well-defined? The function $b(\boldsymbol{\theta})$ must be chosen so that the density integrates (or sums) to 1, i.e., $b(\boldsymbol{\theta}) = \log \int \exp\{\sum_{j=1}^{k} u_j(y)\theta_j + c(y)\}\, dy$. The *natural parameter space* is defined as $\boldsymbol{\theta} = \{\boldsymbol{\theta} \in \mathbb{R}^k | -\infty < b(\boldsymbol{\theta}) < \infty\}$. (Cf., Bickel and Doksum 2001, 58–59).

5. Consider k independent competing risks X_j that have exponential distributions with parameters μ_j, $j = 1, \ldots, k$. Define the lifetime as $Y = \min\{X_1, \ldots, X_k\}$. Use the representation of the exponential distribution as a member of the exponential family to calculate the expectation and variance of Y.

*6. In the case of ordinary regression $\mathbf{Y} = \mathbf{X}\beta + \varepsilon$, where $\varepsilon \sim N(\mathbf{0}, \sigma^2\mathbf{I})$. The likelihood is $(2\pi\sigma^2)^{-n/2}\exp(-(\mathbf{Y} - \mathbf{X}\beta)^T(\mathbf{Y} - \mathbf{X}\beta)/2\sigma^2)$. (a) Show that this can be written in the form $\exp([\mathbf{Y}^T\mathbf{X}\beta - B(\beta)]/\sigma^2 + c(\mathbf{Y}, \sigma) + d(\sigma))$. (b) By differentiating the log-likelihood show that the MLEs for β solve the *normal equations* $\mathbf{X}^T\mathbf{Y} = E[\mathbf{X}^T\mathbf{Y}] = \mathbf{X}^T\mathbf{X}\beta$. (c) The solution is $\hat{\beta} = (\mathbf{X}^T\mathbf{X})^{-1}\mathbf{X}^T\mathbf{Y}$, provided that the inverse exists. This is the *ordinary least squares* (OLS) estimator. It is a *linear* function of Y_i's. (d) Show that $\hat{\beta} \sim N(\beta, \sigma^2(\mathbf{X}^T\mathbf{X})^{-1})$. (e) The variance σ^2 is usually estimated by $\hat{\sigma}^2 = (\mathbf{Y} - \mathbf{X}\hat{\beta})^T(\mathbf{Y} - \mathbf{X}\hat{\beta})/(n - k)$. Show that this is unbiased.

*7. Continuation. If $\varepsilon \sim N(\mathbf{0}, \sigma^2\mathbf{W})$ for some known positive definite matrix $\mathbf{W}$, then the likelihood is $(2\pi\sigma^2)^{-n/2}\ |\mathbf{W}|^{-1/2}\exp(-(\mathbf{Y} - \mathbf{X}\beta)^T\mathbf{W}^{-1}(\mathbf{Y} - \mathbf{X}\beta)/2\sigma^2)$. (a) Show that this can be written in the form $|\mathbf{W}|^{1/2}\exp([\mathbf{Y}^T\mathbf{W}^{-1}\mathbf{X}\beta - B(\beta)]/\sigma^2 + C(\mathbf{Y}, \sigma) + d(\sigma))$. (b) A transformed model $\mathbf{W}^{-1/2}\mathbf{Y} = \mathbf{W}^{-1/2}\mathbf{X}\beta + \mathbf{W}^{-1/2}\varepsilon$ has mean $\mathbf{W}^{-1/2}\mathbf{X}\beta$ and errors $\mathbf{W}^{-1/2}\varepsilon \sim N(\mathbf{0}, \sigma^2\mathbf{I})$. Deduce that the normal equations are $\mathbf{X}^T\mathbf{W}^{-1}\mathbf{Y} = \mathbf{X}^T\mathbf{W}^{-1}\mathbf{X}\beta$, with solution $\hat{\beta} = (\mathbf{X}^T\mathbf{W}^{-1}\mathbf{X})^{-1}\mathbf{X}^T\mathbf{W}^{-1}\mathbf{Y}$ (e.g., Rao 1973, 221). This is the *generalized least squares* (GLS) estimator. (c) Show that the GLS estimator has $\mathrm{Cov}(\hat{\beta}) = \sigma^2(\mathbf{X}^T\mathbf{W}^{-1}\mathbf{X})^{-1}$.

8. Newton's method has the following geometric motivation. Suppose we want to solve the equation $f(x) = 0$, and have a guess x_0 available. If $f(x_0) \neq 0$, we can try to improve the solution by replacing $f(x)$ with its tangent line at $x = x_0$, $L(x) = f(x_0) + f'(x_0)(x - x_0)$. This intersects the x-axis at x_1, $L(x_1) = 0$, so $x_1 = x_0 - f(x_0)/f'(x_0)$ is an updated guess. In (1.10) we seek the solution to the vector equation $\mathbf{f}(\beta) = \mathbf{0}$, where $\mathbf{f}(\beta) = \mathbf{U} - E[\mathbf{U}] = \mathbf{X}^T\mathbf{Y} - \partial/\partial\beta B(\beta)$. The tangent line is replaced by a first-order Taylor series expansion about a trial value $\beta_{(i)}$, $\mathbf{L}(\beta) = \mathbf{f}(\beta_{(i)}) + \partial/\partial\beta^T \mathbf{f}(\beta_{(i)})(\beta - \beta_{(i)})$. Setting $\mathbf{L}(\beta_{(i+1)}) = \mathbf{0}$ we find $\beta_{(i+1)} = \beta_{(i)} - [\partial/\partial\beta^T\, \mathbf{f}(\beta_{(i)})]^{-1}\mathbf{f}(\beta_{(i)}) = \beta_{(i)} + [\partial^2/\partial\beta\partial\beta^T B(\beta_{(i)})]^{-1}(\mathbf{U} - E_{(i)}[\mathbf{U}])$.

9. Show that (1.11) and (1.12) are equivalent to (1.10).
*10. Show that the *hat matrix* $\mathbf{H} = \mathbf{X}(\mathbf{X}^T\mathbf{X})^{-1}\mathbf{X}^T$ (not to be confused with the Hessian!), is symmetric ($\mathbf{H} = \mathbf{H}^T$) and idempotent ($\mathbf{H} = \mathbf{H}^2$) and consequently the ith diagonal element, h_{ii}, is between 0 and 1. Let $\hat{\boldsymbol{\beta}}$ denote the OLS estimate of $\boldsymbol{\beta}$ in the model $\mathbf{Y} = \mathbf{X}\boldsymbol{\beta} + \varepsilon$, with $\text{Var}(\varepsilon) = \sigma^2\mathbf{I}$, and define $\hat{\mathbf{Y}} = \mathbf{X}\hat{\boldsymbol{\beta}}$. Show that the covariance matrix of the residual vector $\mathbf{Y} - \hat{\mathbf{Y}}$ equals $\sigma^2(\mathbf{I} - \mathbf{H})$. Let $\hat{\boldsymbol{\beta}}_{(i)}$ denote the OLS estimate when the ith observation is not used in the fitting, and define $\hat{\mathbf{Y}}_{(i)} = \mathbf{X}\hat{\boldsymbol{\beta}}_{(i)}$. Notice that the prediction of Y_i is now $\mathbf{x}_i\hat{\boldsymbol{\beta}}_{(i)}$, where $\mathbf{x}_i$ denotes the ith row of $\mathbf{X}$. Show that $\hat{Y}_i = (1 - h_{ii})\mathbf{x}_i\hat{\boldsymbol{\beta}}_{(i)} + h_{ii}Y_i$, so that the derivative of $\hat{Y}_i$ with respect to Y_i equals h_{ii} (Welsch 1983).
11. Derive the leverages mentioned in Example 1.3.
*12. To motivate Cook's distance, note that numerator of Cook's distance is $(\hat{\mathbf{Y}} - \hat{\mathbf{Y}}_{(i)})^T(\hat{\mathbf{Y}} - \hat{\mathbf{Y}}_{(i)})$. Consider $\mathbf{Y} \sim N(\mathbf{X}\boldsymbol{\beta}, \boldsymbol{\sigma}^2\mathbf{I})$, where the rank of $\mathbf{X}$ is k and $\text{Cov}(\hat{\boldsymbol{\beta}}) = \sigma^2(\mathbf{X}^T\mathbf{X})^{-1}$. Show that $(\hat{\boldsymbol{\beta}} - \boldsymbol{\beta})^T(\mathbf{X}^T\mathbf{X})(\hat{\boldsymbol{\beta}} - \boldsymbol{\beta})/\sigma^2 \sim \chi^2$ distribution with k degrees of freedom. Therefore, $(\hat{\boldsymbol{\beta}} - \boldsymbol{\beta})^T(\mathbf{X}^T\mathbf{X})(\hat{\boldsymbol{\beta}} - \boldsymbol{\beta})/k\hat{\sigma}^2 \sim F_{k,n-k}$, the F distribution with k and $n - k$ degrees of freedom.
13. Consider two probabilities $0 < q_j < 1$, for $j = 0, 1$. Define $\text{RR} = q_1/q_0$ and $\text{OR} = \{q_1/(1 - q_1)\}/\{q_0/(1 - q_0)\}$. Assume that $q_1 = 2q_0$ and plot both RR and OR as functions of q_0 for $0 < q_0 < 1/2$.
*14. The concept of a "saturated model" is a bit tricky. Suppose we toss a coin independently n times, and the chance of "heads" is $0 < p < 1$. Consider two cases. First, suppose we only know that the total number of heads is y. Then, we would base inference on the binomial model $Y \sim \text{Bin}(n, p)$, and assume that $Y = y$ is the observed value. Second, suppose the outcome of the ith toss is y_i and we know the ordered outcomes $(y_1, \ldots, y_n)$. In this case we would have a vector of random variables $(Y_1, \ldots, Y_n)$, where the $Y_i \sim \text{Ber}(p)$ are independent, and $(Y_1, \ldots, Y_n) = (y_1, \ldots, y_n)$ is the observed value. The two models are usually equally informative, but the deviances calculated under the two models differ, because they correspond to different saturated models. In the former case $\ell^* = \log\{[n!/(y!(n - y)!)](y/n)^y((n - y)/n)^{n-y}\}$, whereas in the latter case $\ell^* = 0$. This shows that deviance is not appropriate as a general measure of lack of fit.
*15. Consider the model $Y \sim \text{Ber}(p)$. In logistic regression the mean $E[Y] = p$ is mapped to the linear predictor $\mathbf{X}^T\boldsymbol{\beta}$ by a canonical link function $\text{logit(p)} = \mathbf{X}^T\boldsymbol{\beta}$. Alternative mappings are provided by (*i*) the *probit link*, $\Phi^{-1}(p) = \mathbf{X}^T\boldsymbol{\beta}$, where $\Phi(x) = (2\pi)^{-1/2}\int_{-\infty}^{x}\exp(-z^2/2)\,dz$; (ii) *complementary log-log link* $\log(-\log(1 - p)) = \mathbf{X}^T\boldsymbol{\beta}$; (iii) *identity link* $p = \mathbf{X}^T\beta$, etc. To motivate (ii), consider a follow-up period [0, 1] and assume that the cumulative hazard ((2.5) of Chapter 4; and (7.1)) of a waiting time T of an individual with covariates $\mathbf{X}$ is $\Lambda(1)\exp(\mathbf{X}^T\boldsymbol{\beta})$ at the end of the period. Define $Y = 1$, if $T \leq 1$, and $Y = 0$ otherwise. Show that $\log(-\log(1 - p)) = \alpha + \mathbf{X}^T\boldsymbol{\beta}$, where $\alpha = \log(\Lambda(1))$ can be absorbed into the constant term of the model.
16. Carry out the logistic regression of the two 2×2 tables suggested at the end of Example 2.4. Is an interaction term needed?

*17. Consider a model $Y_i \sim \text{Ber}(p_i)$, where $\text{logit}(p_i) = \alpha_0 + \alpha_1 x_i$, $i = 1, \ldots, n$. Suppose $Y_i = 1$ indicates that i dies (recovers from an illness) and x_i is i's level of exposure (amount of medicine), so we are modeling a *dose-response* relationship. The problem of *inverse dose-response* asks for a dose $x = x(c)$ such that the probability of death (recovery) is some predetermined value $0 < c < 1$. (*a*) Write $c^* = \text{logit}\,(c)$, and deduce that an estimator of the dose is $\hat{x}(c) = (c^* - \hat{\alpha}_0)/\hat{\alpha}_1$. (b) Note that if x is the true value, then $L(x) = \hat{\alpha}_0 + \hat{\alpha}_1 x - c^* \sim N(0, V(x))$ asymptotically, where $V(x) = v_{00} + 2x v_{01} + x^2 v_{11}$ and $v_{ij} = \text{Cov}(\hat{\alpha}_i, \hat{\alpha}_j)$, $i, j = 0, 1$ are the elements of matrix (1.13). Using the result $L(x)^2/V(x) \sim \chi_1^2$ deduce a second degree polynomial in x whose roots give the 95% confidence interval for $x(c)$. (c) Alternatively, if x is the true value, we must have $\alpha_0 + \alpha_1 x = c^*$, so $\alpha_0 = -\alpha_1 x + c^*$. Thus, we can write $\text{logit}(p_i) = c^* + \alpha_1(x_i - x)$, $i = 1, \ldots, n$. This model can be fitted for any x by offsetting c^* to get the profile likelihood $\hat{\ell}_0(x)$. Deduce that an alternative 95% confidence interval is of the form $\{x | 2(\hat{\ell}_1 - \hat{\ell}_0(x)) < 3.841\}$.

18. Show that if we add the term $\gamma(x - t)$ into the model $\log(\lambda_{xt}) = \alpha_0 + \alpha_1 x + \beta t$, then the model parameters are not identifiable.

19. Consider the data of Example 3.6. Fit a model that has a separate slope for age for employed and unemployed. Check the residuals of the model. Are there indications of remaining lack of fit?

20. Consider the following data on the incidence of occupational diseases in Finland in 1983, by industry and sex:

	Reported Cases		Population At Risk (in 1000's)	
Industry	Males	Females	Males	Females
1. Agriculture	160	183	139	116
2. Forestry	116	2	54	3
3. Man. of Consumer Goods	194	371	49	93
4. Man. of Wood and Paper Prod.	575	167	112	56
5. Metal Industries, Mining	850	211	160	47
6. Other Manufacturing	284	92	70	30
7. Building, Construction	633	20	164	19
8. Trade	87	64	120	148
9. Restaurants, Hotels	2	25	10	47
10. Traffic	212	26	131	49
11. Finance, Real Estate	14	21	51	85
12. Public Admin., Defense	132	42	64	72
13. Other Social Services	75	142	80	315
14. Other Services	59	21	38	45

A topic of concern is whether the risk of occupational diseases differs among males and females. A comparison of crude rates by sex may be confounded by the fact that males and females work in different industries. Use Poisson regression, indirect standardization, (3.6) and (3.7), and direct standardization (3.5), to study the relative risk between males and females.

21. Suppose the number of deaths in age $x = 0, 1, \ldots, \omega$, during year $t = 1, 2, \ldots, T$ are Poisson distributed, $D_{xt} \sim \text{Po}(\theta_t \mu_x K_{xt})$, where K_{xt} is the person years, the μ_x's are a set of known standard mortality rates, and θ_t is an unknown relative risk parameter of the year t. A linear estimator of θ_t is of the form $Y_t = \Sigma_x c_x D_{xt}$, where the c_x's are some weights. The estimator is unbiased if $E[Y_t] = \theta_t$. Incorporate the condition of unbiasedness using

Lagrange multipliers, and show that the minimum variance linear unbiased estimator of θ_t is obtained by choosing $c_x = 1/\Sigma_u \mu_u K_{ut}$ for all x. Deduce then that the standardized mortality ratio is the minimum variance linear unbiased estimator of the relative risk.

*22. As in Exercise 21, suppose the number of deaths are of the form $D_{xt} \sim \text{Po}(\theta_t \mu_x K_{xt})$, where the K_{xt}'s are the person years, the μ_x's are known standard rates, and θ_t's are unknown parameters to be estimated. Show that $D_t = \Sigma_x D_{xt}$ is a sufficient statistic for θ_t. Conclude with the help of the Rao-Blackwell theorem (cf., DeGroot 1987, 373) that as a function of the sufficient statistic, the standardized mortality ratio $Y_t = D_t / \Sigma_u \mu_u K_{ut}$ is a minimum variance unbiased estimator of θ_t. This result is stronger than that of Exercise 21, because no restriction to linear estimators is needed, and its derivation is simpler, since no real calculations are needed - once one knows Rao-Blackwell!

*23. Consider the likelihood equation (1.8) in the form $\mathbf{X}^T\mathbf{Y} = \mathbf{X}^T\mathbf{E}[\mathbf{Y}]$. In the Poisson regression case $Y_i \sim \text{Po}(\exp(\mathbf{X}_i^T \beta)K_i)$, $i = 1, \ldots, n$, we noted that they can be solved by resorting to an offset term. Alternatively, define a vector $\mathbf{M}$ with the ith element equal to Y_i/K_i, and $\mathbf{K} = \text{diag}(K_1, \ldots, K_n)$. Multiply the likelihood equation from the right by $\mathbf{K}^{-1}$ to get $\mathbf{X}^T\mathbf{M} = \mathbf{X}^T\text{E}[\mathbf{M}]$. Writing $\mathbf{W} = \text{diag}(E[\mathbf{M}])$ we get that $\text{Cov}(\mathbf{M}) = \mathbf{W}\mathbf{K}^{-1}$. Thus, an alternative numerical method is to base the estimation on rates, and multiply the weights by $1/K_i$'s in iteration.

24. In Finland, the state provides support to municipalities for health and social care, using allocation formulas. A 1992 law stipulated that support for health care should be proportional to the product of population size and "level of illness" in the municipality. As a measure of level of illness, the SMR as defined in (3.6), was adopted, with x = age and t = municipality. (a) Do you think mortality is a good measure of illness? (b) Suppose you are in the Municipal Board. What kind of incentive does this formula give you, if you are considering whether to improve the health care of the elderly? (c) The median population size of a municipality is approximately 5,000. Suppose 1% of the population is expected to die annually. What is the coefficient of variation of the allocation, from year to year, in a municipality of median size, if deaths from three consecutive years are used to calculate the SMR? Having considered the three issues you will understand why the law was subsequently changed.

25. Assume that $Y_i \sim \text{Po}(\mu_i)$, $i = 0, 1$, are independent, and define $Y = Y_0 + Y_1$. (a) Show that conditionally on $Y = y$, we have $Y_0 \sim \text{Bin}(y, \mu_0/(\mu_0 + \mu_1))$. (b) Using this, show that the probability of finding $Y_0 < 3$ in Example 3.2 is 0.9993, provided that H_0: $\lambda_0 = \lambda_1$ holds. This a direct way of confirming the significance of the excess risk.

26. Derive the ML estimators of the loglinear model parameters for the capture-recapture experiment discussed in Section 3.4.

27. Continuation. Show by a direct calculation that $\hat{N} = Y_{11} + Y_{10} + Y_{01} + \hat{\lambda}_{00}$ is equal to the classical dual systems estimator, as defined in Section 6 of Chapter 2.

28. Consider the negative binomial distribution as parametrized in Section 4. Derive the values of the gamma parameters α_i and β_i as functions of μ_i and σ^2.

29. Equivalently with (5.4), we have $\text{Var}(\hat{N}) = N(1 - p_1)(1 - p_2)/(p_1 p_2)$. Substitute estimators $\hat{p}_1 = m/n_2$, $\hat{p}_2 = m/n_1$ into this to get the variance estimator first derived in Exercise 11 of Chapter 2.
*30. Consider a 2-way contingency table with expected counts $E[Y_{ij}] = \lambda_{ij}$. (a) Under a loglinear main effects model $\log(\lambda_{ij}) = \mu + \alpha_i + \beta_j$ with conditions $\Sigma_i \alpha_i = 0 = \Sigma_j \beta_j$ the model contains $1 + (I - 1) + (J - 1) = I + J - 1$ parameters. Therefore, $IJ - I - J + 1 = (I - 1)(J - 1)$ degrees of freedom remain. (b) Under a full interaction model $\log(\lambda_{ij}) = \mu + \alpha_i + \beta_j + \gamma_{ij}$ with conditions $\Sigma_j \gamma_{ij} = 0$ for each $i = 1, \ldots, I$, and $\Sigma_i \gamma_{ij} = 0$ for each $j = 1, \ldots, J$ the model becomes saturated, so the number of new additional parameters must be $(I - 1)(J - 1)$. To see this directly, note that the first set of conditions introduces I conditions for the γ_{ij}'s and the second set introduces J additional conditions. However, one of the latter conditions is superfluous, because the first I conditions already imply that $\Sigma_{ij} \gamma_{ij} = 0$. Thus, the number of new free parameters introduced is $IJ - I - J + 1 = (I - 1)(J - 1)$. (c) Under the association model $\log(\lambda_{ij}) = \mu + \alpha_i + \beta_j + \varphi \nu_i \eta_j$ with conditions $\Sigma_i \nu_i = 0 = \Sigma_j \eta_j$ and $\Sigma_i \nu_i^2 = 1 = \Sigma_j \eta_j^2$, there are two conditions for both υ_i's and η_j's, so $(I - 2) + (J - 2)$ parameters are free to vary. One more degree of freedom is lost due to φ. Hence, the total number of new free parameters is $I + J - 3$.
31. Consider the capture-recapture model (5.5). Show that conditioning on $n_{1i} = 1$, we have $u_{1i} = 1 - m_i$, where $m_i \sim \text{Ber}(p_{2i})$. Thus, the parameters $\boldsymbol{\beta}_2$ of (5.5) can be estimated by applying ordinary logistic regression to first capture. Correspondingly, taking $n_{2i} = 1$, we may use $m_i \sim \text{Ber}(p_{1i})$ to estimate $\boldsymbol{\beta}_1$.
32. Differentiate the loglikelihood (7.6) with respect to the (scalar) parameter β. From this expression you can see that each $X_{(i)}$ actually has a Bernoulli distribution. What is the probability of success? Calculate also the second derivative and check that it gives the correct Bernoulli variance.
33. Continuation. The so-called *log-rank test* for the hypothesis $H_0 : \beta = 0$ can derived from the results of Exercise 32 by setting $\beta = 0$. The (score) test statistic is $\ell'(0)/(-\ell''(0))^{1/2}$. Show that it is of the form: sum of independent Bernoulli variables minus their expectation, divided by the standard deviation of the sum. Therefore, it has an asymptotic standard normal distribution.
34. Prove the result of Example 7.5.
35. Continuation. Consider matched sets of individuals $R_{(i)}, i = 1, \ldots, n$. Suppose a subset $A_i \subset R_{(i)}$ has $\#A_i = n_i$ cases, and $R_{(i)} \backslash A_i$ consists of non-cases. Such data can arise from a case-control study in which the cases A_i are matched with some controls, and they together form the set $R_{(i)}$, or it can arise from a cohort study in which individuals are first matched (e.g., by residence) into sets $R_{(i)}$ and during the follow-up those in A_i happen to fall ill. Show that by conditioning on the number of cases in $R_{(i)}$, in both cases the likelihood is

$$L_{(i)} = \exp\left(\sum_{i \in A_i} \mathbf{X}_i^{\mathrm{T}} \boldsymbol{\beta}\right) \Bigg/ \sum_{B_i \subset R_{(i)}, \#B_i = n_i} \exp\left(\sum_{i \in B_i} \mathbf{X}_i^{\mathrm{T}} \boldsymbol{\beta}\right).$$

36. (Continuation) Suppose $n_i = 1$ for all sets i with $\#R_{(i)} = 2, i = 1, \ldots, n$. I.e., we have n case-control pairs. Based on the above likelihood, show (the otherwise mind-boggling result) that conditional logistic regression can be run using an ordinary logistic regression program by creating a data set in which there are n observations (data rows), for each observation the outcome variable is 1 ("success"), the explanatory variables are the differences between the case's explanatory variables and the control's explanatory variables, and there is no constant term.

*37. Following the notation of Section 7, let $Z_{(i)} = k$ if individual k withdrew at time $T_{(i)}$. Denote the history up through the ith withdrawal by $H_i = \{T_{(1)}, Z_{(1)}, \delta_{(1)}, \ldots, T_{(i)}, Z_{(i)}, \delta_{(i)}\}$. The full likelihood is $L(H_n)$. Note that $L(H_i|H_{i-1}) = P(Z_{(i)}|H_{i-1}, \delta_{(i)}, T_{(i)}) \times P(\delta_{(i)}, T_{(i)}|H_{i-1})$, and hence

$$L(H_n) = \prod_{i=1}^{n} L(H_i|H_{i-1})^{1-\delta_i} \prod_{i=1}^{n} P(\delta_{(i)}, T_{(i)}|H_{i-1})^{\delta_i} \prod_{i=1}^{n} P(Z_{(i)}|H_{i-1}, \delta_{(i)}, T_{(i)})^{\delta_i}.$$

The first product involves censoring only. The second product involves the times of non-censored withdrawal, which under (7.1) do not provide information about $\boldsymbol{\beta}$. Under (7.1), the components of the last product are given by (7.4). Assume that the proportional hazards model holds, and show that the partial likelihood (7.5) is derived by ignoring the first two products above. If the covariate vectors vary with time, $\mathbf{X}_{(k)} = \mathbf{X}_{(k)}(t)$, then they can be included in the definition of H_i and a similar expression can be derived; see Cox and Oakes (1984, Section 8.4).

38. Consider formula (2.7) of Chapter 4. Prove (8.1) by first splitting the integral into an integral from 0 to t, and an integral from t to ∞. Then, majorize the integrand on $[0, t]$ by 1, and on (t, ∞) by $p(x)/p(t)$. Note that the inequality is strict unless $p(t) = 1$.

39. Use a computer to generate realizations of a random walk of length 20. Stop each random walk if it reaches the level $r = -5$. Calculate the expectation of the state of the walks that have not been stopped at epochs $n = 1, 5, 10, 15, 20$. How do the expectations behave as a function of n?

40. Prove formally that the Box-Muller method produces two independent variables with standard normal distributions.

41. Simulation of logistic regression. Generate values of explanatory variables $X_i \sim N(\mu, \sigma^2)$ for $i = 1, \ldots, n$. Then, generate uniforms $U_i \sim U[0, 1]$ and calculate $p_i = \exp(\beta_0 + \beta_1 X_i)/(1 + \exp(\beta_0 + \beta_1 X_i)$ using, e.g., $n = 30$, $\beta_0 = -0.1$, and $\beta_1 = 1.0$. Now generate the observations $Y_i = 1$ if $U_i \leq p_i$, otherwise let $Y_i = 0$.

42. Generate a sample from a spatial Poisson process into a unit square such that the expected number of points is 100. I.e., pick a value Y from Po(100), and locate Y points into unit square by picking independently each x-coordinate and each y-coordinate from $U[0, 1]$. Does the point pattern correspond to your idea of complete randomness?

6
Multistate Models and Cohort-Component Book-Keeping

In this chapter we develop some theory and notation for multistate life tables and general linear growth models. Life tables are synthetic calculations that are intended to summarize the overall implications of period transition rates in populations with one or more states defined by region, marital status, labor force participation, etc. We will provide a formulation that takes duration (i.e., time spent in a state) into account.

As anticipated in Chapter 4, when the generation of births at constant rates of fertility is added to a life table population, a theory of stable populations follows. Life table calculations also provide the "engine" on which the cohort-component population forecasts are based. The matrix model we emphasize is often called a *Leslie model*, in honor of Leslie (1945). However, Bernardelli (1941) and Lewis (1942) had earlier considered the matrix formulation. Cannan (1895) had used the equivalent arithmetic already, and many European states and the U.S. had used the arithmetic in the 1920's and 1930's (cf., DeGans 1999). Therefore, a more neutral name seems to be in order, and we will refer to the linear growth model.

Calculations concerning population evolution are used in economic contexts such as pension planning, disability insurance, assessment of health care costs etc. Often, relevant statistics can be calculated from the population numbers and rates, so they can be viewed as functions of population numbers, or *demographic functionals*. Multistate models are also connected to Markov chains that are used to describe state transitions in many branches of science.

Section 1 presents multistate life tables in a probabilistic context analogous to that of Chiang (1968). An application to Finnish nuptiality data is described, and a model for simple disability insurance is formulated. Section 2 defines the linear growth model and develops aspects of classical stable population theory and the so-called weak ergodicity. In Section 3 we open the multistate system to external migration and consider alternate ways of parametrizing migration flows. Section 4 defines the concepts of demographic functional and functional forecasts. In Section 5 we examine some details of the linear growth model and population renewal at the level of individual ages. In Section 6 we will mention applications of Markov chains to an ecological population.

1. Multistate Life Tables

1.1. Numerical Solution Using Runge-Kutta Algorithm

Define $I(x) = 1$ if an individual is alive in age $x \geq 0$ and $I(x) = 0$ otherwise. The probability of surviving to age x can be written as $p(x) = E[I(x)]$. In equation (2.2) of Chapter 4, the probability of survival was shown to satisfy the differential equation $p'(x)/p(x) = -\mu(x)$ in terms of the hazard. The equation was solved analytically in (2.4). We will show below that (2.2) has an analogue in the multidimensional case. Although the multidimensional case does not allow an explicit analytical solution, except in special cases, (2.2) can be solved numerically without recourse to the analytical solution. The solution is based on a standard method for first order differential equations, the so-called fourth order Runge-Kutta method (e.g., Press et al. 1992, 710–714), which we now describe.

Consider a differential equation $y' = f(x, y)$, where y is to be solved as a function of x subject to a known starting value $y_0 = y(x_0)$. The simplest method for getting an approximate numerical solution to the equation is to determine a step size $h > 0$, set $x_n = x_{n-1} + h$, and determine the approximations $y_{n+1} \approx y_n + hf(x_n, y_n)$, $n = 0, 1, 2, \ldots$ This is *Euler's method*. It uses information about the derivative of y only at the beginning of each interval $[x_n, x_{n+1}]$. One can try to improve on Euler's method by getting a better estimate of the derivative in the interval. The fourth order *Runge-Kutta method* uses four estimates of the derivative, one at the beginning, one at the end, and two in the middle. The algorithm is:

$$y_{n+1} = y_n + (a_1 + 2a_2 + 2a_3 + a_4)/6, \tag{1.1}$$

where $a_1 = hf(x_n, y_n)$, $a_2 = hf(x_n + h/2, y_n + a_1/2)$, $a_3 = hf(x_n + h/2, y_n + a_2/2)$, and $a_4 = hf(x_n + h, y_n + a_3)$. The coefficients of the a_i have been chosen so that the method is accurate to the fourth degree, i.e., the error is $O(h^5)$ as defined in Chapter 1.

Example 1.1. Runge-Kutta Illustration. Suppose $\mu(x) = \mu > 0$ for $x \geq 0$ and use the fourth order Runge-Kutta method for solving $p'(x) = -\mu p(x)$ subject to $p(0) = 1$. The exact solution is $p(x) = \exp(-\mu x)$. We have $y_0 = 1; a_1 = -\mu h$; $a_2 = -\mu h(1 - \mu h/2)$; $a_3 = -\mu h(1 - \mu h/2 + (\mu h)^2/4)$; and $a_4 = -\mu h(1 - \mu h + (\mu h)^2/2 - (\mu h)^3/4)$. Therefore, $y_1 = 1 - \mu h + (\mu h)^2/2! - (\mu h)^3/3! + (\mu h)^4/4!$, or the first step is equal to the fourth order Taylor series approximation to the true value of $\exp(-\mu h)$. By taking h small enough, we can achieve any degree of accuracy. ◊

To apply the Runge-Kutta method to (2.2) of Chapter 4, we take $y(x) = p(x)$ and $f(x, y) = -\mu(x)p(x)$. The starting value is $p(0) = 1$. For most ages we take $h = 1$, but for age 0 we may take two steps, first $h = 28/365$ corresponding to neonatal mortality, and the second step size is $h = 1 - 28/365$. The values $\mu(1)$, $\mu(1.5)$, $\mu(2)$, $\mu(2.5)$ can be estimated, e.g., using methods discussed in Section 2.4 of Chapter 4. For the first year of life procedures based on Example 2.11 of Chapter 4 may be applied, for example.

1.2. Extension to Multistate Case

Suppose now that there are J states. An individual is born into a state, and may later move to another state, move back, etc. For example, a person is born into never married state and may later marry, become divorced or widowed, remarry, etc. Labor force participation, migration, and even acquisition of skills and knowledge are other examples of transition among states. Some basic references to this area are Rogers (1975), Rees and Wilson (1977), Land and Rogers (1982), ter Heide and Willekens (1984). More recent contributions include Schoen (1988), Gill and Keilman (1990), Van Imhoff (1990), Ekamper and Keilman (1993), and Rogers (1995).

Define an indicator vector $\mathbf{I}(x) = (I_1(x), \ldots, I_J(x))^T$ for $x \geq 0$ such that $I_j(x) = 1$ if the individual is in state j at age x and $I_j(x) = 0$ otherwise. Define $\mathbf{e}_j$ as a J-component column vector of all zeroes except a 1 in the jth position; for example, $\mathbf{e}_2 = (0, 1, 0, \ldots, 0)^T$. Set $\mathbf{p}_j(x) = E[I_j(x)]$ and $\mathbf{p}(x) = (p_1(x), \ldots, p_J(x))^T$, so $\mathbf{p}(x) = E[\mathbf{I}(x)]$ gives the probabilities that the individual is in each of the states $j = 1, \ldots, J$ at age x. We assume that the individual changes state according to the following rules:

(1) If $\mathbf{I}(x) = \mathbf{e}_j$, i.e., the individual is in state j at age x, then, independently of the individual's earlier history, the probability of moving to state $i \neq j$ before age $x + h$ is $\nu_{ij}(x)h + o(h)$, where $\nu_{ij}(.) \geq 0$ is continuous.
(2) The probability of two or more transitions in a period of length $h > 0$ is $o(h)$.

We call the functions $\nu_{ij}(.)$ *hazards* or *transition intensities*.

Consider the probability $p_i(x + h)$ that individual is in state i in age $x + h$. We can express the probability in terms of the probabilities $p_j(x)$. There are three cases. The individual either was in i already and did not leave, was in some other state and moved to i, or made two or more transitions. Therefore, we can write,

$$\begin{aligned} p_i(x+h) = {} & \left(1 - \sum_{j \neq i} \left\{\nu_{ji}(x)h + o(h)\right\}\right) p_i(x) \\ & + \sum_{j \neq i} \left\{\nu_{ij}(x)h + o(h)\right\} p_j(x) + o(h). \end{aligned} \tag{1.2}$$

As in the case of (2.1) of Chapter 4, divide (1.2) by h, rearrange terms, and let $h \to 0$, to get for each $i = 1, \ldots, J$ that

$$p_i'(x) = -\sum_{j \neq i} \nu_{ji}(x) p_i(x) + \sum_{j \neq i} \nu_{ij}(x) p_j(x). \tag{1.3}$$

In matrix form (1.3) can be written as

$$\mathbf{p}'(x) = \boldsymbol{\nu}(x)\mathbf{p}(x), \tag{1.4}$$

where the left hand side is a vector of the derivatives and $\boldsymbol{\nu}(x) = (\nu_{ij}(x))$ is a $J \times J$ matrix, where for $i \neq j$ the elements $\nu_{ij}(x)$ are as defined in condition (1) above, but for $i = 1, \ldots, J$ we take

$$\nu_{ii}(x) = -\sum_{j \neq i} \nu_{ji}(x), \tag{1.5}$$

the negative of the hazard of leaving state i in age x. Notice that (1.4) is the multistate counterpart of (2.2) of Chapter 4. Let us first consider two special cases.

First, the single cause of death case can be described as a two-state model with states "alive" ($j = 1$) and "dead" ($j = 2$). The latter state is *absorbing*, or it has $\nu_{12}(x) = 0$ for all $x > 0$. If we write $\nu_{21}(x) = \mu(x)$, as before, then we have

$$\boldsymbol{\nu}(x) = \begin{bmatrix} -\mu(x) & 0 \\ \mu(x) & 0 \end{bmatrix}. \tag{1.6}$$

In this case, $p_1(x)$ is given by (2.4) and (2.5) of Chapter 4, and $p_2(x) = 1 - p_1(x)$.

Second, assume that $\boldsymbol{\nu}(x) \equiv \boldsymbol{\nu}$. By a direct calculation one can show that

$$\mathbf{p}(x) = \left\{ \sum_{i=0}^{\infty} (x\boldsymbol{\nu})^i / i! \right\} \mathbf{p}(0) \tag{1.7}$$

satisfies the equation $\mathbf{p}'(x) = \boldsymbol{\nu}\mathbf{p}(x)$ (e.g., Gantmacher 1959; Schoen 1988, 72–73). The matrix in brackets on the right hand side of (1.7) actually defines the *exponential function with matrix argument $x\boldsymbol{\nu}$.*

In fact, a slightly more general case can also be handled analytically. Suppose that the $\boldsymbol{\nu}(x)$ are simultaneously diagonalizable, i.e., we can write $\boldsymbol{\nu}(x) = \mathbf{U}\boldsymbol{\gamma}(x)\mathbf{V}^T$, where $\mathbf{V}^T\mathbf{U} = \mathbf{I}$, and $\boldsymbol{\gamma}(x) = \text{diag}\,(\gamma_1(x), \ldots, \gamma_J(x))$ has the eigenvalues of $\boldsymbol{\nu}(x)$. Note that $\mathbf{V}^T = \mathbf{U}^{-1}$ and the columns of $\mathbf{U}$ contain the eigenvectors of $\boldsymbol{\nu}(x)$ normalized in some manner (cf., Rao 1973, 42–43). In other words, the spectral decompositions of the matrices $\boldsymbol{\nu}(x)$ are such that the matrices $\mathbf{V}$ and $\mathbf{U}$ do not depend on x. (We will discuss spectral decomposition further in Section 2.2.) Define, for $j = 1, \ldots, J$,

$$\Gamma_j(x) = \exp\left(\int_0^x \gamma_j(s)\, ds \right) \tag{1.8}$$

and let $\boldsymbol{\Gamma}(x) = \text{diag}(\Gamma_1(x), \ldots, \Gamma_J(x))$. Now, the solution of (1.4) is simply

$$\mathbf{p}(x) = \mathbf{U}\boldsymbol{\Gamma}(x)\mathbf{V}^{\mathbf{T}}\mathbf{p}(0). \tag{1.9}$$

To confirm that (1.4) is satisfied, note that $\boldsymbol{\nu}(x)\mathbf{p}(x) = \mathbf{U}\boldsymbol{\gamma}(x)\mathbf{V}^{\mathbf{T}}\mathbf{U}\boldsymbol{\Gamma}(x)\mathbf{V}^{\mathbf{T}}\mathbf{p}(0) = \mathbf{U}\boldsymbol{\gamma}(x)\boldsymbol{\Gamma}(x)\mathbf{V}^{\mathbf{T}}\mathbf{p}(0) = \mathbf{p}'(x)$.

Returning to the case where $\boldsymbol{\nu}(x) \equiv \boldsymbol{\nu}$ are constant, write $\boldsymbol{\gamma}(x) = \boldsymbol{\gamma}$. Then we have $\Gamma_j(x) = \exp(\gamma_j x)$ and $\boldsymbol{\Gamma}(x) = \boldsymbol{\Gamma}(1)^x$. We now illustrate how the spectral decomposition can be used to calculate the right hand side of (1.7) (cf., Hoem and Funck Jensen 1982, 179).

Example 1.2. A Three-State Labor Force Model. Suppose we have only three states: Employed ($j = 1$), Unemployed ($j = 2$), and Dead ($j = 3$), with transition intensities

$$\boldsymbol{\nu}(\boldsymbol{x}) = \begin{bmatrix} -0.08 & 0.05 & 0 \\ 0.06 & -0.07 & 0 \\ 0.02 & 0.02 & 0 \end{bmatrix}. \tag{1.10}$$

In other words, life expectancy is $1/0.02 = 50$ years, irrespective of working status; the probability of becoming unemployed is about 6% each year, and the probability of getting a job is about 5% for an unemployed, per year. Consider a person who is employed at the start of the study (or $x = 0$), so $\mathbf{I}(0) = (1, 0, 0)^T$ is the initial state. Note that under (1.10), the third (absorbing) component of the vector $\mathbf{p}(x)$ does not influence the evolution of the first two in (1.4), so it can be omitted in the following calculation. Using a software package with linear algebra capabilities, such as Matlab or MATHEMATICA, one can calculate the spectral decomposition of the 2×2 upper left corner of (1.10) as

$$\begin{bmatrix} -0.08 & 0.05 \\ 0.06 & -0.07 \end{bmatrix} = \begin{bmatrix} -0.707107 & -0.640184 \\ 0.707107 & -0.768221 \end{bmatrix} \begin{bmatrix} -0.13 & 0 \\ 0 & -0.02 \end{bmatrix} \times \begin{bmatrix} -0.771389 & 0.642824 \\ -0.710023 & -0.710023 \end{bmatrix}. \tag{1.11}$$

The middle matrix on the right has eigenvalues on the diagonal, the columns of the first matrix on the right are the corresponding eigenvectors, and the last matrix on the right is the inverse of the first. Using the starting value $\mathbf{p}(0) = (1, 0)^T$, the solution (1.9) gets the form

$$\mathbf{p}(x) = \begin{bmatrix} 0.545455e^{-0.13x} + 0.454545e^{-0.02x} \\ -0.545455e^{-0.13x} + 0.545455e^{-0.02x} \end{bmatrix}. \tag{1.12}$$

In general the decomposition (1.11) might involve complex eigenvalues and eigenvectors, but the solution (1.12) is always real. Note that the second component of (1.12) is a nonmonotone function of x. ◊

Apart from the special cases, there is no analytical solution to (1.4). A formal solution in terms of the so-called *product integral* is available (cf., Gantmacher 1959; Andersen et al. 1993, 88–95), but for a numerical solution we can work directly with (1.4). A number of methods for solving it have been proposed (e.g., Schoen 1988, 75). Other than the constant hazards assumption, the most popular is based on the assumed linearity of the solution. As noted by Rogers (1995, 96), this can be viewed as a generalization of the linearity assumption in the single region case (Example 2.2 and Exercise 9 of Chapter 4).

Example 1.3. Hazards Producing a Linear Solution. Suppose that (1.4) has a solution of the form $\mathbf{p}(x) = (\mathbf{I} + x\mathbf{B})\mathbf{a}$ that has (componentwise) $\mathbf{0} \le \mathbf{p}(x) \le \mathbf{1}$ for $x \in [0, 1]$. Then we must have $\mathbf{p}(0) = \mathbf{a}$ with $\mathbf{0} \le \mathbf{a} \le \mathbf{1}$. Also $\mathbf{p}'(x) = \mathbf{Ba}$, so we must have $\boldsymbol{\nu}(x)(\mathbf{I} + x\mathbf{B})\mathbf{a} = \mathbf{Ba}$. As any J linearly independent vectors $\mathbf{a}$ with $\mathbf{0} \le \mathbf{a} \le \mathbf{1}$ actually span the whole space $\mathbb{R}^J$, it follows that $\boldsymbol{\nu}(x)(\mathbf{I} + x\mathbf{B}) = \mathbf{B}$, or $\boldsymbol{\nu}(x) = \mathbf{B}(\mathbf{I} + x\mathbf{B})^{-1}$ provided that the inverse exists. Hence, we have $\mathbf{B} = \boldsymbol{\nu}(0)$. The linearity assumption may provide a reasonable numerical approximation in many situations, but Hoem and Funck-Jensen (1982, 198–200) point out several short-comings. ◊

Given that closed-form analytical solutions are not to be had, for practical computation we will resort to the Runge-Kutta method (1.1). This method will easily extend to handle time-varying covariates. To solve the system of differential equations (1.4), in vector notation $\mathbf{y}' = \mathbf{f}(x, \mathbf{y})$, we substitute $\mathbf{y} = \mathbf{p}(x)$ and $\mathbf{f}(x, \mathbf{y}) = \boldsymbol{\nu}(x)\mathbf{p}(x)$. A technical issue that comes up is that the algorithm does not automatically ensure that $\mathbf{1}^T\mathbf{p}(x) = 1$ or that $\mathbf{0} \le \mathbf{p}(x) \le \mathbf{1}$. Adjustments to satisfy these conditions can be made during each round of Runge-Kutta iteration. If the problems are severe a shorter step size may be adopted.

We note that (1.1) can be started at any age x_0 by taking an arbitrary starting value for $\mathbf{p}(x_0)$, such as $\mathbf{p}(x_0) = \mathbf{e}_j$ for some j, and solving for $\mathbf{p}(x)$ when $x > x_0$. Any life table quantity can thus be obtained. For example, suppose an individual is born into one of states $j = 1, \ldots, J$ with probabilities given by the components of the vector $\mathbf{p}(0)$. In analogy with (2.7) of Chapter 4, the vector of expected years spent in different states over his or her life time is

$$\mathbf{e}_0 = \int_0^\infty \mathbf{p}(x)\, dx, \tag{1.13}$$

where the integration is performed element by element. In the case of Example 1.2 the life expectancy of 50 years becomes divided into two parts: 26.9 years spent working and 23.1 years unemployed. To verify, note that the integral of the first component of (1.12) over x in $(0, \infty)$ equals $0.545455/0.13 + 0.454545/0.02 = 26.9231$, for example.

For life table construction we need conditional life expectancies by state. Define ${}_z\mathbf{p}_j(x) = E[\mathbf{I}(x + z)|\mathbf{I}(x) = \mathbf{e}_j]$, i.e., it is the vector of probabilities of being in different states in age $x + z$, given that the person was in state j at exact age x. This vector of probabilities can be calculated for any z using Runge-Kutta, taking ${}_0\mathbf{p}(x) = \mathbf{e}_j$ as the initial value. We can then define the vector of state-specific remaining life expectancies, conditionally on $\mathbf{I}(x) = \mathbf{e}_j$, as

$$\mathbf{e}_j(x) = \int_0^\infty {}_z\mathbf{p}_j(x)\, dz. \tag{1.14}$$

This is a multi-state generalization of the e_x defined by formula (2.8) of Chapter 4.

In multistate forecasting, considerations similar to those discussed in Section 3 of Chapter 4 apply. Let $k_j(t)$ be the density of population in age t in state j. The expected survivors to different states from those who were in state j in age $[x, x + 1)$ one year earlier are given by

$$\int_x^{x+1} {}_1\mathbf{p}_j(t) k_j(t)\, dt. \tag{1.15}$$

Generalizing (3.6) of Chapter 4, we may define the vector of average survival probabilities to age $[x+1, x+2)$ as

$$
{}_1\bar{\mathbf{p}}_j(x) = {}_1\mathbf{p}_j(x+1)\frac{2k_j(x+1)+k_j(x)}{3(k_j(x+1)+k_j(x))} + {}_1\mathbf{p}_j(x)\frac{2k_j(x)+k_j(x+1)}{3(k_j(x+1)+k_j(x))}. \tag{1.16}
$$

Define $K_{j,x}$ as the size of the population of state j who are in age $[x, x+1)$ at a given moment. Then, the vector of expected survivors in age $x+1$ one year later is

$$
\sum_{j=1}^{J} {}_1\bar{\mathbf{p}}_j(x) K_{j,x}. \tag{1.17}
$$

For each x we go over the states, and then move to $x+1$.

1.3. Duration-Dependent Life Tables

As above, we consider states $j = 1, \ldots, J$ with an indicator vector $\mathbf{I}(x) = (I_1(x), \ldots, I_J(x))^T$, where $I_j(x) = 1$ if an individual is in state j in age $x \geq 0$ and $I_j(x) = 0$ otherwise. Defining $\mathbf{p}(x) = E[\mathbf{I}(x)]$, we have a vector of probabilities of being in different states. A *multistate life table* is simply a set of tabulated values of $\mathbf{p}(x)$ and some of its functionals, such as (1.14). The overall aim of the table is to summarize the transition conditions of a chosen time period. Unfortunately, tabulating such probabilities and state-specific expected waiting times is cumbersome when starting ages and states vary.

Another aspect that sets a multistate life table apart from the single state life table is the possible presence of population heterogeneity associated with past event history. Heterogeneity may, in principle, arise from any aspect of past state transitions, as illustrated in Section 4.3.3 of Chapter 4.

1.3.1. Heterogeneity Attributable to Duration

In this section we will develop a theory that can take certain aspects of duration into account. By *duration* we may refer to the total time spent in a given state, to the length of the last visit in a given state, or more generally, to any positive functional of the sojourn times in a given state, such as those given by (7.9) of Chapter 5.

Example 1.4. Remarriage Probability Varies with Time Spent Non-married. Figure 1 shows how the average relative risk of remarriage is related to duration since end of marriage for those whose marriage ended due to divorce and for those who became widowed, among women in Finland in 1998. (Here, the baseline against which relative risk is measured is average intensity of marriage in a given age; in Figure 1 average relative risk by duration is obtained by averaging such relative risk estimates over age.) For the divorced the relative risk of a new marriage declines rapidly. This is consonant with the notion that finding a new spouse is

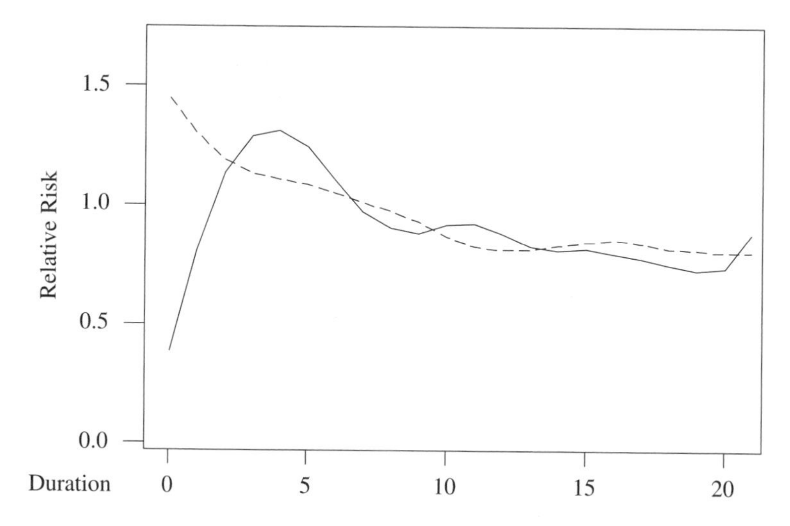

FIGURE 1. Average Relative Risk of Remarriage Among Widowed (Solid) and Divorced (Dashed) as a Function of the Duration of Widowhood and Divorce, Respectively.

often a cause of divorce. For the widowed the relative risk is below 1 for short durations, but increases to about three in durations of 3–4 years, and declines to one thereafter. That is, the effect of duration is *not* multiplicative between the two populations. Although we do not show the details here, we note that among the widowed the relative risk is roughly the same in each age. However, among the young divorced the relative risk of a new marriage increases with the duration, so a multiplicative model incorporating age and duration is not appropriate among the divorced. These examples illustrate the limitations of the proportional hazards model. ◊

1.3.2. Forms of Duration-Dependence

It is difficult (but not impossible, cf., Wolf 1988) to accommodate duration effects into the calculation of life tables analytically, because we have a case of time varying covariates (cf., Section 7 of Chapter 5). It is easier to resort to simulation. If proportional hazards are appropriate, one can estimate duration effects via Poisson regression or via Cox regression (cf., Sections 3 and 7 of Chapter 5). Or, more general hazard models can be used that allow for the interaction of duration and age. Given the hazard estimates, one can simulate state transitions individual by individual. In this manner a collection of state transition paths can be formed. It is then a matter of simple arithmetic to estimate relevant probabilities and expectations. We will now describe both some theoretical and practical issues that come up when implementing a multistate model[1].

[1] Based on our experiences in developing the C++ program MTABLE at the University of Joensuu.

The starting point is the differential equation $\mathbf{p}'(x) = \boldsymbol{\nu}(x)\mathbf{p}(x)$ in (1.4). Define $\mathbf{D}(x) = (D_1(x), \ldots, D_J(x))^T$ as the vector of durations at age x. At least two possible concepts of duration seem relevant. One can choose $D_j(x)$ either as *time ever spent in j by age x* or as *time spent during the current visit*[2] *in j by age x*. The usual Cox model for the effect of duration assumes that

$$\nu_{ij}(x, \mathbf{D}(x)) = \nu_{0ij}(x)\exp(\boldsymbol{\beta}_{ij}^T\mathbf{D}(x)), \tag{1.18}$$

where $\nu_{0ij}(x)$ is the baseline intensity of those with $\mathbf{D}(x) = \mathbf{0}$. Note that $\boldsymbol{\beta}_{ij}$ is a vector. If only the duration $D_j(x)$ of the current sojourn is relevant, then a general proportional hazards model assumes that there are functions $g_{ij}(.) \geq 0$ such that

$$\nu_{ij}(x, \mathbf{D}(x)) = \nu_{0ij}(x)g_{ij}(D_j(x)). \tag{1.19}$$

A general duration-dependent intensity model assumes that the intensities are of the form $\nu_{ij}(x, d)$ with $0 \leq d \leq x$, and the intensity for a person with duration $d = D_j(x)$ at age x is $\nu_{ij}(x, D_j(x))$.

A possible problem in the proportional hazards formulations derives from the imbalance in the data. To simplify, suppose that only the duration d of current sojourn matters. Omitting dependency on i and j, the model (1.18) is equivalent to a main effects log-linear model $\log \nu(x, d) = \alpha_x + \beta_d$. While this may be a realistic model in some situations it is good to remember that our intuition from ordinary 2-way analysis of variance does not carry over, as such, to this case, because for ages $x = 0, \ldots, \omega$ the possible values of duration are also $d = 0, \ldots, \omega$, but we have to have $d \leq x$. Thus, estimates of β_d for short durations depend on most ages, but estimates for long durations depend only on the oldest ages.

1.3.3. Aspects of Computer Implementation

The model (1.4) is in continuous time, so in principle an unlimited number of state transitions are possible during a time unit. We can always approximate the process by taking the time unit small enough so that the possibility of more than one transition can be ignored. Suppose an individual starts at exact age $x = 0, 1, \ldots, \omega - 1$, at state j. First we use the Runge-Kutta method to calculate the vector of probabilities ${}_1\mathbf{p}_j(x) = E[\mathbf{I}(x+1)|\mathbf{I}(x) = \mathbf{e}_j]$. Then we select the state at $x + 1$ randomly using ${}_1\mathbf{p}_j(x)$.

If a transition to k from state j occurs, then the time spent in j must be specified. As a first approximation we may choose the time of transition from a uniform distribution $U[0, 1]$. This is equivalent to the assumption of Example 1.3. To refine, one could use information concerning the derivative of the solution at the end points of the interval (Exercise 5). If the randomly chosen state at $x + 1$ is also j, then one time unit is added to the time spent in j.

Repeating the above procedure we obtain a path consisting of state transitions and their times of occurrence. One can keep track of such characteristics of the

[2] This particular case is a so-called *age-dependent semi-Markov model*, i.e., transition intensities depend on state, age, and duration of current sojourn (Mode 1985, 244–245).

paths that are of interest and store them for further processing. Using the output one might wish to answer following types of questions:

(i) given that the person is in state j in age t, what is the probability that he or she is in state k at age $u > t$;
(ii) given that the person is in state j in age t, what is the distribution of time the person spends in state k by age $u > t$;
(iii) given that the person is in state j in age t, what is the distribution of the waiting time until next entry to state k;
(iv) given that the person is in state j in age t and enters state k in some age, what is the probability that he or she exits state k via state $h \neq k$?

In all cases the answer can be numerically determined from a simulated probability distribution of the variable of interest. Summary measures such as the expectation, the standard deviation, or tail probabilities can also be calculated based on the distribution.

1.3.4. Policy Significance of Duration-Dependence

Exposure distributions or *duration distributions* can have considerable significance in social policy. Consider long-term unemployment, for example. The chance of becoming unemployed may depend on population heterogeneity. Some people may find work (or loose a job) more easily than others because their knowledge, skills, and attitudes. On the other hand, being unemployed (or getting a job) may be due to luck. If the chance of finding a new job decreases with the duration of unemployment, bad luck may accumulate (cf., discussion of "randomness and predestination" at the end of Section 2.4 of Chapter 4 and Section 8 of Chapter 5). In the first case, remedial training might be an effective measure for improving the job opportunities of the unemployed. In the latter case remedial measures may not help, and the unemployed might be best helped with insurance mechanisms, as in the case of disability, for example. Exposure distributions from duration-dependent multistate life tables can show us whether chance alone could explain the observed exposure distributions.

1.4. Nonparametric Intensity Estimation

The estimation of the transition intensities is challenging because a multistate population with J states can logically have up to $J(J-1)$ transition flows for each age. Each flow may have idiosyncratic characteristics (e.g., mortality as compared to the remarriage of widows). Different methods may turn out to be optimal for each. For a general discussion, see Hoem and Funck Jensen (1982), and Andersen et al. (1993). We present two nonparametric approaches that rely on local linearity (Section 2.4 of Chapter 4) and kernel smoothing (Section 9 of Chapter 5). More general graduation methods are discussed by Keyfitz (1977, Ch. 10) and nonparametrics by Härdle (1990) and Green and Silverman (1994). The nuptiality example of Section 1.5 provides the background for our discussion of the general duration-dependency model. Duration refers to duration during current sojourn and is truncated to the nearest lower integer.

Let $N_{ij}(t, d)$ be the number of transitions from state j to i, in exact age that belongs to interval $[t, t+1)$, given that duration in the beginning of the year was in $[d, d+1)$, $d = 0, \ldots, t$. Let $K'_j(t, d)$ be the number of individuals in the age $\times$ duration category in the beginning of the year and $K''_j(t, d)$ the number of individuals in the age $\times$ duration category at the end of the year. Person years can then be approximated as $K_j(t, d) = (K'_j(t, d) + K''_j(t, d))/2$, and the corresponding *o/e* rate is $\nu_{ij}(t, d) = N_{ij}(t, d)/K_j(t, d)$. Given the large number of pairs (t, d), the *o/e* rates may be unstable. Computation of local averages can often provide a smoother estimate.

Preliminary analyses suggest that in the flows we consider age effects are larger than duration effects. Therefore, we will adopt a two-stage estimating strategy, trying first to get the age effects right under as few assumptions as possible. A separate estimation of duration effects under smoothness assumptions is presented afterwards. Let us write $\nu_{ij}(t, d) = \nu_{ij}(t)\psi_{ij}(t, d)$, where

$$\sum_{d=0}^{t} \psi_{ij}(t, d) K_j(t, d) \Big/ \sum_{d=0}^{t} K_j(t, d) = 1. \tag{1.20}$$

Thus, $\nu_{ij}(t)$ is the *average intensity* at t, and $\psi_{ij}(t, d)$ is the *relative risk* at duration d. Define $N_{ij}(t) = \Sigma_d N_{ij}(t, d)$, so that $v_{ij}(t) = N_{ij}(t)/K_j(t)$ is an *o/e* rate.

Consider exact ages $t = 1, 2, \ldots, \omega - 1$. Emulating the approach of Section 2.4 of Chapter 4, consider the interval $[t-1, t+1)$. Suppose that the average rate and the population density are locally linear. One can then deduce (Exercise 8) that the estimator

$$\hat{\nu}_{ij}(t) = \frac{N_{ij}(t-1) + N_{ij}(t)}{K_j(t-1) + K_j(t)} - \frac{2(\nu_{ij}(t-1) - \nu_{ij}(t))(K_j(t-1) - K_j(t))}{3(K_j(t-1) + K_j(t))}. \tag{1.21}$$

corrects for both linear effects at exact age t. Having estimates available at exact values of t, we can use any interpolation technique (such as the Karup-King formula, cf. Shryock and Siegel 1976, 554) to estimate the ages $t + 0.5$.

One way to estimate the relative risk parameters is to use kernel smoothing. Fix t and d. Using a Gaussian kernel with smoothing parameter $h > 0$, we obtain the following estimate for the relative risk at $d = 0, \ldots, t$, as compared to the average risk at t,

$$\hat{\psi}_{ij}(t, d|h) = \sum_{s=d}^{\omega} \nu_{ij}(s, d) K_j(s, d) \exp\left(-\frac{(s-t)^2}{2h^2}\right) \Big/ \sum_{s=d}^{\omega} \hat{\nu}_{ij}(s) K_j(s, d) \exp\left(-\frac{(s-t)^2}{2h^2}\right). \tag{1.22}$$

Since $\nu_{ij}(t, d)K_j(t) = N_{ij}(t, d)$, the estimator is of the form "observed count $\div$ expected count", or it is a nonparametric form of indirect standardization (Section 3.3 of Chapter 5). For a given d, (1.22) weights *o/e* rates in different ages according to how far they are from t, and by person years. Conditioning on h, a rough confidence interval for $\psi_{ij}(t, d|h)$ can be obtained by estimating $\text{Var}(\nu_{ij}(t, d))$ by $\nu_{ij}(t, d)/K_j(t, d)$.

Cross-validation can be used to choose h (e.g., Härdle 1990). Define predicted relative risk at t and d, $\tilde{\psi}_{ij}(t, d|h)$, by (1.22) with *the summation restricted* in both numerator and denominator *to* $s = d, \ldots, \omega$ with $s \neq t$. Define the corresponding predicted residuals as

$$e_{ij}(t, d|h) = N_{ij}(t, d) - \tilde{\psi}_{ij}(t, d|h)\hat{\nu}_{ij}(t)K_j(t, d). \tag{1.23}$$

A *cross-validation estimator* of the smoothing parameter is a value of h that minimizes the sum of squared predicted residuals for some set of values of (t, d). In the application of Section 1.5 discussed next we searched for a value $h = h(t)$ for each t, that minimizes the sum,

$$\sum_{d=0}^{t} e_{ij}(t, d|h)^2, \tag{1.24}$$

for example.

1.5. Analysis of Nuptiality

What is the *probability that a marriage ends in a divorce*? As a multiple decrement process a person's marriage can end in a divorce or upon death of either spouse. In popular press, one frequently sees estimates relating the number of divorces to the number of new marriages in a given year. This practice can be approximate at best, since (a) current divorces do not come from the same cohorts as the current marriages, and (b) both past divorces and marriages influence the measure. Statistical agencies sometimes calculate a "probability of divorce" in year t by adding the fractions of those marriages formed during each of the years $y < t$ that ended by divorce during year t. For example, the official statistics of Finland use this measure, and around year 2000 the probability of a marriage ending in a divorce is claimed to be about 50%. The measure is a bit analogous to the total fertility rate (cf., Shryock and Siegel 1976, 346) but, unfortunately, patterns of past divorces can bias this measure.

We have analyzed the nuptiality of the Finnish women in 1998 (using the program MTABLE). The states of the system are Single, Married, Divorced, Widowed, and Dead (cf., Figure 2).

The total number of person years coming from the four living states were $N = 601{,}100 + 1{,}004{,}000 + 234{,}800 + 269{,}000 = 2{,}108{,}900$. With five states there are potentially $5 \times 4 = 20$ flows, but in the case of nuptiality, only nine

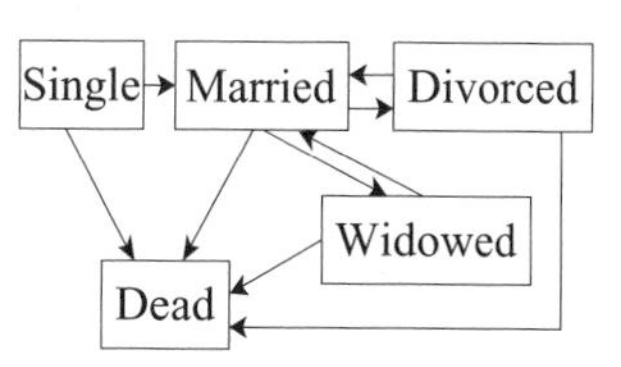

FIGURE 2. Possible State Transitions in Nuptiality Processes.

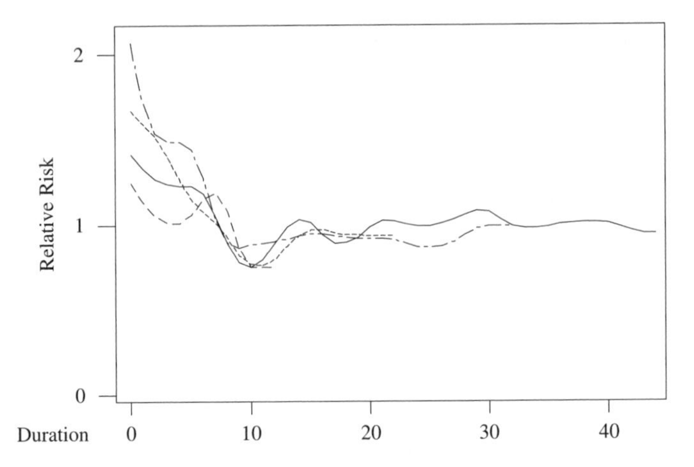

FIGURE 3. Relative Risk of Death Among Married as a Function of the Duration of Marriage: Average (Solid), in Age 30 (Dashed), in Age 40 (Dotted), and in Age 50 (Dash-Dotted).

are logically possible. Except for the flow from Single to Married, the intensities may depend separately on age and on duration. As discussed in Example 1.4, a proportional hazards assumption is not appropriate for all flows. Our results are based on the general duration-dependent intensity model.

Data on state transitions were available from year 1998, by age $x = 17, \ldots, 99$ and duration $d = 0, \ldots, x - 17$. The estimation consisted of three steps. (1) Estimates of average intensity were calculated with (1.21) for exact ages $x = 17, 18, \ldots, 100$, based on data from the two neighboring ages, when available. (2) For each age, estimates of relative risk (1.22) were calculated. The smoothing parameter was determined by minimizing (1.24) for each age. Values were restricted to range $2 \leq h \leq 10$ on *a priori* grounds. A comparison to estimates obtained with fixed values $h = 5.0$ and $h = 7.5$ showed that the estimates of transition intensities were insensitive to the exact value of the smoothing parameter. (3) The relative risk estimates were further smoothed across duration (using RSMOOTH of Minitab) for each age.

Consider mortality (cf., Figure 1 of Chapter 4). For the divorced and the widowed the duration effects (not shown) are relatively small, but we see in Figure 3 that for the married there are systematic effects. Short marriage durations are associated with high relative risk of mortality. The effect is more pronounced in older ages than younger ages. Since most of the marriages occur in ages 20–30, the finding is consonant with the notion that those who marry atypically late initially experience a relatively high level of mortality which then declines as the duration of marriage increases.

An analysis of the intensity of widowhood (or equivalently of husband's death) has a similar pattern, but the dependency on duration is even stronger (details not shown). Since spouses are of a roughly similar age, this indicates, that male mortality is similarly associated with the duration of marriage. We speculate that marriage can act as a selection mechanism that first tends to select those who are

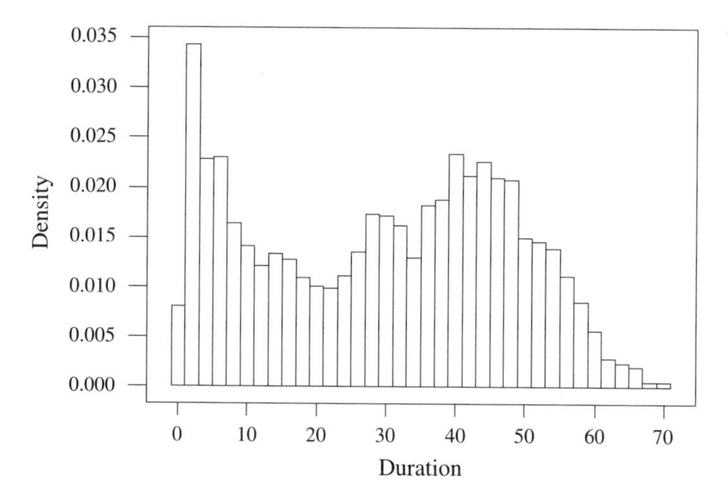

FIGURE 4. Distribution of Time Spent in the Divorced State, if Ever Divorced, for a Single at Age 17.

relatively healthy, due to genetics or life style, but does not provide much additional protection. The genetic make-up or life style of those who are left out or divorce may entail greater risks of a kind that a later marriage may reduce.

Returning to the question of the probability that a marriage ends in a divorce, we can simply repeatedly begin a nuptiality history in age 17 in the Single state, calculate the number of times entry into Marriage occurs, calculate the number of times entry into Divorce occurs, and divided the latter by the former. This *life table probability of divorce* comes out 39%, considerably less than the official figure of 50%.

To illustrate other statistical characteristics, consider the time a woman will spend in the divorced state, conditionally on her becoming divorced at all. Figure 4 has a simulated probability distribution for the time spent in divorce. We see that the distribution is (essentially) bimodal. This is also a multiple decrement phenomenon, in which the first mode is primarily due to those who remarry soon after the divorce. The latter mode is primarily due to those who do not remarry, but exit the state of Divorce via death.

1.6. A Model for Disability Insurance

To indicate the broad applicability of the simulation approach to the multistate setting, consider a model for disability insurance. For a general discussion see Haberman (1999); here we consider a highly stylized setting. Suppose there are $J = 4$ states: $j = 1$ Employed; $j = 2$ Unemployed or outside the labor force but able to work; $j = 3$ Disabled; $j = 4$ Dead. Consider an individual born into state $j = 2$ at time t, who is in state $\mathbf{I}(x)$ at $t + x$. Suppose the salary of the individual in age x is of the form $s(x, d)$ given that he or she has worked $d \leq x$ years in his or her life time. A fraction $0 < c < 1$ is paid as a premium for disability insurance.

Instead of a fixed benefit, suppose that the benefit is equal to $b(d)$, if the number of years worked is d when the entry to the state of disability occurs. How should c be determined if the interest rate at time $t+x$ is $\rho(t+x)$?

Suppose the times of entry into Employed are $0 \leq Y_1 < Y_2 < \cdots$ with respective durations Z_i. Suppose the cumulative duration or employment before the ith entry is H_i, with $H_1 = 0$ and $H_i = Z_{i-1} + \cdots + Z_1$ otherwise. At birth, the discounted value of the entire salary is

$$S = \sum_{i=1}^{\infty} \int_0^{Z_i} s(Y_i + x, H_i + x) \exp\left(- \int_{Y_i}^{Y_i+x} \rho(u)\, du \right) dx. \tag{1.25}$$

Similarly, suppose the ages of entry into Disability are $0 \leq X_1 < X_2 < \cdots$. with durations $D_1, D_2, \ldots$ and H_i^* years worked before the ith entry. Then, the total value of the discounted benefits is

$$B = \sum_{i=1}^{\infty} b(H_i^*) \int_{X_i}^{X_i+D_i} \exp\left(- \int_0^x \rho(u)\, du \right) dx. \tag{1.26}$$

Since the times of entries to and exists from the various states are random, both S and B are random variables. The integrals involving interest rates in (1.25) and (1.26) can be evaluated numerically.

To calculate the expectations of S and B, we can independently generate paths $i = 1, \ldots, N$, calculate the value S_i of (1.25) and B_i of (1.26) for each, and then take the averages $\bar{S} = (S_1 + \cdots + S_N)/N$ and $\bar{B} = (B_1 + \cdots + B_N)/N$. Equating the two expected values, we can determine the premium as fraction $c = \bar{B}/\bar{S}$. Much more complex benefit, salary, and payment schemes can be accommodated in a similar manner. In addition, we may let the interest rates $\rho(x)$ to be random.

2. Linear Growth Model

2.1. Matrix Formulation

The book-keeping of population change can be based on several slightly different ways of data collection. Rather than pursue generality, we will give one set of definitions that will be consonant with the estimation theory of Chapter 4. We first define how time, age, and region are to be understood. Then, we proceed to develop the necessary arithmetic in matrix form.

We will assume that the *same units are used for age and time*. Typically the unit will be one year. Sometimes forecasters wish to enter less data by using five-year age groups (or ages 0, 1–4, 5–9, 10–14, ...). The theory we present assumes that such data have been interpolated into one year age-groups. The population of year t will refer to the population existing at a single point in time. We will assume this is the *beginning* of the year, or January 1, year t. (Note that some countries use the end of the year in their official statistics!) The *jump-off population* will be the population of year $t = 0$. This is the population that one wishes to treat as the latest known population. The vital rates of year t (relating to births, deaths, and migration) will refer to time $[t, t+1)$. The first forecasted births, deaths, and migrations will then occur during year $t = 0$, and the first forecasted population will be that of year $t = 1$.

Age $x = 0$ refers to those whose exact age is in the interval $[0, 1)$, age $x = 1$ refers to the interval $[1, 2)$ etc. The *highest possible age* is denoted by ω, and it refers to the open-ended interval $[\omega, \infty)$. Therefore, there are $\omega + 1$ ages in all. Births are attributed to women only. The *lowest age of childbearing* is α, and the *highest age of childbearing* is β (cf., Section 4.2 of Chapter 4). We will assume that $0 < \alpha < \beta < \omega$.

Population sizes of year t are denoted by a vector of the form

$$\mathbf{V}(t) = (\mathbf{V}(0, t)^T, \ldots, \mathbf{V}(\omega, t)^T)^T. \tag{2.1}$$

Three different interpretations will be given to the vector depending on the context. First, suppose we have a female population of a single region. In that case $\mathbf{V}(x, t)$ is a scalar giving the number of women in age x. Second, suppose we have a population consisting of both males and females. Then, $\mathbf{V}(x, t) = (V_1(x, t), V_2(x, t))^T$, where $V_1(x, t)$ is the number of females in age x and $V_2(x, t)$ is the number of males in age x. Third, suppose we have a closed system consisting of males and females from regions $j = 1, \ldots, J$. We can then write $\mathbf{V}(x, t) = (\mathbf{V}_1(x, t)^T, \mathbf{V}_2(x, t)^T)^T$, where $\mathbf{V}_1(x, t) = (V_{11}(x, t), \ldots, V_{1J}(x, t))^T$ and $V_{1j}(x, t)$ is the number of females in age x, in region $j = 1, \ldots, J$. In analogy, we write for males $\mathbf{V}_2(x, t) = (V_{21}(x, t), \ldots, V_{2J}(x, t))^T$.

The cohort-component arithmetic of all three cases can be written in matrix form as

$$\mathbf{V}(t+1) = \mathbf{R}(t)\mathbf{V}(t), \tag{2.2}$$

once the matrix $\mathbf{R}(t)$ has been properly defined. The assumption required for (2.2) to hold is that, in each case, the population is closed. An extension allowing for migration will be given below. We will call (2.2) the *linear growth model*.[3]

Define $\mathbf{R}(t)$ in terms of blocks, $\mathbf{R}(t) = (\mathbf{R}(x, y, t))$, where $x, y = 0, 1, \ldots, \omega$. In all cases $\mathbf{R}(x, y, t) = \mathbf{0}$, unless $x = 0$ and $\alpha \le y \le \beta$; or $y = x - 1$; or $x = y = \omega$. In other words, the matrices are of the form (cf., Feeney 1970),

$$\mathbf{R}(t) = \begin{bmatrix} \mathbf{0} & \cdots & \cdots & \mathbf{0} & \mathbf{R}(0,\alpha,t) & \cdots & \mathbf{R}(0,\beta,t) & \mathbf{0} & \cdots & \mathbf{0} \\ \mathbf{R}(1,0,t) & \mathbf{0} & \cdots & \cdots & \cdots & \cdots & \cdots & \cdots & \cdots & \mathbf{0} \\ \mathbf{0} & \mathbf{R}(2,1,t) & \mathbf{0} & \cdots & \cdots & \cdots & \cdots & \cdots & \cdots & \mathbf{0} \\ \mathbf{0} & \mathbf{0} & \mathbf{R}(3,2,t) & \mathbf{0} & \cdots & \cdots & \cdots & \cdots & \cdots & \mathbf{0} \\ \vdots & \vdots & \vdots & \vdots & \vdots & \vdots & \vdots & \vdots & \vdots & \vdots \\ \mathbf{0} & \cdots & \cdots & \cdots & \cdots & \cdots & \cdots & \mathbf{0} & \mathbf{R}(\omega,\omega-1,t) & \mathbf{R}(\omega,\omega,t) \end{bmatrix}. \tag{2.3}$$

[3] In time series analysis the same term is sometimes used differently, to describe a state-space model with a linear trend (e.g., Chatfield 1996, 184).

In the case of female population we would have $\mathbf{R}(0, x, t) =$ expected number of girls, born during t per woman in age x, that survive to the beginning of next year; $\mathbf{R}(x, x-1, t) =$ proportion of survivors from age $x-1$ at t to age x at $t+1$; and $\mathbf{R}(\omega, \omega, t) =$ proportion of survivors in age ω.

If males are included we would have

$$R(0, x, t) = \begin{bmatrix} R_1(0, x, t) & 0 \\ R_2(0, x, t) & 0 \end{bmatrix}, \tag{2.4}$$

where $R_1(0, x, t) =$ expected number of girls, born during t per woman in age x, that survive to the beginning of next year, and $R_2(0, x, t) =$ expected number of boys, born during t per woman in age x, that survive to the beginning of next year. For survival we would have

$$R(x, x-1, t) = \begin{bmatrix} R_1(x, x-1, t) & 0 \\ 0 & R_2(x, x-1, t) \end{bmatrix}, \tag{2.5}$$

where $R_1(x, x-1, t)$ gives the female proportion of survivors from age $x-1$ to x during t, and $R_2(x, x-1, t)$ gives the corresponding proportion for males. $\mathbf{R}(\omega, \omega, t)$ is defined analogously.

Finally, in the multiregional case $\mathbf{R}(0, x, t)$ is a $2J \times 2J$ matrix consisting of four blocks, as in (2.4). Each block is a $J \times J$ matrix. The matrix $\mathbf{R}_1(0, x, t)$ has the form

$$\mathbf{R}_1(0, x, t) = \begin{bmatrix} R_{111}(0, x, t) & R_{112}(0, x, t) & \dots & R_{11J}(0, x, t) \\ R_{121}(0, x, t) & R_{122}(0, x, t) & \dots & R_{12J}(0, x, t) \\ \vdots & \vdots & \vdots & \vdots \\ R_{1J1}(0, x, t) & R_{1J2}(0, x, t) & \dots & R_{1JJ}(0, x, t) \end{bmatrix}, \tag{2.6}$$

where $R_{1ij}(0, x, t)$ expected number of girls born to women in age x in region j during t that are alive in region i at the end of the year. Matrices $\mathbf{R}_2(0, x, t) = (R_{2ij}(0, x, t))$ for boys are similarly defined. The remaining two blocks are $J \times J$ matrices of all zeroes. For survival, $2J \times 2J$ matrices of the form (2.5) are defined where $J \times J$ matrices $\mathbf{R}_1(x, x-1, t)$ have the (i, j) elements $R_{1ij}(x, x-1, t) =$ proportion of women in age $x-1$ in region j at t that survive to region i at the end of the year, as in (2.6). Definitions for males are similar.

Assuming that we have an estimate of the jump-off population $\hat{\mathbf{V}}(0)$ and that we have forecasts $\hat{\mathbf{R}}(t)$ for $t = 0, \dots, T-1$, then the *cohort-component forecast* of $\mathbf{V}(T)$ is simply

$$\hat{\mathbf{V}}(T) = \hat{\mathbf{R}}(T-1) \cdots \hat{\mathbf{R}}(0)\hat{\mathbf{V}}(0). \tag{2.7}$$

We conclude with three comments relating to the generation of births in computer simulations. First, it is common that births are generated using age-specific fertility rates. In all cases the probability of a child's survival to the end of the year must be accounted for. Second, if the forecast is based on *o/e* rates, then the proper multiplier is the number of person years during the year rather than the population in the beginning of the year. In practice, survival of women can be simulated and then person years can be calculated. Thus, a correct calculation can be made.

However, when this is done, (2.2) does not exactly represent the actual calculation. Third, it is conventional to attribute births to women only. Logically, they could equally well be attributed to men, but women appear to be preferred for ease of data collection. From this perspective we are using a so-called *female dominance* model. This is a particular solution to the so-called *two-sex problem* that is particularly relevant when, instead of births, one considers how the incidence of new marriages is best to be modeled (e.g., Goodman 1967, McFarland 1972, Pollard 1975, Schoen 1988).[4]

Example 2.1. Two-Sex Problem. Fix a calender year and let $Y_{xy} \sim \text{Po}(\lambda_{xy} K_{xy})$ be the number of marriages among females of age x and males of age y. Suppose there are N_x females and M_y males at risk of marriage. The intensity of marriage in the two ages is estimated as $\hat{\lambda}_{xy} = Y_{xy}/K_{xy}$, but how should we think about K_{xy}? Suggestions include $K_{xy} = N_x$ (female dominance); $K_{xy} = M_y$ (male dominance); $K_{xy} = (N_x + M_y)/2$ (arithmetic mean); $K_{xy} = (N_x M_y)^{1/2}$ (geometric mean); $K_{xy} = N_x M_y/(N_x + M_y)$ (harmonic mean), etc. No suggestion has found universal acceptance, however. Empirical evidence shows that there are "marriage circles" defined by socio-economic factors and adopted life style, within which spouses are typically found (Henry 1972, Bozon and Heran 1989). This heterogeneity is not explicitly considered in the classical proposals. Thus, one model may be a good approximation in one cultural or geographic setting but another model may be better in another (Alho, Saari and Juolevi 2000). ◊

2.2. *Stable Populations*

In Section 2.2.2 of Chapter 4 we introduced the concept of stable population in connection with life tables. For some purposes, such as forecasting, stable population theory is relatively unimportant because, unrealistically, it assumes that the vital rates remain constant over time. Yet, the concepts of asymptotic growth rate and asymptotic age-distribution are useful for understanding the long-term implications of current rates. We will now develop the stable population theory in the multistate case, based on the matrix representation (2.2).

Suppose we have $\mathbf{R}(t) = \mathbf{R}$ for all $t = 0, 1, 2, \ldots$, where $\mathbf{R}$ is a real-valued $m \times m$ matrix of the form (2.3). In case of a female population, $m = \omega + 1$; in case of a two-sex population we have $m = 2(\omega + 1)$; and in case of a J region population we have $m = 2J(\omega + 1)$. The matrix $\mathbf{R}$ has m eigenvalues γ_i and m linearly independent right eigenvectors $\mathbf{w}_i \neq \mathbf{0}$ that satisfy the equation $\mathbf{R}\mathbf{w}_i = \gamma_i \mathbf{w}_i$. Since $\mathbf{R}$ is not symmetric, it has separate linearly independent left eigenvectors $\mathbf{u}_i \neq \mathbf{0}$ such that $\mathbf{u}_i^T \mathbf{R} = \gamma_i \mathbf{u}_i^T$. Define $\mathbf{\Gamma} = \text{diag}(\gamma_1, \ldots, \gamma_m)$, $W = [\mathbf{w}_1, \ldots, \mathbf{w}_m]$, and $\mathbf{U} = [\mathbf{u}_1, \ldots, \mathbf{u}_m]$. A left and a right eigenvector that correspond to different eigenvalues are orthogonal, and they can be normalized so that $\mathbf{U}^T \mathbf{W} = \mathbf{I}$. It then follows that $\mathbf{R}$ has the *spectral decomposition* $\mathbf{R} = \mathbf{W}\mathbf{\Gamma}\mathbf{U}^T = \gamma_1 \mathbf{w}_1 \mathbf{u}_1^T + \cdots + \gamma_m \mathbf{w}_m \mathbf{u}_m^T$ (cf., Rao 1973, 43–44; Karlin and Taylor 1975, 540–542). The eigenvalues satisfy

[4] The problem is also central in enterprise demography, when mergers of firms are modeled.

the characteristic equation $|\mathbf{R} - \gamma \mathbf{I}| = 0$. This is a polynomial of order m of γ, with m real or complex roots that are the eigenvalues. No special properties are required of $\mathbf{R}$ for these results to hold.

Suppose now that all fertility rates for ages $\alpha \leq x \leq \beta$ and all transition rates (relating to survival and migration) for ages $0 \leq x \leq \beta$ are strictly positive. To carry through the technical argument we now make a detour. For the moment, let us exclude all males, and all females in ages $x > \beta$, from consideration. That is, we delete all elements relating to them from the vectors $\mathbf{V}(t)$ and the matrix $\mathbf{R}$, so that, e.g., in the case of a single region female population $\mathbf{R}$ has $\beta + 1$ rows and columns. Since $\alpha < \beta$ the strict positivity of the rates implies that from some power j on, all elements of the reduced matrix $\mathbf{R}^k$, $k > j$, are strictly positive. The so-called Perron-Frobenius theorem (Gantmacher 1959, Karlin and Taylor 1975, 542 ff) tells us then that $\mathbf{R}$ has a unique, strictly positive eigenvalue, say γ_1, such that $\gamma_1 > |\gamma_i|$ for $i > 1$. The corresponding right and left eigenvectors can also be chosen real and nonnegative. Using the spectral decomposition one can then show that $(\mathbf{R}/\gamma_1)^k \to \mathbf{w}_1 \mathbf{u}_1^T$, as $k \to \infty$. It follows that for large k we have the asymptotic approximation

$$\mathbf{R}^k \mathbf{V}(0) \sim \gamma_1^k \mathbf{w}_1 \left\{ \mathbf{u}_1^T \mathbf{V}(0) \right\}, \tag{2.8}$$

where $\sim$ means that the elementwise ratios of left hand side and right hand side converge to 1. We see that in the long run the initial population $\mathbf{V}(0)$ influences only the level of population via the scalar $\mathbf{u}_1^T \mathbf{V}(0)$. The *asymptotic age-distribution* is determined by $\mathbf{w}_1$ (when normalized so the elements sum to one), and the annual *asymptotic (or intrinsic) growth rate* is given by $\log(\gamma_1)$. The fact that the asymptotic age-distribution and growth rate do not depend on the initial age-distribution is called the *ergodicity* of the process. Note that the right hand side of (2.8) defines a stable population, i.e., a population that grows exponentially and whose age-distribution does not change (cf., Section 2.2.2 of Chapter 4).

Having established the result for the female population in age $x \leq \beta$, we can extend it to older females by noting that the surviving women in any age $x > \beta$ are (in this deterministic treatment) a constant fraction of those in age $= \beta$. Hence, their number will asymptotically also grow/decline exponentially. Assuming that the female life expectancy is finite, we see that a representation of the form (2.8) holds for females of all ages. Males can similarly be accommodated because the expected number of male births is a constant multiple ($= \kappa/(1 + \kappa)$) in terms of the notation of Chapter 4) of the female births, so they, and the numbers of male survivors, will also grow exponentially. This completes the proof of the asymptotic behavior of the population when fertility and mortality rates do not change over time. As shown by Keiding and Hoem (1976) the results go through in a probabilistic context as well when proportions are interpreted as probabilities and the average number of children per woman is interpreted as a statistical expectation.

Although the assumption of unchanging transition rates is crude, the cohort-component book-keeping, and the corresponding linear growth model, were important in the theory of population forecasting. Exponential and logistic models used earlier for the total population had the drawback that they either lead to an increase or to a decrease, forever. In contrast, a population may have unchanging transition rates, a positive current growth rate, but a negative intrinsic growth rate.

2.3. Weak Ergodicity

It is clear that if the matrices $\mathbf{R}(t)$ change over time, there is no guarantee of a particular long-term growth rate nor that there would necessarily be an age distribution that the population might tend to. However, a more subtle asymptotic property does hold. Subject to regularity conditions *any two population vectors will become proportional* if subjected to the same sequence of matrices $\mathbf{R}(t)$. We give the main ingredients of the result here, but leave the details into complements.

Suppose we have $n \times n$ matrices $\mathbf{A}(t) = (a_{ij}(t))$, $t = 0, 1, 2, \ldots,$ that all have a strictly positive element in at least one location on every row. Let two sets of vectors $\mathbf{X}(t) = (X_1(t), \ldots, X_n(t))^T$ and $\mathbf{Y}(t) = (Y_1(t), \ldots, Y_n(t))^T$ evolve according to $\mathbf{X}(t+1) = \mathbf{A}(t)\mathbf{X}(t)$ and $\mathbf{Y}(t+1) = \mathbf{A}(t)\mathbf{Y}(t)$ from some strictly positive starting values $\mathbf{X}(0)$ and $\mathbf{Y}(0)$. It follows that all elements of $\mathbf{X}(t)$'s and $\mathbf{Y}(t)$'s are strictly positive for all t. Consider the following ratios $M_t = \max\{X_i(t)/Y_i(t)|i = 1, \ldots, n\}$ and $m_t = \min\{X_i(t)/Y_i(t)|i = 1, \ldots, n\}$. Clearly, $M_t \geq m_t$, but note that $M_t = m_t$ only if the vectors $\mathbf{X}(t)$ and $\mathbf{Y}(t)$ are proportional. Matrix multiplication by a positive matrix has the following *contraction property*,

$$m_t \leq X_i(t+1)/Y_i(t+1) \leq M_t, \tag{2.9}$$

for all $i = 1, \ldots, n$. It follows that M_t's form a non-increasing sequence that has a limit $M_t \to M$ as $t \to \infty$, and m_t's form an non-decreasing sequence with limit $m_t \to m \leq M$ as $t \to \infty$. The limits can be shown to be equal provided, for example, that the following two conditions hold. First, the positive elements in the matrices $\mathbf{A}(t)$ always occur in the same locations, are bounded from above, and bounded away from zero. I.e., there are constants $0 < a < A$ such that for those elements with $a_{ij}(t) > 0$ we actually have $a \leq a_{ij}(t) \leq A$ (e.g., LeBras 1977; Caswell 2001, 375). Second, there is an integer $j > 0$ such that all elements of any j-fold product of $\mathbf{A}(t)$ matrices are strictly positive.

We can translate this result in demographic terms as follows. Consider the linear growth model (2.2) and assume that all transition rates and fertility rates are bounded away from zero and bounded above. Then, two multistate population systems that are subject to the same sequence of matrices $\mathbf{R}(t)$ will have asymptotically the same distribution by age, sex and region, although the common distribution may change over time and the population has no fixed asymptotic growth rate. This is the so-called *weak ergodicity property* of demography. Intuitively, it can be interpreted as saying that all populations will eventually "forget" their earlier age-distributions. The current age-distribution depends on past rates only.

Another way to think about the result is that a product of non-negative matrices $\mathbf{P}(t) \equiv \mathbf{R}(t) \ldots \mathbf{R}(0)$ resembles increasingly a matrix of rank $= 1$, in the sense that there is a sequence $\mathbf{M}(t)$ of matrices of rank 1 such that the difference $\mathbf{P}(t) - \mathbf{M}(t) \to \mathbf{0}$ as $t \to \infty$. (This can happen even though the rank of the product would be n for all t!) Therefore, the population at $t = 0$ influences the asymptotic total size of the population, but not its age distribution. The age distribution changes as a function of $\mathbf{R}(t)$'s, as does the rate of growth.

3. Open Populations and Parametrization of Migration

3.1. Open Population Systems

The multistate linear growth model of a closed population system describes all in- and out-migration flows within the J states. That is, there are $J(J-1)$ transition flows by age and sex. Although this is, in principle, the most satisfactory way to handle state transitions, it is often hard to apply in practice since the number of flows that must be considered can be very large. Along with the difficulty of data collection and the lack of international standards, these considerations have led to the use of a various shortcut procedures.

The simplest way to handle migration is to make assumptions about the net number of migrants by age and sex, for each future year. The method is appealing if in-migration is large and out-migration is small. Under those circumstances changes in population size do not have an important effect on out-migration, so not much would be gained by considering out-migration via transition intensities. In-migration typically cannot meaningfully be analyzed via such intensities, because "the rest of the world" is a very heterogeneous risk population, and changes in its size and composition may have little to do with migration into the area of interest.

We formulate the net-migration model by opening a system of J regions to the rest of the world. Parallel to the definition of $\mathbf{R}$ in Section 2.1, define $\mathbf{N}(x,t) = (\mathbf{N}_1(x,t)^T, \mathbf{N}_2(x,t)^T)^T$, where $\mathbf{N}_1(x,t) = (N_{11}(x,t), \ldots, N_{1J}(x,t))^T$ and $N_{1j}(x,t)$ is the net-number of female migrants from the rest of the world in age x, to region $j = 1, \ldots, J$. Similarly, write for males $\mathbf{N}_2(x,t) = (N_{21}(x,t), \ldots, N_{2J}(x,t))^T$. Then, define $\mathbf{N}(t) = (\mathbf{N}(0,t)^T, \ldots, \mathbf{N}(\omega,t)^T)^T$, and replace formula (2.2) by

$$\mathbf{V}(t+1) = \mathbf{R}(t)\mathbf{V}(t) + \mathbf{N}(t). \tag{3.1}$$

Starting from time $t = 0$, the evolution of the population system to time $T > 0$ follows the equation

$$\mathbf{V}(T) = \left\{\prod_{t=0}^{T-1} \mathbf{R}(t)\right\} \mathbf{V}(0) + \sum_{k=0}^{T-1} \left\{\prod_{t=k+1}^{T-1} \mathbf{R}(t)\right\} \mathbf{N}(k), \tag{3.2}$$

where the products are "backward" as in (2.7), and a matrix product with no elements (this occurs when $k = T-1$) is defined as an identity matrix. When $J = 1$, the model (3.2) describes a single region, two-sex population that is open to migration.

3.2. Parametric Models

Consider the internal flows among the J regions. There are several intermediate models of out-migration rates. Notably, Rogers (1986) has used the so-called double exponential model to describe the level and age-structure of migration intensity using ten parameters. Others have used data-analytic techniques (e.g., Van Imhoff et al. 1997, Lin 1999, Willekens 1999). We will briefly outline two approaches of the latter type.

3.2.1. Migrant Pool Model

The *migrant pool model* uses out-migration rates that are not destination specific. One first forecasts the total number ("pool") of out-migrants from all regions. In-migrants are then obtained by redistributing the migrant pool back to the regions according to some forecasted shares. Statistically this means that destination is independent of the origin, or that we have a log-linear model representation of net migration from j to i,

$$R_{sij}(x+1, x, t) = \exp(\alpha_{si}(x, t) + \beta_{sj}(x, t)), \tag{3.3}$$

where $s = 1$ for females and $s = 2$ for males. Even simpler versions are obtained by taking the parameters to be age-independent, for example $\alpha_{sj}(x, t) \equiv \alpha_{sj}(t)$ or $\beta_{sj}(x, t) \equiv \beta_{sj}(t)$. The parameters of the loglinear model can be estimated using Poisson regression. However, due to the independence assumption one can directly estimate the outmigration rates, and the shares, and do the multiplication.

The migrant pool model requires J out-migration rates, and J shares, for each age and sex. If J is large, then a considerable reduction in the number of parameters is achieved, compared to the full set of $J(J-1)$ interstate flows. For example, Finland produces forecasts of the population of approximately $J = 450$ municipalities, so the model of full flows would have about 200,000 parameters for each age and sex, whereas the pooled model only has 900. On the other hand, if $J = 2$, no savings are achieved.

3.2.2. Bilinear Models

It is well-known that the intensity of migration is heavily age-dependent in a way that is rather similar in most regions. Bilinear models of the type discussed in Chapter 5 provide a description of age patterns.

Consider the following three ($J = 3$) regions of Finland: the Helsinki region (consisting of cities of Helsinki, Espoo, Vantaa, Kauniainen); North-Eastern Finland (Lappland, North Carelia, and Kainuu); and the remaining West-Central Finland. Helsinki region has typically gained migrants, and North-Eastern Finland has lost. There are six flows. For sexes $s = 1, 2$, consider a bilinear model of the form

$$R_{sij}(x+1, x, t) = \mu_{sij}(t) + \gamma_s(x) + \alpha_{si}(t)\nu_s(x) + \beta_{sj}(t)\eta_s(x) + \varepsilon_{sij}(x, t), \tag{3.4}$$

where $E[\varepsilon_{sij}(x, t)] = 0$. For interpretation and identifiability, we may assume, for example, that $\Sigma_x \gamma_s(x) = \Sigma_x \nu_s(x) = \Sigma_x \eta_s(x) = 0$; $\Sigma_x \gamma_s(x)\nu_s(x) = \Sigma_x \gamma_s(x)\eta_s(x) = \Sigma_x \nu_s(x)\eta_s(x) = 0$; and $\Sigma_x \nu_s(x)^2 = \Sigma_x \eta_s(x)^2 = 1$ for $s = 1, 2$ separately. Then, $\mu_{sij}(t)$ would determine the overall level of the intensity from j to i during year t, and $\gamma_s(x)$ would determine the dependence of migration intensity on age, and the remaining two terms would represent interactions between

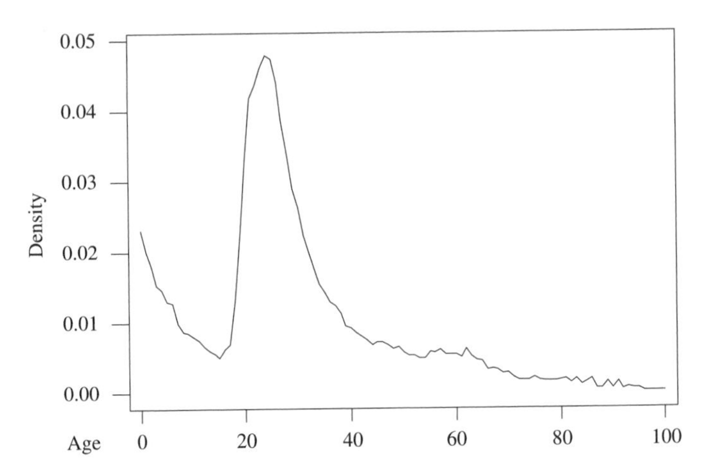

FIGURE 5. Average Density of Male Migration in Finland, Across Three Regions, During 1987–1997.

flows and age. Consider the males. Figure 5 provides the average distribution of migration intensity, across the six flows and 11 years of observation. Principal components were used to estimate the vectors (or "factors" in the terminology of factor analysis; e.g., Afifi and Azen 1979, 324–325) $(\nu_s(0), \ldots, \nu_s(\omega))^T$ and $(\eta_s(0), \ldots, \eta_s(\omega))^T$, see Figure 6. The solid curve depicting $\nu_s(x)$'s accounts for 67% of the variation around the mean, and the dashed curve depicting $\eta_s(x)$'s adds 6%, for a total of 73%. We see from the solid curve that the most important aspect of deviations from average, is in terms of how much of migration is concentrated in ages 19–29 as opposed to ages < 10, 30–40, and 60–70. A large positive

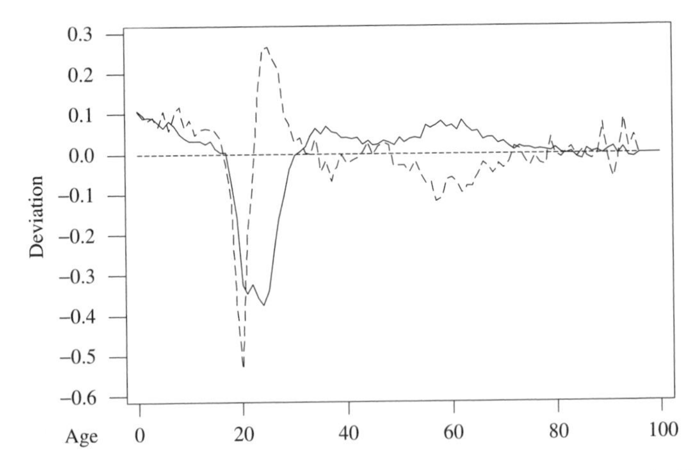

FIGURE 6. Two Most Important Patterns of Deviation from Average Age Distribution of Migration Intensity.

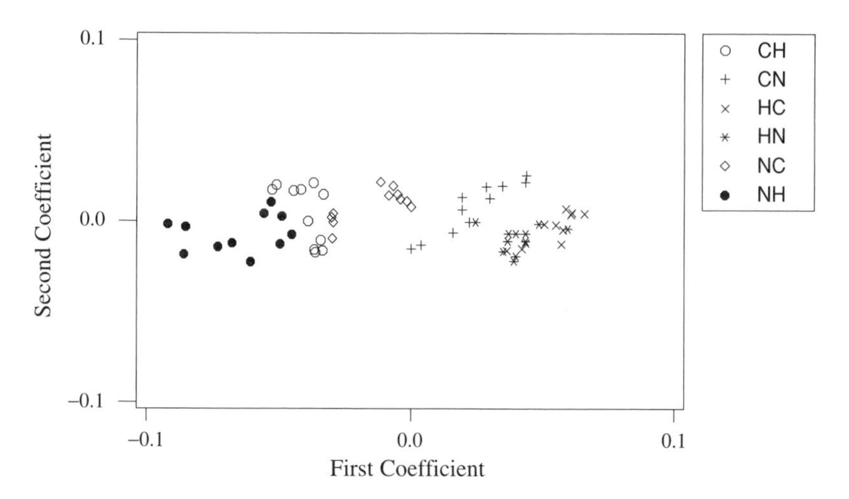

FIGURE 7. Coefficients of Deviations from the Mean for the Six Flows (H = Helsinki, CN = West-Central, N = North-East), During 1987–1997.

(negative) coefficient for this pattern in a given year for a given flow would indicate that there were relatively few (many) males in ages 19–29 in that flow. The second most important way the flows differ is in terms of how many 18–21 year olds have moved as opposed to 24–27 year olds. The younger age bracket coincides with the beginning of higher education and/or leaving military service, and the latter with family formation and seeking of permanent employment. Figure 7 shows the coefficients (or "factor loadings") $(\alpha_{ijS}(t), \beta_{ijS}(t))$ as points on a plane for years $1987, \ldots, 1997$, for each of the six flows. Although the evolution of time has not been indicated in the plot, we note that for some flows the age pattern has changed in a regular manner (notably flow CN, or the flow from West-Central to North-East) but in others changes have been more erratic.

4. Demographic Functionals

The notion of a multistate population system is motivated by two types of considerations. First, we may primarily be interested in the size of the total population, but disaggregating the population by state other than age and sex may be helpful in formulating the forecasts of the vital rates. For example, we might wish to disaggregate the population by ethnic categories and marital status for the purpose of analyzing either fertility or survival, if it is known that fertility, mortality, and migration depend heavily on ethnicity.

Second, the states may be of direct interest by themselves. For example, we may be interested in marriage patterns on their own right; we may wish to analyze trends in unemployment, etc. In these applications, the possible differences in the vital rates of the different states may be of secondary interest, and the states may be viewed as functions of the total populations via the prevalence rates of the states by age and sex.

More generally, we define a *demographic functional* as a function of either a population vector or a vector of vital rates. Since both vectors can be viewed as functions of age, we are speaking of a function of function, or functional. The function may also be random given the total population vector or vital rates.

Example 4.1. Marriage Prevalence as a Functional. Let $\pi_{sj}(x, t)$ be the fraction of those in age x at time t, in region j, of sex s, who belong to a specific subpopulation, e.g., those in the Married state. Then, $\pi_{sj}(x, t)$ is called the *prevalence* of marriage. The total married female population at time t in region j is then the following demographic functional,

$$\sum_{x=0}^{\omega} \pi_{1j}(x, t) V_{1j}(x, t). \tag{4.1}$$

Forecasting (4.1) involves two sources of uncertainty: how accurately can we forecast the vector $\mathbf{V}_{1j}(t)$, and how accurately can we forecast the corresponding (random) vector of prevalences $\boldsymbol{\pi}_{1j}(t)$. The approach that analyses multistate problems via prevalence rates is sometimes called *Sullivan's method*. For reasons similar to the ones discussed in Section 4.3.3 of Chapter 4, prevalence rates are actually complicated functions of past transitions between the states, so care is needed in their application. ◊

Example 4.2. Life Expectancy as a Functional. The remaining life expectancy e_x, as defined in (2.8) of Chapter 4, is a nonrandom, nonlinear functional of the age-specific mortality rates. We can view its forecast as a functional forecast. ◊

Example 4.3. Age Dependency Ratio. One of the most useful functions of age-distributions is the so-called *age dependency ratio*. It is usually defined as the ratio of the population in ages <15 or >64 to those who are in ages 15–64. Therefore, conditionally on the population vector its value is a fixed (i.e., nonrandom), nonlinear function of the population vector. The age dependency ratio gives a rough indication of how many dependents each person in working age must support. ◊

Example 4.4. A Relation Between Prevalence and Incidence. In the folklore of epidemiology the following argument concerning prevalence is sometimes given. Suppose a population of size N is composed of those D who are diseased and $N - D$ who are not. Let the average duration of the disease be d and let the incidence of disease be ν. Then, we should have $D =$ (number of new cases per year) $\times$ (average duration) $= \nu(N - D)d$. The prevalence of disease is $p = D/N$. Then we have that $p/(1 - p) = \nu d$, or *prevalence odds* $=$ incidence $\times$ duration. For the argument to hold, one has to assume that (i) the population being studied is stationary, and (ii) incidence and expected duration of illness are uncorrelated as functions of age (cf., Alho 1992c). Both assumptions may fail (e.g., intensities of most flows of Section 1.5 depend heavily on age leading to a possible violation of (ii)), so the formula is a rough approximation only. ◊

5. Elementwise Aspects of the Matrix Formulation

The matrix formulation of Section 2 is helpful in showing the broad outlines of population renewal. However, examination of some of the elementwise relationships provides additional insights. We consider first survival in a closed multi-state setting, and then the renewal of female births in a single state case.

Consider the number of individuals of sex s in region j, who are in age $x \geq t$ at time t. They were in age $x - t$ at jump-off time $t = 0$, so their number is

$$V_{sj}(x, t) = \sum_{i_0=1}^{J} \sum_{i_1=1}^{J} \cdots \sum_{i_{t-1}=1}^{J} V_{s,i_0}(x - t, 0) \exp\{r_{s,i_1,i_0}(x - t + 1, x - t, 1) + \cdots + r_{s,j,i_{t-1}}(x, x - 1, t - 1)\}. \tag{5.1}$$

In later chapters we will treat the elements of the matrices $\mathbf{R}(t)$ as random variables. In the single region case ($J = 1$) the sum reduces to a single exponential term, so the stochastic analysis of survival involves merely a sum in the log-scale. However, when $J > 1$, we have a sum of J^t terms (this can be a large number: e.g., when $J = 2$, and $t = 50$, we have $2^{50} \approx 10^{15}$ summands), and no transformation reduces (5.1) into a linear form exactly. Taylor-series approximations can be provided, but loss of accuracy cannot be avoided.

Assume now that $J = 1$, and consider the youngest female age-group during year $t > \beta$. At that time all women giving birth have, themselves, been born after the jump-off year. It follows that for $j = 1$ we can write

$$V_{1j}(0, t) = \sum_{x=\alpha}^{\beta} V_{1j}(0, t - x) \exp\{r_{1j}(1, 0, t - x) + \cdots + r_{1j}(x, x - 1, t - 2) + r_{1j}(0, x, t - 1)\}. \tag{5.2}$$

This is called a *renewal equation* for the youngest age, because it expresses the value of year t in terms of the values of past years $t - x$. Under the assumption of constant vital rates, one can solve the renewal equation to determine the asymptotic growth rate of the population defined in Section 2.2. (In this case the exponential terms of (5.2) comprise the net maternity function appearing on the right hand side of (4.4) of Chapter 4.) We will come back to this in Section 5.1 of Chapter 9.

6. Markov Chain Models

When individuals move from state to state in a multistate demographic system, they create migration histories that can be described probabilistically. The simplest such model is the *Markov chain* in which an individual moves in discrete time among a finite or countably infinite number of states and the probability of moving at step n from state j to state k only depends on j and k, and not what states the individual had visited prior to n (e.g., Çinlar 1975).[5] The theory of Markov chains is related to the theory of stable populations, as discussed in Section 2.2. Instead of pursuing those topics we provide an ecological example that uses both Markov chain ideas and capture-recapture techniques to analyze a multistate population system.

Example 6.1. Metapopulation of Butterflies. Consider butterflies that live in J meadows. Each meadow may be too small to sustain a separate population, but migrants from other meadows may regenerate a population that has become extinct due to a storm, for example. A population consisting of such communicating subpopulations is called a *metapopulation* in ecology. The situation is of ecological interest, because human intervention may alter the pattern of meadows and forest land and pose a threat to the butterflies (Wahlberg, Moilanen, and Hanski 1996). The parameters of ecological interest include the probability of death within a meadow and the probability of death during migration. These are hard to estimate because it is impracticable to keep track of all butterflies in an experimental situation. Instead, ecologists use capture-recapture techniques to study the population.

Assume that during days $t = 1, \ldots, T$ a total of N butterflies have been captured and marked. This generates a capture history of locations $s_1, \ldots, s_{n_i}$ and times $t_1 < \cdots < t_{n_i}$, for each captured butterfly $i = 1, \ldots, N$, where n_i is the number of captures. Movements of butterflies can be viewed as having no memory: the probability of leaving a meadow for another at time t depends only on the meadow the butterfly is in, not on the path before t. Therefore, a Markov chain model is appropriate. Let $j = 1, \ldots, J$ correspond to different meadows. Define a $J \times J$ matrix of transition probabilities $\mathbf{P} = (p(j, k))$ with

$$p(j, k) = P(\text{state is } k \text{ at time } t + \text{state is } j \text{ at time } t). \tag{6.1}$$

These probabilities depend on mortality during the transition, and mortality while in a meadow. For each t there is a set of meadows $B(t)$ in which catches were made with capture probabilities $0 < \rho_j(t) < 1$ for $j \in B(t)$. These probabilities are primarily influenced by the weather. We omit the complex details, but note that the probability of the observed path can then be expressed in terms of the transition matrix $\mathbf{P}$ and the capture probabilities $\rho_j(t)$. As discussed in Hanski, Alho, and Moilanen (2000) it is natural to let transition probabilities to depend on the area of the meadows, their mutual distances, and mortality, via parametric models. The object is to estimate $\mathbf{P}$ and the capture probabilities. In this application it is impracticable to calculate the derivatives of the likelihood function. However, the maximization can be carried out using global optimization methods such as *simulated annealing* that rely on a stochastic search of the parameter space (Press et al. 1992, 436ff). In fact, Markov chain theory provides a practical method for carrying the search (cf., Ripley 1987, 181–182). ◊

---

[5] For example, the random walk model used to describe leadership duration in Section 8 of Chapter 5 is Markov chain with states $\{r, \ldots, 0, 1, 2, \ldots\}$.

Exercises and Complements (*)

1. Show that if $\mu(t) = b/(1 - bt)$ for $t \in [0, 1]$, then the Runge-Kutta method with $h = 1$ produces an exact solution for $p(1)$.
2. Use the Runge-Kutta method to solve numerically the value of the survival function $p(t)$ for $t = 0, 1, \ldots, 100$, when the force of mortality is given by the Gompertz-Makeham law with $A = 0.00376$, $R = 0.0000274$, and $\alpha = 0.104$. Compare the result to the exact value obtained by integrating the hazard.
3. Consider a four state system with states employed ($j = 1$), unemployed ($j = 2$), outside workforce ($j = 3$), and dead ($j = 4$). Being absorbing, the last state can be left out. Use the spectral representation to calculate $\mathbf{p}(t)$ for $t = 0, 1, \ldots, 20$ when the constant transition intensities are given by the matrix

$$\begin{bmatrix} -0.08, & 0.03 & 0.10 \\ 0.02 & -0.07 & 0.10 \\ 0.04 & 0.02 & -0.22 \end{bmatrix},$$

   and the person starts from outside the workforce.
4. Continuation. Calculate the expected years spent in different states (during [0, 20]) in the setting of Problem 3.

*5. Consider a function $y(x)$, $x \in [0, 1]$, such that $y(0) = 0$, $y(1) = 1$, $y'(0) = \beta$, and $y'(1) = \gamma$. Determine a, b and c so that function $z(x) = ax + bx^2 + cx^3$ has $z(x) = y(x)$ and $z'(x) = y'(x)$ at $x = 0, 1$. Interpreting $y(x) = E[I(x) = \mathbf{e}_k | I(0) = \mathbf{e}_j]/E[I(1) = \mathbf{e}_k | I(0) = \mathbf{e}_j]$ the values for β and γ are available from the Runge-Kutta output. Neglecting the possibility of more than one transition one might then impute the time of departure from j as $z^{-1}(U)$, where $U \sim U[0, 1]$. This solution is only feasible if $z(x)$ turns out to be monotone.

6. Consider the setting of Example 1.3 with $\mathbf{p}(x) = (\mathbf{I} + x\mathbf{B})\mathbf{a}$ and $\boldsymbol{\nu}(x) = \mathbf{B}(\mathbf{I} + x\mathbf{B})^{-1}$ for $x \in [0, 1]$. Suppose we estimate transition intensities by *o/e* rates, say, $\boldsymbol{\nu}(1/2) = \hat{\boldsymbol{\nu}}$. Then, deduce from the latter equation the estimate $\hat{\mathbf{B}} = (\mathbf{I} - (1/2)\hat{\boldsymbol{\nu}})^{-1}\hat{\boldsymbol{\nu}}$. Substitute into the first equation to get $\hat{\mathbf{p}} = (\mathbf{I} - (1/2)\hat{\boldsymbol{\nu}})^{-1}(\mathbf{I} + (x - 1/2)\hat{\boldsymbol{\nu}})\mathbf{a}$. (cf., Rogers and Ledent 1976).
7. What is the *average age at retirement*? As in the case of mean age at childbearing (cf., Section 4.2 of Chapter 4), different answers to this question can be given depending on what the goal of the calculation is. First, one can simply calculate the average age at retirement of those who retire in a given year. This may be what is wanted, but this average depends on the sizes of the earlier birth cohorts, and on the earlier transitions to the state of retirement, so it is certainly not a pure period summary of transition intensities. How can a multistate model be used to define the concept?
8. Consider a transition $j \to i$, but omit the indices from $N_{ij}(t)$ and $K_j(t)$ to simplify the notation. Suppose the density of population is $k(s) = k_0 + k_1(t - s)$

and the average rate is $\nu(s) = \nu_0 + \nu_1(t - s)$ for $s \in [t - 1, t + 1)$. Deduce that

$$K(t-1) + K(t) = \int_{t-1}^{t+1} k(s)\,ds = 2k_0;$$

$$N(t-1) + N(t) = \int_{t-1}^{t+1} \nu(s)k(s)\,ds = 2\nu_0 k_0 + 2\nu_1 k_1/3.$$

Use the estimates

$$\nu_1 = v(t) - v(t-1); \quad k_1 = K(t) - K(t-1)$$

to obtain the estimator (1.21) for ν_0.

*9. A "quick and dirty" way to assess the statistical significance of multistate life table summaries is as follows. Consider (1.13), and suppose first that we observe a cohort of size N under no censoring. In this case, we estimate the components of (1.13) by $\bar{T}_j$ = average time spent in $j = 1, \ldots, J$. Let V_j = variance of the times spent in j, so the standard error is $(V_j/N)^{1/2}$. Second, instead of cohort data, suppose we have period data that come from a stationary population of size N. In this case we could repeatedly generate samples of size N using the estimated transition intensities, and perform the same calculations as for a cohort. These *bootstrap replications* would give us an estimate of the sampling distribution of (1.13). Our proposal is to use the above period data procedure, even if the data do not come from a stationary population, and to call standard errors calculated in this way *stationary equivalent standard errors* or *SESE*'s. In this case we determine the birth rate of the stationary population underlying simulation so that N = person years lived in the population from which the data came. (a) Can you see why $\bar{T}_j$ and V_j can be estimated from any number ($\neq N$) of simulations rounds? (b) When would you expect SESE's to be too small, or too large? (Hint: think of younger and older age-distributions than the stationary one.)

10. Consider eigenvectors $\mathbf{R}\mathbf{w}_i = \gamma_i \mathbf{w}_i$ and $\mathbf{u}_j^T \mathbf{R} = \gamma_j \mathbf{u}_j^T$ with $\gamma_i \neq \gamma_j$. Show that $\mathbf{u}_j^T \mathbf{w}_i = 0$.

11. Consider a female population in two regions ($J = 2$). Suppose the female population in age $x = \beta$ is exponentially increasing with rate γ, or

$$V_{1j}(\beta, t) = V_{1j}(\beta, 0)\exp(\gamma t),$$

for $j = 1, 2$. Suppose the probability that a person in age β in region i survives to be of age $x > \beta$ in region j is $p_{ji}(x, \beta)$ for $i = 1, 2$ and $j = 1, 2$. Show that the $V_{11}(x, t)$ and $V_{12}(x, t)$ also evolve exponentially at rate γ.

12. Consider a female population, closed to migration, that has constant fertility and mortality rates. Restrict attention to ages $x = 0, \ldots, \beta$. Suppose the limits of childbearing ages are $\alpha = 2$ and $\beta = 4$. The matrix $\mathbf{R}$ is of the form

$$\mathbf{R} = \begin{bmatrix} 0 & 0 & * & * & * \\ * & 0 & 0 & 0 & 0 \\ 0 & * & 0 & 0 & 0 \\ 0 & 0 & * & 0 & 0 \\ 0 & 0 & 0 & * & 0 \end{bmatrix},$$

where $*$ denotes some strictly positive fertility rate (on first row) or survival probability (on first subdiagonal). Show that there is a power j such that all elements of $\mathbf{R}^k$ with $k > j$ are strictly positive. (Hint: One way to do this is to replace $*$ by, e.g., 1, and to carry out the multiplications with a computer.)

13. Consider a matrix $\mathbf{R} = (r_{ij})$ with $i = 0, \ldots, \beta$ and $j = 0, \ldots, \beta$. Suppose the elements $r_{0j} = f_j$ are strictly positive for $j = \alpha, \ldots, \beta$. Similarly, assume that the elements $r_{i+1,i}$ are strictly positive. All other elements of $\mathbf{R}$ are zero. Consider the eigenvalue problem,

$$\mathbf{R}\mathbf{w} = \lambda \mathbf{w},$$

where $\mathbf{w} = (w_0, \ldots, w_\beta)^T$ is non-zero vector. Define

$$p_x = \prod_{i=1}^{x} r_{i,i-1}.$$

Show first that if λ is an eigenvalue, then the corresponding eigenvector has the form $w_x = c p_x / \lambda^x$ for $x = 1, \ldots, \beta$ and $w_0 = c$ is some constant.

14. Using this, show that λ must satisfy the polynomial equation,

$$\lambda^{\beta+1} = \sum_{x=\alpha}^{\beta} f_x p_x \lambda^{\beta-x}.$$

Note that the coefficients $f_x p_x$ on the right hand side are the discrete version of the net maternity function (provided that only female births are considered in f_x!).

15. By considering values $\lambda > 0$, show that a positive, real solution to the polynomial equation of Exercise 13 exists. To show that it is unique requires more work (cf., Keyfitz 1977, 48).

16. Solve the polynomial equation of Exercise 3 numerically (using the secant method, Newton's method, or by using existing software) for a data set of your country.

17. *Exponential population growth.* Suppose population at time t is $V(t)$. Assume that its growth rate satisfies the differential equation $V'(t)/V(t) = r(t)$. If $V(0) = A$, show that for $t < 0$,

$$V(t) = A \exp\left(\int_0^t r(s)\, ds \right).$$

18. *Logistic population growth*. Suppose the population growth rate satisfies the equation $V'(t)/V(t) = r(t)(M - V(t))/M$, where $M > 0$ is some constant. Show that if $V(0) = A < M$, then by defining $B = (M - A)/A$ we get

$$V(t) = M \exp\left(\int_0^t r(s)\,ds\right) \Big/ \left(B + \exp\int_0^t r(s)\,ds\right).$$

*19. Prove the relationship (2.9) by showing that

$$X_i(t+1)/Y_i(t+1) = \sum_{j=1}^{n} w_{ij}(t)X_j(t)/Y_j(t),$$

where $w_{ij}(t) = a_{ij}(t)Y_j(t)/\Sigma_h a_{ih}(t)Y_h(t)$.

*20. Continuation. Suppose the non-zero elements of matrices $\mathbf{A}(t)$ are located in fixed locations in such a way that for some $j > 1$ any j-fold product $\mathbf{A}(t + j - 1)\mathbf{A}(t + j - 2)\cdots\mathbf{A}(t) \equiv \mathbf{B}(t) = (b_{ij}(t))$ has only strictly positive elements (cf., Exercise 12). Then, (2.9) holds for the subsequences $\mathbf{X}^*(t+1) = \mathbf{A}^*(t)\mathbf{X}^*(t)$ and $\mathbf{Y}^*(t+1) = \mathbf{A}^*(t)\mathbf{Y}^*(t)$, where $\mathbf{A}^*(t) = (a^*_{ij}(t)) = \mathbf{B}(tj)$ for $t = 0, 1, 2, \ldots$, and the starting values are $\mathbf{X}^*(0) = \mathbf{X}(0)$ and $\mathbf{Y}^*(0) = \mathbf{Y}(0)$. I.e., we are picking every j^{th} vector from the original sequences. (a) Show that if the non-zero elements in $\mathbf{A}(t)$ satisfy $0 < a \le a_{ij}(t) \le A$, then there are constants $0 < a^* < A^*$ such that $a^* < a^*_{ij}(t) < A^*$. (b) Define $w^*_{ij}(t) = a^*_{ij}(t)Y^*_j(t)/\Sigma_h a^*_{ih}(t)Y^*_h(t)$, and show that $0 < c^*/n < w^*_{ij}(t)$, where $c^* = (a^*/A^*)^2$. (Hint: conclude from $\mathbf{Y}^*(t) = \mathbf{A}^*(t)\mathbf{Y}^*(t-1)$ that $A^*\Sigma_h Y^*_h(t-1) > Y^*_j(t) > a^*\Sigma_h Y^*_h(t-1)$.)

*21. Continuation. Define M^*_t and m^*_t for the $X^*(t)$ and $Y^*(t)$ processes as for the original ones.

(a) Show that $M^*_{t+1} - m^*_{t+1} = M'_{t+1} - m'_{t+1}$, where

$$M'_{t+1} = \max_i \sum_{j=1}^{n}(w^*_{ij}(t) - c^*/n)\frac{X^*_j(t)}{Y^*_j(t)};$$

$$m'_{t+1} = \min_i \sum_{j=1}^{n}(w^*_{ij}(t) - c^*/n)\frac{X^*_j(t)}{Y^*_j(t)}.$$

(b) Show first that

$$M'_{t+1} < M^*_t \sum_{j=1}^{n}(w^*_{ij}(t) - c^*/n) = M^*_t(1 - c^*),$$

and then that $m'_{t+1} > m^*_t(1 - c^*)$.

(c) Conclude that $M^*_{t+1} - m^*_{t+1} < (M^*_t - m^*_t)(1 - c^*)$. Since $0 < c^* < 1$, this proves that the limits of M^*_t and m^*_t, and hence those of M_t and m_t, are equal. This proof of weak ergodicity is due to LeBras (1977).

*22. Consider a single region. An alternative to additive net migration is to use the so-called *census survival rates* or *census survival probabilities* in place of ordinary survival proportions. The idea is that one corrects the mortality rate (and birth rates) to reflect the net effect of migration in each age.

23. Suppose the transition probabilities (6.1) of a Markov chain are given by a $J \times J$ matrix $\mathbf{P}$. Suppose each state can be reached in one step from any state. Check that a column vector of J ones is a right eigenvector of $\mathbf{P}$ corresponding to eigenvalue 1. Note that the jth row of the product $\mathbf{P}^k$ gives the k-step transition probabilities of the chain. Using the Perron-Frobenius theorem, show that 1 is the largest eigenvalue and there is a J-vector $\mathbf{u} = (u_1, \ldots, u_J)$ such that $u_j > 0$ is the probability that the chain is in state j for large k *irrespective of the state it has started from.* This is an ergodic property of Markov chains. The u_j's determine the *invariant distribution* of the chain when they are normalized to sum to 1.

7 Approaches to Forecasting Demographic Rates

Statistical prediction theory accepts, as a starting point, that error cannot be avoided. The best forecast is the one that minimizes error according to the chosen criterion. This is in contrast with the "crystal ball" usage, in which it is assumed that forecasting is possible only when the future can be seen clearly, without error. We believe that the statistical outlook has much to offer to demography. In particular, recognizing uncertainty leads towards its quantification. This aids in decision making by helping us to prepare for realistic future alternatives in a systematic or at least thoughtful manner.

In this chapter we develop a conceptual basis for the discussion of statistical aspects of demographic time series, and provide guidance to the critical use of time series models in demography. The emphasis will be on simple models rather than theoretical generality. In Section 1 we discuss the basic building blocks of time series models. In Section 2 we refine the models by allowing for intermediate levels of autocorrelation. Section 3 discusses the various ways nonconstant means can be handled. Then, in Section 4 we discuss models for processes whose variance changes over time.

1. Trends, Random Walks, and Volatility

A collection of random variables Y_t where t belongs to some index set is called a *stochastic process*[1]. Earlier, the assumption of independence was natural in many applications. For example, in Chapter 5 we used random variables $Y_1, Y_2, \ldots, Y_n$ to represent observations coming from different individuals (or different age-groups, different sexes etc.). Here, we associate the observed value $Y_t = y_t$ to time t, so the random variables can be used as a probabilistic model for a *time series*. This creates a natural ordering for the variables, and many forms of dependence can be entertained.

[1] The random variables are assumed to be defined on the same probability space.

If $Y_t = \varepsilon_t$, where $\varepsilon_1, \varepsilon_2, \ldots, \varepsilon_n$ is an i.i.d. sequence of random variables with $E[\varepsilon_t] = 0$ and $\mathrm{Var}(\varepsilon_t) = \sigma_\varepsilon^2$, then (especially in engineering literature) one often speaks of *white noise* or a *white noise process*.[2]

Define $Z_t = Y_1 + \cdots + Y_t = \varepsilon_1 + \cdots + \varepsilon_t$, for $t = 1, 2, \ldots, n$, and $Z_0 = 0$. This is a *random walk*. It is characterized by the fact that the first differences, or *increments*, $Z_t - Z_{t-1} = \varepsilon_t$, form an independent sequence. Suppose we have observed the process Z_t for $t = 1, \ldots, n$, and we would like to forecast its future values. Since the increments $\varepsilon_{t+1}, \varepsilon_{t+2}, \ldots$ are independent of $Z_t, t \leq n$, and they have mean zero, the minimum mean squared error forecast is the latest observed value of Z_n, forever after.

Random walks have long been used as models for stock prices, because in efficient markets stock prices should be unpredictable (e.g., Bachelier 1900; Taqqu 2001; Bernstein 1998). In continuous time the corresponding model is called *Brownian motion*. It has been used as a model for the erratic movement of particles in liquids, where collisions with other particles occur continuously. We will present evidence in Example 4.1 of Chapter 8 that a random walk also provides a serviceable approximation for the (logarithm of the) total fertility rate in industrialized countries. This provides us intuition concerning the relationship between period and cohort fertility.

Example 1.1. Cohort Fertility Is Smoother. Figure 1, dashed line, is a realization of a process $T_t = 1.7 \times \exp(Z_t)$, where Z_t is a random walk with the standard deviation of the unit increment $\sigma_\varepsilon = 0.06$. (Motivation for this particular choice will be provided in Example 4.1 of Chapter 8.) At $t = 0$ the process starts at 1.7.

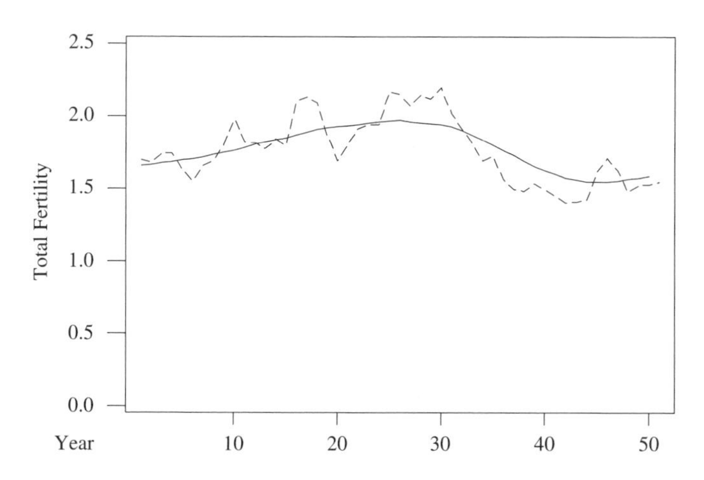

FIGURE 1. Hypothetical Cohort (Solid) and Period (Dashed) Fertility Under a Pure Period Random Walk Model.

[2] If made audible via a transmitter, the process sounds like noise you hear in between stations on radio.

The process T_t represents the period total fertility rate. The solid curve is moving average of the series, with weights $w_i > 0$, $w_{15} + \cdots + w_{49} = 1$. The weights used in the graph correspond to the distribution of total fertility to single years of age, as estimated for 1985 in Italy; cf., Example 4.1. Thus, the solid curve can be interpreted as the cohort total fertility rate. (For an example of an observed cohort total fertility series, see Figure 7 of Chapter 4.) The curves have been matched so that the cohort value has been plotted for the year when the cohort is of age = 28, the mean of the fertility distribution. That is, the solid curve can be defined as $C_t = w_{15}T_{t-13} + \ldots + w_{49}T_{t+21}$. We find that the cohort curve is much smoother than the period curve although, by construction, *all variation is due to period effects*. ◊

In principle, the example could be turned around so that period fertility would be represented as a weighted average of cohort fertility. However, in the absence of period effects it would be difficult to imagine why cohorts in their different phases of childbearing might coordinate their timing to produce the observed variations in period fertility.

The example shows that the relative smoothness of the cohort curve is to be expected even when there are no cohort effects. It is certainly plausible that the cohort point of view is useful in understanding the childbearing decisions of the couples. However, in order to be able to capitalize on the regularities of the cohort fertility in forecasting, more is needed than mere smoothness!

Let us now take $\mu \neq 0$, and define first $Y_t = \mu + \varepsilon_t$, and then $Z_t = Y_1 + \cdots + Y_t = t\mu + \varepsilon_1 + \cdots + \varepsilon_t$, for $t = 1, 2, \ldots, n$. The Z_t process is a *random walk with a drift*. For $\mu > 0$ this process tends to wander up and for $\mu < 0$ it tends to wander down. We see that an assumption of nonzero mean for the increments actually induces a *linear trend* into the summed series, $E[Z_t] = t\mu$. In long-term analysis of stock prices it is necessary to take into account the fact that stocks have appreciated at an average rate of several percent per year. Thus, a rough approximation of the development of a stock's price would be to assume that in t years' time the current price will be multiplied by a factor $\exp(t\mu + \varepsilon_1 + \cdots + \varepsilon_t)$, $\mu > 0$. In contrast, in the analysis of mortality we typically observe declines that are interrupted by plateaus or even increases. Thus, a model of the same type with $\mu < 0$ may provide a serviceable approximation for many ages. In both cases, it is not simply the value of μ that is of interest, but also the value of σ_ε^2, or the *volatility*, because it determines how much the process tends to wander around the trend. In fact, since the sum of i.i.d. terms with mean zero and finite variance is (subject to regularity conditions) approximately normally distributed, the change in value has an approximate log-normal distribution. Therefore, *if* the values of μ and σ_ε are known, and the process starts from value V at $t = 0$, then the probability is approximately 95% that the process is within limits $V\exp(t\mu \pm 1.96\sigma_\varepsilon t^{1/2})$ at $t > 0$. This is an example of a *prediction interval*, i.e., an interval that has a prescribed probability of containing the value of a random variable. (In contrast, a confidence interval is a random interval with a prescribed probability of including a constant, such as a mean.)

In addition to giving rise to random walks, white noise provides a basis for simulating arbitrarily correlated variables. To see this, define $\varepsilon = (\varepsilon_1, \ldots, \varepsilon_n)^T$ a vector of i.i.d. variables with $\sigma_\varepsilon^2 = 1$. Let $\mathbf{\Sigma}$ be an arbitrary covariance matrix. The Cholesky decomposition gives us a way to find a lower triangular matrix $\mathbf{C}$ such that $\mathbf{\Sigma} = \mathbf{CC}^T$. It follows that a vector $\mathbf{Y} = \mathbf{C}\varepsilon$ has covariance matrix $\mathbf{\Sigma}$, because $\text{Cov}(\mathbf{Y}) = \mathbf{C}\text{Cov}(\varepsilon)\mathbf{C}^T = \mathbf{\Sigma}$. Note that the lower triangularity implies that any Y_t depends on ε_i, $i = 1, \ldots, t$, but not on $i > t$.

Example 1.2. Cholesky Decomposition. Suppose the target covariance $\mathbf{\Sigma}$ and the Cholesky matrix are of the form

$$\mathbf{\Sigma} = \begin{bmatrix} 1 & \varphi & \varphi^2 \\ \varphi & 1 & \varphi \\ \varphi^2 & \varphi & 1 \end{bmatrix}, \quad \mathbf{C} = \begin{bmatrix} c_{11} & 0 & 0 \\ c_{21} & c_{22} & 0 \\ c_{31} & c_{32} & c_{33} \end{bmatrix}, \tag{1.1}$$

where $|\varphi| < 1$. Write $c = (1 - \varphi^2)^{1/2}$, for short. By a direct matrix multiplication one can show that a solution is $c_{11} = 1$, $c_{21} = \varphi$, $c_{31} = \varphi^2$, $c_{22} = c$, $c_{32} = \varphi c$, $c_{33} = c$. (Note that the decomposition is only unique up the sign of the diagonal terms.) Consider the transformed values $\mathbf{Y} = \mathbf{C}\varepsilon$. We find that $Y_1 = \varepsilon_1$, $Y_2 = \varphi\varepsilon_1 + c\varepsilon_2$, $Y_3 = \varphi^2\varepsilon_1 + \varphi c\varepsilon_2 + c\varepsilon_3$. One consequence of these relationships is that we can write $Y_t = \varphi Y_{t-1} + c\varepsilon_t$ for $t = 2$ and $t = 3$. This is an example of the so-called autoregressive processes that will be discussed in more detail in the next section. ◊

2. Linear Stationary Processes

In the 1920's, 1930's, and 1940's, when demographers were developing the cohort-component forecasting system, probabilists developed foundations for the so-called stationary processes. This theory was based on a linear transformation of white noise, much the same way as the Cholesky decomposition was used above. Although the main features of the theory were essentially perfected by the beginning of the 1950's (cf., Doob 1953), their practical application in statistics did not become standard until the publication of the monograph by Box and Jenkins in 1970 (second edition 1976, third 1994). Early examples of their use in demography include Saboia (1974, 1977). In this section we will develop the theory with two primary purposes in mind. First, we want to be able to discuss the strengths and limitations of basic time series techniques. Second, we will establish a number of formulas regarding the prediction errors of such processes that will later be useful in the description of qualitative aspects of errors of different types of forecasts. For details about practical modeling, and time series analysis in general, we refer to standard textbooks such as Box and Jenkins (1976), Chatfield (1996), or Harvey (1989).

2.1. Properties and Modeling

2.1.1. Definition and Basic Properties

Let $\ldots, Y_{-1}, Y_0, Y_1, Y_2, \ldots$ be a (doubly infinite) sequence of random variables. As above, we associate the observed value $Y_t = y_t$ with time. A particular realization $\ldots, y_{-1}, y_0, y_1, y_2, \ldots$ of the process is called a *sample path*. Suppose the i.i.d. sequence $\ldots, \varepsilon_{-1}, \varepsilon_0, \varepsilon_1, \varepsilon_2, \ldots$ with $E[\varepsilon_t] = 0$ and $\mathrm{Var}(\varepsilon_t) = \sigma_\varepsilon^2$ is white noise. As in the case of Cholesky decomposition (Example 1.2), let us assume that each Y_t can be written in the form

$$Y_t = \psi_0 \varepsilon_t + \psi_1 \varepsilon_{t-1} + \psi_2 \varepsilon_{t-2} + \cdots, \tag{2.1}$$

where $\psi_0 = 1$, and the series of the absolute values of ψ_j's converges. The process ε_t is also called an *innovation process*, because its values generate the Y_t's.[3] The process (2.1) is called a *linear process*, because each Y_t is a linear function of the innovation process. Since the expectation of each term on the right hand side of (2.1) is zero, it follows that $E[Y_t] = 0$ for all t. In practice, processes (2.1) are used for centered data (i.e., for variables from which the estimated mean has been subtracted) so the assumption of mean zero is not a limitation. If the estimated mean is imprecise, e.g., if the number of observations is too small, the theory is only an approximate guide.

The variance of Y_t is finite, and of the form

$$\mathrm{Var}(Y_t) = \sigma_\varepsilon^2 \sum_{j=0}^{\infty} \psi_j^2, \tag{2.2}$$

for all t. More generally, we have that

$$\mathrm{Cov}(Y_t, Y_{t+k}) = \sigma_\varepsilon^2 \sum_{i=0}^{\infty} \psi_i \psi_{i+k} \tag{2.3}$$

for all t, and $k \geq 0$.

We have observed that the mean of the process Y_t does not change over time. Moreover, since the *autocovariance* (2.3) only depends on the lag k (not on t), the process is called *stationary* (in the wide sense).

Define $\gamma_k = \mathrm{Cov}(Y_t, Y_{t+k})$. The *autocorrelation function* of the process is given by $\rho_k = \gamma_k/\gamma_0$ for $k = 0, 1, 2, \ldots$. When data for $t = 1, \ldots, n$ are available, autocovariance is usually estimated by the *sample autocovariance* $c_k = \Sigma_t (Y_t - \bar{Y})(Y_{t+k} - \bar{Y})/n$, where $\bar{Y} = (Y_1 + \cdots + Y_n)/n$ and the summation is over $t = 1, \ldots, n-k$. Autocorrelation is estimated by the *sample autocorrelation* $r_k = c_k/c_0$.

---

[3] From a mathematical point of view the ε_t's form an orthonormal basis of a vector space (Hilbert space) on which each of the Y_t is defined, with coordinates given by the ψ_j's. For most aspects of the theory, an assumption of uncorrelatedness of the innovations would suffice.

Autocorrelation is a useful tool in the identification of a linear model. Unfortunately, as a rule of thumb, the standard error of the first sample autocorrelation is approximately $n^{-1/2}$ (e.g., Box and Jenkins 1976, 34–36). A time series must have at least 50–100 observations to allow for a somewhat precise estimate of the autocorrelations. This in itself is a strong reason for considering *parsimonious models*, i.e., models with a small number of parameters.

2.1.2. ARIMA Models

We now define a subclass of linear processes that depend on a small number of parameters. An advantage is the availability of relatively objective methods of identifying a model from the class.

Example 2.1. MA(q) Processes. Assume $\psi_q \neq 0$, and $\psi_j = 0$ for $j > q$. Then, (2.1) defines a *moving average process of order* q, which is usually denoted as $MA(q)$. Written with the customary symbolism $\psi_1 = -\theta$, the MA(1) process is of the form $Y_t = \varepsilon_t - \theta\varepsilon_{t-1}$, for example. Its variance is $\text{Var}(Y_t) = \sigma_\varepsilon^2(1+\theta^2)$, and its autocorrelation function is zero except $\rho_1 = -\theta/(1+\theta^2)$. An MA(2) process is usually written as $Y_t = \varepsilon_t - \theta_1\varepsilon_{t-1} - \theta_2\varepsilon_{t-2}$, etc. As a limiting case, taking $q = 0$ we obtain the white noise discussed in Section 1. $\Diamond$

Moving averages are frequently used in demography and economics to smooth out random variation. Suppose, for example that $D_t \sim \text{Po}(\mu_t K_t)$ is the number of deaths in year t (in a given age range, in a given area), where μ_t is the hazard and K_t is the number of person years. Define $m_t = D_t/K_t$ as the observed mortality rate. Using 5 years on both sides to estimate the local level for year t we get the smoothed value

$$\hat{m}(t) = \sum_{j=-5}^{5} w_j m_{t-j}, \tag{2.4}$$

where $w_j > 0$ and $w_{-5} + \cdots + w_5 = 1$. Then the smoothed values are essentially moving average processes, and as such autocorrelated. To illustrate the possible consequences of smoothing, suppose that $\mu_t \equiv \mu$ and $K_t \equiv K$ for all t. In this case $E[m_t] = \mu$ and $\text{Var}(m_t) = \mu/K$. Suppose the deaths during different years are independent with $\mu = 0.01$ and $K = 10{,}000$, so 100 deaths are expected every year. Let $w_j = 1/11$. Figure 2 has a graph of such a process for $t = 1, \ldots, 100$. We see that smoothing creates artificial waves in the plot of the estimate even though the underlying time series values are i.i.d. This is called a *Slutsky effect* in recognition of the pioneering work of Slutsky (1927).

Example 2.2. AR(1) Processes. An *autoregressive process of order 1*, or an $AR(1)$ process, satisfies the recursive equation,

$$Y_t = \varphi Y_{t-1} + \varepsilon_t, \tag{2.5}$$

where $|\varphi| < 1$. Using the recursion (2.5) for $t-1$, and substituting back in, we get that $Y_t = \varepsilon_t + \varphi\varepsilon_{t-1} + \varphi^2 Y_{t-2}$. Continuing in this manner we get after n steps that $Y_t = \varepsilon_t + \varphi\varepsilon_{t-1} + \cdots + \varphi^n\varepsilon_{t-n} + \varphi^{n+1}Y_{t-n-1}$. Since $|\varphi| < 1$, the last term

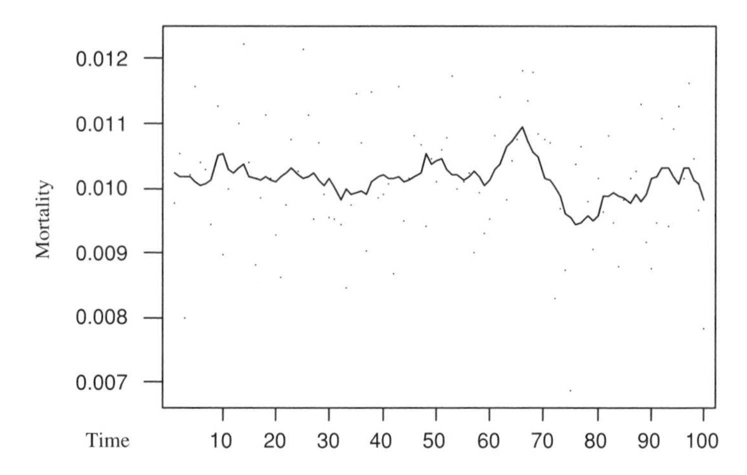

FIGURE 2. Hypothetical Mortality Rates and a Moving Average Estimate of their Level.

converges to zero, as $n \to \infty$. Thus, an AR(1) process is obtained by taking $\psi_j = \phi^j$ for all j, in (2.1). Note that the assumption $|\varphi| < 1$ guarantees that the variance (2.2) is finite. In fact, $\text{Var}(Y_t) = \sigma_\varepsilon^2/(1-\varphi^2)$ and $\text{Cov}(Y_t, Y_{t+k}) = \sigma_\varepsilon^2 \varphi^k/(1-\varphi^2)$. It follows that the autocorrelation function is $\rho_k = \varphi^k$ for all $k = 0, 1, 2, \ldots$ Thus, in contrast with the MA(1) process, whose autocorrelation is zero after one lag, the current value of the AR(1) process is correlated with *all* earlier (and future) values. We can interpret ε_t as a one-step ahead prediction error, because if Y_{t-1} is known we predict Y_t by φY_{t-1}. ◊

In analogy with (2.5) one can define the general *autoregressive process of order p*, or *$AR(p)$*, by the recursion, $Y_t = \varphi_1 Y_{t-1} + \cdots + \varphi_p Y_{t-p} + \varepsilon_t$, where $\varphi_p \neq 0$.[4] To provide a compact description, it is customary to define a *back shift (or lag) operator B* such that $BY_t = Y_{t-1}$, $B^2 Y_t = Y_{t-2}$ etc. We can define a polynomial operator $\Phi(B) = 1 - \varphi_1 B - \cdots - \varphi_p B^p$. Then, the AR($p$) process can be written as $\Phi(B)Y_t = \varepsilon_t$.

To guarantee that such a recursive process has a representation (2.1) (i.e., that it defines a stationary process with a finite variance) the coefficients φ_j must be such that the roots of the polynomial equation $\Phi(B) = 0$ are strictly greater than 1 in absolute value. For example, when $p = 1$, we have $1 - \varphi B = 0$, or $B = 1/\varphi$, so the condition is satisfied in Example 2.2. In this case we have $(1 - \varphi B)Y_t = \varepsilon_t$, or $Y_t = (1 - \varphi B)^{-1}\varepsilon_t = (1 + \varphi B + \varphi^2 B^2 + \cdots)\varepsilon_t$.

Define another operator $\Theta(B) = 1 - \theta_1 B - \cdots - \theta_q B^q$. Then, the MA($q$) process of Example 2.1 can be written as $Y_t = \Theta(B)\varepsilon_t$. An *autoregressive moving average process*, or *ARMA(p, q)* process, is defined by the equation $\Phi(B)Y_t = \Theta(B)\varepsilon_t$. For example, when $p = q = 1$, we get the ARMA(1,1) process

[4] This notion generalizes further to *vector-valued autoregressive* (*VAR*) processes, in which the coefficients are matrices (cf., Chatfield 1996, Ch. 12).

$Y_t - \varphi Y_{t-1} = \varepsilon_t - \theta\varepsilon_{t-1}$. The ARMA(2,2) process is usually written as $Y_t - \varphi_1 Y_{t-1} - \varphi_2 Y_{t-2} = \varepsilon_t - \theta_1\varepsilon_{t-1} - \theta_2\varepsilon_{t-2}$ etc.

It is clear from the defining recursive equation of the AR(p) processes that ε_t can be expressed in terms of the Y_{t-j}'s for $j \geq 0$. To guarantee the same for the MA(q) processes, and ARMA(p, q) processes in general, we must require that the roots of the polynomial equation $\Theta(B) = 0$ are greater than one in absolute value. Such processes are called *invertible*. In the case of MA(1) process, this means that we must have $|\theta| < 1$, for example.

A final piece in the description of ARMA(p, q) processes is to tie up the representation $\Phi(B)Y_t = \Theta(B)\varepsilon_t$ with (2.1). Define a power series $\Psi(B) = 1 + \psi_1 B + \psi_2 B^2 + \cdots$, so (2.1) can be written as $Y_t = \Psi(B)\varepsilon_t$. The representation (2.1) of ARMA(p, q) processes is obtained by equating the two power series $\Psi(B) = \Phi(B)^{-1}\Theta(B)$. In the case of ARMA(1,1) process we get $\psi_j = (\varphi - \theta)\varphi^{j-1}$ for $j > 0$, for example. We see that the ARMA(p, q) processes are a subclass of linear processes such that $\Psi(B)$ is a ratio of two polynomials.

The concept of *ARIMA*(p, d, q) models, or *autoregressive integrated moving average* models, is obtained by assuming that the d-fold difference of the process follows an ARMA(p, q) model. For example, suppose Y_t follows an ARMA(p, q) model, and define $Z_t = Y_0 + \cdots + Y_t$. In this case Z_t is the summed, or *integrated*, version of Y_t, and we have that $(1 - B)Z_t = Y_t$. Therefore, Z_t follows the ARIMA($p, 1, q$) model. Furthermore, if $X_t = Z_0 + \cdots + Z_t$, then $(1 - B)X_t = Z_t$ and $(1 - B)^2 X_t = Y_t$, so X_t is an ARIMA($p, 2, q$) process etc.

Example 2.3. EWMA Processes. Consider an ARIMA(0,1,1) model of the form $(1 - B)Z_t = \varepsilon_t - \theta\varepsilon_{t-1}$, where $0 < \theta < 1$. With some algebra, one can show that $Z_t = \varepsilon_t + m_{t-1}$, where

$$m_{t-1} = (1 - \theta)(Z_{t-1} + \theta Z_{t-2} + \theta^2 Z_{t-3} + \cdots) \tag{2.6}$$

can be viewed as the "level" of the process at time t. Since the weights $(1 - \theta)\theta^j$, $j = 0, 1, \ldots$ sum to 1 and fall off exponentially, this estimate of level is often called *exponentially weighted moving average*, or *EWMA*. We see from (2.6) that $m_t = (1 - \theta)Z_t + \theta m_{t-1}$, so for $0 < \theta < 1$, the estimate of the level is updated as a weighted average of the new observation and previous estimate. Substituting in $Z_t = \varepsilon_t + m_{t-1}$ we see that the updating equation can also be expressed as $m_t = (1 - \theta)\varepsilon_t + m_{t-1}$. This is the so-called *error-correction form* of the updating formula. Even before the systematic development of the theory of ARMA models by Box and Jenkins, the EWMA method had evolved into a forecasting method on its own right (cf., Muth 1960). In this approach, a forecast of Z_{t+1} is m_t, because the future error ε_{t+1} has mean zero and is independent of the past observations. From the error correction form we see that in general, the forecast is $\hat{Z}_{t+k} = m_t$. In estimating m_t one often uses judgment to select the parameter θ rather than estimate it from the data. In this case it is customary to call $1 - \theta$ as the *smoothing parameter*. One way to think about the smoothing parameter is that it determines the weighting involved in the computation of the local level (2.6). If we have a (subjective) view of how far back the data are relevant in the determination of the local level, then a value may possibly be determined. Chatfield (1996, 70) notes that values of the smoothing parameter in the range from 0.1 to 0.3 are often preferred. An illustration will be given in Figure 6. ◊

2.1.3. Practical Modeling

The first step in modeling is to *plot the data*. This reveals if there are unusual observations that may have a large influence on estimation. Sometimes the unusual observations are data errors that should be corrected before proceeding further. At other times they may be real, but reflect unusual aspects of the process. Examples include peaks in mortality or fluctuations in fertility caused by wars, epidemics or famines; level shifts in population data caused by changes in national or other administrative borders; or discontinuities caused by changes in migration or naturalization policies.

Whether the series varies around a fixed mean with a constant variance often can be seen from the plot. Note that apart from social, economic, or political factors, the volatility of a demographic process may change simply as a consequence of population growth because the variance of a binomial or Poisson variable is proportional to the expected value of the number of events.

In addition to the plot, one would typically compute the autocorrelation function. We see from (2.3) that the autocorrelation of all linear processes (2.1) must eventually converge to zero because the absolute convergence of the series of ψ_j's implies that $\psi_j \to 0$ as $j \to \infty$. In contrast, if the series has a polynomial trend then, depending on the length of series, the lag, and the order of the polynomial, many types of persistent fluctuating patterns can manifest themselves. Thus, if the autocorrelations do not approach zero quickly, then the series may be best approximated by a nonstationary model.[5]

For example, a visual inspection of the sex ratio at birth in Figure 6 of Chapter 4 suggests that the process does not have a constant mean. This shows up in the autocorrelation function. It starts from 0.52 at lag $= 1$, and then declines in roughly monotone manner, but at lag $= 51$ we still observe a value as high as 0.23. The latter value appears to be statistically significant because there are $n = 250$ observations. (If the kth autocorrelation is approximately $\varphi^{|k|}$, then the variance of an autocorrelation beyond the first is approximately $n^{-1}(1+\varphi^2)(1-\varphi^2)^{-1}$ (Box and Jenkins, 1976, 35), and the estimated standard error is about 0.08.) In contrast, the autocorrelations of the first differences begin from -0.41 at lag $= 1$, and remain small in absolute value, with one value at 0.15 and the rest much smaller. A comparison of parsimonious ARIMA(p, 1, q) models shows that an ARIMA(0,1,1) model provides an approximate representation for the series.

[5] Nonlinear models are also an alternative. They are capable of representing different behavior when the series is at a relatively high level as compared to being at a relatively low level; when it is increasing as compared to decreasing, etc. (Complement 15) Existing models appear to have been mostly motivated by economic considerations (e.g., Granger and Teräsvirta 1993), but they may eventually provide useful alternatives for demographic data, as well.

Example 2.4. Vital Processes Appear Nonstationary. We analyzed the logarithm of white age-specific fertility rates in 1921–1988 in ages 14,15,..., 46, and the logarithm of mortality rates for males and females in 1940–1988 in ages 1, 2, 3, 4, 5–9, 10–14,..., 80–84, 85+ in the U.S. Based on plots and the study of autocorrelations we concluded that *all series appeared nonstationary* (see also Lee and Tuljapurkar 1994; Lee 1974). The autocorrelations did not approach zero, as they should for a linear process. We then looked for the smallest d such that the d-th difference both looked stationary in a plot and had an autocorrelation that did approach zero fairly quickly. Fertility had to be differenced twice to remove persistent patterns from autocorrelations in ages 19–44. Mortality had to be differenced twice for stationarity in ages 30–49 for males and in ages 20–49 for females. For other rates differencing once was sufficient. The sample first-autocorrelations r_1 of the first differences of the U.S. fertility series mentioned above varied from -0.24 to 0.75. with average $= 0.41$. For the first differences of the mortality rates we had $-0.39 \leq r_1 \leq 0.53$ with male average -0.02 and female average -0.03. The analysis indicates that while there are opportunities for ARMA modeling of the first differences of these series, the representations may be approximate only. $\diamondsuit$

Once a stationary looking series is found, one tries to identify an ARMA(p, q) model for it. Although there is no theoretical limit for the values of p and q, it is relatively rare that demographically meaningful models would have $p + q > 3$, when annual data are used. (Monthly data displaying seasonality are a different matter that will not be discussed here.) Even values $p = 3$ or $q = 3$ yield models that are rarely interpretable, because they imply an independent influence from year $t - 3$ on the value of the process at year t, even when one controls for the values of the process in years $t - 1$ and $t - 2$. (This effect can be quantified in terms of partial autocorrelations; Complement 12.) In any event, it is advisable to fit at least all of the remaining models and to compare them based on the residual sum of squares, the significance of the parameter estimates, and estimated residuals, much the same way ordinary regression models are identified.

Sometimes there is a peak in autocorrelation at a lag k that defies explanation. Although such peaks can theoretically arise from infinitely many ARMA(p, q) processes, it sometimes happens that the correlation is due to a small number, possibly just one, pair of observations k steps apart, (Y_t, Y_{t-k}) for some t. Such pairs may be difficult to detect from the plot of the series itself. A useful diagnostic tool for investigating this possibility is to make a so-called *lag-plot* with lag k, i.e., a plot of the pairs (Y_t, Y_{t-k}) for all t. We will illustrate this in Section 2.2.2.

As a practical example of the application of the ARIMA models we will consider the annual growth rate of the U.S. population in 1900–1999. The population is the so-called *mid-year population*, or the population as of July 1, each year.[6] In

[6] The data are from Population Estimates Program, Population Division, U.S. Census Bureau, Internet Release Date: April 11, 2000, Revised date: June 28, 2000, http://eire.census.gov/popest/archives/pre1980/popclockest.txt.

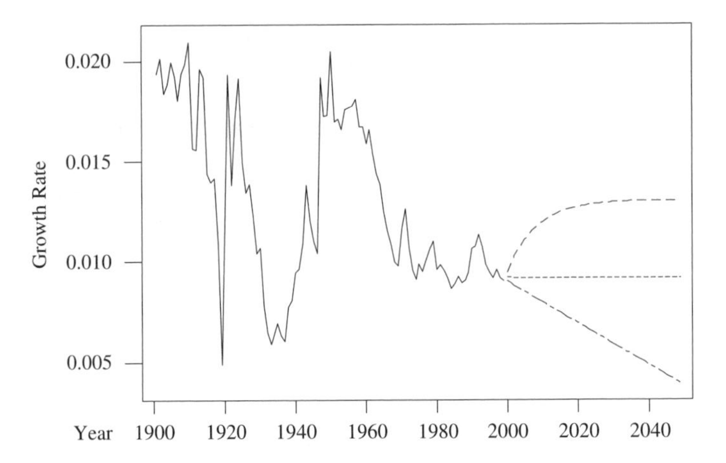

FIGURE 3. The Growth Rate of the U.S. Population in 1900–1999, and Three Forecasts: AR(1) (dashes) and ARIMA(2,1,0) with (dot-dashes) and without a Constant Term (short dashes).

1900–1949 the figures exclude Alaska and Hawaii. Thus, there is a level shift from 1949 to 1950. The population comprises the national resident population (or *de jure* population) except that in years 1917–1919 and 1940–1979 the armed forces overseas have been included. This has the effect of smoothing the growth rate, notably around 1917–1919. Although adjustments could be made, we chose not to do so because their effect would be minor.

Define V_t as the size of the population in year t. Then, $\log(V_{t+1}/V_t)$ is the growth rate from t to $t+1$. Figure 3 has a plot of the growth rate of the U.S. population for 1900–1999, together with three point forecast that will be discussed at the end of this example. The plot shows that the series has a declining trend. The nonstationarity shows up in the autocorrelation function, which declines roughly linearly from 0.85 at lag $= 1$ to -0.37 at lag $= 25$. A plot suggests that the first differences vary around a constant mean. (In Section 4.1 we will see that the variance is not constant, however.) The first seven autocorrelations are -0.122, -0.372, 0.255, 0.149, -0.248, -0.140, 0.278. Beyond lag $= 7$ the correlations are < 0.2 in absolute value. Lag-plots (not shown) indicate that the negative autocorrelation at lag 2 and the positive autocorrelation at lag 7 are largely due to outliers (e.g., declines in 1918–1919 and 1945 coupled with increases in 1920–1921 and 1947). Thus, the best fitting ARIMA model need not be best model for forecasting purposes. We will come back to this issue later, but proceed now with the data as they are.

Since the growth rate is the first difference of log population sizes, an ARMA(p, q) model for the first difference of the growth rate is the same as an ARIMA$(p, 1, q)$ model for the log population size. Slight differences in numerical output may occur, however, depending on how the endpoints of the series are handled in estimation. Various ARIMA$(p, 1, q)$ models were fitted. Based on residual checks, models ARIMA(0,1,1), ARIMA(1,1,0), and ARIMA(1,1,1) are not acceptable. ARIMA(2,1,0) fits better than ARIMA(0,1,2), and just about equally well as ARIMA(0,1,3). Adding autoregressive parameters does not help. Although (as we will see) the first autoregressive coefficient is not significant, ARIMA(2,1,0) is a reasonable choice within this class of models.

We used Minitab to carry out the analyses. Let Y_t be the rate of change. The estimated model is $Y_t - Y_{t-1} = -0.1644(Y_{t-1} - Y_{t-2}) - 0.3901(Y_{t-2} - Y_{t-3}) + \varepsilon_t$, if the mean of the differences is assumed to be zero. The estimated standard error of both autoregressive parameters is 0.0939, so the two P-values are 0.083 and 0.000, respectively. If we allow a nonzero mean by adding a constant term to the model, we get the estimates $Y_t - Y_{t-1} = -0.0001618 - 0.1687(Y_{t-1} - Y_{t-2}) - 0.3937(Y_{t-2} - Y_{t-3}) + \varepsilon_t$, instead. (Note that the constant is *not* the mean of the differences itself, when the model includes autoregressive terms; Exercise 13.) The estimated standard error of the constant term is 0.00020 corresponding to a P-value of 0.430.

In a time series setting, the MLE's are usually calculated under a normal assumption. Even when the assumption is true the MLE's are typically biased to some extent and their estimated standard errors are based on approximations that may not be accurate in small samples. A version of the bootstrap method discussed in Chapter 3, the so-called *parametric bootstrap* can be used to investigate both aspects once a model has been fit (Efron and Tibshirani 1993; cf. Section 8.2 of Chapter 3). The maximum likelihood estimation procedure gives us a set of estimated residuals. In this application of the parametric bootstrap, we can sample with replacement from the set of estimated residuals, use the sampled values as innovations, and generate realizations (sample paths) from the estimated model with the same number of observations as the original series. (Thus the procedure is valid even if the normality assumption is not true.) We produced 1,000 such realizations and re-estimated the ARIMA(2,1,0) model with the constant for each one. This produced 1,000 estimates of the constant γ and the autoregressive parameters φ_1 and φ_2 that can be used to estimate the joint sampling distribution of (γ, φ_1, φ_2). The bootstrap estimates of standard errors of the autoregressive parameters were 0.0935 and 0.0943, so they were essentially identical with the estimate given by Minitab. Similarly, the bootstrap estimate of the standard error of the constant term was 0.00020. In this case the two analyses agreed.

Figure 3 also shows three forecasts of the series. Stationary ARMA(p, q) models do not seem appropriate for the series based on the unacceptable fit, but we have included a forecast made with an AR(1) model to show the effect of using a stationary model for a series that obviously is nonstationary. The other two forecasts are based on an ARIMA(2,1,0) model either with or without a constant term. We see that the AR(1) based forecast continues smoothly from the last observed value to the historical (1900–1999) mean. The ARIMA(2,1,0) without a constant term produces essentially the same forecast as a random walk model. After small initial wiggles it runs parallel to the time axis. The model with a constant term estimates the average rate of change in the growth rate, and assumes the linear change to continue. We will comment on the difference of the latter two models in Example 3.3.

2.2. *Characterization of Predictions and Prediction Errors*

2.2.1. *Stationary Processes*

Suppose we make a forecast for Y_{t+k} at time t. From (2.1) we can write the future values as $Y_{t+k} = F_k(t) + E_k(t)$, where

$$E_k(t) = \psi_0\varepsilon_{t+k} + \psi_1\varepsilon_{t+k-1} + \cdots + \psi_{k-1}\varepsilon_{t+1}, \tag{2.7}$$

and

$$F_k(t) = \psi_k\varepsilon_t + \psi_{k+1}\varepsilon_{t-1} + \cdots \tag{2.8}$$

If the ψ_j's are known, then we know the value of $F_k(t)$ at time t for an invertible ARMA(p, q) process, but $E_k(t)$ is independent of the past and has mean $= 0$. It follows that $F_k(t)$ is the minimum mean-squared-error forecast of Y_{t+k}.[7] Note that *error = forecast − true value*. Hence, $E_k(t)$ is the negative of the forecast error.

Since $E_k(t)$ is independent of $F_k(t)$, its distribution is the same, both conditionally given the past of the process until time t, and unconditionally. To put it in another way, (apart from the problem of identifying and estimating a model for the process) the accuracy of the forecast is independent of the particular sample path the process has followed until time t. Intuitively, this means that the "forecastability" of the linear process is assumed not to depend on history or to change over time.

In practice, the ψ_j's must be estimated from data so $F_k(t)$ is only known up to estimation and specification error. Although such errors can be large, in this section they will be ignored.

Letting $k \to \infty$ in (2.8), we see that the forecast function of all stationary processes of type (2.1) converges to zero (or to the mean when the estimated mean is added back), because the ψ_j's converge to zero. This shows that the analysis of the autocorrelation structure is primarily useful in relatively short term forecasting. In the longer term the value of the mean is decisive.

Suppose now that we use (2.8) to make two forecasts at time t, one for time $t + k$, the other for time $t + k + h$ with $k, h \geq 0$. From (2.7) one can deduce that

$$\mathrm{Cov}(E_k(t), E_{k+h}(t)) = \sigma_\varepsilon^2 \sum_{j=0}^{k-1} \psi_j\psi_{j+h}. \tag{2.9}$$

It follows that, when $F_k(t)$ is known, the covariance structure of the forecast error does not depend on the time t at which the forecast is made. When this is the case, we will write E_k instead of $E_k(t)$. In typical applications the mean of a process must be estimated from the data and the correlation analysis is carried out on

[7] Geometrically, we may view $F_k(t)$ as the projection of Y_{t+k} on the subspace spanned by $(\varepsilon_t, \varepsilon_{t-1}, \ldots)$. The projection is orthogonal, because $F_k(t)$ and $E_k(t)$ are uncorrelated.

centered data. In forecasting the mean is added back in. Denote the variance of the mean estimate by ${\sigma_\mu}^2$. In forecasting k steps ahead, we see that $\mathrm{Var}(\mathrm{E}_k)/\sigma_\mu^2 \rightarrow \mathrm{Var}(Y_t)/\sigma_\mu^2$, as $k \rightarrow \infty$, so σ_μ^2 is of the same order of magnitude as $\mathrm{Var}(E_k)$, and error in the estimation of the mean always remains a factor of uncertainty for all lead times.

Example 2.5. Standard Error Under AR(1) Residuals. Estimation error depends on the autocorrelation structure of the process. Suppose we have observations $Z_t = \mu + Y_t$, where $Y_t = \varphi Y_{t-1} + \varepsilon_t$. That is, Z_t is an AR(1) process with mean μ. Suppose we have observation at $t = 1, \ldots, n$, and we take $\hat{\mu} = (Z_1 + \cdots + Z_n)/n$. What is the standard error of the mean? We have that $\mathrm{Var}(Z_1 + \cdots + Z_n) = n\sigma_Z^2 + 2\{(n-1)\varphi\sigma_Z^2 + (n-2)\varphi^2\sigma_Z^2 + \cdots + \varphi^{n-1}\sigma_Z^2\} \approx n\sigma_Z^2(1+\varphi)/(1-\varphi)$ for large n, so the standard error is approximately $\sigma_Z[(1+\varphi)/n(1-\varphi)]^{1/2}$. We see that the higher the correlation φ, the higher the standard error. For example, if $\varphi = 0.9$, then the standard error is over 4 times bigger than under independent random sampling. ◊

Denote by $\rho(X, Y)$ the correlation between any two variables X and Y. Then, (2.9) leads to the well-known result (Box and Jenkins 1976, 160)

$$\rho(E_k, E_{k+h}) = \sum_{j=0}^{k-1} \psi_j \psi_{j+h} \Bigg/ \left\{ \sum_{i=0}^{k-1} \psi_i^2 \sum_{l=0}^{k+h-1} \psi_l^2 \right\}^{1/2}. \tag{2.10}$$

Example 2.6. Correlations of Forecast Errors For AR(1) Processes. In the case of an AR(1) process, $\psi_{k+j} = \phi^k \psi_j$, so the forecast of Y_{t+k} is $\hat{Y}_{t+k} = \varphi^k Y_t$ for $k = 1, 2, \ldots$ From (2.9) we see that, if φ is known, the theoretical variance of the forecast error is $\sigma_\varepsilon^2(1-\varphi^{2k})/(1-\varphi^2)$. From (2.10) we find that

$$\rho(E_k, E_{k+h}) = \varphi^h \left[\frac{1-\varphi^{2k}}{1-\varphi^{2k+2h}} \right]^{1/2}. \tag{2.11}$$

For large k the correlation is approximately φ^h. For large h the correlation approaches zero. ◊

2.2.2. *Integrated Processes*

Consider an integrated process Z_t that is related to a stationary process Y_t (as defined in (2.1)) via the first differences $Y_t = Z_t - Z_{t-1}$. Suppose we know the values of Z_{t+j} for $j = 0, -1, -2, \ldots$ and we want to forecast Z_{t+k}, for $k = 1, 2, \ldots$ We can always write

$$\begin{aligned} Z_{t+k} &= Z_t + Y_{t+1} + \cdots + Y_{t+k} \\ &= Z_t + F_1(t) + E_1(t) + \cdots + F_k(t) + E_k(t). \end{aligned} \tag{2.12}$$

Therefore, if we ignore the estimation error in the ψ_j's, the optimal forecast is $\hat{Z}_{t+k} = Z_t + F_1(t) + \cdots + F_k(t)$, and the negative of the forecast error is $Z_{t+k} - \hat{Z}_{t+k} = E_1(t) + \cdots + E_k(t) \equiv E_{(k)}$. Although the error depends on t, its moments do not, and we suppress the dependency in our notation. We see from (2.7) that the E_j's are all linear combinations of ε_{t+h}'s with $h = 1, \ldots, k$, so the forecast error is independent of the forecast. A direct calculation yields the result,

$$\operatorname{Cov}(E_{(k)}, E_{(k+h)}) = \sigma_\varepsilon^2 \sum_{i=1}^{k} \left(\sum_{j=0}^{k-i} \psi_j \right) \left(\sum_{l=0}^{k+h-i} \psi_l \right) \tag{2.13}$$

for $k, h \geq 0$. Note that both inner sums of the ψ_j's are bounded in absolute value. It follows that $\operatorname{Var}(E_{(k)})$ is of the order of magnitude k, or $O(k)$, if the parameter estimation error is ignored.

Example 2.7. Correlations of Forecast Errors for Integrated AR(1) Processes. Suppose that the first differences follow an AR(1) process. In this case $F_k(t) = \varphi^k Y_t$, where $Y_t = Z_t - Z_{t-1}$. Therefore, the forecast function $\hat{Z}_{t+k} = Z_t + \varphi(Z_t - Z_{t-1})(1 - \varphi^k)/(1 - \varphi)$ has the asymptotic value $Z_t + \varphi(Z_t - Z_{t-1})/(1 - \varphi)$, as $k \to \infty$. In demographic forecasting $|\varphi|$ is often small, so the asymptotic value tends to be close to the current value. The second moments of the forecast error of are of the form

$$\operatorname{Cov}(E_{(k)}, E_{(k+h)}) = \frac{\sigma_\varepsilon^2}{(1-\varphi)^2} \left[k - (\varphi + \varphi^{h+1}) \frac{1 - \varphi^k}{1 - \varphi} + \varphi^{h+2} \frac{1 - \varphi^{2k}}{1 - \varphi^2} \right]. \tag{2.14}$$

Because the partial sums in (2.13) are all positive for AR(1) first differences, it is easy to show that *the covariance is positive* for $|\varphi| < 1$. We see from (2.14) that the covariance of the forecast error is asymptotically proportional to the shorter lead time, k. Hence, in contrast with the AR(1) case of Example 2.6, the variance increases without a bound. We have $\rho(E_{(k)}, E_{(k+h)}) \to 1$, when h is fixed and $k \to \infty$, and $\rho(E_{(k)}, E_{(k+h)}) \to 0$, when k is fixed and $h \to \infty$. As φ tends to 0, the autocorrelations $\rho(E_{(k)}, E_{(k+h)})$ tend to $(k/(k+h))^{1/2}$, which is the autocorrelation function of a random walk. ◊

Taken together, Examples 2.4 and 2.7 support the conclusion that *the autocorrelations of the forecast errors of the demographic vital rates must typically be positive and high.* This limits the accuracy of empirical estimates of past forecast errors.

For another qualitative insight, consider (2.14) with $h = 0$. Note that under an AR(1) model for the process increments we have $\operatorname{Var}(Y_t) = \sigma_\varepsilon^2/(1 - \varphi^2)$, so for large k we have $\operatorname{Var}(E_{(k)}) \approx k \times \operatorname{Var}(Y_t) \times (1 + \varphi)/(1 - \varphi)$. Thus, an approximation for the variance of the forecast error can be obtained based on a simple random walk model, only then the empirical variance of the process of increments must be multiplied by $(1 + \varphi)/(1 - \varphi)$.

Example 2.8. Standard Error and Random Error. Suppose now that forecasting is carried out using an estimated mean of the differences Y_t. Denote the variance of the mean estimate by σ_μ^2. The mean of the Y_t's introduces a linear trend into the forecast function with a slope equal to the mean. Therefore, the variance of the estimated linear trend at lead time k is $k^2\sigma_\mu^2$, or it is $O(k^2)$. A comparison with (2.13) and (2.14) with $h = 0$ shows that in long-term forecasting based on differenced series the uncertainty concerning the mean always eventually dominates in the overall forecasting error. ◊

We omit the details but note that if Z_t would be a *twice-integrated* version of Y_t (or $Y_t = Z_t - 2Z_{t-1} + Z_{t-2}$), then we have the result,

$$\begin{aligned}&\mathrm{Cov}(E_{[k]}, E_{[k+h]})\\ &\quad = \sigma_\varepsilon^2 \sum_{i=1}^{k} \left(\sum_{j=0}^{k-i} (k - i + 1 - j)\psi_j \right) \left(\sum_{l=0}^{k+h-i} (k + h - i + 1 - j)\psi_l \right), \end{aligned} \tag{2.15}$$

where $E_{[k]}$ denotes the forecast error of Z_t at lead time k. Thus, the variance is $O(k^3)$ for a twice integrated process compared to $O(k)$ for a once integrated process. If an estimated nonconstant mean of the second differences is used in forecasting, then a second degree polynomial trend is introduced into the forecast function. Its variance is $O(k^4)$, so eventually the uncertainty of the trend estimates exceeds that of the random part, just as for once-integrated processes. For these models the width of the prediction intervals is $O(k^2)$, so the intervals open up like a trumpet, as compared to the tulip shape we have for random walks, for example. Unless twice differenced processes are constrained in some way, this result alone precludes their use in many demographic applications.

Figure 5 of Chapter 4 has a graph of the total fertility rate $T(t)$ of Finland. Here we analyze the (post demographic transition) period 1920–1996 that is given in Figure 4, together with 50% prediction intervals for 1997–2025. The series is obviously nonstationary. This is confirmed by a very slowly declining autocorrelation function. We took $Z_t = \log(T(t))$ as the variable to be analyzed. This guarantees the positivity of all results, but, more importantly, it transforms changes into relative scale, which seems reasonable given the large variation in the level of the total fertility rate. Based on a graph, the first differences $Y_t = Z_t - Z_{t-1}$ appear reasonably stationary, except that the zig-zag pattern of the years 1940–1945 visible in Figure 4 produces a corresponding zig-zag pattern in the first differences. This war period[8] is clearly different from the rest. The first two autocorrelations of the differenced series are -0.365 and 0.433. Figure 5A has a lag-plot corresponding to the first autocorrelation. We see that the negative value is due to three outliers. These relate to the war years. In fact, if the points are removed for which $Y(t)$ has $t = 1941$–1944, the first correlation changes from -0.365 to 0.411. In contrast, Figure 5B shows that while the outliers caused by the war are influential at lag 2 also, they are much more in accordance with positive autocorrelation of the remaining values. Removing the points for which $Y(t)$ has $t = 1941$–1945 actually

[8] The war in Finland started in 1939, there was an interim peace from March 1940 to June 1941, and the war continued until 1944.

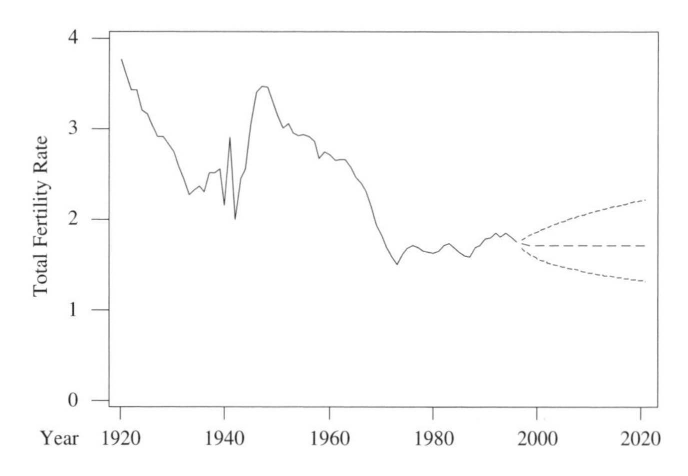

FIGURE 4. Total Fertility Rate of Finland in 1920–1996, and its Forecast for 1997–2021 with 50% Prediction Intervals.

reduces the second autocorrelation from 0.433 to 0.269. It seems clear that identifying an ARIMA model from data that are dominated by war time outliers is not an appropriate approach to forecasting. Therefore, we smoothed the values of the war years 1940–1942 using RSMOOTH.

A graph of the first differences of the adjusted series shows that there is still a large outlier due to the peak of the baby-boom in 1947, but there is no obvious basis for changing this value. After some experimentation we found that ARIMA(1,1,0) gives the best fit among parsimonious models although it would still be rejected on a formal test of the residuals. Thus, we have a model

$$Z_t - Z_{t-1} = \varphi(Z_{t-1} - Z_{t-2}) + \varepsilon_t, \tag{2.16}$$

where $\hat{\varphi} = 0.4984$.

Minitab also gives the estimate $\sigma_\varepsilon^2 = 0.001626$ for the innovation variance, estimated from the residuals of the fitted model. Here, we have to pause. Motivated by forecasting considerations, we have reduced the variability of the process, so using residuals from the smoothed series underestimate past uncertainty. An alternative is to use the fitted values of the adjusted series to estimate the innovation variance from the original observations. Doing this yields the estimate $\sigma_\varepsilon^2 = 0.002902$, or the estimate is nearly doubled. Which estimate is preferable? There is no unequivocal answer, but we note that the difference of the two estimates is due to the war time fluctuations. *Conditioning* on the assumption that there will be no similar fluctuations during the forecast period, we may use the smaller estimate in our illustration.

The last two values of the total fertility rate were $T(1995) = 1.81$ and $T(1996) = 1.76$, with logs $Z_{1995} = 0.59333$ and $Z_{1996} = 0.56531$. Therefore, the last observed difference was $Y_{1996} = -0.028013$. It follows that the point forecast of Z_{1996+k}

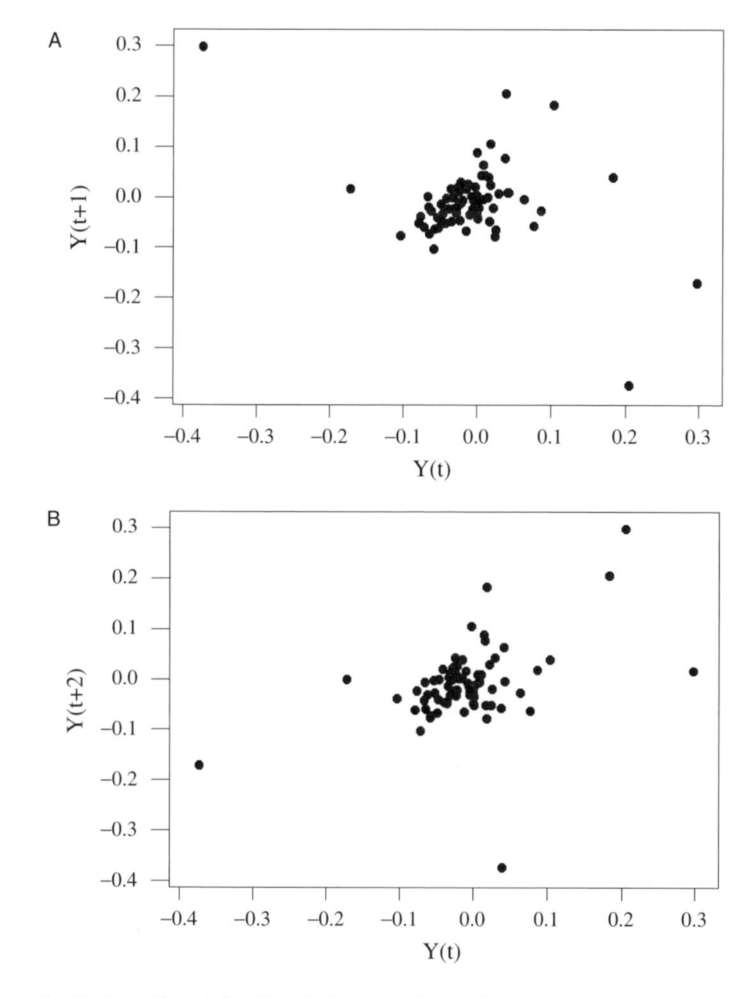

FIGURE 5. (A) Lag-Plot of the First Differences $Y(t)$ at Lag 1.
(B) Lag-Plot of the First Differences $Y(t)$ at Lag 2.

is $\hat{Z}_{1996+k} = 0.56531 + (-0.028013)\{0.4984 + 0.4984^2 + \cdots + 0.4984^k\}$. The point forecast depicted in Figure 4 is $\hat{T}(1996 + k) = \exp(\hat{Z}_{1996+k})$ for $k = 1, \ldots, 25$. The variance of forecast error $\text{Var}(E_{(k)})$ has been calculated using formula (2.14) with $h = 0$, $\varphi = 0.4984$, and $\sigma_\varepsilon^2 = 0.001626$. The 50% prediction intervals are of the form $\exp(\hat{Z}_{1996+k} \pm 0.6745 \times \text{Var}(E_{(k)})^{1/2})$, based on a normal approximation for the distribution of $\hat{Z}$. Although the prediction intervals are symmetric in the log-scale, the exponentiation transforms them into asymmetric ones. We may note some additional aspects of the prediction intervals. First, the estimated uncertainty of the one-step-ahead forecast is quite high relative to the low level of variability observed since the 1970's. This points to a change in volatility. Second, since $[0.002902/0.001626]^{1/2} \approx 1.34$, if we were to use the larger estimate of innovation variance, the intervals would be approximately 1/3 wider.

2.2.3. Cross-Correlations

For future reference we also need results corresponding to the cross-correlations between forecast errors of different processes. Suppose, therefore, that in addition to Y_t given by (2.1), there is another stationary process $Y_t' = \psi_0'\varepsilon_t' + \psi_1'\varepsilon_{t-1}' + \cdots$. Let the innovation processes ε_t and ε_t' have correlation $\rho(\varepsilon_t, \varepsilon_t') = \delta$ and $\rho(\varepsilon_t, \varepsilon_{t+k}') = 0$ for $k \neq 0$. The forecast errors E_k and E_{k+h}' of the two processes have the cross-covariances (cf., (2.9))

$$\mathrm{Cov}(E_k, E_{k+h}') = \delta\sigma_\varepsilon\sigma_\varepsilon' \sum_{j=0}^{k-1} \psi_j \psi_{j+h}'. \tag{2.17}$$

It follows from the Cauchy-Schwartz inequality that the correlation between the prediction errors is *less* in absolute value than the innovation correlation δ even for $h = 0$. Letting $k \to \infty$ in the above formula yields a formula for $\mathrm{Cov}(Y_t, Y_t')$. Hence, an inspection of cross-correlations gives an indication of what the cross-correlations of prediction errors look like.

Similar formulas for the prediction errors of the once and twice integrated processes Z_t, can be obtained from the autocovariance formulas (2.13) and (2.15), if we replace σ_ε^2 by $\delta\sigma_\varepsilon\sigma_\varepsilon'$ and ψ_l by ψ_l'.

These findings lead us to the following methodological remark. Official demographic forecasts typically assume a perfect (positive or negative) correlation between the forecast errors of different vital processes. This is a very restrictive assumption, because even under the current highly simplified setting it can only be valid if (a) the innovations are perfectly correlated, and (b) the processes have identical autocorrelation structures. As we will show in more detail in Chapter 8, in demographic applications neither condition holds.

3. Handling of Nonconstant Mean

Several approaches are available for modeling nonconstant trends. One is differencing the time series one or more times, as we did for the U.S. growth rate and the Finnish total fertility rate, above. Another is to explicitly estimate a smooth trend function using parametric functions, splines, or some form of moving averages. A third possibility is to use a stochastic representation for the trend, and estimate it based on the model.

3.1. Differencing

We consider here the implications of differencing for the forecasts obtained. Suppose we find that the series Z_t is nonstationary, but the first differences

$Y_t = Z_t - Z_{t-1}$ appear to be stationary around a mean $\mu \neq 0$. Let us assume that Z_t is the last observed value, and we want to forecast Z_{t+k} for some $k > 0$. We can write $Z_{t+k} = Z_t + Y_{t+1} + \cdots + Y_{t+k}$. Suppose an AR(1) process with parameter φ describes the centered differences $Y_{t+j} - \mu$ well. Then, as shown in Example 2.6, the best forecast of $Y_{t+j} - \mu$ is $\varphi^j(Y_t - \mu)$. It follows that the best forecast of Z_{t+k} is

$$\hat{Z}_{t+k} = Z_t + k\mu + (Y_t - \mu)\sum_{i=1}^{k} \varphi^i, \tag{3.1}$$

where $Y_t = Z_t - Z_{t-1}$. We see that the presence of μ produces a linear trend $k\mu$ in the forecast function. The trend eventually dominates, because the sum on the right hand side converges to $\varphi/(1 - \varphi)$, as $k \to \infty$.

Example 3.1. Forecasting a Random Walk with a Drift. Note that if $\varphi = 0$, or the first differences are uncorrelated, then Z_t is a random walk process with a drift (if $\mu \neq 0$), and the forecast consists of the *jump-off* or starting value Z_t and a linear trend $k\mu$. The constant term μ would normally be estimated from the data. Suppose the observations were made at times $0, 1, \ldots, n$. Then, the average of the differences is $(Y_1 + \cdots + Y_n)/n = (Z_n - Z_0)/n$, which is the slope of a line between the first and the last observation. Therefore, the forecast function (3.1) is simply a *line that goes through the first and last data points*, $(0, Z_0)$ and (n, Z_n). ◊

The above result provides a quick way to produce a forecast that approximates those obtained from more complex ARIMA$(p, 1, q)$ models that incorporate a constant. The model has been successfully applied in mortality forecasting by Lee and Carter (1992), for example. Often, however, when a differenced series is analyzed its mean is *assumed* to be zero. We did so in the analysis of the Finnish fertility, for example. Indeed, Box and Jenkins (1976, 194) suggest that one should not include a nonzero constant term into the model "unless evidence to the contrary presents itself". This may be a wise course in many fields of application but, the choice can have a major effect on demographic forecasts. In most cases, we suggest one examine the effect of including a constant, to see how it changes the forecast function. The decision to include or not to include the constant can be the single most important aspect of the eventual forecast.

Example 3.2. Trend in Finnish Fertility up to 1930. In Alho (2000) we analyzed the forecast of the Finnish population made by Modeen (1934). ARIMA modeling was applied to historical fertility data from 1776–1925 published by Turpeinen (1978). Modeen did not have access to such data, nor did he have the modern statistical technology available, but it is of interest to see if that would have made a difference. The series of the total fertility rate is nonstationary, and an ARIMA$_H$(0,1,1) model was found to give a serviceable approximation to the data. The constant term was not significant at a 0.05 level, but its inclusion had a marked effect on the point forecast. In retrospect we know that including the constant term would have produced a better forecast for the next 50 years than leaving it out. ◊

3.2. Regression

An alternate way of handling the mean is to directly estimate it using polynomials or other smooth functions. We consider a general case here for use in Section 3.3 of Chapter 8 and in Chapter 9, but note that in practice the most common choice is a first degree polynomial. Suppose we have observed a process Z_t for $t = 1, \ldots, n$, and we want to forecast it for $t = n + 1, \ldots, n + m$. Let us assume that the trend of the process is given by a function $f(.)$ such that

$$Z_t = f(t) + \varepsilon(t) \tag{3.2}$$

where $E[\varepsilon(t)] = 0$, at least for $t = 1, \ldots, n + m$. Suppose there are some known functions $f_j(.)$ such that

$$f(t) = \sum_{j=1}^{k} \beta_j f_j(t), \tag{3.3}$$

where the β_j's are parameters to be estimated. To represent the model in a matrix form, define first $\varepsilon_1 = (\varepsilon(1), \ldots, \varepsilon(n))^T$, $\varepsilon_2 = (\varepsilon(n + 1), \ldots, \varepsilon(n + m))^T$, and $\varepsilon = ({\varepsilon_1}^T, {\varepsilon_2}^T)^T$, and then $\mathbf{Z}_1 = (Z(1), \ldots, Z(n))^T$, $\mathbf{Z}_2 = (Z(n + 1), \ldots, Z(n + m))^T$, and $\mathbf{Z} = (\mathbf{Z}_1^T, \mathbf{Z}_2^T)^T$. Let $\mathbf{X}_1$ be an $n \times k$ matrix with $f_j(i)$ as the (i, j) element, and let $\mathbf{X}_2$ be an $m \times k$ matrix with $f_j(n + i)$ as the (i, j) element. Define the matrix $\mathbf{X} = (\mathbf{X}_1^T, \mathbf{X}_2^T)^T$, the vector of parameters $\boldsymbol{\beta} = (\beta_1, \ldots, \beta_k)^T$, and the covariance matrices $\Sigma_{ij} = E[\varepsilon_i {\varepsilon_j}^T]$ for $i, j = 1, 2$. Then, our past and future data can be written in the form $\mathbf{Z} = \mathbf{X}\boldsymbol{\beta} + \varepsilon$, where $\mathrm{Cov}(\mathbf{Z}) \equiv \boldsymbol{\Sigma}$ is of the form

$$\boldsymbol{\Sigma} = \begin{bmatrix} \Sigma_{11} & \Sigma_{12} \\ \Sigma_{21} & \Sigma_{22} \end{bmatrix}. \tag{3.4}$$

Suppose (3.4) is known. Then, the minimum variance unbiased prediction of $\mathbf{Z}_2$ based on $\mathbf{Z}_1$,

$$\hat{\mathbf{Z}}_2 = \mathbf{X}_2\hat{\boldsymbol{\beta}} + \Sigma_{21}\Sigma_{11}^{-1}(\mathbf{Z}_1 - \mathbf{X}_1\hat{\boldsymbol{\beta}}), \tag{3.5}$$

where $\hat{\boldsymbol{\beta}} = (\mathbf{X}_1^T \Sigma_{11}^{-1}\mathbf{X}_1)^{-1}\mathbf{X}_1^T \Sigma_{11}^{-1}\mathbf{Z}_1$, is the *generalized least squares (GLS) predictor* (e.g., Vinod and Ullah 1981; Chapter 5, Complement 7).

In practice the covariance matrix $\boldsymbol{\Sigma}$ would have to be estimated under a parametric model such as ARMA. Then, the prediction may no longer be unbiased or have minimum variance. Under the assumption of normality it continues to be a maximum likelihood estimator, provided that maximum likelihood is used to estimate the covariance matrix. This can be accomplished in practice by an iterative application of GLS estimation and ARMA modeling of the residuals.

We can write the forecast (3.5) as $\mathbf{L}\mathbf{Z}_1$, where

$$\mathbf{L} = \mathbf{X}_2\left(\mathbf{X}_1^T\Sigma_{11}^{-1}\mathbf{X}_1\right)^{-1}\mathbf{X}_1^T\Sigma_{11}^{-1} + \Sigma_{21}\Sigma_{11}^{-1}\left(\mathbf{I} - \mathbf{X}_1\left(\mathbf{X}_1^T\Sigma_{11}^{-1}\mathbf{X}_1\right)^{-1}\mathbf{X}_1^T\Sigma_{11}^{-1}\right). \tag{3.6}$$

Notice that the prediction error can be written in matrix form as $\hat{\mathbf{Z}}_2 - \mathbf{Z}_2 = [\mathbf{L}, -\mathbf{I}]\mathbf{Z}$, where $\mathbf{I}$ is an $m \times m$ identity matrix. It follows that (ignoring the estimation error in $\boldsymbol{\Sigma}$) we can write the covariance matrix of the prediction error in the form

$$\mathrm{Cov}(\hat{\mathbf{Z}}_2 - \mathbf{Z}_2) = \boldsymbol{\Sigma}_{22} - \boldsymbol{\Sigma}_{21}\mathbf{L}^T - \mathbf{L}\boldsymbol{\Sigma}_{12} + \mathbf{L}\boldsymbol{\Sigma}_{11}\mathbf{L}^T. \tag{3.7}$$

We see that (3.7) does not depend on $\mathbf{Z}_1$ in any way. Therefore, (apart from the identification and estimation of $\boldsymbol{\Sigma}$) the distribution of the forecast error is independent of the segment of the sample path we have observed, just as in ARIMA forecasts. However, if desired, the covariance matrix $\boldsymbol{\Sigma}$ may be chosen so that the variance of errors changes over time.

Example 3.3. Alternative Time Series Forecasts of the U.S. Growth Rate. We saw in Figure 3 that the growth rate of the U.S. population can be reasonably well modeled with an ARIMA(2,1,0) model. However, whether or not one includes a constant term has major implications for the forecast. We can produce a forecast of population based on a starting value (from 1999) and a forecast of the growth rate, and compare the results to a full cohort-component forecast produced by Lee and Tuljapurkar (1994). We label the Lee-Tuljapurkar forecast by *LT*, the AR(1) forecast that assumes stationarity by *AR*, the ARIMA(2,1,0) without a constant term by *ARI*, and the ARIMA(2,1,0) with a constant term by *ARC*. For comparison we include a forecast produced by a simple random walk (*RW*), and a forecast obtained by fitting a linear trend to growth rates using ordinary least squares (*REG*). In other words, the last model is of type (3.2) with $k = 2$, $f_1(t) = 1$ and $f_2(t) = t$, and $\boldsymbol{\Sigma} = \mathbf{I}$, an identity matrix. The results (in millions) are the following (they deviate slightly from Table 1 of Alho and Spencer (1997) due to different data used):

Year	*LT*	*AR*	*ARI*	*ARC*	*RW*	*REG*
2030	336.3	397.3	362.5	343.8	360.4	350.1
2050	371.5	516.5	435.5	379.0	431.5	396.0

Keeping LT as a gold standard, we find that AR forecasts are implausibly high. The forecasts ARI and *RW* are almost indistinguishable, and further away from LT than either ARC or REG. The latter two are close to the much more elaborate LT forecast. The closeness does not appear accidental, in light of findings by Keyfitz and Stoto (cf., Section 1.3 of Chapter 8) that simple forecasts often worked as well as complex ones. ◊

3.3. Structural Models

A third possibility for the handling of nonconstant means is to use so-called structural models, in which the trend is modeled stochastically (Harvey 1989). We will illustrate this approach by two examples.

Example 3.4. Stochastic Local Level Process. Suppose the model is defined via the equations

$$Y_t = \mu_t + \eta_t;$$
$$\mu_t = \mu_{t-1} + \xi_t, \tag{3.8}$$

where $\eta_t \sim N(0, \sigma_\eta^2)$ are i.i.d. and independent of the i.i.d. sequence $\xi_t \sim N(0, \sigma_\xi^2)$. In this model the "local level" μ_t is a random walk, so the model represents a nonstationary series. One way to estimate the "local level" is the following. Note that (3.8) implies that $Y_t - Y_{t-1} = \eta_t - \eta_{t-1} + \xi_t$. Consider the right hand side of this as a process indexed by t. Its mean is zero for all t, and its variance is the same for all t. By our assumptions, observations that are two or more steps apart are uncorrelated, but two consecutive observations have the correlation $\rho_1 = -\sigma_\eta^2/(2\sigma_\eta^2 + \sigma_\xi^2)$. Thus, the right hand side is actually an MA(1) process. Writing the differences in the MA(1) form: $Y_t - Y_{t-1} = \varepsilon_t - \theta\varepsilon_{t-1}$, we get that $\rho_1 = -\theta/(1+\theta^2)$. It follows that the *signal-to-noise ratio* $\sigma_\xi^2/\sigma_\eta^2$ uniquely determines θ. The converse is also true provided that $\theta \geq 0$. Provided that one or the other can be estimated, we can estimate the "local level" at t with the *exponential smoother* obtained by substituting our estimate of θ into the definition of m_{t-1} in (2.6). This also provides the forecast for all future values. ◊

Example 3.5. Stochastic Linear Trend Process. Consider the model

$$Y_t = \mu_t + \eta_t;$$
$$\mu_t = \mu_{t-1} + \beta_{t-1}, \tag{3.9}$$
$$\beta_t = \beta_{t-1} + \nu_t,$$

where Y_t is the observed value of the process, μ_t is the "local level" that changes roughly linearly with the slope β_t. The i.i.d. innovation processes $\eta_t \sim N(0, \sigma_\eta^2)$ and $\nu_t \sim N(0, \sigma_\nu^2)$ are assumed to be independent. As in the previous example, one can show that this process corresponds to an ARIMA(0,2,2) model. Suppose we start the process from $t = 0$ with some initial values for the level μ_0 and slope β_0. It follows that we have $Y_t = \mu_0 + t\beta_0 + (t\nu_1 + (t-1)\nu_{t-1} + \cdots + \nu_t) + \eta_t$. This means that the process is a sum of a deterministic linear trend, an integrated random walk, and an independent sequence of errors of observation. Even though such a series is severely nonstationary, it can have demographic applications if σ_ν^2 is small. ◊

4. Heteroscedastic Innovations

As noted in Section 2.2.1, the theoretical forecast error of ARIMA and other stationary models does not depend on the particular sample path observed so far nor does it depend on the time at which the forecast is being made. (By theoretical forecast error we mean (2.8), which is the error when the forecast is (2.7) with known ψ_k's.) Thus, the forecastability of the process does not vary over time

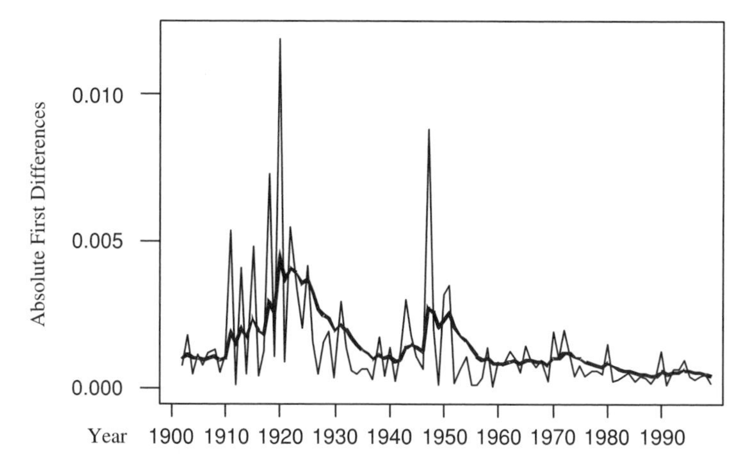

FIGURE 6. Absolute First Differences of the U.S. Growth Rate in 1900–1999, and an Exponentially Smoothed Trend Estimate.

or across sample paths. In stock option trading it has been observed that stock prices appear to be more variable at some times than at others. In other words, their volatility changes over time. We will present an example (Figure 6) that demographic processes also may display changing volatility.

4.1. *Deterministic Models of Volatility*

We noted in Section 2.1.2, the volatility of the vital processes may change simply as a consequence of increasing (or decreasing) population size. Other reasons for change can be traced to improved control over child bearing and ability to alleviate the effect of bad harvests, weather, or epidemics. Inasmuch as such changes can be explained it would seem reasonable to acknowledge them in future forecasts. The simplest way this can be done is in terms of parametric or nonparametric models of variance.

Figure 6 illustrates the issue with U.S. growth rate data. The absolute values of the first differences imply that the volatility of the growth process was much higher during the first half of the century than during the second. In Figure 6 we have used exponential smoothing (EWMA) to estimate their local level (cf., Example 3.4) with a smoothing parameter $= 0.2$ (corresponding to $\theta = 0.8$). In long term population forecasting judgment often is used to assess whether the future will be more of less volatile than the past. In such applications estimates such as those of Figure 6 can provide a starting value for the volatility.

The possibility of changing volatility has other implications for practical modeling. Consider a process of the form (2.1) with independent innovations and $E[\varepsilon_t] = 0$, but with $\mathrm{Var}(\varepsilon_t) = \kappa_t \sigma_\varepsilon^2$. To ensure that the variance of the process is finite, assume that each of the sets $\{\kappa_t, \kappa_{t-1}, \ldots\}$ is bounded for all t. Define $\boldsymbol{\psi}_k = (\psi_k, \psi_{k+1}, \ldots)^T$ for any $k = 0, 1, 2, \ldots$ and $\boldsymbol{\varepsilon}_t = (\varepsilon_t, \varepsilon_{t-1}, \ldots)^T$

for any $t = \ldots, -2, 1, 0, 1, 2, \ldots,$ so $Y_t = \psi_0{}^T \varepsilon_t$. Defining a diagonal matrix $\boldsymbol{\kappa}_t = \text{diag}(\kappa_t, \kappa_{t-1}, \ldots)$, we can write $\text{Cov}(Y_t, Y_{t+k}) = \sigma_\varepsilon^2 \psi_0{}^T \boldsymbol{\kappa}_t \psi_k$. Unlike (2.3) this depends on t. The correlation between Y_t and Y_{t+k} is $\rho_t(k) = \psi_0{}^T \boldsymbol{\kappa}_t \psi_k / [\psi_0{}^T \boldsymbol{\kappa}_t \psi_0 \psi_0{}^T \boldsymbol{\kappa}_{t+k} \psi_0]^{1/2}$. Formula (2.9) for the prediction error covariance gets the form,

$$\text{Cov}(E_k(t), E_{k+h}(t)) = \sigma_\varepsilon^2 \sum_{j=0}^{k-1} \kappa_{t+k-j} \psi_j \psi_{j+h}. \tag{4.1}$$

Hence, the error variances and covariances depend on the time at which the forecast has been made.

Example 4.1. A Heteroscedastic Process with Time Invariant Autocorrelations. Suppose the errors are exponentially increasing, $\kappa_t = e^{\alpha t}$ for some $\alpha \geq 0$. In this case $\boldsymbol{\kappa}_t = e^{\alpha t} \boldsymbol{\kappa}_0$ for any s. It follows that $\rho_t(k) = \rho_0(k)$ for any t. It is an example of a heteroscedastic process that has a constant mean, and an autocorrelation function that is invariant over time. ◊

We conclude that even though the study of the autocorrelation function is a useful tool in determining whether or not a process is stationary, Example 4.1 demonstrates that one cannot reliably use the autocorrelation function (nor any summary statistic that is a function of the autocorrelation function) as the *sole* means of making that decision. Plots are essential.

4.2. Stochastic Volatility

The approach of Section 4.1 relies on an unconditional form of heteroscedasticity, i.e., the variance of the process may change over time but this change is assumed to be the same for all sample paths. By allowing for path dependency, we may obtain a vast number of flexible models. Such models have proven to be especially useful in finance, where massive amounts of time series data must be handled in real time.

In these models changes of the innovation variance are modeled using some stochastic process, much the same way structural models can be used to describe nonconstant means (Engle 1982; for a review, see Bollerslev, Chou, and Kroner, 1992). We can express the *autoregressive conditional heteroscedasticity (ARCH(q))* model of Engle (1982) in our notation by assuming that the values κ_t depend on past squared innovations ε_t^2 according to

$$\kappa_t = \mu + \alpha_1 \varepsilon_{t-1}^2 + \cdots + \alpha_q \varepsilon_{t-q}^2, \tag{4.2}$$

where $\mu > 0$, $\alpha_i \geq 0$. Under (4.2), small (large) squared innovations lead to small (large) κ_t, so innovations of a similar size tend to cluster, on a sample path basis. Still, unconditionally, the processes may have constant variances. These models have been generalized in many ways, to the so-called *generalized ARCH*, or *GARCH* processes, for example. Although their applicability in demographic settings is still an open question, Keilman, Pham, and Hetland (2002) have shown that they can be used to an advantage in some situations. It is clear that demographic time-series can be heteroscedastic, but it is not clear what will turn out be the simplest representation for that.

Exercises and Complements (*)

1. In Example 1.2 we considered a lower triangular Cholesky decomposition. (a) Derive the corresponding representation for an *upper triangular* matrix $\mathbf{C}$. (b) Note that the resulting process for Y_t is essentially identical to that of Example 1.2 but with the time reversed. (c) Note that the same result can be obtained directly from (1.1) by reversing the order of Y_t's. This observation is important in practical modeling because it shows that *a linear process must look similar in all relevant respects whether we let time run forwards or backwards.*

*2. In general, we may think of an $n \times m$ matrix $\mathbf{C} = (c_{ij})$ as a mapping that relates to any $i = 1, \ldots, n$ and any $j = 1, \ldots, m$ a number c_{ij}. Matrix operations, such as multiplication, can also be defined in terms of i and j, so we can consider infinite dimensional matrices. Suppose $\mathbf{C}= (c_{ij})$ is such that on the row $i = \ldots, -1, 0, 1, 2, \ldots$ and column $j = \ldots, -1, 0, 1, 2, \ldots$ we have that $c_{ij} = \psi_{i-j}$ for $j \leq i$, and $c_{ij} = 0$ otherwise. Define correspondingly infinite dimensional vectors $\varepsilon = (\ldots, \varepsilon_{-1}, \varepsilon_0, \varepsilon_1, \varepsilon_2, \ldots)^T$ and $\mathbf{Y}= (\ldots, Y_{-1}, Y_0, Y_1, Y_2, \ldots)^T$. Then, we can write (2.1) exactly in the Cholesky form of Example 1.2, or $\mathbf{Y} = \mathbf{C}\varepsilon$.

3. Show that (2.3) holds.
4. (a) Show that the variance of an MA(1) process is $\text{Var}(Y_t) = \sigma_\varepsilon^2(1 + \theta^2)$. (b) Show that the autocorrelation function of an MA(1) process is zero except that $\rho_1 = -\theta/(1 + \theta^2)$. (c) Show that an MA(2) process has two non-zero autocorrelations, and derive their formulas.
5. Consider (2.4). To assess whether or not the "waves" one may detect in a smoothed series could be due to chance, compute the variance of (2.4) under the assumption that the process has a fixed mean, or $D_t \sim \text{Po}(\mu K_t)$. Use general weights w_j. Under a normal approximation we conclude that if the waves are within ± 2 standard deviations from the mean, they may well be due to the Slutsky effect alone.
6. Derive the variance, autocovariance, and autocorrelation functions of an AR(1) process.
7. Show that (2.5) holds by substituting for the AR(1) process Y_t its representation (2.1).
8. Show that the ψ_j–weights of an ARMA(1,1) are of the form $\psi_j = (\varphi - \theta)\varphi^{j-1}$ for $j > 0$.
9. (a) Show that if $y_t = a + bt$, then first difference is $y_t - y_{t-1} = b$, and (b) if $y_t = a + bt + ct^2$, then the second difference (i.e., difference of differences) is $2c$.

10. Fit an ARIMA model to the logarithm of the total fertility rate of an industrialized country (that has at least 50 years worth of data) in a post demographic transition period.
11. Show that (2.6) holds.
*12. Fitting an AR(k) process to a series should have the last regression coefficient zero, if there is no independent effect from time $t-k$ to time t, given the values of the intermediate years. Under stationarity, both the variable to be explained Y_t, and the last explanatory variable Y_{t-k}, can be explained equally well using the intermediate variables $Y_{t-1}, \ldots, Y_{t-k+1}$. Therefore, the *partial correlation* between Y_t and Y_{t-k}, when controlling for $Y_{t-1}, \ldots, Y_{t-k+1}$, can be estimated by regressing Y_t on $Y_{t-1}, \ldots, Y_{t-k}$ and taking the coefficient of the last term as the estimate at lag $k \geq 1$. This is helpful, especially for choosing the order of an AR(p) process.
13. Consider an AR(p) process around a mean μ of the form $Y_t - \mu = \varphi_1(Y_{t-1} - \mu) + \cdots + \varphi_p(Y_{t-p} - \mu) + \varepsilon_t$. Write the model using a constant term γ, in the form $Y_t = \gamma + \varphi_1 Y_{t-1} + \cdots + \varphi_p Y_{t-p} + \varepsilon_t$. Show that the constant satisfies the relationship $\gamma = \mu(1 - \varphi_1 - \cdots - \varphi_p)$.
*14. Consider the model $Y_t = \mu + \varphi Y_{t-1} + \varepsilon_t$, for $t = 1, \ldots, n$, where the independent innovations are normally distributed. Conditioning on Y_1 Dickey and Fuller (1981) considered the hypothesis $H_0 : \mu = 0$ and $\varphi = 1$, or that the process is a random walk. The principle of likelihood ratio testing (cf., Section 3 of Chapter 1) leads one to consider the statistic

$$R = \sum_{t=2}^{n}(Y_t - \hat{\mu} - \hat{\varphi}Y_{t-1})^2 \Big/ \sum_{t=2}^{n}(Y_t - Y_{t-1})^2,$$

where $\hat{\mu}$ and $\hat{\varphi}$ are the least squares estimators given Y_1, and small values indicate deviation from the null. (An equivalent "F test" type of statistic can also be used.) The distribution of R can be determined by simulation under H_0:
(i) generate i.i.d. values $\varepsilon_t \sim N(0, 1)$ for $t = 2, \ldots, n$;
(ii) set $Y_0 = 0$, and then $Y_t = Y_{t-1} + \varepsilon_t$, for $t = 2, \ldots, n$;
(iii) calculate $\hat{\mu}$ and $\hat{\varphi}$ and store the corresponding R.
Repeating the steps (i)–(iii), say 10,000 times, we can approximate the sampling distribution of R. The value of R computed from the empirically observed data can then be compared to the left hand tail of the distribution to determine a P-value. This is an example of a so-called *unit root test*. An extension in which H_0 specifies a random walk with a drift can similarly be handled (Dickey and Fuller 1981).
*15. *Regime switching*. Consider a model

$$Y_t = \mu + \varphi Y_{t-1} + (\mu' + \varphi' Y_{t-1})\Phi((Y_{t-1} - \mu'')/\sigma) + \varepsilon_t,$$

where $\Phi(.)$ is the c.d.f. of $N(0, 1)$ distribution. When $Y_{t-1} - \mu'' \to -\infty$, the model approaches the form $Y_t = \mu + \varphi Y_{t-1} + \varepsilon_t$. When $Y_{t-1} - \mu'' \to +\infty$, the model approaches the form $Y_t = (\mu + \mu') + (\varphi + \varphi')Y_{t-1} + \varepsilon_t$. Or the model is capable of representing different behavior when it is at a relatively low level and a relatively high level. The parameter σ regulates the speed of change from one regime to the other. This *smooth transition regression* is an example of a nonlinear time series model (Granger and Teräsvirta 1993, 38–39).

16. (a) Verify that in Example 2.6 we have that $\hat{Y}_{t+k} = \varphi^k Y_t$ for $k = 1, 2, \ldots$ (b) Derive (2.11) from (2.10).
17. Show that (2.13) holds, and derive (2.14) by substitution.
18. Consider formula (2.15). Suppose that the second differences of a process are an independent sequence, so $\psi_j = 0$ for $j > 0$. Show that we have then

$$\rho(E_{[k]}, E_{[k+h]}) = \frac{k(k+1)(2k+3h+1)}{[k(k+1)(2k+1)(k+h)(k+h+1)(2k+2h+1)]^{1/2}}.$$

19. Consider two integrated processes. Emulate Example 2.7 to get

$$\mathrm{Cov}(E_{(k)}, E_{(k+h)'}) = \frac{\delta\sigma_\varepsilon\sigma_{\varepsilon'}}{(1-\varphi)(1-\varphi')}\left[k - \frac{\varphi(1-\varphi^k)}{1-\varphi} - \frac{\varphi'(1-(\varphi')^{k+h})}{1-\varphi'} + \frac{(\varphi')^h(1-(\varphi\varphi')^k)}{1-\varphi\varphi'}\right].$$

Asymptotically the corresponding crosscorrelations are of the form $\rho(k, k+h) = \delta(k/(k+h))^{1/2}$.

20. Derive the forecast function (3.1).
21. Verify the formula for the first autocorrelation of the differences of the process (3.8). Solve signal-to-noise ratio in terms of θ, and θ in terms of the signal-to-noise ratio.
22. Show that the second differences of the process (3.9) form an MA(2) process and derive equations that connect the variances of the innovation processes to those of the moving average parameters.
23. Show that (4.1) holds.
24. Consider the model of Example 4.1. Suppose $\psi_j = \varphi^j$ with $|\varphi| < 1$. Show that $\psi_0^T \kappa_t \psi_k = \varphi^k e^{\alpha t}/(1-\varphi^2 e^{-\alpha})$, $\rho_0(k) = (\varphi e^{-\alpha/2})^k$, and $\mathrm{Var}(E_k(t)) = \sigma_\varepsilon^2 e^{\alpha(t+k)}(1-\varphi^{2k}e^{-\alpha k})/(1-\varphi^2 e^{-\alpha})$ for $k = 1, 2, \ldots$. We see that the theoretical forecast error variance is an exponential function of both the jump-off time t and the lead time k.

*25. ARIMA models may produce prediction intervals that eventually become too wide for a vital rate X_t. A logistic transformation $Y_t = \log((X_t - L)/(U - X_t))$ with $U > L$, constrains X_t to remain in $[L, U]$. Assume a random walk model $Y_t \sim N(0, t\sigma^2)$. Choose any two values $L < L' < U' < U$, and consider the probability that $X_t > U'$ or $X_t < L'$, or equivalently $Y_t > U^* = \log((U' - L)/(U - U'))$ or $Y_t < L^* = \log((L' - L)/(U - L'))$. Show that $P(L^* < Y_t < U^*) \to 0$, when $t \to \infty$. Conclude that X_t will eventually be "absorbed" close to U or L.

8
Uncertainty in Demographic Forecasts: Concepts, Issues, and Evidence

Demographic forecasting is historical activity both in terms of methodology and accuracy: to forecast forward and to predict the accuracy of our forecast, we look backward. If the vital rates follow closely their past trends, accurate forecasting is feasible, but increased fluctuations of the rates usually implies rapidly increasing forecast errors. Consider, for example, the forecasts of the U.S. total fertility rate. The forecasts assumed that fertility would stay roughly at the latest observed level. Therefore, those made in the early 1950's and 1970's were accurate for a few years, when the level of fertility remained fairly constant for a decade, whereas the forecasts made in the 1940's, when fertility rose, and in the 1960's, when it declined, were grossly in error (cf. Figure 5 of Chapter 4). More generally, Stoto (1983) found that the major determinant of forecast accuracy was the time at which the forecast was made. Keyfitz (1981, 581–582) credits Lee (1980) with the following analogy.

> "Think of a number of marksmen, all equally competent, facing a target that moves about erratically. Some will do better than others, not because of differences in competence, but because they were fortunate enough that the target stood still when they fired, while others had the bad luck to shoot just before the target moved."

Although the theoretical models available to the forecaster improve over time, this does not necessarily lead to substantially more accurate forecasts. For example, improved socio-economic analyses have increased our understanding of determinants of change in mortality, fertility, and migration, and improved statistical methods allow ever more complicated models to be estimated. Therefore, controlling for the difficulty of forecasting at any given time, one might expect the forecast accuracy to improve over time. However, to effectively utilize the improved theoretical models, one must be able to accurately identify and forecast the determinants of change, and that has proved challenging.

The recognition of both the varying forecastability and the historical character of forecasting methodologies has led many to reject the notion of forecasting altogether. In the United States, for example, the official forecasters of population talked about "forecasts" in the late 1940's (Whelpton, Eldridge, and Siegel 1947 and U.S. Census Bureau 1949), but when the gross errors caused by the baby-boom became evident, the terminology was first switched to "illustrative projections" (U.S. Census Bureau 1958), and later to "projections" (e.g., U.S. Census Bureau 1964, 1984; Day 1993). Our view of these terminological distinctions is the same as that of Harold Dorn (1950) who wrote: "Predictions, estimates, projections, forecasts; the fine academic distinction among these terms is lost upon the user of demographic statistics. So long as numbers which purport to be possible future populations are published they will be regarded as forecasts or predictions, irrespective of what they are called by demographers who prepare them." Indeed, it is difficult to understand, why a national statistical agency would publish anything but the most likely future alternative as the middle variant of their projection.[1]

We will follow Dorn in interpreting the forecaster's task. Producing population forecasts that are highly uncertain can still have value, as the forecast may draw attention to looming public policy issues that would otherwise be neglected. At the beginning of the 21st century, many industrialized countries have not adequately prepared for the retirement of the baby-boom generations that will occur during 2015–2025. Even inaccurate forecasts demonstrate the unpreparedness. The shortfall in retirement funding is uncertain, however, and quantification of the uncertainty can improve the development of public policy. For example, some wishing to avoid investment in retirement funding will try to point to low alternative forecasts and say the problem is small. With an assessment of the probability distribution of forecast error, the public policy debate can distinguish unlikely alternatives from probable ones, and if the forecast is very uncertain, flexible adaptive strategies can be sought to allow for modification as the real path of the future unfolds (Chapter 11).

In Chapter 7 we discussed statistical models for demographic time series and showed how they can be used to quantify forecast uncertainty. Here, we take the demographic tradition and demographic data as starting points, and try to establish "stylized facts" or "boundary conditions" for demographic forecasting that need to be acknowledged. Section 1 discusses how assumptions have been traditionally formulated in cohort-component forecasting. These principles were first formulated in a unified way by Pascal K. Whelpton. Section 2 considers dimensionality problems that arise in mortality forecasting. In Section 3 we will discuss conceptual issues regarding forecast errors. This involves error concepts and classifications, the interpretation of probabilities, the feedback effects of forecasts, the role of expert judgment, and conditional forecasting. In Section 4 we discuss the practical specification of error, including modeling error. We then discuss the measurement of correlations of vital processes and their forecast errors in Section 5. This is a new area of demographic research where relatively little is known so far.

---

[1] Examples of statistical agencies having tried to avoid such an interpretation by publishing an *even* number of variants (e.g., four) have proved dismal. The users have quickly averaged the middle two to produce the most likely figure!

1. Historical Aspects of Cohort-Component Forecasting

1.1. Adoption of the Cohort-Component Approach

As discussed in Chapter 6, *cohort-component forecasting* is an elaboration of the fundamental book-keeping identity: (population at time $t + 1$) = (population at time t) + (births during t) − (deaths during t) + (net-migration during t), in which the book-keeping is done by age and sex. Cannan (1895) first prepared a cohort-component forecast for England and Wales. By the end of the 1920's such forecasts had also been made for the Soviet Union by Tarasov in 1922 (DeGans 1999, 96), for the Netherlands by Wiebols (1925), for Sweden by Wicksell (1926), for Italy by Gini (1926), for Germany by Statistisches Reichsamt (1926), for France by Sauvy (1928), and for the United States by Whelpton (1928). Many details about the early forecasts, especially from the Dutch perspective, can be found in DeGans (1999).

One reason for the increased interest in developing new methods of population forecasting in the early decades of the 20th century appears to have been declining fertility, especially in cities (Fleischhacker, DeGans, and Burch 2003), although in the case of the Netherlands, overpopulation was a concern (DeGans 1999, 24). In Germany, Burgdörfer (1932, 32), an author associated with national socialism, characterized Berlin as an "infertile city" and predicted that the "two-child system" would lead to a population decline. In Sweden, left-leaning social scientists Myrdal and Myrdal (1934, 87–88, 94) believed that improved contraception was the cause of declining fertility. They thought that decline would continue in the foreseeable future. These widely held views posed problems to the earlier methods of forecasting. For example, in the first forecast of Finland, Modeen (1934) criticized the logistic model introduced by Verhulst (1838) and later popularized by Pearl and Reed (1920), and Yule (1925), because the logistic model (together with the simpler exponential model) always predicted growth (or decline), but could not incorporate a change from growth to decline.

1.2. Whelpton's Legacy

In the United States the cohort-component method was pioneered by Pascal K. Whelpton. In a sequence of papers (Whelpton 1928, Thompson and Whelpton 1933, Ch. X, Whelpton 1936, and Whelpton, Eldridge, and Siegel 1947) he developed a unified program for population forecasting. Whelpton realized that book-keeping by age and sex would not necessarily make the resulting forecast for the total population more accurate, but at the very least it would provide more information to the user. Even more importantly, he articulated many of the central problems in the methodology of formulating assumptions for the vital rates. This will be the topic of the remainder of the section. We will use the meticulously compiled forecast report Whelpton et al. (1947) as the primary source material. Unless otherwise noted, the quotes below are from the report.

In discussing the "hypothetical mortality trends in the United States, 1945–2000", Whelpton decided to make three alternative sets of mortality assumptions, designated as "high mortality", "low mortality", and "medium mortality". "The first represents the smallest declines in the age-specific death rates that seem probable, the second the largest declines that are considered reasonable, and the third a position approximately midway between the extremes."

Whelpton's methodological ideas are well summarized by the following paragraph that discusses the way the high, middle, and low variants are to be made:

"With each of these assumptions it is possible to extrapolate past trends according to some formula and arrive at hypothetical death rates for any future year. An alternative procedure is to consider past trends and the likelihood of future changes, form an opinion as to the percentage reduction in death rates to be expected by a given future year, and obtain rates for the intervening years by interpolation. The former method may seem to have the advantage of being less influenced by personal bias, nevertheless the personal element would remain in the choice between two or more formulas fitting past trends equally well but giving different results for the future. More important, the extrapolation of past trends according to such formulas might lead to future rates which would seem incompatible with present knowledge regarding causes of death and means of controlling them. After some experimentation with both methods, the second alternative was chosen as the more desirable for the purpose at hand."

We see that Whelpton objects to the use of mathematical extrapolation methods because they do not rid us of the subjectivity inherent in model choice and because he fears they may produce results that are contrary to "present knowledge". This is essentially the same reasoning most producers of official forecasts still use. For example, the U.S. Office of the Actuary has followed Whelpton's ideas almost literally, in that they have used target values for the reduction of age-specific mortality rates in their forecasts (Section 2.2).

Whelpton used essentially similar reasoning to reject mathematical extrapolations in the forecasting of fertility. Although these elements of his methodology have also become standard procedures in many statistical offices, the unfortunate fact is that Whelpton's forecast for fertility was among the most erroneous ever made. He missed the baby-boom. Whelpton assumed that the U.S. total fertility rate of the white women would decline from 2.42 in 1945 to 2.06 in 1960, but in reality it rose to 2.90 by 1946 and to 3.53 by 1960! The increase of 0.48 child per woman during 1945–1946 was the biggest observed during the 20th century. Recognizing that Whelpton was one of the very best demographers of his time, we may look at Whelpton's reasoning more closely, to see if there is anything we can learn for the future.

To set his fertility variants Whelpton first looked at the historical trends in the United States. He had native white age-specific fertility series available for 1920–1945, and nonwhite age-specific series for 1930–1945. He complemented these short series by statistics on children under 5 years of age per women in age 20–44 years of age for the census years 1800–1940, by nine major statistical divisions of the United States (New England, Middle Atlantic, East North Central, West North Central, South Atlantic, East South Central, West South Central, Mountain, Pacific). Considerable attention was paid to corrections for underenumeration in the censuses. He then compared the changes in the number of children ever born among the white and nonwhite female population by age and marital status in 1910 and 1940. After that he studied annual birth rates of women by parity (years 1920–1945 for whites, 1930–1945 for nonwhites), and changes in age-specific birth rates in the nine major divisions in 1918–1921, 1929–1931, and 1939–1941. After the detailed study of the past U.S. trends Whelpton compared the U.S. gross reproduction rates (see Section 4.2 of Chapter 4) to other countries (Norway, Sweden, Finland, Denmark, Netherlands, England and Wales, France, Germany, Czechoslovakia, Austria, Portugal, Italy, Hungary, Poland, Bulgaria, South Africa, Australia, New Zealand, Japan) during "early years" (mostly 1870's), "shortly after World War I" (mostly early 1920's), and "shortly before World War II" (mostly late 1930's). Whelpton summarized the experience of Western countries with reliable data as follows: "A long-time downward trend in fertility has been the almost universal rule. Upswings have occurred but rarely, and have been relatively small and of short duration."

After the historical comparisons Whelpton discussed "causes in the long-time decrease in fertility in the United States" with a view of formulating opinions regarding the long-time future trend in fertility. Five hypotheses considered were: "(1) a less favorable marriage rate, (2) a rise in the proportion of pregnancies ending in a miscarriage or stillbirth, (3) the greater frequency of illegal abortions, (4) an increase in sterility or low fecundity, and (5) an increase in the voluntary limitation of family size". After a detailed discussion Whelpton concludes that "the great preponderance of evidence" indicates that the voluntary limitation is the most significant cause. Whelpton then went on to discuss "causes of short-time changes in birth rates" that he thought would be "helpful in estimating the probable fertility during the next few years". He analyzed the effect of war and economic prosperity on nuptiality and birth rates. The overall conclusion was that "the factors which will primarily determine the long-term future trend of fertility will be (1) the speed with which the pattern of effective family planning is adopted by additional groups of the population and (2) the number of children that couples decide to have". So far, we find no fault in Whelpton's analyses. On the contrary, their meticulous detail far surpasses what one commonly sees in more recent forecast reports.

What finally went wrong is related to Whelpton's assessment of the desired family size. He believed that the past extension of effective family planning would rapidly continue as a consequence of war time shifts of population. In particular "millions of women and girls who might never have sought employment in time of peace took jobs in offices, stores, and factories. These changes have tended to bring people with a regional or family background of high fertility into contact with those having a background of low fertility. Such contacts disseminate more widely the knowledge of effective measures of family planning and the point of view that leads to their use." Clearly, Whelpton believed (like Myrdal and Myrdal in Sweden) that the forces of modernization connected with urbanization and women's increased participation in the labor force were in operation, and would prevail. Whelpton was misled. Although "a high degree of economic prosperity plus war time psychology resulted in a substantially larger number of births during 1942–1945 than was expected", Whelpton believed this was a short term fluctuation. We know from other sources (Beale 2004) that a factor in this assessment was the apparent change in the timing of births. The observed rise in fertility was disproportionately due to first births and interpreted as delayed child-bearing deferred during the Great Depression of the 1930's. It was thought that this could not continue, but contrary to the expectation both cohort and period measures of fertility rose rapidly after the war.

Finally, and most interestingly, Whelpton was one of the first developers of surveys concerning desired family size (e.g., Whelpton and Kiser 1946, 1947). He used data collected in by the American Institute of Public Opinion which shows that in 1941 the desired family size was 2.97 children per family, but in 1945 it was 3.30 children per family. Whelpton wrote: "The change in opinions from 1941 to 1945 could mean that there will be a tendency toward larger families in the future. It seems more probable, however, that it reflects the psychology and economic conditions of the war and that a survey a few years later will elicit replies which are more like those of 1941 than 1945." Thus, Whelpton used a theoretical argument to reject some empirical evidence he saw.[2]

Had Whelpton accepted the desired family size data, his forecast would have been accurate for about twenty years. Now, even his high forecast variant that assumed the 1945 level to persist was *too low by approximately one child per woman* during the same period. Whelpton's middle forecast of 1.9 of the total fertility rate of the year 2000 was much more accurate!

1.3. Do We Know Better Now?

Some forecasters believe that advances in demographic research have led us to understand changes in childbearing much better than in Whelpton's time. Examples include the use of cohort and duration approaches, instead of the period approach that Whelpton used, as a basis of fertility forecasting. Unfortunately, they have not led to improvements in accuracy.

Example 1.1. Cohort Approach to Fertility Forecasting. In 1964 the U.S. Census Bureau started to use completed cohort fertility a basis for forecasting age-specific fertility. The rationale was that cohort fertility corresponds to actual childbearing whereas period fertility is a synthetic concept. Characteristically, part of the data used was compiled by Whelpton earlier. For the year 1980 the high variant of the 1964 forecast for the total fertility rate was 3.44 and the low variant was 2.59. The actual rate was 1.9. As discussed in Chapter 7 the relative smoothness of the cohort rate does not mean that it is necessarily the relevant quantity to forecast, because it needs to be disaggregated into age-specific rates by assumptions concerning the timing of childbearing, simultaneously in all ages. In addition, one has to consider

[2] As noted in Bongaarts and Bulatao (2000, 93) and Hendershot and Placek (1981), fertility intention data has been of variable predictive value in the forecasting of completed fertility during the post World War II era.

cohorts that have just started child bearing, or who will do so in the future. Their completed fertility will be known in the next 30 years or later, and it may have little to do with the fertility decisions of those whose completed fertility is known. ◊

Example 1.2. Effect of Marriage Duration on Fertility. Keilman (1990, 65–66) describes changes in Dutch forecasts of fertility during 1967–1970. Prompted by the poor results of the 1965 forecast a working group consisting of forecasters of the Netherlands' Central Bureau of Statistics, planners of the National Physical Planning Agency and Central Planning Bureau, and some prominent academic demographers recommended that fertility be forecasted for marriage cohorts by the duration of marriage. Presumably, the idea was that child bearing would follow in some predictable way the life course of a couple. Although the conceptual analysis underlying the change was sophisticated, the forecasting results were poor. As a result, in the official forecasts made in the 1980's marriage duration was abandoned, and age (retaining the cohort perspective) was reintroduced. More generally, Keilman and Kučera (1991) found that methodology had little impact on accuracy of national forecasts by the Netherlands and the Czechoslovak Socialist Republic. ◊

Example 1.3. Was the Baby-Boom a Unique Phenomenon? It is sometimes thought that the baby-boom that occurred (depending on the country) from the late 1940's until the 1960's was a unique event and that we should not expect equal surprises unless something corresponding to World War II were to occur. However, as mentioned in Chapter 4 already, the role of war was not at all clear in the creation of the boom. Moreover, fertility changed in the Mediterranean countries during 1985–1995 from the total fertility rate of over 2 to 1.3–1.4, or *in relative terms* by as much as it did during the baby-boom. Just like the baby-boom, this change was missed by official forecasts. ◊

The examples show that developments in fertility have repeatedly taken even the best experts by surprise. Surprisingly, forecasting mortality has been of comparable difficulty.

Example 1.4. Trend Extrapolation Versus Judgment. In Alho (1990c) we compared the accuracy of official forecasts of mortality to extrapolations based on ARIMA models. The directly age-standardized female mortality (cf., Section 3.3 of Chapter 5) in the U.S. during 1920–1986 was considered. The rate started from about 0.022 and declined to about 0.006. No segment of the series looked stationary. In order to prevent the forecasts from being implausibly high or low, it was assumed that the rate must remain in the interval [0.002, 0.03]. A logit-transformation was applied to the rate $r(t)$ of year t, so the transformed rate was of the form $w(t) = \log((r(t) - 0.002)/(0.03 - r(t)))$. Simple trend forecasts from an ARIMA(1,1,0) model with a constant were calculated. Official forecasts up to the year 1986, with jump-off years 1950, 1955, 1965, and 1977, were matched by ARIMA(1,1,0) forecasts with data up to the jump-off year. The ARIMA extrapolations were more accurate in three cases and the official forecast was more accurate for the jump-off year 1965. For males the first two official forecasts were more accurate, the last two less accurate, than the ARIMA extrapolations. Overall, the official forecasts tended to overshoot the future mortality, whereas the extrapolations tended to be too low. Lee and Miller (2001) have provided evidence that Lee-Carter method outperformed the official forecasts for life expectancy. ◊

An entirely different approach to forecasting the vital rates is to consider them in an economic framework (cf., Schultz 1981). Econometricians (McDonald 1979, 1980, 1981; Butz and Ward 1979) have experimented with dynamic stochastic models in which a demographic variable (such as yearly births or the total fertility rate) is explained directly in terms of its correlatedness with economic variables. Wheeler (1984) has similarly modeled population growth in developing countries. As noted by Land (1986, 898–899), it has proven difficult to find persistent statistical relationships of this sort and even when such relationships exist it is difficult to forecast the economic variables with enough accuracy to improve the demographic forecasts. To illustrate some of the difficulties, we consider the effect of an extreme economic shock.

Example 1.5. Counterintuitive Data on Economic Shocks and Demographics. In 1991–1993 Finland went through an economic shock comparable to the Great Depression of the 1930's. In the following table we present data from 1988–2000, on the change in gross domestic product (GDP), unemployment rate, total fertility rate (TFR), male life expectancy (e_0), and net migration (NET).

Year	Change in GDP (%)	Unemployment (%)	TFR	e_0	NET (1,000)
1988	4.9	4.5	1.69	70.7	1.3
1989	5.7	3.1	1.78	70.9	3.8
1990	0.0	3.2	1.79	70.9	7.1
1991	−7.1	6.6	1.79	71.3	13.0
1992	−3.3	11.7	1.85	71.7	8.5
1993	−1.1	16.3	1.81	72.1	8.4
1994	4.0	16.6	1.85	72.8	2.9
1995	3.8	15.4	1.81	72.8	3.3
1996	4.0	14.6	1.76	73.0	2.7
1997	6.3	12.7	1.75	73.4	3.7
1998	5.3	11.4	1.70	73.5	3.4
1999	4.0	10.2	1.74	73.7	2.8
2000	5.7	9.8	1.73	74.1	2.6

As one would expect, the decrease in production led to an increase in unemployment, with a lag of approximately two years. If anything *fertility rose* slightly, *life expectancy increased faster*, and *net migration was higher* during the depression, than at other times. All these developments are counterintuitive from a common sense point of view, but perhaps less so from a historical perspective, since population growth and economic growth appear not to have been systematically related (Simon 1977, 47). ◊

Keyfitz (1982, 744–746) discusses other reasons preventing theory from further improving forecasts. He found that simplistic forecasts of population size to be much less accurate than published official forecasts, but the latter were similar in accuracy to simple forecasts (Keyfitz 1981, 588–599). Similarly, a large empirical study focusing on population forecasts prepared by the U.S. Bureau of Census and by the U.N. using the cohort-component method found that "for projections of total population size, simple projection techniques are more accurate than more complex techniques" (Stoto 1983, 13). Thus, although it is easy to choose a forecasting method that will work poorly in a given situation, once the obviously poor methods are excluded, it is difficult to choose the best of the remaining, competing methods. A large empirical study of forecasting methods in a variety of settings concluded that

> "forecasting accuracy depends on the type of data and the forecasting situation considered... As a consequence any monolithic approach to forecasting has been eliminated as a practical alternative... Furthermore, the empirical evidence indicates that forecasting accuracy can often be achieved through simple methods. (Makridakis et al. 1984, vii–viii)

In fact, given the difficulty of model choice, combining forecasts that have been made based on different principles is an appealing idea (cf., Clemen 1989).

2. Dimensionality Reduction for Mortality

Cohort-component forecasting of population may require forecasts for a hundred or more age-specific mortality rates for each sex. As noted in Example 2.4 of Chapter 4, deaths can further be analyzed by cause. To allow for a meaningful use of time-series techniques, some form of dimensionality reduction is desirable. Fortunately, there are regularities in mortality change that allow for simplification, and it turns out that unless causes of death are of interest in themselves, it is often not necessary to consider them in forecasting. These are the topics we address here. Classical techniques, not to be discussed, include model life table techniques (e.g., United Nations 1983) and actuarial graduation procedures (e.g., Keyfitz 1977, Heligman and Pollard 1980).

2.1. *Age-Specific Mortality*

Let $\mu(x, t)$ be the age-specific mortality rate in age x during year $t \geq 0$ (we suppress sex in the notation for simplicity). Consider a class of models,

$$\mu(x, t) = \exp(\alpha(x) + \beta(x, t)), \tag{2.1}$$

where $\beta(x, 0) = 0$ for all x. The *rate of change* for mortality is $\partial/\partial t \log \mu(x, t) = \partial/\partial t\, \beta(x, t)$. Assume first that $\beta(x, t) = \xi(t)$. In that case we have a loglinear, proportional hazards model whose parameters can be estimated under a Poisson assumption, for example. A *constant rate of change* model would take $\xi(t) = \delta t$ with $\partial/\partial t\, \beta(x,t) = \delta$, but we know from Section 2.2.3 of Chapter 4 that such a simple model is not likely to hold.

A bilinear model uses $\beta(x,t) = \delta(x)\xi(t)$. If $\xi(t) = t$, then $\beta(x,t) = \delta(x)t$, and we can interpret the constant $\delta(x)$ as the rate of change in mortality in age x. The model $\mu(x,t) = \exp(\alpha(x) + \delta(x)t)$ is of a standard loglinear form that can be estimated via Poisson regression. A forecast for future rates is then simply $\hat{\mu}(x,t) = \exp(\hat{\alpha}(x) + \hat{\delta}(x)t)$. As discussed in Section 6 of Chapter 5, in the more general case with $\beta(x,t) = \delta(x)\xi(t)$, maximum likelihood estimation under a Poisson assumption or a normal assumption (principal components) is still feasible. Then, the forecast of future mortality would be of the form $\hat{\mu}(x,t) = \exp(\hat{\alpha}(x) + \hat{\delta}(x)\hat{\xi}(t))$, where the forecast $\hat{\xi}(t)$ would have to be obtained through other means, such as ARIMA modeling. For an investigation of the accuracy of models of this type, see Bell (1997).

We conclude with a remark on smoothing. Since the parameters $\alpha(x)$ and $\delta(x)$ are typically estimated from data, they may have to be smoothed before use to avoid erratic variations in neighboring ages. In particular, suppose $\delta(x+1) < \delta(x) < 0$ for some x. Then, for t large enough we will always have $\mu(x+1,t) < \mu(x,t)$ under the model $\mu(x,t) = \exp(\alpha(x) + \delta(x)\xi(t))$, regardless of $\alpha(x)$, if $\xi(t) \to \infty$. For unsmoothed δ(x)'s such effects can appear fairly quickly.

Example 2.1. Rates of Mortality Decline in Europe. Figure 1 plots estimates of $\hat{\delta}(x)$ by age from eleven European countries (Austria, Denmark, Finland, France, Germany, Italy, Netherlands, Norway, Sweden, Switzerland, U.K) for females and for males during a 30-year period ending between 1997–2001 for which the data were available. The average rates of decline were computed from ages 0, 1–4, 5–9, ..., 95–99. The lower end of the age interval is indicated in the figure. For

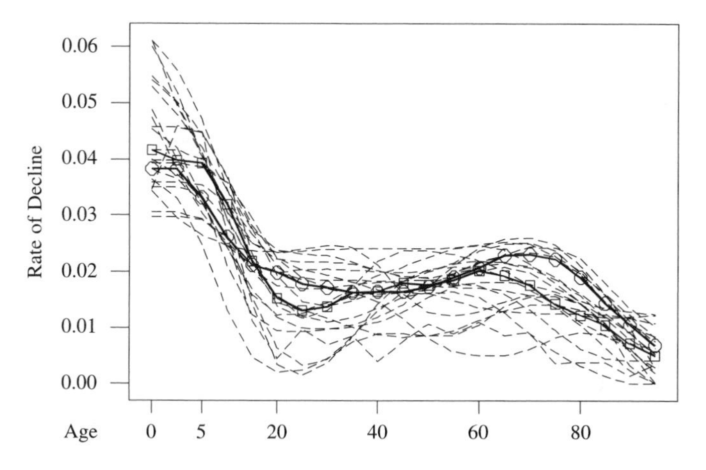

FIGURE 1. Smoothed Rate of Decline in Age-Specific Mortality for Females and Males and its Median Across 11 European Countries, for Females (Circle), and for Males (Square).

forecasting purposes the rates of decline were smoothed (using RSMOOTH) and restricted to be positive. We make no effort to distinguish the countries here, but concentrate instead on the median values and the variability around the medians. Notice that mortality has continued to decline the fastest in the lowest ages. In those ages in which most of the deaths occur, the decline for females has exceeded the decline for males. There is a fair amount of variation across the countries and one has to be concerned that the mortalities of the different countries do not drift too far apart in a forecast. Based on Figure 2 of Chapter 4 we see that during 1880–1990 the rate of decline in ages around 70 has been about 0.01 per year in Finland. Figure 1 shows that in many countries *more* gains have been made in those ages during the past 30 years. This suggests that the nature of mortality improvement has gradually changed. ◊

2.2. *Cause-Specific Mortality*

Consider deaths as classified by cause $k = 1, \ldots, K$. For example, the U.S. Office of the Actuary has used $K = 9$ primary causes of death, with heart diseases, cancer, and vascular diseases the most important ones (Example 2.4 of Chapter 4). In this case, the age-specific mortality is the sum of cause-specific mortalities, $\mu(x, t) = \mu_1(x, t) + \cdots + \mu_K(x, t)$. For each cause, the Office of the Actuary has postulated a target rate of change by cause τ_k. In a forecast $t = 1, \ldots, 25$ years ahead (Wade 1987) a smooth curve (cf., Andrews and Beekman 1987, 21) was used to connect the initial rate of change $\zeta_k(x)$ in age x to the target. Define $\gamma_k(x) = \zeta_k(x) - \tau_k$. Then, the resulting model for cause $k = 1, \ldots, K$ can be written as

$$\beta_k(x, t) = \tau_k t + \operatorname{sgn}(\gamma_k(x)) \sum_{s=1}^{t} \log(1 + |\gamma_k(x)| 10^{(6s-31)/25}) 10^{(31-6s)/25}, \qquad (2.2)$$

where $\operatorname{sgn}(z) = 1$ for $z \geq 0$ and $\operatorname{sgn}(z) = -1$ for $z < 0$. The complex expression is designed to lead to a smooth change from the initial rate of change to the target rate of change. However, (2.2) can actually be approximated fairly well with a second degree polynomial of t (Alho and Spencer 1990a, 213–214; 1990b, 611). The targeting approach followed by the Office of the Actuary is quite similar in spirit to the one suggested by Whelpton (Section 1.2).

The targets used in practice have been much closer to each other both across age and cause than are the empirical estimates at jump-off time (Alho and Spencer 1990a). This leads one to suspect that the cause-specific analysis has only been partially relevant for the specification of the forecast. Yet, it is clear that different causes of death of death could be treated differentially (e.g., Van den Berg Jeths et al. 2001). We will now discuss both theoretical and practical issues that arise when this is attempted.

The analysis of trends in cause-specific mortality is complicated by changes in the *International Classification of Diseases* (ICD). Although efforts are made to ensure continuity by dual coding of a single year's data, or *bridge-coding*, inevitable discontinuities and more gradual changes may occur. The revisions typically are more refined than their predecessors. For example, in the 10th revision of the ICD, or ICD-10, there are 8,000 categories for cause of death, whereas there were 5,000 in ICD-9. For results of a bridge-coding exercise between ICD-10 and ICD-9, see Anderson et al. (2001).

Apart from the data problems, it is often thought that if the trends of mortality due to different causes are different, then the causes should be analyzed separately. To see that this need not be the case, assume that the trend of mortality in a given age (we suppress age in the notation) during t, due to cause $k = 1, \ldots, K$ is of the form

$$\mu_k(t) = \sum_{j=1}^{s} \beta_j(k) f_j(t), \tag{2.3}$$

where the $f_j(.)$'s are known functions and the $\beta_j(k)$'s are parameters. The age-specific mortality rate is then of the form

$$\mu(t) = \sum_{j=1}^{s} \beta_j f_j(t), \tag{2.4}$$

where $\beta_j = \beta_j(1) + \cdots + \beta_j(K)$. We see that the trend of the sum is of the same form as the trends of the components. It follows that one would not expect there to be much difference in the forecast accuracy of the two approaches provided that the linear models (2.3) hold for each cause. Nevertheless, exceptions may occur.

Example 2.2. Emerging Cause of Death. Assume a polynomial model $f_j(t) = t^j$. If the degree s of the polynomial in (2.3) depends on k, then an emerging cause with a small current share of deaths may have a high value of s. In such a case we might erroneously identify a too small value for s when using a model for the total mortality (2.4). In long term forecasting this could make a difference. ◊

This example illustrates the possible advantage of disaggregation by cause. However, it can be turned around. Consider a cause of death that represents a small fraction of all deaths. Suppose the recorded number of deaths is rapidly increasing for that cause due to improving classification of deaths by cause. Initially, new diseases are infrequently diagnosed and deaths due to them are allocated to other causes. With better recognition more cases are found and a rapid spread of the disease may be predicted. We might call this an *early detection bias*. When the diagnostic practices become established, the recorded incidence levels approach the actual incidence. In the case of AIDS, for instance, these considerations may have been more relevant than the possibility mentioned in Example 2.2.

From a statistical perspective it is clear that *if* the data are correct, then one cannot lose efficiency by analyzing different causes jointly, instead of one by one. However, even here for the benefits of the joint analysis to materialize, special circumstances must prevail. Suppose the trends of the cause-specific time series are of the form (2.3). Assume that their errors are built up of innovations that are contemporaneously cross-correlated, so the processes themselves are crosscorrelated (e.g., as in (2.17) in Chapter 7). Then, the GLS estimators of the parameters $\beta_j(k)$ are the same whether the causes are analyzed jointly or separately and, similarly, the predictions of the future values of the processes are the same (Alho 1991). A condition that could lead to improvements under joint prediction is essentially that one of the series serves as a *leading indicator* for the others, i.e., the innovations of the series could be used to predict the innovations of the other series. Another condition would be if judgement could be more effectively used in forecasting deaths by cause than in forecasting the aggregate.

3. Conceptual Aspects of Error Analysis

3.1. Expected Error and Empirical Error

Recall that *error* = *forecast* − *true value*. We can use the concept of error after the future has unfolded, and we know how accurate the forecast turned out to be. However, for an error analysis to be really useful, we need to be able to characterize future uncertainty beforehand, at the time a forecast is made.

By *expected error* we refer to error as assessed at the time a forecast is made, before the future unfolds. By *empirical error* we refer to errors as assessed after the future has unfolded and the attained values of the process being forecasted have become observed.[3] The user of population forecasts wants to know the accuracy ahead of time and so is primarily interested in the expected error of a current forecast. The empirical errors of past forecasts are primarily useful if they help us either to improve forecasting methodology or to estimate the expected error. If future errors can be assumed to be similar (or at least not dramatically larger) than past errors, then the past errors provide us directly with estimates of the error to be anticipated in the future.

A key element in expected error is that it is always *model based*. If we mis-specify the model, the error assessment may be wrong. If the mis-specification is due to overfitting, an underestimation of expected error may occur. On the other hand, consider fitting an ARMA model to a once or twice differenced data series. Even if the model fits well, and leads to a small residual variance, the severe nonstationarity of the model can lead to forecast intervals that eventually cover values that are, in Whelpton's words, "incompatible with present knowledge". In such a case a model-based expected error may exceed empirical error.

3.2. Decomposing Errors

3.2.1. Error Classifications

Hoem (1973) classified sources of forecast inaccuracy into three main categories: (a) estimation and registration errors; (b) errors due to random fluctuations; and (c) erroneous trends in the mean vital rates. The first category refers to errors in parameter estimates, errors in basic data (on jump-off population and vital rates),

---

[3] It is customary to call these as *ex ante* and *ex post* errors. According to the *Oxford English Dictionary*, "ex post" is an abbreviation of "ex postfacto", meaning 'from what is done afterwards'. The etymology of "ex ante" is hazier.

and rounding errors. The second comprises the inherent stochasticity of the vital rates (e.g., binomial or Poisson variation, and random variation in their expectations). The third category involves various forms of model mis-specification (such as unincorporated gradual change or gross shifts of level). Keilman (1990) gave a similar list. In Alho (1990c, 523) we looked at the classification from the perspective of statistical modeling and defined the following four categories:

"(1) model mis-specification: the assumed parametric model is only approximately correct;
(2) errors in parameter estimates: even if the assumed parametric model would be the correct one, its parameter estimates will be subject to error when only finite data series are available;
(3) errors in expert judgment: an outside observer may disagree with our judgments or 'prior' beliefs about the parameters of the model;
(4) random variation, which would be left unexplained even if the parameters of the process could be specified without any error: since any mathematical model is only an approximation, one would expect there to be random variation."

We note that the four sources depend conceptually on each other. For example, random variation gives rise to estimation error, and errors of judgment may be equivalent to model mis-specification. Data errors fall in this classification under category (2). They require separate stochastic modeling. An example of this is the probabilistic assessment of error in census data that will be given in Chapter 10. Note also that (3) need not be the only source of error in judgmental forecasts. For example, estimation errors belonging to category (2) may influence the error of the forecast during the first years, and the classes (1)–(4) may all be applicable.

In practice, we have found that often the most important category of error is either model mis-specification or error of judgment. Any forecast must implicitly or explicitly choose the degree to which the future trend will continue the past trend, and the degree to which future variation about the trend will resemble past variation. As summarized by the following example, different choices can lead to drastically different forecasts.

Example 3.1. Sensitivity to Assumptions. Alternative cohort-component projections made around 1990, for the U.S. population in 2050, range from about 280 million to 507 million (U.S. Census Bureau 1992), and even 553 million (Ahlburg and Vaupel 1990). These projections are all scenario-based and their diversity reflects alternative assumptions rather than residual error or error in estimated coefficients. Pflaumer (1992) used two alternative ARIMA models for total U.S. population in 2050. One yielded a point forecast of 402 million, the other 557 million. The sensitivity to assumptions is also indicated by the fact that consecutive forecasts often show greater variance than the population they are trying to predict. Thus, the median Census Bureau forecast for 2050 increased in one year from 383 million to 392 million (U.S. Census Bureau 1992; Day 1993), an amount which exceeds the forecasted annual change even under their highest growth scenario.◊

3.2.2. Alternative Decompositions

A more precise discussion of the components (1)–(4) is feasible for formal models. Let μ_t denote the trend of a time series at time t and let ε_t denote the random deviation of the future value about its trend. The future value can be written as $X_t = \mu_t + \varepsilon_t$. Let $\mu_t(\beta)$ denote a parametric model for μ_t, and let $\mu_t(\hat{\beta})$ be a forecast of X_t based on estimated values of the parameters. For example, Lee and Carter (1992, 661) used the model $\mu_t(\beta) = \beta_0 + \beta_1 t$ for forecasts of the log of the mortality rate for a particular age group in the U.S. The forecast error $\mu_t(\hat{\beta}) - X_t$ is equal to the sum of three terms, $(\mu_t(\beta) - \mu_t) + (\mu_t(\hat{\beta}) - \mu_t(\beta)) - \varepsilon_t$. The first term represents model mis-specification (1), the second reflects errors in the estimated parameters of the model (2), and the third reflects random variation (4). Finally, a forecaster holding other prior views might have derived an estimator $\tilde{\beta}$ for the parameters. In that case we could further decompose $\mu_t(\hat{\beta}) - \mu_t(\beta) = (\mu_t(\tilde{\beta}) - \mu_t(\beta)) + (\mu_t(\hat{\beta}) - \mu_t(\tilde{\beta}))$. Here the first term reflects estimation error (2) conditionally on the other prior views, and the second is due to a difference in judgment (3).

Other decompositions are possible. For example, the sum of the first two terms is the error in the estimated trend. We have shown in Chapter 7 (Example 2.8 in particular) that random variation is important in the short run, but error in the estimated trend often dominates in long-range forecasts. In Section 2.2 we noted that the U.S. Office of the Actuary has used a model approximately equal to $\mu_t(\beta') = \beta_0' + \beta_1' t + \beta_2' t^2$. If the Office of the Actuary's specification were correct, or $\mu_t = \mu_t(\beta')$, the model error for a linear forecast would be $(\beta_0' - \beta_0) + (\beta_1' - \beta_1)t + \beta_2' t^2$. Even if we had $\beta_0 \approx \beta_0'$ and $\beta_1 \approx \beta_1'$, so the two models agreed for the recent time periods and for the near future, the model error would be approximately $\beta_2' t^2$. The standard error arising from the estimation of β_1 is linear in t in this example, implying that model mis-specification dominates estimation error in long-range forecasts.

3.3. Acknowledging Model Error

Model error is a central component of forecast error, but it is rarely discussed in statistics texts. Chatfield (1996) is an exception. In Alho and Spencer (1985) we applied the approximately linear models of Sacks and Ylvisaker (1978) to demographic forecasting in order to account for model error in the prediction intervals. A more ambitious synthesis via model averaging is discussed by Draper (1995), but see also Tukey (1995). For an application of these ideas in epidemiology, see Volinsky et al. (1997). Here we discuss the topic in terms that are readily applicable in demographic forecasting.

3.3.1. Classes of Parametric Models

We discuss first model error in the context of time series regression (Section 3.2 of Chapter 7). Consider a time series $Z_t = f(t) + \varepsilon(t)$ that has been observed for $t = 1, \ldots, n$. The goal is to predict the process for $t = n + m$, for $m = 1, 2, \ldots$ Consider a single value of m. By a *model* of $f(t)$ for $t = 1, 2, \ldots, n, n + m$, we mean a class of functions with the domain $A_m = \{1, \ldots, n, n+m\}$. For example, (3.3) of Chapter 7 defines such a class: M = all linear combinations of the functions $f_1(.), \ldots, f_k(.)$. To fix ideas, we begin by assuming $n > k$.

Consider two cases. First, if $f(.)$ does not belong to M, the model M is erroneous. The degree of error can be measured in different ways. A simple method is to use $\tilde{f}(t) - f(t)$ as the *model error for prediction* at $t = n + m$, where $\tilde{f}(t)$ is an estimate of $f(t)$ that would have been obtained if there had been no error $\varepsilon(t)$, $t = 1, \ldots, n$. If $\tilde{f}(t)$ is based on least squares fit of M to $f(1), \ldots, f(n)$ with $n \geq k$, then we can write $\tilde{f}(t) = (f_1(t), \ldots, f_k(t))(\mathbf{X}_1^T\mathbf{X}_1)^{-1}\mathbf{X}_1^T(f(1), \ldots, f(n))^T$, where $\mathbf{X}_1$ is an $n \times k$ matrix with $f_j(i)$ as the (i, j) element, as in Section 3.2 of Chapter 7.

Second, if M contains $f(.)$ there is no model error for any lead time m. If we add functions to M, then the enlarged model, say M_1, also contains $f(.)$. However, if a large number of variables were added relative to the number of observations n, then eventually $k > n$, the resulting statistical estimates may become unstable, and model error reappears. (For an extreme example, if M_1 is the class of all functions with domain A_m, there is no model error but the model is useless, as it leads to the same practical estimates as if we had no model at all.) One could attempt to measure the degree of model error for prediction by the asymptotic bias in a setting in which $n \to \infty$ and $k \to \infty$ (e.g., Portnoy 1988), but given our aims, we will not pursue this matter further.

The above discussion will lead to different measures of model error for different future years m. One should not be too surprised by this. For example, incorrectly choosing the order of a polynomial in regression would lead to errors that depend on m. As noted in Example 2.2, emerging causes can make such a choice especially difficult in mortality forecasting.

One way to take model error into account in the calculation of prediction intervals is to estimate $f(n+m)$ under alternative plausible models M_j, $j = 1, \ldots, J$. Denote the corresponding estimates by $\hat{f}(n+m; j)$. Suppose one of the models, say, M_1 is the correct one. Then, $\hat{f}(n+m; 1)$ is a "model-unbiased" estimate of $f(n+m)$. It follows that $|\hat{f}(n+m; j) - \hat{f}(n+m; 1)|$ is an approximation to the absolute value of the model error for prediction of M_j. Suppose the M_i is the preferred model. In reality we do not know which model is the correct one, but can use

$$B_i(m) = \max\{|\hat{f}(n+m; j) - \hat{f}(n+m; i)||j = 1, \ldots, J\} \tag{3.1}$$

as a conservative estimate of bias. The variance estimate $V(m) = \mathrm{Var}(\hat{f}(n+m; i))$ obtained from the preferred model i could then be replaced by the mean squared error $V(m) + B_i(m)^2$ in the calculation of two-sided prediction intervals, for example (cf., Cochran 1977, 12–15). Although (3.1) depends on the set of plausible models being entertained, the calculation of (3.1) even under just two alternative models may be enough to alert the forecaster that model error is a real possibility.

3.3.2. Data Period Bias

Let us continue to assume that we have data from a data period $t = 1, \ldots, n$. A frequent problem a forecaster may face is, should all the data be used in forecasting.

There are two complementary points of view to this problem. The first has to do with length of the data period n relative to the lead time m.

Data on Finnish fertility since 1776 are available (Figure 5 of Chapter 4). In Section 2.2.2 of Chapter 7 we used only data starting from 1920 because the earlier part cannot structurally be fit with the same ARIMA model as the latter part. Some may argue that one should only concentrate on even the shorter period since 1973 because the nature of the series may have changed again. We would have no disagreement with that view if the intention were to forecast only 5 years ahead. However, basing a forecast 25 years ahead on a data period of about 25 years would not be prudent. We can see from the figures that periods of 25 years have had idiosyncratic features that are only revealed against a longer background. Thus, models based on a short data period may be seriously in error. To summarize, our practical advice is that *one should always have a longer data period than the forecast period, and preferably two to three times as long*.

The second aspect is that even if the order of magnitude of the base period n (relative to lead time m) is not at issue, the specific choice can be hard to make. We suspect that often convenience rather than factors related to series itself dictate the choice. Still, alternative data periods will lead to alternative forecasts and alternative assessments of model error. In Alho and Spencer (1997) we introduced a practical method of taking such data period biases into account. The method is based on the same idea as (3.1). For concreteness, suppose ARIMA(p, d, q) models are being entertained. Define M_j as the estimate obtained from the data period $t = j, j + 1, \ldots, n$. Depending on the application we might want to use different values of p, d and q for different j. Or, we might keep those fixed and just vary j to get different parameter estimates.

For illustration, consider the U.S. growth rate (Figure 3 of Chapter 7). Concentrate on the decline in growth rate. Suppose one believes that the rate declines, but cannot decide exactly which of the periods starting from $j = 1900, 1901, \ldots,$ 1949, and ending at 1999, to take as a basis. A plausible compromise is to take the average over the starting years as the preferred estimate. This decline is $= 6.34 \times 10^{-5}$ per year. It determines "i" in (3.1). A histogram of the absolute values of (3.1) is given in Figure 2. We see that the maximum error is, in this case, about three times the size of the point estimate. Consonant with the fact that the average was used to get the preferred estimate, a less conservative approach is as follows. Suppose all starting values are viewed as equally likely to be correct. Then, the histogram would actually represent equally likely values of the bias, and the mean of the absolute values might be a compromise. In this case the mean of the absolute errors is 5.64×10^{-5}, still almost as big as the point estimate. This analysis suggests that it is not possible to get a reliable estimate of the future population growth rate by analyzing the growth series alone. For more accuracy, other information must be brought to bear.

3.4. *Feedback Effects of Forecasts*

In the previous sections we have not taken into account the possibility that forecasts have feedback effects that would directly influence their accuracy. Although

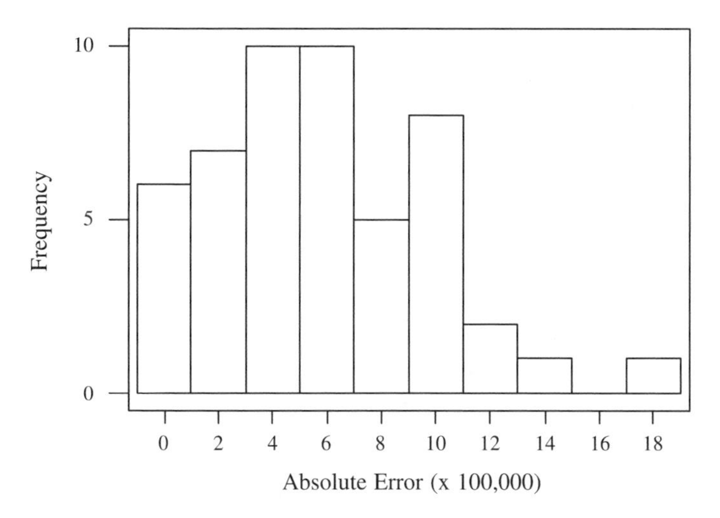

FIGURE 2. Distribution of Absolute Errors of Decline in Growth Rate.

decisions concerning additional births, health behavior, or moving from one place to another, are made by individuals, a classical view is that such decisions depend on social or community level values and economic conditions that have some coercive force over the individuals (Durkheim 1937). One use of forecasts is to influence such values. For example, as noted in Section 1, in many European countries cohort-component forecasts were made in the 1920's and 1930's with the specific motivation of fighting against imminent population decline. In other words, the intention was to produce a forecast that would make itself false, or *self-defeating*.

Self-fulfilling forecasts are also a possibility. In energy policy, for example, forecasts of increasing demand are used to justify the building of new power plants. The resulting increase in supply keeps prices in control, thus allowing increased consumption of energy. In demography, forecasts of increasing net migration may be used to justify the build-up of infrastructure (e.g., native tongue teaching in schools, training of social workers, provision of entry-level housing etc.) to receive future migrants, and this may lead to an increased inflow.

The possibility that forecasts may perform a feedback function from the past vital processes via behavior modification back to the vital processes are a reason to question the possibility of a meaningful statistical analysis of demographic forecasts and forecast errors. We recognize that such feedback mechanisms are possible, but point out that influencing people in this manner is harder than one might think. Attempts to influence fertility in the industrialized countries suggests that the policies typically have had relatively little effect (I.N.E.D. 1976, Ekert 1986).[4] Even in the case of immigration, government policies may be changed by

[4] Even the pro-natalist policies of the national socialist regime in Germany, in the 1930's, had only a temporary effect on fertility.

external events. In the United States, for example, legislated immigration quotas have frequently been exceeded when political and economic conditions have led to an unexpected influx of illegal immigrants.

Example 3.2. Planning Optimism. In the 1970's, in Europe, there was much optimism about the possibilities of social planning. In Finland, the government decided to replace "bystander's forecasts" of population that incorporate no assumptions about specific policies, by "participant's forecasts" in which the state would harmonize social policies on the regional level in such a way that future population would actually follow a *population plan* (Väestöennusteryhmä 1973). Conceptual models for the work were sought from the regional input-output tables and other planning tools developed in Sweden, Norway, Italy, France, the Netherlands, the United Kingdom, West-Germany, and the Soviet Union. These models attempted to give a system-theoretic picture of the regional economies, regional populations and their change (Talousneuvoston aluejaosto 1972). Despite the enthusiasm of the planners, and the seeming rationality of the plans, people simply ignored them. As the discrepancy between plans and subsequent development became large enough, the whole concept of population plans was abandoned. $\Diamond$

Our tentative conclusion is that while forecasts may lead to changes in demographic behavior, the large scale effects are probably indirect, via long chains of changes in attitudes, social norms, institutions etc. Empirical examples of significant short term feedback influences in national level forecasts are hard to find.

3.5. *Interpretation of Prediction Intervals*

3.5.1. *Uncertainty in Terms of Subjective Probabilities*

In philosophical literature it is shown that probabilities can be given numerous interpretations that sometimes conflict (e.g., Kyburg 1970, Jeffrey 1983). We do not discuss them in any generality, but note that for the communication of stochastic population forecasts to users, some intuitively understandable interpretation is needed.

In general, a forecaster must be prepared to describe a *stochastic* or *probabilistic forecast* as representing his or her subjective views of the likelihood of future developments. Since forecasting is typically a group effort, the forecast must actually correspond to the consensus view of the group. Moreover, a reputable team of forecasters typically tries to present evidence and arguments to show that statistical modeling was done efficiently and provided a good fit to the data, and judgment was exercised in a defensible manner. Thus, reputable forecasts are constrained in many ways by peer criticism or the prospect of rejection by potential users. Whether the "author" of a forecast is an individual or a group, the probabilities that are published are intended to correspond to the author's views in a very specific sense. This will be taken up next, using the machinery of set theory.

Consider a non-empty set Ω of elements, one and only one of which will occur. The set of possible events is taken to be a collection $\mathscr{F}$ of subsets of Ω with certain properties: (i) the sure thing is an event, i.e., $\Omega \in \mathscr{F}$; (ii) the complement of any event is also an event, i.e., if $A \in \mathscr{F}$ then its complement $A^c \in \mathscr{F}$; and (iii) if A and B are both events, then "either A or B" is an event, i.e., if $A \in \mathscr{F}$ and $B \in \mathscr{F}$, then their union $A \cup B \in \mathscr{F}$.[5]

Referring to Figure 4 of Chapter 7, the subsets could describe childbearing in 2020. For example, if $A =$ "the total fertility rate in year 2020 is > 2.21", then $A^c =$ "the total fertility rate in year 2020 is ≤ 2.21". (Their union $\Omega = A \cup A^c =$ "the total fertility rate in year 2020 is > 2.21, or ≤ 2.21" is an event that is certain to occur.) If $B =$ "the total fertility rate in year 2020 is < 1.32", then, $(A \cup B)^c =$ "the total fertility rate in year 2020 is in the interval [1.32, 2.21]" etc. Using the so-called De Morgan rules, one can show that the intersection can be expressed in terms of unions and complements. (Note that $A^c \cap B^c = (A \cup B)^c$ in the example at hand, for example.) This means that we have a simple set theoretic language available with operators corresponding to "not" (complement), "or"(union), "and" (intersection) to form expressions for new events.

If $P(A)$ is the probability of an event $A \in \mathscr{F}$, then it satisfies the rules (iv) $P(\Omega) = 1$; and (v) if $A \cap B = \emptyset$ for $A, B \in \mathscr{F}$ then $P(A \cup B) = P(A) + P(B)$. It follows from these that $P(A^c) = 1 - P(A)$ for $A \in \mathscr{F}$ also holds. In our example, based on the numbers underlying the figure (see page 214), we would have $P(A) = P(B) = 1/4$, for example, so $P((A \cup B)^c) = 1/2$. How should such quantitative probabilities be interpreted?

Major contributions to the theory of subjective probabilities were Ramsey (1926), de Finetti (1931, 1937, 1974), and Savage (1954). A textbook treatment of the theory is given in Fine (1973) and a philosophically oriented but mathematically rigorous treatment is given in Jeffrey (1983); see also Howson and Urbach (1993). Continuing with the class $\mathscr{F}$ of events that satisfies (i)–(iii), suppose there is an ordering relationship "$\preceq$" such that (a) it is not true that $\Omega \preceq \emptyset$; (b) *comparability* holds: either $A \preceq B$ or $B \preceq A$ for any $A, B \in \mathscr{F}$; (c) *monotonicity* holds: if $A \cap C = \emptyset$ and $B \cap C = \emptyset$, then $A \cup C \preceq B \cup C$ if and only if $A \preceq B$; (d) *transitivity* holds: if $A \preceq B$ and $B \preceq C$, then $A \preceq C$ for $A, B, C \in \mathscr{F}$. Subject to further conditions one can prove that corresponding to such a relationship there exists a unique probability P satisfying (iv)–(v) such that $A \preceq B$ if and only if $P(A) \leq P(B)$. The conditions are satisfied, for example, if for any n the set Ω can be partitioned into n subsets $D_1, \ldots, D_n \in \mathscr{F}$ that are equally likely (i.e., both $D_i \preceq D_j$ and $D_j \preceq D_i$ hold for $i, j = 1, \ldots, n$).[6]

The relationship "$\preceq$" is intended to correspond to a *qualitative* (or *comparative*) *probability*: $A \preceq B$ means that "B is at least as likely as A" (e.g., Savage 1954, 30). The interpretation of (a) is that a certain event should be strictly more likely than an impossible event. The conditions (b)–(d) can then be interpreted as characterizing the beliefs or an individual who is "rational" in the sense of being able to compare any events of interest, thinks of probabilities in an additive manner, and is consistent

[5] For technical reasons (iii) is usually given for countable unions.

[6] For a discussion and an alternative formulation in terms of "fine" and "tight" conditions, see Savage (1954, 37–38).

in thinking. (The notion of rationality will further be discussed in Section 4.2 of Chapter 12.) The partitioning condition for quantification says that there are equally likely events that can be used as a yardstick to measure the probabilities of other events. This would be true if, say, an unlimited number of coin-tossing experiments could be included into $\mathscr{F}$. If our views of the world are more "coarse" (e.g., so that a partition is available up to some finite value of n only), it may only be possible to determine P to some degree of accuracy.

The classical result shows that an individual's degrees of belief can be represented in terms of quantitative probability statements. However, as different individuals may hold conflicting views, this opens up the possibility of conflicting probability statements that are simultaneously true. This is, indeed, the case. However, an approximate consensus view can arise under quite general circumstances, if rational individuals are presented information in an unbiased manner. To indicate how this can come about, consider the following classical example.

Example 3.3. Achieving Approximate Consensus on Probabilities. Consider two individuals R and S. R thinks a coin is biased, so a chance of getting heads is about 0.1 and perhaps a lot less. He is not quite sure, however, and the standard deviation around the expected value could be about 0.1. Define $f(p) = p^{\alpha-1}(1-p)^{\beta-1}$ for $p \in [0, 1]$ and $\alpha > 0$ and $\beta > 0$. Let $B(\alpha, \beta) = \int f(p)dp$. Then, the *beta distribution* $\mathrm{Be}(\alpha, \beta)$ (e.g., DeGroot 1987, 294–296) has the density $f(p)/B(\alpha, \beta)$. R's views can then possibly be represented by, say, Be(1,9), because this distribution has expectation $\alpha/(\alpha+\beta) = 0.1$ and variance $\alpha\beta/[(\alpha+\beta)^2(\alpha+\beta+1)] \approx 0.09^2$. Suppose S has opposite views that can be represented by Be(9,1) with the mean 0.9. In both cases the probabilities reflect both what the individuals perceive as likely, and their uncertainty about the most likely value. How could one get them to come to a consensus? Suppose the true probability of heads is actually $p_0 = 0.3$. We arrange a coin tossing experiment for R and S and observe X heads in n independent tosses of the coin. The number of heads has a binomial distribution, so the probability is proportional to $p^X(1-p)^{n-X}$. Given the prior views $\mathrm{Be}(\alpha, \beta)$ a rational person would compute the posterior distribution that is proportional to the product of the prior and the likelihood of the data,[7] that is, proportional to $p^X(1-p)^{n-X}p^{\alpha-1}(1-p)^{\beta-1} = p^{\alpha+X-1}(1-p)^{\beta+n-X-1}$. We notice that this integrates to $B(\alpha+X, \beta+n-X)$, so the posterior view must be represented by the distribution $\mathrm{Be}(\alpha+X, \beta+n-X)$, whose mean is $(\alpha+X)/(\alpha+\beta+n) = (\alpha/n + X/n)/(\alpha/n + \beta/n + 1)$. By the law of large numbers, $X/n \to p_0$ as $n \to \infty$, so the mean converges to the right value. For the variance we have $(\alpha+X)(\beta+n-X)/[(\alpha+\beta+n)^2(\alpha+\beta+n+1)] \to 0$. Thus, the individual eventually learns the true value and becomes certain about his or her belief! Figure 3 gives simulated trajectories of the expected values of for R and S in one such experiment. ◊

[7] This is what an idealized "rational" individual would do. As discussed by Edwards (1982) and Starmer (2000) opinions can be more resistant to change, in practice.

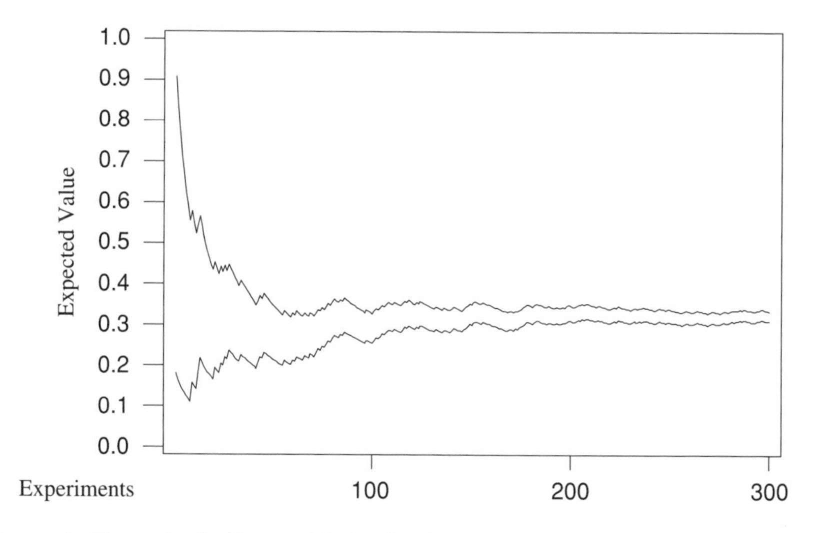

FIGURE 3. Change in the Expected Value for the Probability of Heads in a Sequence of Coin Tossing Experiments for an Individual with a Prior Expectation of 0.9 (Upper) and an Individual with a Prior Expectation of 0.1 (Lower).

The classical results involve highly idealized individuals, whose abilities in introspection surpass what we are normally capable of. Techniques of elicitation have been developed to discover dormant beliefs in a person, who has not consciously thought about a particular matter, or who outright denies being capable of expressing his or her views in this manner. A popular method is to pose the problem in terms of betting; for general discussion of other methods see Kadane and Wolfson (1998) and for a demographic application see Daponte, Kadane, and Wolfson (1997).

Example 3.4. Elicitation of Probabilities via Betting. Consider the event "the total fertility rate in year 2020 is in the interval [1.32, 2.21]" that we assign a probability of 0.5 based on a time-series analysis underlying the intervals in Figure 4 of Chapter 7. A person who truly believes in the assessment should be willing to pay 1 unit for a gamble in which he or she would win 2 units or more, in case the true total fertility is in 2020 is inside the interval, because then: $\textit{expected winning} - \textit{cost} \geq 0.5 \times 2 - 1 = 0$.[8] However, suppose the person thinks that the chances are $p > 0.5$ that the future value will be in the interval. Then, he or she should be willing to pay 1 unit for a gamble that would only pay as little as $1/p$. Conversely, if the person accepts a gamble in which the winnings are 1.5 units

[8] Due to risk aversion (e.g., Arrow 1971, Chapter 3) a somewhat higher value than 2 would often be needed. For example, a person may prefer (and not be indifferent) to receiving 1 unit with certainty rather than accepting a lottery ticket with equal probabilities of payoffs 0 units and 2 units. This topic is discussed in Section 4.2 in Chapter 12.

or more, then the subjective probability that would make this gamble rational must be $p \geq 1/1.5 = 2/3$. Experiments of this type have been used to assess the uncertainty of migration forecasts (Alho 1998). ◊

In practice, views of individuals or groups may violate the conditions (b)–(d) in various ways (e.g., Kahneman and Tversky 1982). This need not invalidate the general approaches or interpretations outlined here, but care has to be exercised in any elicitation.

"When we speak of belief in common life, we always mean that we consider the object of belief more likely than not; the state of mind in which we rather reject than admit, we call *un*belief. When the mind is quite unbalanced either way, we have no word to express it, because the state is not a popular* one...

* Many minds, and almost all uneducated ones, can hardly retain an intermediate state. Put it to the first comer, what he thinks on the question whether there be volcanoes on the unseen side of the moon larger than those on our side. The odds are, that though he has never thought of the question, he has a pretty stiff opinion in three seconds." (de Morgan 1847, 182–183)

Moreover, empirical evidence using pairwise comparison interview techniques indicates that there are severe limits to our abilities to maintain transitivity when questioned repeatedly about a value of an item of interest (e.g., Alho, Kangas, and Kolehmainen 1996). Although transitivity can always be imposed using a number of methods, the result may be sensitive to the method used. This suggests that we may have to satisfied with less precision in the quantification of probabilities than we might wish.

3.5.2. Frequency Properties of Prediction Intervals

Even if a forecasting group can agree on a particular quantification of the expected error, users of prediction intervals want the intervals to possess frequentist interpretations, e.g., 95% prediction intervals should contain the future value in 95% of the cases, not much more, not much less (i.e., the intervals should be *externally calibrated*). Unfortunately, due to the high autocorrelations of forecast errors (cf., Chapter 7), the empirical validation of prediction intervals is difficult. Autoregressive models provide a simple example.

Example 3.5. Assessing Prediction Intervals for ARIMA Forecasts. Consider an ARIMA$(p, d, 0)$ model. Its forecast function is determined by the $p + d$ last observations. Suppose that a forecast k steps ahead is made at time t and a $100(1 - \alpha)$ level prediction interval is computed. Define $X = 1$, if the observation at $t + k$ is included in the interval, otherwise $X = 0$. Then, observations during the time segment $[t - p - d + 1, t + k]$ determine X. The length of the segment is $k + p + d$. Suppose also that we have n consecutive, non-overlapping segments available, and we calculate a k-step ahead forecast for the last observation of the segment using the $k + d$ first observations in each segment, in turn. Define $X_i = 1$ if the last observation was in the interval for segment $i = 1, \ldots, n$ and otherwise $X_i = 0$. Since the experiments during different segments are independent, the laws of large numbers entail that $(X_1 + \cdots + X_n)/n \to 1 - \alpha$, as $n \to \infty$. Or, in the long run the intervals cover the true value with the right frequency. However, in this argument a data series of length $(k + p + d)n$ is reduced to a sequence of n observations only. For large k, there may be very few independent observations, or none at all. In practice, we would use all sequences of length $k + p + d$ to assess the coverage probabilities of the intervals, but the high correlation of the corresponding X indicators means that the increase in information can be much less than $(k + p + d)$-fold. ◊

3.6. Role of Judgment

3.6.1. Expert Arguments

A statistical examination of past developments provides a relatively neutral starting point for a forecast. Although subjectivity is always involved in the choice of a statistical model, the principles of simplicity or parsimony (cf., Section 2.1.1 of Chapter 7) and consistency with the data often lead to a small set of models that any competent modeler would consider plausible. However, even if a relatively objective basis for model choice exists, the chosen models may suffer from shortcomings.

First, statistical models do not explicitly include notions of causality or understanding.[9] As pointed out by Whelpton, it may happen that the models produce forecasts that conflict with other information we may possess about the vital processes. For example, a time-series model may lead to forecasts or prediction intervals (of life expectancy or total fertility rate, for example) that are implausibly high or implausibly low in view of past experience. An expert may point this out, and suggest how the analysis should be adjusted in light of such knowledge.

Second, statistical analyses tend to emphasize long-term developments. However, we may have knowledge of emerging factors that are believed to have an effect on the trends in the future even though such effects have not been apparent in the past (cf., Example 2.2). For example, knowledge of changes in smoking behavior may suggest that mortality trends will change in the future. Again an expert may point this out, and suggest how the forecast should be adjusted in light of such knowledge.

Third, statistical models typically assume that the uncertainty of forecasting is similar in the future to what it has been in the past. Or, if changing volatility is allowed, one has to specify a mechanism of change that operates in the future as it did in the past (e.g., Section 4 of Chapter 7). Yet, demographic processes may undergo periods of relative calm and relative turbulence for reasons that can be explained. An expert may point this out, and suggest how a forecast should be adjusted in light of this.

The three cases mentioned above do not exhaust the ways in which judgment may be exercised to adjust model-based forecasts to better correspond to reality.

---

[9] E.g., the well-known *Granger causality* says that two time-series do not show causal dependence if knowing the past values of the second series does not help in predicting the first, in the mean squared sense (Granger 1969; Wiener 1956). Or, the notion of causality is reduced to a formal property of conditional expectations. For extensions, see, e.g., Chamberlain (1982) and Florens and Mouchart (1982).

However, to preserve the intended interpretation of the forecasts, in all cases a *careful argumentation is necessary when adjustments are made.* Given the relatively low predictive power of our social science theories (e.g., Example 1.5 above), such an argumentation can rarely be conclusive. But if no arguments are given, then the resulting forecast may appear arbitrary. We give three stylized arguments that appear legitimate to us.

Example 3.6. Mortality Differences Across Countries. In the early 1950's the female life expectancy was 72.4 in Denmark and 73.3 in Sweden. The two countries were leading the world at that time. In 2002 the corresponding life expectancies were 79.7 and 82.6. Both countries lagged behind Japan with a female life expectancy of 84.3. During a 50 year period the advantage of Sweden over Denmark grew from 0.9 years to 2.9 years. It is thought that life style factors (smoking, alcohol use) explain much of the change. Since these are factors that can be influenced by government activities (information, improved health care systems), it is reasonable to expect that the Swedish advantage will not continue to grow indefinitely, and it may even begin to shrink. ◊

Example 3.7. Fertility in the Mediterranean Countries. The decline of period fertility to an unprecedented low level in Italy (1.24 in 2000) and Spain (1.26 in 2000) would lead to a higher level of childlessness than suggested by fertility surveys. This suggests that some degree of recovery may take place in the coming 10–20 years. ◊

Example 3.8. Migration to Germany. After the fall of Soviet power, and the unification of East and West Germany, migration into Germany became a practical possibility for a pool of German speakers who would have liked to migrate even earlier. As the pool gradually becomes depleted, it is likely that net-migration will decline to a lower level than that observed in the 1990's. ◊

The practical difficulties observed in connection with the elicitation of probabilities suggest that it is much harder to come up with meaningful uncertainty statements using judgment alone than to argue for a particular point forecast. These difficulties are compounded by the well-known phenomenon of *expert overconfidence* (Kahneman, Slovic, and Tversky 1982, Part VI). An expert may be in a particularly tight spot when asked to express his or her uncertainty concerning a topic he or she is supposed to be an expert on! A possible way to circumvent such awkward situations is to approach uncertainty in relative terms. In the spirit of Examples 3.6–3.8, judgment may well be useful in an assessment of whether future uncertainty should be viewed as being bigger, equal, or smaller than uncertainty in the past. Statistical modeling can provide an estimate of the past level.

3.6.2. Scenarios

As far as we know, the use of scenarios originates from military applications during the Cold War, in the 1950's and 1960's (cf., Kahn 1962, 150–153; quotations below are from this source). At that time, scenarios were devised as aids to thinking about events that are not only "unpleasant" but also "unexperienced". Among other things the scenarios "call attention, sometimes dramatically and persuasively, to the large range of possibilities that must be considered"; "force the analyst to deal with details and dynamics that he might more easily avoid"; and "illuminate the interaction of psychological, social, political, and military factors". To be plausible they must "relate at the outset to some reasonable version of the present, and must throughout relate rationally to the way people could behave". Thus, the scenarios are very much based on causal thinking, and use ideas of continuity to try to make the "unthinkable" future analyzable.

Thus, scenarios involve not only what is likely to happen, but also alternatives we might not otherwise be able to, or might not wish to see. This is very much in the same spirit as we have approached forecasting. However, while it is clear that if we contemplate the course of a thermonuclear war, we cannot have much empirical basis for formulating probabilities concerning the future outcomes, in demography the situation is different as we have perhaps the longest and most systematic body of historical evidence of any social science.

3.6.3. Conditional Forecasts

When new policies are contemplated, one might wish to forecast their consequences. In this case we may not have direct evidence upon which to base a forecast, and we may have to condition on particular actions being taken with more or less well specified consequences. Although all forecasts are conditional on what was observed in the past, we define a *conditional forecast* as a forecast that is conditional on the occurrence of some future event.

Suppose Y is a criterion variable of interest, such as some demographic intensity measure (fertility, mortality, migration etc.), and suppose a policy maker wants to influence its value. Write $Y = m_Y + \varepsilon_Y$, where $E[\varepsilon_Y] = 0$ and assume that the policy maker can create a control variable Z such that the controlled version of Y is $Y_Z = Y - Z$. We call Y_Z an *adaptive scenario*, because it explicitly conditions on a policy being adopted that produces a value for Z in the future (Alho 1997). Whatever the value of Z, the distribution of Y_Z can then be interpreted as the conditional distribution of Y given Z. In the simplest case, we may assume that $Z = \alpha + \beta\varepsilon_Y + \varepsilon$, where ε is independent of ε_Y with $E[\varepsilon] = 0$. Here α and β are parameters that the policy maker can choose within some limits. They influence both the mean and variance of the variable to be controlled. The role of ε is to represent unexpected disturbances that are caused by the introduction of Z. Indeed, $\text{Var}(Y_Z) = (1 - \beta)^2\text{Var}(Y) + \text{Var}(\varepsilon)$, so the adaptive scenario may even be more uncertain than the uncontrolled Y. Under this model it is possible to make assumptions about future policies, incorporate them into forecasts, and still retain the notion that such scenarios are uncertain.

4. Practical Error Assessment

To assess the uncertainty of future population, we need to look at the past forecastability of the vital rates and the accuracy of past forecasts. We do not have to accept that future forecasts will be exactly as accurate as the past forecasts, but we have to be prepared to defend our models and assumptions if we do not so assume. By looking at the way vital rates have been forecasted in the past, we may learn a great deal about why errors were made in the past, to what extent they might be avoided, and how large we might expect them to be in the future. In Section 4.1 we will define commonly used error measures. In Section 4.2 we show how baseline forecasts can be used to provide error assessments. Section 4.3 discusses the modeling of errors of the U.N. world forecasts using a random effects model.

4.1. Error Measures

Suppose $X > 0$ is a random variable representing future population size, the future level of fertility etc. Let T be its forecast, which is based on past data. Then, *forecast error* is $\varepsilon = T - X$. The *absolute error* is $|\varepsilon|$, the *squared error* is ε^2, and the *relative error* is ε/X. To characterize the level of error over a set of forecasts, one typically conditions on the realized value of X. In this case, the *mean absolute error* (*MAE*) is $E[|\varepsilon|]$, the *mean squared error* (*MSE*) is $E[\varepsilon^2]$, the *mean relative error* (*MRE*) is $E[\varepsilon/X]$, and the *mean absolute relative error* (*MARE*) is $E[|\varepsilon|/X]$, for example. The *bias of the forecast* is $B(X) = E[\varepsilon]$. The various measures are estimated from data by their sample averages. For example, if we have a set of values $X_i, i = 1, \ldots, n$ with forecasts $T_i, i = 1, \ldots, n$, then MARE would be estimated by $(1/n)\Sigma_i|T_i - X_i|/X_i$.[10]

The variance of the forecast error is $\text{Var}(\varepsilon) = E[\varepsilon^2] - E[\varepsilon]^2 = E[\varepsilon^2] - B^2$, so the mean squared error is of the form,

$$MSE = \text{Var}(\varepsilon) + B^2. \tag{4.1}$$

Other error measures account for bias, as well, but only the mean squared error has this elegant decomposition. If the interest centers on understanding how past errors came about, both the variance and the bias are of interest. However, if we intend to use empirical measures of past errors in an assessment of future uncertainty, then the future bias would be unknown and using $\text{Var}(\varepsilon)$, instead of MSE, can lead to an underestimation of the level of uncertainty. The moments in (4.1) can also be taken conditionally on either T or X, depending on the desired interpretation.

In his study of Dutch forecast errors Keilman (1990) established several qualitative results that have emerged in many other studies since (e.g., Keilman 1998; Bongaarts and Bulatao 2000, Chapter 2). Perhaps the single most important finding was to show that the MRE of population size has depended heavily on age. Fertility has been overestimated to the extent that over a 15 year forecast period the MRE of the age-group 0–4 has been approximately 0.28, or 28%. Similarly, survival in old-age has been underestimated, especially for females, so that the MRE of age-group 85+ has been approximately −0.15, or −15%, over a 15 year ahead forecast period (Keilman 1990, 83). This illustrates how errors in different

[10] A frequently used measure is the *mean absolute percentage error* (*MAPE*) $= 100 \times$ *MARE* that is estimated by $100 \times (1/n)\Sigma_i|T_i - X_i|/X_i$.

age ranges have compensated for each other. The forecast for the total population has been much more accurate.

Empirical estimates of error typically show that uncertainty increases with lead time (e.g., Keilman 1990, 105). However, examples such as Whelpton's forecast of the U.S. total fertility rate (Section 1.2) show that for a given forecast it may well happen that the errors first increase and then start to decrease. Occasional examples of this type occur, if the estimates are based on a small number of observations. We conclude that some form of error modeling (using time-series or other statistical models) is preferable to the direct use of error measures if the intention is to use characterize expected error. This is supported by the fact that for many countries very few past forecasts are available, and there is no country for which a statistically reliable estimate of past forecast error is available for lead times above 50 years. For many applications (e.g., pensions) forecasts up to 50 years or more are, nevertheless, needed.

4.2. Baseline Forecasts

As a potential remedy to the difficulties of empirical error estimation, in Alho (1990c) we suggested that naive or baseline forecasts be used to obtain *omnibus error assessments*, i.e., assessments that capture all sources of error simultaneously.[11] Consider the total fertility rate during the 20th century. In many European countries and the U.S. the rate declined until the 1930's or so. Then, it increased until the 1950's and 1960's and declined after that. As pointed out by Lee (1974) the available forecasts display a remarkable regularity: the forecast has typically been very close to the current value. If the total fertility rate were a random walk, using today's value for all future times would be optimal.[12] Indeed, in industrialized countries a graph of these series (e.g., Figure 4 of Chapter 7) often looks approximately like that of a random walk. We conclude that using the *current value* as the forecast is a simple, reasonable *baseline forecast for fertility* which conceivably can be improved upon, but which is not easy to beat.

A similar argument is available for mortality. In industrialized countries, we have seen a steady decline in mortality during this century, with an occasional plateau in one country or another, but with no major upturns.[13] Official forecasts of mortality have typically assumed that the decline will continue for a while, and then level off. However, as pointed out in Example 1.4 (recall (3.1) of Chapter 7), a simple *baseline forecast for mortality* that would have done as well as (or better than) the official forecasts is to assume that the *recent past rate of decline continues*

---

[11] In economics, naive forecasts are routinely used as benchmarks in the assessment of forecast accuracy, cf., Öller and Barot (2000), for example. In fact, the notion is a generalization of "Theil's *U*" (Theil 1966).

[12] A wider class of models, the *martingales*, are defined by the property that today's value is the optimal forecast, cf., Chung (1974).

[13] In the 1980's and 1990's, the countries of the former Soviet Union experienced increases in mortality that are not compatible with the "stylized facts" we are presenting.

for the next few decades. Again, this is a simple, reasonable forecast that possibly can be beat, but not very easily.

In many industrialized countries net migration has behaved in a rather erratic fashion around a mean, but with many national variations. Thus, a *baseline forecast for net migration* that often can be taken as a starting point is to assume that the *recent average number* will continue to enter (or leave) the country.

Such baseline or *naive* forecasts can be useful in the assessment of the expected error of forecasts, because their empirical accuracy can always be assessed. We may simply make as many naive forecasts as we have past jump-off years available, and calculate the empirical errors. These errors should not be smaller than the errors of the more complex forecasting methods actually used. (If they are, one should consider changing from the complex method to the naive one!) Therefore, if the forecastability of the process does not dramatically deteriorate, the empirical error of the naive forecasts provide useful assessments of the expected error for any other forecasting method that is not less accurate than the naive method.

Naive forecasts cannot replace model-based error estimates (strictly speaking, they are based on certain implicit modeling assumptions themselves!), but they can serve as a useful complement. Since modeling error is an important source of error, it is useful to have available a non-parametric technique that avoids assumptions about parameter structure or distributional form.

Example 4.1. Error Estimates for Fertility Forecasts in Europe. Figure 4 displays empirical estimates of the absolute relative error of naive forecasts for the logarithm of total fertility in six European countries with data ending in 2000 and starting between 1751–1900. In the order of size of error, from the largest to the smallest, they are the Netherlands, Denmark, Norway, Finland, Iceland, and Sweden.

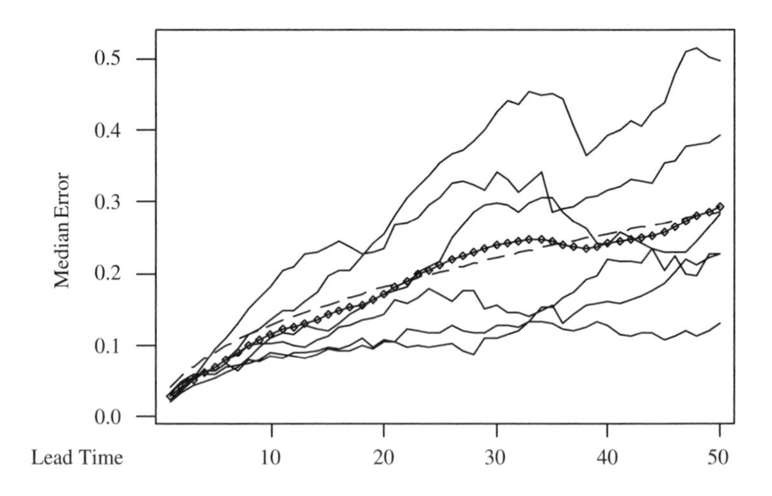

FIGURE 4. Median Relative Error of Fertility Forecast as a Function of Lead Time for Six Countries with Long Data Series, their Average (Circle), and a Random Walk Approximation.

Figure 4 also has the mean of the six countries, as smoothed by the RSMOOTH procedure of Minitab, and the error of a random walk whose volatility closely matches the mean. If the steps of the random walk are normal (Gaussian) with variance (volatility) 0.06^2, then the median of the absolute value of the error is $0.6745 \times 0.06 \times t^{1/2}$, given as the dashed line in Figure 4. To appreciate the order of magnitude, note that at lead time 30 the mean of the relative errors is approximately 0.20. This corresponds to an expected absolute error of about 20%. Under a normal (Gaussian) model of relative error, this corresponds to a relative standard deviation of about 30%. $\diamondsuit$

A study of the autocorrelation functions of the six countries shows some autocorrelation (0.1–0.3) at short lags (cf., Section 2.2.2 of Chapter 7 for the effect of war in Finland). The median of the estimated standard deviations of the first differences is 0.045. An AR(1) process with parameter $\varphi \approx 0.25$ provides a serviceable model of the series. The results of Example 2.7 of Chapter 7 imply that a random walk model produces comparable prediction intervals as an ARIMA(1,1,0) with this correlation structure, if the standard deviation is multiplied by $[(1 + 0.25)/(1 - 0.25)]^{1/2} = 1.29$. Or, the matching scale estimate should be $1.29 \times 0.045 = 0.055 \approx 0.06$, a value we arrived at in Example 4.1 via the error of naive forecasts. Moreover, the overestimation of the level of uncertainty at short lead times when a random walk approximation is used (see Figure 4) essentially vanishes if ARIMA(1,1,0)-based formula (2.14) of Chapter 7 is used. Thus a more refined approximation would be an AR(1) model for the first differences with first autocorrelation = 0.25 and innovation variance 0.045^2. We see that error estimates based on naive forecasts and those deriving from ARIMA models give similar results for these data.

Given the paucity of data, a corresponding analysis cannot be validly carried out for countries with time series 40–50 years long. For short lead times, say, up to 15 years a meaningful analysis can, however, be carried out. One can also argue that the consideration of the remote past is not as relevant as the most recent past. Perhaps the level of uncertainty is less if the most recent period alone is considered? In Europe, the opposite is true, however. During 1960–2000 the errors for 22 European countries are typically *larger* than the estimates obtained for the subset of the countries with long data series. For lead time 15 years, the median error (across countries) is 0.26. This is approximately twice the mean value of Figure 4 for lead time 15. Hence, fertility has been unusually volatile in Europe during the last 40–50 years. In part, the recent high volatility can be attributed to the decline that forms the end of the baby-boom. However, another factor is the emergence of extremely low fertility in Central Europe and the Mediterranean countries.

Example 4.2. Error Estimates for Mortality Forecasts in Europe. In an analysis of data from nine European countries (Austria, Denmark, France, Italy, the Netherlands, Norway, Sweden, Switzerland, and the United Kingdom), we have compared the volatility of mortality in ages 50–54, 55–59, ..., 90–94 with data ending in 2000, and starting at various times, the earliest being the United Kingdom in 1841. The baseline forecast was made by assuming the decline observed during the most

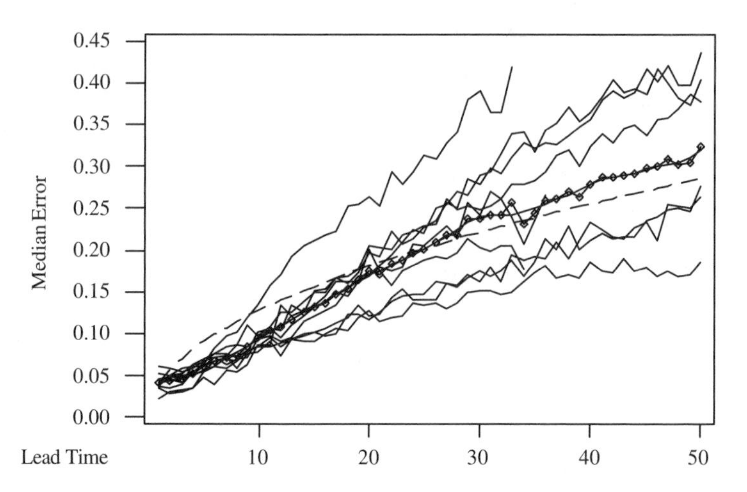

FIGURE 5. Median Relative Error of Mortality Forecast as a Function of Lead Time for Nine Countries with Long Data Series, their Average (Circle), and a Random Walk Approximation.

recent 15 years to continue indefinitely. The data were aggregated over age-groups for each country to provide the median level of relative error for each country, for each lead time. Figure 5 has a plot of the median errors, their average, and a random walk approximation. Surprisingly, a matching level of error is obtained with the same volatility 0.06^2 as for fertility. ◊

From a comparison of Figures 4 and 5 we find that the relative error one can expect in age-specific mortality forecasts is similar to that for total fertility. How can this be reconciled with the generally held view that forecasting mortality is easier? Perhaps a partial answer is that usually survival rather than mortality is meant. If one makes a large error in forecasting a mortality rate that is of the order of 1 percent, then the relative error in the number of survivors is 1/100 of that.

4.3. Modeling Errors in World Forecasts[14]

The U.N., the World Bank, and the U.S. Census Bureau publish cohort-component forecasts for all countries of the world. We will review simple techniques of error modeling, and show how, based on past and current forecasts of the U.N., prediction intervals for the total population size can be derived.

4.3.1. An Error Model for Growth Rates

Let $V(t)$ be the population size in the beginning of the year t. Defining the average growth rate during $[t, t+1)$ as $\rho(t) = \log(V(t)/V(t-1))$, we get that for $t > 0$,

$$V(t) = V(0)\exp(\rho(0) + \cdots + \rho(t-1)) \tag{4.2}$$

[14] This section reviews Appendix F (http://books.nap.edu/books/0309069904/html/index.html) of Bongaarts and Bulatao (2000), and presents some unpublished findings.

To match the available data, we index the jump-off years of interest by $k = 0, 5, 10, 20$ that correspond to calendar years $1970 + k$. Our data come in the form of average growth rates during 5 year intervals. The end points of the intervals will be indexed by $m = 5, 10, 15, 20, 25, 30$ corresponding to calendar years $1970 + m$. The average growth rate during $[m - 5, m)$ is

$$\bar{\rho}(m) = \log(V(m)/V(m-5))/5. \tag{4.3}$$

A major advantage of the cohort-component method is that the effect of age-structure on crude rates can be accounted for. Therefore, assume that the true growth rate during the year t is of the form

$$\rho(t) = c(t) + \Psi(0, t) + \xi(t), \tag{4.4}$$

where $c(t)$ is a function whose values can be forecasted using cohort-component methods; $\Psi(0, t)$ represents gradual deviation from assumed fertility, mortality, and migration rates during $[0, t)$; and $\xi(t)$ represents unpredictable annual perturbations in fertility, mortality, or migration. Assume that the $\xi(t)$'s are i.i.d. with $E[\xi(t)] = 0$. For any $u < t$, define $\Psi(u, t) = \psi(u) + \cdots + \psi(t)$ where the $\psi(t)$'s are i.i.d. with $E[\psi(t)] = 0$. This is our basic model of error.

To estimate the model parameters, let $Y(k, m)$ be the estimated error in the average growth rate for a forecast made at $1970 + k$ for a 5-year period ending at $1970 + m$, where $m > k$ are multiples of 5. This is further influenced by factors $\pi(k)$ that are i.i.d. with $E[\pi(k)] = 0$, representing error in the assumed jump-off value of the growth rate at $1970 + k$. This can reflect data error, the effect of past ξ's, errors of judgment on the average growth rate etc.

We omit most of the technical details below, and concentrate on issues that have the greatest numerical influence on the final estimates.

4.3.2. Second Moments

Defining $\text{Var}(\pi(t)) = \sigma_\pi^2$, $\text{Var}(\psi(t)) = \sigma_\psi^2$, $\text{Var}(\xi(t)) = \sigma_\xi^2$, and by assuming that the sources of error are *independent* of each other, one can deduce (we omit calculations) the representation

$$E[Y(k, m)^2] = \sigma_\pi^2 + (m - k - 14/5)\sigma_\psi^2 + \sigma_\xi^2/5. \tag{4.5}$$

From these moment equations one can estimate σ_ψ^2 and $\sigma_\pi^2 + \sigma_\xi^2/5$ using linear regression on the squared values $Y(k, m)^2$ with m and k as explanatory variables. A further calculation shows that

$$E[(Y(k, m) - Y(k + 5, m))^2] = 2\sigma_\pi^2 + 5\sigma_\psi^2, \tag{4.6}$$

so one can make separate estimates of σ_π^2 and σ_ξ^2.

Assume that the world has regions $i = 1, \ldots, I$, with countries $j = 1, \ldots, n_i$. All symbols are indexed accordingly, $\sigma_{\pi ij}^2 = \text{Var}(\pi_{ij}(t))$, $\sigma_{\psi ij}^2 = \text{Var}(\psi_{ij}(t))$, and $\sigma_{\xi ij}^2 = \text{Var}(\xi_{ij}(t))$. This specification provides for a large number of variance components, and some parametrization was deemed prudent. Assume the model $\sigma_{\pi ij} = c_{ij}\sigma_{\pi i}$, $\sigma_{\psi ij} = c_{ij}\sigma_{\psi i}$, and $\sigma_{\xi ij} = c_{ij}\sigma_{\xi i}$, where c_{ij} is a country specific volatility parameter, and the region specific variance components are identified via the normalizing condition $\sigma_{\pi i}^2 + \sigma_{\psi i}^2 + \sigma_{\xi i}^2 = 1$. Since the relative magnitudes of the normalized components are the same for all countries j within region i, the variance of the forecast error increases with lead time the same way for all countries within a region, but the scales allow different countries within a region to have different levels of variance.

The region specific components were estimated using the normalized errors $y_{ij}(k,m) = Y_{ij}(k,m)/\{\Sigma_{u,v} Y_{ij}(u,v)^2\}^{1/2}$ as data. The moment equations noted above were applied to each country within a region, and the estimates of the country specific variance components were averaged and normalized to sum to 1, which led to estimates of $\sigma_{\pi i}^2$, $\sigma_{\psi i}^2$, and $\sigma_{\xi i}^2$. Let $S_i^2(k,m)$ denote the estimate obtained by substituting these estimates into the right hand side of (4.5). A direct estimate of the scale is then

$$\hat{c}_{ij} = \left\{ \sum_{k,m} Y_{ij}^2(k,m) \Big/ \sum_{k,m} S_i^2(k,m) \right\}^{1/2}. \tag{4.7}$$

For some countries the period from which our data come from may have been unusually volatile or calm. An alternative estimator is a *composite estimator* (cf., Rao 2003, 57) of the form

$$\tilde{c}_{ij} = \gamma_i \hat{c}_{ij} + (1 - \gamma_i)\hat{c}_i, \tag{4.8}$$

where $0 \leq \gamma_i \leq 1$, and

$$\hat{c}_i = \frac{1}{n_i} \sum_{j=1}^{n_i} \hat{c}_{ij}. \tag{4.9}$$

Alternative calculations were carried out using $\gamma_i \equiv \gamma = 1.0, 0.85, 0.70$.

To apply these models, the world was divided into $I = 10$ regions: Region around China and India; Middle East; East Asia (excluding China); Pacific Islands; Western Tropical Africa; Non-tropical and Eastern Tropical Africa; North America and Australia; South and Central America; Southern, Western and Northern Europe; and Former Socialist States around Russia. To be able to aggregate population data across countries in a given region, it was assumed that the correlations are $\text{Corr}(\psi_{ij}(t), \psi_{ih}(t)) = \text{Corr}(\pi_{ij}(t), \pi_{ih}(t)) = \text{Corr}(\xi_{ij}(t), \xi_{ih}(t)) = \rho_i$ for $j \neq h$. It turned out that the correlations within regions were low, with average 0.15. The highest correlation, 0.50, was observed in countries neighboring the former Soviet Union. The observation period includes the break-up of the Soviet Union. Since such upheavals may occur in the future, it was deemed prudent to consider alternative calculations that assume the intraregional correlation to be $\rho = 0.15, 0.375, 0.50$.

There were not sufficient data to estimate the possible correlations across the ten regions. The uncertainty of world forecasts turned out to be very sensitive to these assumptions, however. For example, a modest interregional correlation of 0.1 had the effect of multiplying the standard error estimates for the world as a whole by 1.28, as compared to standard errors that assumed independence. Again, prudence dictates that some allowance for interregional correlation is made.

4.3.3. *Predictive Distributions for Countries and the World*

Suppose one makes a new forecast at a time $k = K$ for the year $K + t$. After some algebra we find that under our model the ratio of the forecast to the true value for country j in region i is

$$\frac{\hat{V}_{ij}(K, K+t)}{V_{ij}(K+t)} = \exp\left[\pi_{ij}(K)t + \sum_{n=1}^{t} n\psi_{ij}(K+t-n) + \sum_{h=0}^{t-1} \xi_{ij}(K+h)\right]. \tag{4.10}$$

We see that the ψ's produce errors whose variance increases with the cube of the lead time, the π's produce errors whose variance increases with the square of the lead time, and the ξ's produce errors whose variance increases proportionally to the lead time.

If all the variance parameters were known, *a priori*, the variance of the relative error would be

$$c_{ij}^2 \left\{t^2\sigma_{\pi i}^2 + (2t+1)(t+1)t\sigma_{\psi i}^2/6 + t\sigma_{\xi i}^2\right\}. \tag{4.11}$$

Estimation error for the variance components was incorporated using bootstrap.

As an illustration, we present the quantiles of the predictive distribution of the *world population* (in millions) corresponding to $\gamma = 0.85$, $\rho = 0.375$, and interregional correlation of 0.1. These figures are based on a jump-off year of 1995.

	Quantiles				
Year	0.025	0.25	0.50	0.75	0.975
2030	7,463	7,910	8,143	8,380	8,900
2050	7,948	8,665	9,050	9,492	10,876

Without an assumption of the interregional correlation of 0.1, a 95% prediction interval in 2050 would have been [8,184, 10,488]. As mentioned in Chapter 1, a recent U.N. forecast for the world in 2050 has a high variant of 10.9 billion and a low variant of 7.7 billion. Therefore, based on the analysis outlined above, the interval can be considered approximately as a 95% prediction interval.

Even though the U.N. interval for the world as a whole appropriately reflects the uncertainty of forecasting, this is due to the perfect correlation assumption implicit in the calculation. The high variant is obtained by adding the high variants for the countries, and the low variant is obtained by adding the low variants for all the countries. The high-low intervals for the individual countries have a much smaller probability of covering the future values. We now present a comparison of the U.N. forecasts to stochastic forecasts made for the U.S. (Lee and Tuljapurkar 1994), for Austria (Hanika, Lutz and Scherbov 1997), for Norway (Keilman, Pham and Hetland 2002), for the Netherlands (DeBeer and Alders 1999), for Finland (Alho 1998), and for Lithuania (Alho 2002a), and to the present estimates. The estimates used below incorporate estimation error via bootstrap but are not composite.

To quantify the uncertainty implied by each forecast we calculated the ratio of the upper end point of a 95% prediction interval to the median forecast, and in the case of U.N. (2001), the ratio of the high forecast to the middle forecast. Table 1 is obtained for lead times $t = 10, 30, 50$, where "U.N. Empirical" refers to estimates obtained with the methods of this section.

A comparison of the U.N. scenario-driven forecasts and careful stochastic forecasts shows that the U.N. intervals are much narrower. They do not give a similarly realistic assessment of uncertainty for the individual countries as they do for the world as a whole.

TABLE 1. The Ratio of the Upper End Point of a 95% Prediction Interval to the Median Forecast in Stochastic Forecasts (Stochastic), and as Derived from the Empirical Analysis of the Past U.N. Forecasts (U.N. Empirical), and the Ratio of the High U.N. Forecast to the Median Forecast for Lead Times 10, 30, and 50.

Country	Lead	U.N.	Stochastic	U.N. Empirical
United States	10	1.017	1.039	1.018
	30	1.069	1.154	1.073
	50	1.152	1.372	1.151
Austria	10	1.003	1.035	1.023
	30	1.024	1.112	1.098
	50	1.074	1.232	1.210
Finland	10	1.005	1.030	1.032
	30	1.029	1.153	1.142
	50	1.087	1.402	1.309
Lithuania	10	1.004	1.047	1.047
	30	1.027	1.155	1.234
	50	1.087	1.307	1.560
Norway	10	1.005	1.040	1.031
	30	1.031	1.190	1.112
	50	1.086	1.450	1.224
The Netherlands	10	1.004	1.023	1.011
	30	1.029	1.110	1.046
	50	1.083	1.200	1.096

A comparison of the careful stochastic forecasts to the present model that uses the past errors of the U.N. forecasts from 1970–1990 as a basis, shows broad agreement. For Austria the results are almost identical. However, the data period appears to have been less volatile for the U.S. and Norway than the much longer time-series material Lee and Tuljapurkar, and Keilman and co-workers, have used. This seems to be the case in the Netherlands, as well, where DeBeer and Alders have viewed the future as more volatile than the past performance of the U.N. forecasts suggests. The difference for Finland at lead time 50 may be due to the same thing. In the stochastic forecast of Finland, fertility in the near future was assumed to have the recent past volatility that is quite low in historical perspective. Later the volatility was assumed to increase to the historical median levels. In the case of Lithuania, the rapidly increasing values of the present analysis may depend on the other, formerly Soviet countries.

We also note that the stochastic forecasts of Austria and the Netherlands that have been constructed by primarily judgmental methods show a markedly lower level of uncertainty than those of the U.S., Norway, Finland and Lithuania that have primarily relied on statistical time-series techniques.

We conclude that the results of the present analysis reflect a short data period, and some results may depend on developments the neighboring countries. Yet, the results have been derived based on a unified empirical methodology that involves judgment in a minimal way. The broad agreement of the results, despite the very different methods used, suggest that the stochastic forecasts are relatively robust. Burdick, Manchester and Bang (2003) come to a similar conclusion in their assessment of stochastic methods in connection with the U.S. Social Security Trust Fund.

On the other hand, the "U.N. Empirical" estimates appear more variable than those coming from more complex stochastic analyses, so our comparison also suggests that composite estimation that borrows strength from regions deemed similar can be beneficial. Table 2 presents estimates for 27 EU/EEA countries, including those that joined the EU in 2004 (Cyprus is omitted due to data problems). The estimates of uncertainty include estimation error and borrowing of strength using composite estimation. The estimates are based on 10,000 simulations. A lognormal approximation can be used to arrive at a prediction interval for the total population, so for example the upper limit of an 80% prediction interval for a lead time $t = 30$ for Poland would be obtained by multiplying a point forecast (such as the one given by the U.N., for example), by $\exp(1.2816 \times 0.071) = 1.095$.

The countries have been ordered according to the relative error at lead time $t = 50$. We see that uncertainty is related to small size (and possibly to the level of migration). Taking logarithms of the relative error at $t = 50$ and of population size, we obtain a scatter plot that appears roughly bivariate normal. The correlation between the logged variables is -0.424 (P-value $= 0.027$), which supports the conclusion of a negative association.

4.4. Random Jump-Off Values

In practical forecasting data problems can sometimes be a major component of uncertainty. Chapter 10 is devoted to the modeling of error in census numbers in the U.S. context. Elsewhere, similar estimates are not typically available and judgment must be used. In Alho and Spencer (1985) we suggested that random jump-off values be used to reflect uncertainty of this type. We will illustrate the issues in the context of a forecast made for Lithuania[15], but note that the

[15] This section uses material from the report Alho J.M. (2001) Stochastic Forecast of the Lithuanian Population 2001–2050. The research was undertaken with support from European Union's Phare ACE programme 1998, Project P98-1023-R. The reasoning reflects what was known around 2000–2001.

TABLE 2. Composite Estimates ($\gamma = 0.85$) of the Standard Deviation of the Relative Error of the Forecast of the Total Population for 27 EU/EEA Countries of 2004 for Lead Times $t = 10, 30, 50$, and Population in 2000.

	Lead Time			
Country	10	30	50	Pop. in 2000
Belgium	0.013	0.040	0.088	10251
Italy	0.014	0.041	0.090	57536
France	0.014	0.042	0.093	59296
Netherlands	0.014	0.042	0.093	15898
Denmark	0.015	0.043	0.096	5322
Iceland	0.015	0.045	0.099	282
Norway	0.016	0.047	0.103	4473
United Kingdom	0.018	0.052	0.116	58689
Finland	0.019	0.056	0.124	5177
Poland	0.015	0.071	0.149	38671
Greece	0.025	0.073	0.162	10903
Germany	0.026	0.077	0.170	82282
Czech. Rep.	0.018	0.085	0.180	10269
Slovakia	0.019	0.086	0.182	5391
Austria	0.030	0.087	0.193	8102
Sweden	0.031	0.090	0.200	8856
Hungary	0.021	0.098	0.206	10012
Lithuania	0.023	0.107	0.227	3501
Spain	0.036	0.105	0.232	40752
Slovenia	0.024	0.111	0.235	1990
Latvia	0.026	0.119	0.252	2373
Estonia	0.027	0.123	0.260	1367
Portugal	0.043	0.126	0.278	10016
Malta	0.047	0.137	0.304	389
Switzerland	0.047	0.139	0.308	7173
Ireland	0.053	0.156	0.346	3819
Luxembourg	0.070	0.205	0.454	435

specification of randomness in the jump-off values was only completed after the forecast had been released. We will concentrate on population size and on old-age mortality.

4.4.1. *Jump-Off Population*

The jump-off population of our forecast was the January 1, 2000 resident population in Lithuania. Official estimates put the total population at 3.699 million, based on an earlier census and vital registration data. The results of the census of 2001 were not released at the time. However, it had been announced that the enumerated population on April 1, 2001, was 3.496 million, a difference of 0.203 million (or 5.5% of the official estimates).

In the absence of a post-enumeration survey (cf., Chapter 10), any adjustment of the census count was deemed speculative, but some reconciliation of the existing estimates was necessary. Based on discussions with Lithuanian experts, the situation was analyzed as follows. First, in 1990–1994 there had been some undocumented emigration of Slavs. Some had worked in the communist party and related institutions; some may have feared for new language requirements; some may have had an economic motive such as cashing in on their newly privatized apartment; yet others may have left simply to join their family. Thus the official statistics for year 2000 were assessed as having been roughly 50 thousand too high, leading to a revised estimate of 3.649 million. On the other hand, it was thought that the Lithuanian census may have suffered from an undercount of possibly 50 thousand inhabitants. This was 1.4% of the census count, a figure comparable to pre-2000 non-black undercounts in the United States (Example 2.2 of Chapter 2). Taking the two factors into account, an adjusted census figure of 3.546 million was taken as the most credible count for the resident population at census time. Based on birth and death registration it was determined that the rate of natural increase was approximately zero during year 2000. It was thought that during the Soviet years net undercount was low, so the difference, 0.103 million, would consist of undocumented emigration to West. Most of this was thought to have happened in 1995–2000. This implied an annual out-migration of about 17,000 inhabitants. Since the census day was April 1, 2001, the population of January 1,2000, was thought to have been about 22,000 thousand inhabitants higher. Our final estimate of the jump-off population was 3.568 million.

How uncertain is the estimate? Under a normal model we could represent the unknown population size by a distribution $N(3.568, \sigma^2)$. While the census count (3.496) could be too high, this seems unlikely. Assuming that the probability is 2.5% that the census is too high, we get $\sigma \approx (3.546 - 3.496)/2 = 0.025$, or 0.7% of population size. An estimate of this type would then have to be translated into a model by age and sex. Presumably, the uncertainty would be the greatest in those ages that would most likely migrate, or most likely be missed in a census count. Young adult males are one such group.

If an option for a random jump-off population is not available in a computer program one is using, a quick way to implement a random jump-off value is to start a stochastic forecast one year earlier and let the uncertainty of survival and/or migration capture the uncertainty of the estimate. Bias may incur, however, if the assumptions concerning the autocorrelation of mortality or migration cannot be tailored to match what is intended.

4.4.2. Mortality

A comparison of the Lithuanian age-specific mortality to that of the Nordic countries showed that the Lithuanian mortality was higher in ages 0–89 for females and in ages 0–79 for males, but *lower* in older ages. For example, in 1999 mortality (per 1,000) in ages 95+ in Lithuania (= LI), and in the Nordic countries (DK = Denmark, FI = Finland, NO = Norway, SE = Sweden) was

Country	LI	DK	FI	NO	SE
Females	240	354	348	391	409
Males	211	453	412	429	495

In other words, the Lithuanian rates were approximately one half of those of the Nordic countries. Another peculiarity was that male mortality was *lower* in Lithuania than female mortality. On the other hand, in 1990 the Lithuanian rates were 373 for females and 409 for males, which was more in line with the rates elsewhere. A comparison of rates for the highest age might be confounded by variations in the age distribution. However, we noted that in age 90–94 the Lithuanian mortality was estimated at 186 for females and 203 for males in 1999, whereas in the Nordic countries (alphabetical order) the female rates were 212, 227, 221, and 225 for females, and 272, 277, 278 and 303 for males. A possible bias in the Lithuanian old-age mortality data might have been caused by underenumeration of deaths, overestimation of population, or overstatement of age (cf., Section 2, Chapter 2). The latter may be judged the most credible.

In conclusion, for forecasting purposes it was decided to replace Lithuanian mortality figures for 2000 by the average of age-specific mortality in the four Nordic countries in ages 90+ for females, and in ages 85+ for males. This has the merit of being simple to explain, but the drawback that there is no simple yardstick for the measurement of uncertainty. An alternative one might have considered is to use (e.g., polynomial) regression to extrapolate current mortality in the oldest ages by using rates in younger ages. This would yield an error estimate automatically.

5. Measuring Correlatedness

From a statistical point of view we can think of much of classical demography as dealing with expected values. Variances are rarely considered and more complex second order characteristics, correlations, often are loosely treated. We will address here three statistical aspects that come up. First, we consider the definition of correlation in a time-series context, then we consider the necessity of using modeling assumption to estimate correlations, and third, we consider the effect of measurement error on estimated correlations.

Consider two time series $X(t)$ and $Y(t)$, and define $\rho(X(t), Y(t)) = \mathrm{Cov}(X(t), Y(t))/\{\mathrm{Var}(X(t))\mathrm{Var}(Y(t))\}^{1/2}$ as their correlation at time t. Since, e.g., $\mathrm{Cov}(X(t), Y(t)) = E[(X(t) - E[X(t)])(Y(t) - E[Y(t)])]$, we see that $\rho(X(t), Y(t))$ measures association when the means $E[X(t)]$ and $E[Y(t)]$ have been subtracted. If the processes are nonstationary, the meaning of the correlation depends on how the mean is viewed. If the mean is nonconstant (as in a regression model), then we can have a situation in which, say, in-migration and fertility go up and down together, but are *not* correlated *if* the association is due concomitant changes of the means. In contrast, if we consider the mean to be constant (e.g., in a random walk the means would be $X(0)$ and $Y(0)$ if we condition on $X(0)$ and $Y(0)$ and are interested in values at $t > 0$) and the fluctuations as purely random, the same data would lead to a finding of a positive correlation.

The second complication arising in a time-series context is the fact that the number of correlation parameters increases faster than available data. To appreciate this, note that with n observations $X(1), \ldots, X(n)$ there are n variances to be considered, but $n(n-1)/2$ covariances. Thus, the number of correlation parameters increases in proportion to the *square* of the number of observations. It follows that some modeling assumption *has* to be made. In fact, one can view the ARIMA theory of Chapter 7 as an attempt at parametrizing autocorrelations with a small number of parameters. The following examples show that some simple parametrizations may provide at least a rough approximation.

Example 5.1. Constant Correlations Across Ages. Consider the logarithms of male and female mortality in five year age-groups 65–69, 70–74, 75–79, 80–84, and 85+ in the U.S. in 1940–1988. Forecasts were produced using each of the years starting from 1945 as a jump-off year, in turn. The data until the jump-off year were used for prediction. The predictions were calculated by fitting an ARIMA(1,1,0) model with a constant term. We have ten cross-correlations for the forecast errors. They vary quite a bit by lead time. The minimum and maximum correlations are the following: 0.35 and 0.60 for lead = 1; −0.08 and 0.55 for lead = 5; −0.14 and 0.47 for lead = 10; 0.26 and 0.91 for lead = 20; −0.06 and 0.78 for lead = 30. Since the correlation estimates for the different lead times are not independent, it is not easy to summarize these data. However, it appears that the correlations are typically positive, with 0.3 or 0.4 the most typical values. A model that assumes a constant correlation (≈ 0.4) between all ages provides an approximation to these data. ◊

Example 5.2. Constant Correlations Across Causes of Death. In Alho and Spencer (1990b, 223–225) we estimated the cross-correlations of the prediction errors of log mortality rates, between causes of death, for the U.S. data from 1973–1985. The lag = 0 correlations for males varied from −0.61 to 0.84 with the average 0.24, and for females they varied from −0.57 to 0.87 with the average 0.18. The distribution of the correlations between the minimum and maximum values was roughly uniform. Again a model of constant correlation (≈ 0.25) provides a rough approximation. (An alternative, however, would be to focus on the aggregate mortality rates rather than the rates by cause, thereby reducing the number of covariances.) ◊

Example 5.3. Uncorrelated Errors for Different Vital Rates. Keilman (1990, Figure 5.1, 83) has demonstrated that the Dutch fertility and mortality forecasts have both been too high since the 1960's. The same is true for many other industrialized countries, such as the U.S., Canada, and the Nordic countries. The common cause for both errors appears to be that the demographers had predicted that the future rates would be close to the existing ones. The forecast errors were determined by the trends of the vital rates. These both happened to be down, causing the overestimates. However, during the 1940–1960 period fertility rose rapidly, but mortality declined. It is clear that nobody was able to correctly forecast the upsurge of fertility at that time. Therefore, an assumption of zero correlation appears plausible. Further evidence of the low level of correlation is presented in Keilman (1997) for the Netherlands and Norway.

It is well known that a negative correlation has existed between mortality and fertility rates in preindustrial conditions, caused by wars, famines, and epidemics (see, e.g., Turpeinen (1978) for the Finnish evidence during 1750–1900). There can be similar fluctuations in the developing countries today. In industrialized countries the reasons for a mortality forecast to fail are different from the reasons for a fertility forecast to fail. This supports an assumption of independence. ◊

Example 5.4. Constant Correlations Across Countries Within a Region. To see how the same vital rates may behave in different countries in the same region, we considered fertility in ages 15–19, 20–24, ..., 40–44, in Denmark, Finland, Norway, and Sweden, in 1970–1991. The rates for age group 15–19 increased dramatically in the early 1970's in all countries. After that they smoothly declined more than 50% to a level that is a bit lower than in 1970. Using current value as a naive forecast for all future years at any jump-off year during the period would have produced highly correlated prediction errors. In age 20–24 relatively smooth declines were observed in all countries. Again, naive forecasts would have had highly correlated prediction errors. For age 25–29 the experiences were more mixed. Finland had an upward trend all through the period, whereas the other countries had a U-shaped pattern. From 1980 on all forecasts would have been too low. For earlier jump-off times this would have been true for Finland, but the other countries would have initially experienced lower fertility than forecasted. In age 30–34 all countries had a U-shaped pattern that ended up a bit higher in 1991 than it started from in 1970. Again, naive forecasts would have had highly correlated forecast errors in all countries, especially after 1980. In age 35–39 the development was U-shaped in Denmark and Norway. In Finland and Sweden the development was more steadily upward. In age 40–44 all countries experienced first a decline and then an increase. The turning points were different, so the signs of the prediction errors of naive forecasts would have depended heavily on the year they were made. Inasmuch as official demographic forecasts resemble naive forecasts, we conclude that in the Nordic countries the forecast errors of fertility can be expected to have positive correlations over the long run, across the countries. However, in the short run the differential timing of changes may produce a more mixed picture. For these countries a model of constant correlation across countries might be appropriate. ◊

A third issue arises when we view trends of vital processes as being random, and the target of estimation is the correlation between trends. In this case the observations contain *measurement error*. For example, suppose that conditionally on hazards λ_X and λ_Y, $X \sim \text{Po}(\lambda_X K_X)$ and $Y \sim \text{Po}(\lambda_Y K_Y)$ are independent, with K_X and K_Y the person years. Suppose we are interested $\rho(\lambda_X, \lambda_Y)$. With only X and Y available, we base our estimation on the o/e rates $m_X = X/K_X$ and $m_Y = Y/K_Y$. Since they are unbiased for the intensities λ_X and λ_Y, we can write $m_X = \lambda_X + \varepsilon_X$ and $m_Y = \lambda_Y + \varepsilon_Y$, where $E[\varepsilon_X|\lambda_X, \lambda_Y] = E[\varepsilon_Y|\lambda_X, \lambda_Y] = 0$, $\text{Var}(\varepsilon_X|\lambda_X, \lambda_Y) = \lambda_X/K_X$, and $\text{Var}(\varepsilon_Y|\lambda_X, \lambda_Y) = \lambda_Y/K_Y$. Here, ε_X and ε_Y are also independent conditionally on λ_X and λ_Y. Using the conditional independence we note first that $E[(\lambda_X + \varepsilon_X - E[\lambda_X + \varepsilon_X])(\lambda_Y + \varepsilon_Y - E[\lambda_Y + \varepsilon_Y])] = \text{Cov}(\lambda_X, \lambda_Y)$. But since $E[(\lambda_X + \varepsilon_X - E[\lambda_X + \varepsilon_X])^2] = \text{Var}(\lambda_X) + E[\varepsilon_X^2]$, and similarly for $\lambda_Y + \varepsilon_Y$, we find that estimates of correlations will be systematically *biased towards zero* (cf.,

Fuller 1987, 7–11). Thus, in the case of small expected counts, when the coefficient of variation of the Poisson count is large (cf., Section 5 of Chapter 4), correlation estimates can be severely biased.

We can attempt to correct the correlation estimate by subtracting the estimated Poisson variance of the count, e.g., $\text{Var}(\varepsilon_X) = \lambda_X/K_X$, from the empirical variance $\text{Var}(X)$, and similarly for Y. In fact, if we have observed the data $(X(t), Y(t)), t = 1, \ldots, n$, leading to *o/e* rates $m_X(t)$ and $m_Y(t)$, then an estimator of $\text{Var}(\lambda_X)$ is

$$\frac{1}{n-1}\sum_{t=1}^{n}(m_X(t) - \bar{m}_X)^2 - \frac{1}{n}\sum_{t=1}^{n} m_X(t)/K_X(t). \tag{5.1}$$

This is unbiased if the counts for different t are independent. We caution, however, that in the case of small expected counts (i.e., when the need for a bias correction is the greatest) the estimate of the second, bias correction term, can be unstable. In this case, the only hope may be to impose additional structure on the problem by assuming a model for the change of rates.

Exercises and Complements (*)

1. Study the cross-correlations of the time series of Example 1.5.

*2. *Co-integration*. Consider the bilinear model for the mortality in age x of year t, $\log \mu(x,t) = \alpha(x) + \delta(x)\xi(t) + \varepsilon(x,t)$, where $\varepsilon(x,t) \sim N(0, \sigma_\varepsilon^2)$ are i.i.d., and $\xi(t)$ is an ARIMA(p,d,q) process for some $d > 0$. Suppose $\delta(x) \neq 0$ for all x. Then each age-specific series is nonstationary, but they have a *common trend* determined by $\xi(t)$. Consider any two ages $x \neq y$. Define a vector-valued process $\mathbf{W}(t) = (\log \mu(x,t), \log \mu(y,t))^T$ and a vector $\mathbf{U} = (1/\delta(x), -1/\delta(y))^T$. It follows that $\mathbf{W}(t)^T\mathbf{U} = \alpha(x)/\delta(x) - \alpha(y)/\delta(y) + \varepsilon(x,t)/\delta(x) - \varepsilon(y,t)/\delta(y)$ is an uncorrelated process with a constant mean. This is a special case of co-integration: a *co-integrating vector* $\mathbf{U}$ removes the common trend(s) from a vector-valued process $\mathbf{W}(t)$ so that the resulting process is, roughly speaking, stationary and invertible. For a rigorous discussion, see Johansen (1995).

3. *De Morgan Rules*. (a) Prove the rule $A \cap B = (A^c \cup B^c)^c$, and (b) deduce from this that $A \cup B = (A^c \cap B^c)^c$. (Hint: define, $1_A(x) = 1$, if $x \in A$, and $1_A(x) = 0$ otherwise. Then, $1_{A^c}(x) = 1 - 1_A(x)$, $1_{A\cap B}(x) = 1_A(x)1_B(x)$.)

4. Consider the total fertility rate, a year from now. A demographer offers you a gamble that costs 1 unit, and in which you get back 4 units, if fertility is within $\pm 2\%$ of the current value, but if it is outside those limits, you get nothing. Infer how likely it is, in the demographer's view, that fertility is within $\pm 2\%$ of the current value.

*5. *Combining forecasts*. Consider two forecasts X_1 and X_2 of some random variable X. Define the forecast errors as $\varepsilon_j = X_1 - X$, $j = 1, 2$, and assume that $E[\varepsilon_j] = 0$. Denote $\text{Var}(\varepsilon_j) = \sigma_j^2$ and $\text{Cov}(\varepsilon_1, \varepsilon_2) = \rho\sigma_1\sigma_2$. Show that any linear combination of the two forecasts in which the first gets the weight κ and the second the weight $1 - \kappa$ is also unbiased, with error $\varepsilon(\kappa) = \kappa\varepsilon_1 + (1 - \kappa)\varepsilon_2$. The error variance is $\mathrm{Var}(\varepsilon(\kappa)) = \kappa^2\sigma_1^2 + (1 - \kappa)^2\sigma_2^2 + 2\kappa(1 - \kappa)\rho\sigma_1\sigma_2$. (a) Differentiate with respect to κ and set the derivative to zero to find the minimum at

$$\kappa = \left(\sigma_2^2 - \rho\sigma_1\sigma_2\right) / \left(\sigma_1^2 + \sigma_2^2 - 2\rho\sigma_1\sigma_2\right).$$

(b) Show that if $\rho = 0$, then the weights are proportional to the inverses of the variances. (c) Show that if $\sigma_1 = \sigma_2$, then $\kappa = 1/2$ irrespective of ρ. (d) What is the minimizing variance?

*6. Consider forecasts of the world population for 2025. I.I.A.S.A. (Lutz 1994), the U.N. (1993), and the World Bank (1992) offered the following values (in millions) as the most likely: 8,955; 8,472; 8,345, respectively. Suppose (for the sake of illustration) that all forecasts have the pairwise correlation of 1/4 and the standard deviation of the error of the I.I.A.S.A. forecast is 1/2 of that of each of the other two. First combine the U.N. and WB forecasts by giving them the weight 1/2. Let the common standard deviation of error of those forecasts be σ. Show that (a) the standard deviation of the error of the combined forecast is $(5/8)^{1/2}\sigma$; (b) the covariance of the combined forecast and the I.I.A.S.A. forecast is $(1/8)\sigma^2$; (c) the correlation of the I.I.A.S.A. forecast and the combined forecast is $(1/10)^{1/2}$; (d) the weight given to the I.I.A.S.A. forecast is 0.8, so the weights given to the other forecasts are 0.1 each; (e) the resulting combined forecast for the world population is 8,846.

7. Show that (5.1) is an unbiased estimator of $\mathrm{Var}(\lambda_X)$ if the counts are independent.

9
Statistical Propagation of Error in Forecasting

In the previous two chapters we have discussed the statistical forecasting of time series as applied to demographic rates. The main goal of this chapter is to show how the separate pieces can be brought together to form a predictive distribution of future population. Indeed, a major purpose of the whole book is to provide sufficient detail about the most important factors needed so that realistic stochastic forecasts can be produced.

An early and largely unrecognized contribution of Törnqvist to stochastic forecasting is discussed first. In Section 2 we define the concept of predictive distribution, and discuss its nature from a frequentist and Bayesian point of view. This includes an introduction to Markov Chain Monte Carlo techniques in a time-series setting. Section 3 discusses the formulation of forecasts as databases and their uses. Some useful parametrizations of the large number of cross-covariances and cross-lagged covariances of forecast errors of vital rates are discussed in Section 4. Analytical models for forecast error and an analytical approach to the propagation of error are presented in Section 5. Section 6 introduces the simulation approach. We conclude in Section 7 by discussing how the results of a simulation experiment can be post-processed to allow alternative interpretations of the results.

1. Törnqvist's Contribution

The first serious attempt to describe population forecasting from a stochastic point of view is, to the best of our knowledge, due to L. Törnqvist[1] (1949) in connection with a forecast he helped the Central Statistical Office of Finland to produce. Törnqvist was professor of statistics at the University of Helsinki and his other work was close to econometrics, notably index number theory (Nordberg 1999) and an early consideration of cost-benefit analysis for statistical data collection (Törnqvist 1948).

---

[1] Readers interested in the developments of computer operating systems might be interested to learn that Leo Törnqvist was grandfather to Linus Torvalds, the creator of the operating system LINUX. Törnqvist introduced the young Linus to the art of computer programming.

In discussing the reasoning behind the forecast variants, Törnqvist (1949, 69–70) suggested that one begin by trying to determine such "primary series" whose values would be constant, apart from random deviations. For example, Törnqvist logistically transformed mortality in 5-year age-groups, estimated the annual rate of change for the transformed values, and considered the rate of change as a primary series. Due to random deviations the observed values of the series had to be considered as "statistical variables". Based on past data one could form a relatively good impression of the deciles of their probability distribution. In order to limit the analysis of past series for practical reasons, Törnqvist concluded that "it seems permissible to determine the deciles more or less subjectively". In some cases he used data from Sweden to get a view of the development that was not obscured by events related to world War II.

In the future, the primary series attains values that can be considered as "random samples" from the distribution. Törnqvist's point forecast was the median of the estimated distribution, i.e., the (estimated) probability is 50% that future value of the process will be below the forecasted value, and the probability is 50% that the future value will be above the forecasted value. He called this the "most likely value". Similarly, he proposed that the low forecast be chosen so the probability is 10% that the future population will be below it, and the high forecast be chosen so the probability is 10% that the future population be above it.

Having decided on the forecast variants for the vital rates, Törnqvist discussed ways of combining them to produce the future population forecast. He thought it reasonable to try all different combinations, but saw it most useful to combine high fertility with high life expectancy, and low fertility with low life expectancy. Although this is in keeping with the practice started by Whelpton and others, it is characteristic of Törnqvist's statistical thinking that he realized that the high forecast would be more "optimistic" for the population size, and the low forecast would be more "pessimistic" for the population size, than the variants for the individual vital rates.[2,3]

A step Törnqvist did not take was to consider methods that would have produced a prediction interval consisting of, say, the first and ninth decile of the population size itself. This would have involved carrying out the *statistical propagation of error* from the vital rates to the corresponding population size.

Törnqvist did his statistical work at approximately the same time Whelpton was completing his contributions in a deterministic frame work. The latter have been very influential while Törnqvist's efforts have mostly gone unnoticed (Hoem 1973 is an exception). Lack of computing facilities and undeveloped theory for carrying out the propagation of error may have been one reason. Moreover, little was known

---

[2] A simple example is this is the following. Suppose X and Y are independent random variables with $N(0, 1)$ distributions. Then the interval $[-1, 1]$ is a 68.3% prediction interval for both. However, since $X + Y \sim N(0, 2)$, the interval obtained by combining the high limits and low limits, or $[-2, 2]$, contains $X + Y$ with probability 84.3%.

[3] The interpretation of optimistic and pessimistic is not universal. Some statistical agencies have called their high forecasts pessimistic and low forecasts optimistic, even though the latter are associated with higher mortality, because the drain on government pensions is larger.

in the 1940's about the empirical errors of forecasts and about the meager improvements in the accuracy of forecasting from advances in demographic theory.

Finally, how did Törnqvist fare as a forecaster of fertility? The rates of year 1947 were the most recent available to him. This was the peak year of the Finnish baby-boom, and the estimated total fertility rate was 3.44. Törnqvist assumed that the rate would rapidly decline, so that for the 5-year period 1951–1955 the most likely value would be 2.36, with an 80% prediction interval [2.18, 2.54]. Or the width of the interval was approximately $\pm 7.6\%$ of the point forecast, for a forecast going approximately 6 years into the future. The interval impressively failed to include the future value, 2.98, which was more than 26% higher than the point forecast. After the initial decline, Törnqvist considered it most likely that fertility would remain roughly constant, so for the period 1996–2000 the most likely value was also 2.36, with an 80% interval of [1.85, 2.83]. This is a 50-year ahead forecast, and the width of the interval is approximately $\pm 21.3\%$ of the point forecast. The observed average value was 1.73, slightly below the 80% range.

Törnqvist thought that the difference between his "optimistic" and "pessimistic" assumptions concerning future fertility was "relatively large". Yet, under a random walk model, both the 6-year ahead forecast and the 50-year ahead forecast would imply a standard deviation of unit increment of approximately 0.024. As discussed in Chapter 8, this is a low value, because more recent analyses support standard deviations as high as 0.06. This explains why the short term intervals were too narrow.

2. Predictive Distributions

Loosely speaking, a predictive distribution of a future vital rate can be defined as its conditional distribution given everything we have learned in the past. We have discussed its interpretation in an informal way in Sections 3.6 and 4.3 of Chapter 8. To make the concept more concrete, we will here consider three special cases that are relevant in demographic applications: time series regression, random walks, and a simple ARIMA model.

In the 1990's there developed a vast literature on the so-called Markov Chain Monte Carlo methods (e.g., Liu 2001, Gelman et al. 1995). These methods were first introduced in physics in the 1940's and 1950's (Metropolis N., Rosenbluth, and Teller 1953). A recursive set of calculations is set up that produces correlated samples from the joint posterior distribution of all parameters. Sections 2.2 and 2.3 show how a particular method, the Gibbs sampler (Gelman et al. 1995, 326–327), can be used in conjunction with simple time series models. We note in passing that similar calculations form the basis of a Bayesian analysis of count data that was discussed in Section 4.3 of Chapter 5.

2.1. *Regression with a Known Covariance Structure*

Consider the regression model defined in (3.2), (3.3), and (3.4) of Chapter 7. For example, we might have $Z_t = f(t) + \varepsilon(t)$ representing the logarithm of a mortality rate with a constant rate of change, $f(t) = \beta_1 + \beta_2 t$. We assume we have a vector of past observations $\mathbf{Z}_1 = \mathbf{X}_1\beta + \varepsilon_1$, and would like to predict future observations $\mathbf{Z}_2 = \mathbf{X}_2\beta + \varepsilon_2$. The GLS estimator $\hat{\beta} = (\mathbf{X}_1^T \Sigma_{11}^{-1}\mathbf{X}_1)^{-1}\mathbf{X}_1^T \Sigma_{11}^{-1}\mathbf{Z}_1$ is the minimum variance unbiased estimator of β. Under normality $\hat{\beta}$ is also the MLE. Formula (3.5) of Chapter 7 gives the minimum variance unbiased prediction for $\mathbf{Z}_2$, with an error characterized by the covariance matrix (3.7).

If β and Σ were known, the conditional distribution of $\mathbf{Z}_2$ given $\mathbf{Z}_1$ would be (e.g., Rao 1973, 522)

$$\mathbf{Z}_2|\mathbf{Z}_1 \sim N\big(\mathbf{X}_2\beta + \Sigma_{21}\Sigma_{11}^{-1}(\mathbf{Z}_1 - \mathbf{X}_1\beta), \Sigma_{22} - \Sigma_{21}\Sigma_{11}^{-1}\Sigma_{12}\big). \tag{2.1}$$

In this case (2.1) could be viewed as a predictive distribution representing the alternative future paths given that we have seen $\mathbf{Z}_1$. However, when β has to be estimated, one would have to consider jointly the sampling distribution of $\hat{\beta}$ and the estimated conditional distribution of $\mathbf{Z}_2$ given $\mathbf{Z}_1$. This is not entirely natural in many time-series applications in which only one sample path is observed.

The Bayesian approach provides an alternative. The idea is to use the language of probability theory to express uncertainty about model parameters, in this case β. Conditionally on β, the density of $\mathbf{Z}_1$ is

$$f(\mathbf{Z}_1|\beta) = c \times \exp\Big(-\frac{1}{2}(\mathbf{Z}_1 - \mathbf{X}_1\beta)^T \Sigma_{11}^{-1}(\mathbf{Z}_1 - \mathbf{X}_1\beta)\Big), \tag{2.2}$$

where $c > 0$ is a normalizing constant that makes the density to integrate to 1. Its exact value will not be needed, and we will use c as a generic symbol in the sequel. To express our uncertain knowledge about β before having seen $\mathbf{Z}_1$ we might, for example, be willing to act as if there is a vector $\mathbf{b}$ and a covariance matrix $\mathbf{S}$ such that $\beta \sim N(\mathbf{b}, \mathbf{S})$. In this case, the *prior density* of β is of the form

$$g(\beta) = c \times \exp\Big(-\frac{1}{2}(\beta - \mathbf{b})^T \mathbf{S}^{-1}(\beta - \mathbf{b})\Big). \tag{2.3}$$

We pause here to comment on the formulation of the prior. Consider the mortality setting mentioned in the beginning, where $f(t) = \beta_1 + \beta_2 t$. The difficulty is that although we may hold prior views about the future level of mortality $f(t)$, it may be difficult to come up with a two dimensional prior for the parameters (β_1, β_2). Therefore, in Alho and Spencer (1985) we represented prior views about $f(t)$ by specifying an additional "datum" at the target year $t = n + m$; the strength of the prior views was reflected in the specification of the variance of the datum, and the datum was taken to be conditionally independent (given β_1, β_2) of the past and future realizations of mortality. This is close to the use of targets that Whelpton favored, but in our formulation the targets are random. In this case, the calculations can all be carried out via mixed estimation introduced by Theil and Goldberger (1961). For an application of this method in old-age mortality, see Alho and Nyblom (1997). Girosi and King (2003) have come to a similar conclusion in their extensive study of cause-specific mortality.

Continuing with the regression example, we note that the conditional density of $\boldsymbol{\beta}$ given $\mathbf{Z}_1$ is $h(\boldsymbol{\beta}|\mathbf{Z}_1) = c \times f(\mathbf{Z}_1|\boldsymbol{\beta})g(\boldsymbol{\beta})$. This follows from the Bayes' Theorem. The density happens to be of a multivariate normal form,

$$h(\boldsymbol{\beta}|\mathbf{Z}_1) = c \times \exp\Big(-\frac{1}{2}\boldsymbol{\beta}^T(\mathbf{X}_1^T\boldsymbol{\Sigma}_{11}^{-1}\mathbf{X}_1 + \mathbf{S}^{-1})\boldsymbol{\beta} + (\mathbf{Z}_1\boldsymbol{\Sigma}_{11}^{-1}\mathbf{X}_1 + \mathbf{b}^T\mathbf{S}^{-1})\boldsymbol{\beta} + c'\Big), \quad (2.4)$$

where c' does not involve $\boldsymbol{\beta}$. Therefore, we have that

$$\boldsymbol{\beta}|\mathbf{Z}_1 \sim N(\tilde{\boldsymbol{\beta}}, \mathbf{M}), \quad (2.5)$$

where the posterior covariance matrix is

$$\mathbf{M} = \left(\mathbf{X}_1^T\boldsymbol{\Sigma}_{11}^{-1}\mathbf{X}_1 + \mathbf{S}^{-1}\right)^{-1} \quad (2.6)$$

and the posterior mean is

$$\tilde{\boldsymbol{\beta}} = \mathbf{M}\big(\mathbf{X}_1^T\boldsymbol{\Sigma}_{11}^{-1}\mathbf{Z}_1 + \mathbf{S}^{-1}\mathbf{b}\big). \quad (2.7)$$

Formulas (2.5), (2.6), and (2.7) define the *posterior distribution* of $\boldsymbol{\beta}$ given the observed data and the prior views expressed in (2.3).

If a point estimate for $\boldsymbol{\beta}$ is desired, the posterior mean is a natural candidate. This is optimal under a quadratic loss function. We see that it is a "matrix weighted average" of $\hat{\boldsymbol{\beta}}$ and $\mathbf{b}$, with weights $\mathbf{M}(\mathbf{X}_1^T\boldsymbol{\Sigma}_{11}^{-1}\mathbf{X}_1)$ and $\mathbf{M}\mathbf{S}^{-1}$. (This is also the origin of the term "mixed estimation" mentioned above although the details of the formulations are slightly different.)

To see how the prior view influences the estimation, write $\mathbf{S} = \kappa\mathbf{S}_0$, where $\kappa > 0$ is a scale parameter. One can show that $\tilde{\boldsymbol{\beta}} \to \mathbf{b}$ and $\mathbf{M} \to \mathbf{0}$, as $\kappa \to 0$, so in the limit we have a case in which $\boldsymbol{\beta}$ is assumed to be completely known, *a priori*, and $\mathbf{Z}_2$ has the distribution (2.1), where $\beta = \mathbf{b}$. On the other hand, suppose that $\kappa \to \infty$. One can show that then $\tilde{\boldsymbol{\beta}} \to \hat{\boldsymbol{\beta}}$, the GLS estimator, and $\mathbf{M} \to (\mathbf{X}_1^T\boldsymbol{\Sigma}_{11}^{-1}\mathbf{X}_1)^{-1}$. In this case nearly nothing is assumed about the regression parameters before seeing the data. The limiting posterior distribution of $\boldsymbol{\beta}$ is the same as the sampling distribution of $\hat{\boldsymbol{\beta}}$, so the Bayesian and frequentist analyses are equivalent. Only now $\boldsymbol{\beta}$ is random rather than $\hat{\boldsymbol{\beta}}$! A similar equivalence result holds in many other settings, as well, when flat priors are used.

More generally, (2.1) gives the conditional distribution of $\mathbf{Z}_2$ for any $\boldsymbol{\beta}$ and $\mathbf{Z}_1$. Therefore, $\mathbf{Z}_2$ has the same conditional distribution as a variable of the form

$$\mathbf{X}_2\boldsymbol{\beta} + \boldsymbol{\Sigma}_{21}\boldsymbol{\Sigma}_{11}^{-1}(\mathbf{Z}_1 - \mathbf{X}_1\boldsymbol{\beta}) + \boldsymbol{\xi}, \quad (2.8)$$

where $\boldsymbol{\xi} \sim N(\mathbf{0}, \boldsymbol{\Sigma}_{22} - \boldsymbol{\Sigma}_{21}\boldsymbol{\Sigma}_{11}^{-1}\boldsymbol{\Sigma}_{12})$ is independent of both $\boldsymbol{\beta}$ and $\mathbf{Z}_1$. We can uncondition with respect to $\boldsymbol{\beta}$ by using its posterior distribution (2.5). Since (2.8) is linear in $\boldsymbol{\beta}$ the resulting conditional distribution given $\mathbf{Z}_1$ is still a normal distribution, $\mathbf{Z}_2|\mathbf{Z}_1 \sim N(E[\mathbf{Z}_2|\mathbf{Z}_1], \mathrm{Cov}(\mathbf{Z}_2|\mathbf{Z}_1))$, where $E[\mathbf{Z}_2|\mathbf{Z}_1] = \mathbf{X}_2\tilde{\boldsymbol{\beta}} + \boldsymbol{\Sigma}_{21}\boldsymbol{\Sigma}_{11}^{-1}(\mathbf{Z}_1 - \mathbf{X}_1\tilde{\boldsymbol{\beta}})$ from (2.1), and where

$$\mathrm{Cov}(\mathbf{Z}_2\,|\,\mathbf{Z}_1) = \big(\mathbf{X}_2 - \boldsymbol{\Sigma}_{21}\boldsymbol{\Sigma}_{11}^{-1}\mathbf{X}_1\big)\mathbf{M}\big(\mathbf{X}_2^T - \mathbf{X}_1^T\boldsymbol{\Sigma}_{11}^{-1}\boldsymbol{\Sigma}_{12}\big) + \boldsymbol{\Sigma}_{22} - \boldsymbol{\Sigma}_{21}\boldsymbol{\Sigma}_{11}^{-1}\boldsymbol{\Sigma}_{12} \quad (2.9)$$

based on (2.8). This is the formal Bayesian *predictive distribution* of $\mathbf{Z}_2$. One can also show that the covariance (2.9) converges to the covariance (3.7) of Chapter 7 when $\kappa \to \infty$. Thus, the Bayesian interpretation of the frequentist predictive distribution is that it corresponds to a formulation in which very little, or nothing is assumed about the parameters, *a priori*.

Example 2.1. Posterior of an AR(1) Process with Known Autocorrelations. Consider an AR(1) process around a mean μ, $Z_t - \mu = \varphi(Z_{t-1} - \mu) + \varepsilon_t$, with $\varepsilon_t \sim N(0, \sigma^2)$ i.i.d. Suppose the observed values are $Z_1, \ldots, Z_n$. In this case we set $\mathbf{Z}_1 = (Z_1, \ldots, Z_n)^T$, $\mathbf{X}_1 = \mathbf{1}$, and $\mathbf{\Sigma}_{11} = (\sigma^2 \varphi^{|i-j|}/(1 - \varphi^2))$. In other words, here μ takes the role of β. We have that $\hat{\mu} = (\mathbf{1}^T \mathbf{\Sigma}_{11}^{-1} \mathbf{1})^{-1} \mathbf{1}^T \mathbf{\Sigma}_{11}^{-1} \mathbf{Z}_1$ is a weighted average of the observations. Suppose we have a prior $\mu \sim N(b, S^2)$. Define $C = (\mathbf{1}^T \mathbf{\Sigma}_{11}^{-1} \mathbf{1} + S^{-2})^{-1} \mathbf{1}^T \mathbf{\Sigma}_{11}^{-1} \mathbf{1}$. Then, the posterior mean is a simple weighted average, $\tilde{\mu} = C\hat{\mu} + (1 - C)b$. In particular, if $\varphi = 0$, we have $C = (n/\sigma^2 + S^{-2})^{-1} n/\sigma^2$ corresponding to the familiar result that the optimal weights are proportional to the inverses of the variances. The posterior variance of μ is then $(n/\sigma^2 + S^{-2})^{-1}$. Furthermore, the best prediction of Z_{n+k} is $\hat{Z}_{n+k} = \tilde{\mu} + \varphi^k (Z_n - \tilde{\mu})$. $\Diamond$

Example 2.2. Conditional Likehood Errors of an AR(1) Process. A slightly modified version of the AR(1) likelihood is obtained by noting that if φ is known, then the forecast errors one step ahead are $Z_t - \varphi Z_{t-1}$ i.i.d. $\sim N((1 - \varphi)\mu, \sigma^2)$. Conditioning on Z_1 a likelihood for the remaining observations is obtained. The conditional MLE for μ is simply the average divided by $1 - \varphi$, or $\hat{\mu} = \{(Z_n - \varphi Z_{n-1}) + \cdots + (Z_2 - \varphi Z_1)\}/\{(n - 1)(1 - \varphi)\}$. This can be coupled with a prior, as in Example 2.1. $\Diamond$

The Bayesian model and the error classification of Chapter 8, Section 3.2.1 are related. The posterior uncertainty (2.6) represents error in parameter estimates (2), and the covariance matrix in (2.1) represents unpredictable residual error. Disagreements concerning the prior (2.3), either in terms of mean, variance, or distributional form, would be an example of error of expert judgment (3). Note that the above, highly simplified analysis does not incorporate modeling error at all.

2.2. Random Walks

In the previous section we assumed that the second moments of the processes of interest were known. Here, we will consider the estimation of variance and mean simultaneously. The results are classical (e.g., Box and Tiao 1973), but we will tailor them to a time series context.

Suppose we have i.i.d. observations $\varepsilon_i \sim N(0, \sigma^2)$, $i = 1, \ldots, n$. For analytical convenience it is customary to reparametrize the model via the *precision* $\tau = 1/\sigma^2$. Then, the density of the data is $c \times \tau^{n/2} \exp(-\tau \Sigma_i \varepsilon_i^2/2)$. Suppose that the prior distribution of τ has a density of the form $c \times \tau^{\alpha-1} \exp(-\tau\beta)$, a gamma distribution $G(\alpha, \beta)$ with mean α/β and variance α/β^2. Then, the posterior density of τ

is of the form $c \times \tau^{\alpha+n/2-1} \exp(-\tau(\beta + \Sigma_i \varepsilon_i^2/2))$. This is a gamma distribution $G(\alpha + n/2, \beta + \Sigma_i \varepsilon_i^2/2)$. Using the posterior mean of τ, we get the Bayes estimate $\tilde{\sigma}^2 = (\beta/2 + \Sigma_i \varepsilon_i^2)/(n + \alpha/2)$. If n is large relative to α and β, this is close to the MLE $\hat{\sigma}^2 = \Sigma_i \varepsilon_i^2/n$.

Example 2.3. Predictive Distribution of a Random Walk. Consider a random walk $Y_t, t = 0, 1, \ldots, n$, that starts from a known value Y_0. Then, $Y_t - Y_{t-1} = \varepsilon_t \sim N(0, 1/\tau)$ are i.i.d. Assuming τ has a prior density $G(\alpha, \beta)$, τ has the posterior distribution given above. To derive numerically the predictive distribution for the future values $Y_{n+k}, k = 1, \ldots, m$, of the process, we can

(i) sample a value of τ from its posterior distribution $G(\alpha + n/2, \beta + \Sigma_i \varepsilon_i^2/2)$;
(ii) generate i.i.d. values $\varepsilon_{n+k} \sim N(0, 1/\tau), k = 1, \ldots, m$, using the new value for the variance;
(iii) calculate $Y_{n+k} = Y_n + \varepsilon_{n+1} + \cdots + \varepsilon_{n+k}$.

By repeating the steps (i)–(iii), we get a set of simulated vectors $(Y_{n+1}, \ldots, Y_{n+m})$, so we can estimate the predictive distribution to any degree of accuracy (where we take the model specifications as given). ◊

Example 2.4. Predictive Distribution of a Random Walk with a Drift. A random walk $Y_t, t = 0, 1, \ldots, n$, with a drift μ has increments $Y_t - Y_{t-1} = \varepsilon_t + \mu$, with $\varepsilon_t \sim N(0, 1/\tau)$, that are i.i.d. Suppose we have independent priors $\mu \sim N(b, S^2)$ and $\tau \sim G(\alpha, \beta)$. Conditionally on the observed increments the precision and drift are no longer independent. However, suppose we know μ. Then, we can get the posterior of τ for given μ using the results above by identifying $\varepsilon_t = Y_t - Y_{t-1} - \mu$. On the other hand, suppose we know τ, then (cf., Example 2.1) we have that $Y_t - Y_{t-1} \sim N(\mu, 1/\tau), t = 1, \ldots, n$ are i.i.d. Therefore, the posterior of μ for given τ is $N(\tilde{\mu}, (n\tau + S^{-2})^{-1})$, where $\tilde{\mu} = C\hat{\mu} + (1 - C)b$ with $\hat{\mu} = (Y_n - Y_0)/n$. In general, a *Gibbs sampler* can be set up by taking a sample of one parameter given the others, then taking a sample of the next variable given the first and the others etc. In our case, we can take samples from the joint posterior of (τ, μ) by (cf., Williams 2001, 268)

(i) taking a sample $\tau_{(1)}$ from the posterior of τ given some arbitrarily chosen value of μ, such as the mean;
(ii) taking a sample $\mu_{(1)}$ from the posterior of μ given $\tau = \tau_{(1)}$;
(iii) by taking a sample $\tau_{(2)}$ from the posterior of τ given $\mu = \mu_{(1)}$ etc. This produces a sequence of samples $(\tau_{(i)}, \mu_{(i)}), i = 1, 2, \ldots$ A predictive distribution of the future values of the process can then be generated as in Example 2.3:
(iv) corresponding to a sampled pair $(\tau_{(i)}, \mu_{(i)})$ generate a sequence of i.i.d. innovations $\varepsilon_{n+k} \sim N(0, 1/\tau_{(i)}), k = 1, \ldots, m$;
(v) calculate the values $Y_{n+k} = Y_n + \mu_{(i)}k + \varepsilon_{n+1} + \cdots + \varepsilon_{n+k}$.

Repeating the procedure many times allows us to estimate the predictive distribution to the accuracy desired. ◇

Like a Markov Chain, the iterative steps (i)–(iii) always start from the most recent values of the parameter vector. Thus, the approach is called a *Markov Chain Monte Carlo* method.

Initially, the values produced by a Gibbs sampler depend on the chosen starting values. However, it is possible to prove that under regularity conditions (cf., Robert and Casella 1999, 296) an ergodicity result similar to that of the finite state Markov chains (or stable populations) holds even in the case of continuous posterior densities. Therefore, the sampler is first run for several hundred or thousand times during the so-called *burn-in period*. Only after that can the generated values be viewed as samples from the joint posterior.

Choosing the length of the burn-in period is a nontrivial problem. Some of the basic difficulties can be well understood from a demographic perspective. Suppose we have a multistate model representing municipalities in an archipelago. Suppose that the probability is low that one moves from one island to another, although one may move with high probability between municipalities within any given island. Gibbs sampling is analogous to simulating the movement of an individual from municipality to municipality. A simulated individual may never move out of the initial island in a finite number of steps, so the individual's path may end up describing a single island only. Or, even if a change of island occurs, one or more islands of the archipelago can still be left without visits by the time the simulation ends. A practical problem in Gibbs sampling (and other Markov Chain Monte Carlo methods) is that we may not know the setting well enough to be sure that our parameter space does not look like an archipelago with hard to reach islands. Software has been developed to aid in deciding the burn-in period and studying the convergence to an invariant distribution (e.g., Best, Cowles and Vines 1995).

2.3. *ARIMA(1,1,0) Models*

Going beyond random walks, possibly the simplest integrated process is ARIMA$(1, 1, 0)$. The complexity of the details of the predictive distribution calculations increases rapidly when autocorrelation has to be considered. Still, the basic principles are similar to those we used for random walks.

Consider first a process Y_t such that the increments $Y_t - Y_{t-1} \equiv Z_t$ form an AR(1) process around a mean. More precisely, assume that $Z_t - \mu = \varphi(Z_{t-1} - \mu) + \varepsilon_t$, with $\varepsilon_t \sim N(0, 1/\tau)$ i.i.d. As before, assume we have priors $\mu \sim N(b, S^2)$ and $\tau \sim G(\alpha, \beta)$ for some constants b, S, α, and β. For the remaining correlation parameter, assume a uniform prior $\varphi \sim U(-1, 1)$. Based on the observed values of $Y_t, t = 0, 1, \ldots, n$, we can deduce the increments $Z_t, t = 1, \ldots, n$. For simplicity, condition on Z_1. Then, a Gibbs sampler can be based on the following conditional distributions. (i) Conditionally on φ and τ, the posterior distribution of μ is $N(C\hat{\mu} + (1 - C)b, ((n-1)(1-\varphi)^2\tau + S^{-2})^{-1})$, where $C = (n-1)(1-\varphi)^2$

$\tau((n-1)(1-\varphi)^2\tau + S^{-2})^{-1}$ and $\hat{\mu}$ is as given in Example 2.2. (ii) Conditionally on φ and μ, the posterior distribution of τ is $G(\alpha + (n-1)/2, \beta + \Sigma_t \varepsilon_t^2/2)$, where $\varepsilon_t = Z_t - \mu - \varphi(Z_{t-1} - \mu)$ for $t = 2, \ldots, n$. (iii) Conditionally on μ and τ the posterior distribution of φ is of the form $c \times \exp(-\Sigma_t\{Z_t - \mu - \varphi(Z_{t-1} - \mu)\}^2\tau/2)$ for $\varphi \in (-1, 1)$.

Unlike the other cases, the conditional posterior of φ is not immediately obvious. *Rejection sampling* (cf., Ripley 1987, 60–62; Press et al. 1992, 290–296) can be used in this and many other situations. Note first that the summation in the exponent is a second degree polynomial in φ, so for $\varphi \in (-1, 1)$ the posterior must be proportional to the density of $N(\tilde{\varphi}, W)$, where $\tilde{\varphi} = \Sigma_t(Z_t - \mu)(Z_{t-1} - \mu)/\Sigma_t(Z_{t-1} - \mu)^2$ and $W = \{\tau\Sigma_t(Z_{t-1} - \mu)^2\}^{-1}$. A value from the posterior can now be sampled in two steps. First, pick a candidate value φ' from $N(\tilde{\varphi}, W)$. If it is in $(-1, 1)$, we accept φ'. If it is not, we reject φ', and pick another candidate and check if it can be accepted. We continue until an accepted value is found. The accepted values obtained in this manner are samples from the posterior of φ given μ and τ.

The approach can be extended to other ARIMA$(p, d, 0)$ processes, but the details can be complex due to stationarity conditions. For a general approach that does not rely on a conditional likelihood, see Chib and Greenberg (1994).

3. Forecast as a Database and Its Uses

In the past, population forecasts have typically been published in book form. The user has had to wade through pages of small print to find the information he or she is looking for. Having to do this three times (for the middle, high and low variants) certainly hinders appreciation of the uncertainty of the forecast. The book format is even less suitable for presentation of a predictive distribution. Quantiles of the predictive distribution of population aggregates are not simply obtained from the corresponding quantiles of the distributions of the components that are not perfectly correlated. In Alho and Spencer (1991) we proposed that population forecasts should be implemented in a computerized database form, instead.

A *database* can be defined as a collection of data files and a collection of computer programs that are capable of storing, updating, and extracting data from the files. In the case of population forecasting this would mean that sufficient information concerning the forecast is stored, so that the predictive distribution of a user's choice can be output. An important aspect of the database concept is that one would want to get the answers in real time.

The database approach is intended to bring the predictive distribution to the user's desk. This is important in policy settings, where the role of statistical information is complex. Policy preferences frequently are formed on the basis of preferences for certain actions, with only a loose relation to the true state of nature that statistics attempt to estimate or describe. When alternative forecasts are available but their probabilities are unstated, policy makers are pretty free to choose the forecast that best agrees their preferred policy and to criticize forecasts opposing their preferences. Such criticism deflects the policy debate away from the real issues (different values and different assumptions concerning the relation of policy choices to outcomes) and towards a supposedly value-free disagreement, namely what is the future population going to be like. If a predictive distribution shows that the alternative forecasts are about equally likely, such a debate is unenlightening, however. If the predictive distributions shows that one forecast is more likely than the other, then debate might move to consider probabilities of different outcomes, where the probabilities take into account both the error distribution of the forecast(s) and the probability distribution of the outcome conditional on the policy choice. Knowing, in real time, a realistic predictive distribution, can expose the source of the policy disagreement and lead to better argumentation.

Another benefit from predictive distributions is that they emphasize the sequential revision aspects of policy making. As the future unfolds, more becomes known, and adjustments to policy may be called for. The need for such revisions may be anticipated when the expected error of the forecasts is large. Providing an explicit assessment of uncertainty helps protect against overconfidence in a forecast and helps protect against the use of low probability scenarios as rebuttals to more likely forecasts. In many applications, such as the design of pension programs, the predictive distribution can help one evaluate the riskiness of alternative strategies (cf., Chapter 11). Population size may be a relatively minor source of uncertainty in some of those calculations but a major source in others. It may be hard to tell which situation prevails without a realistic assessment of the uncertainty of the future population.

Two basic approaches for the construction of a forecast database are available. In the analytical approach one stores the point forecast and descriptions of forecast errors. Programs are written that approximate the variances of forecast errors for the aggregates of the user's choice using linearizing transformations. This will be discussed in Section 5. The other approach relies on simulation, in which samples are taken from the predictive distribution and stored. Other programs can then read selected stored values and produce statistical summaries from them. This approach will be discussed in Sections 6 and 7. Under either approach, a difficulty is presented by the large number of cross-covariances and cross-lagged covariances. A way around this issue is to parametrize the covariances, as we discuss next.

4. Parametrizations of Covariance Structure

In Section 4.1 we consider the problem of estimating the variance of a sum of random variables. Motivated by the general considerations, in Section 4.2 we define a scaled model of error that is closely linked to simple random walk theory, for use in propagation of error in population forecasts. Section 4.3 tackles the issue of models for covariances for errors in migration forecasts; such models are especially useful because the number possible covariances is large yet the information for estimating them typically is weak.

4.1. Effect of Correlations on the Variance of a Sum

In analyzing cohort-component forecasts we continually deal with various sums of random variables. For example, the total fertility rate of a future year t is a sum of possibly *cross-correlated* age-specific fertility rates. The forecast error of any age-specific vital rate in age x can be viewed as accruing annually, so it is a sum of *autocorrelated* annual terms. In cohort survival we calculate sums of age-specific mortality rates that are correlated over age and time. The population itself as aggregated over age is a sum. More generally, we may be interested in linear combinations of variables with positive coefficients, e.g., the disabled population as aggregated over age according to age-specific prevalence rates. To put the different types of sums into perspective we start by approaching the problem abstractly.

Let $\varepsilon_1, \ldots, \varepsilon_n$ be random variables with $\text{Var}(\varepsilon_i) = s_i^2$ ($s_i > 0$) and $\text{Cov}(\varepsilon_i, \varepsilon_j) = \rho_{ij} s_i s_j$. Let $S_{\text{ind}}^2 = s_1^2 + \cdots + s_n^2$ denote the variance of the sum of the ε_i's under independence, and $S_{\text{dep}}^2 = (s_1 + \cdots + s_n)^2$ the variance of the sum under perfect dependence (i.e., $\rho_{ij} = 1$). Finally, let

$$S^2 = \sum_{i,j=1}^{n} \rho_{ij} s_i s_j \tag{4.1}$$

be the exact variance. Defining the (weighted) average correlation as

$$\bar{\rho} = \sum_{i \neq j} \rho_{ij} s_i s_j \Big/ \sum_{i \neq j} s_i s_j, \tag{4.2}$$

a simple calculation shows that $S^2 = (1 - \bar{\rho}) S_{\text{ind}}^2 + \bar{\rho} S_{\text{dep}}^2$. Clearly, if $\bar{\rho}$ is non-negative, then $S_{\text{ind}}^2 \leq S^2 \leq S_{\text{dep}}^2$. If we have a good guess at $\bar{\rho}$, then we can estimate S^2 by an appropriate linear combination of S_{ind}^2 and S_{dep}^2.

Consider now a single age-specific vital rate. In this case the ε_i's may represent the annual changes of the rate, and the goal is to derive an approximation to the variance of the rate during a future year n.

Example 4.1. Independence, AR(1), and Perfect Dependence. Suppose the ε_i's have $s_i^2 = s^2$ with an AR(1) structure $\rho_{ij} = \rho^{|i-j|}$, where $|\rho| < 1$. In this case $S_{\text{ind}}^2 = ns^2$, $S_{\text{dep}}^2 = n^2 s^2$, and one can show that $S^2 = (2\rho^{n+1} - n\rho^2 - 2\rho + n)s^2/(1 - \rho)^2$. Furthermore, $\bar{\rho} = 2\rho(\rho^n - n\rho + n - 1)/[n(n-1)(1-\rho)^2]$. Asymptotically $S^2/S_{\text{ind}}^2 \sim (1+\rho)/(1-\rho)$, so S^2 is much closer to S_{ind}^2 than S_{dep}^2. ◊

Example 4.1 can be extended to represent the total fertility rate. In this case, the ε_i's would correspond to n age-specific fertility rates of a given year, although the variances of the error terms should then depend on age.

Example 4.2. Error in a Cohort Survival Setting. Consider cohort survival from age x to age $x + n$. Let ε_i be the deviation of the mortality rate from its mean in age $x + i - 1$, $i = 1, \ldots, n$. Then, the sum of ε_i's is the relative deviation in the number of survivors to age $x + n$. Suppose we have $\varepsilon_i = \varepsilon_{i1} + \cdots + \varepsilon_{ii}$, where the ε_{ij}'s are the error increments for the age-specific rate of age $x + i - 1$. Let us assume that (a) the variances of the increments are homogeneous and equal to s^2; (b) for a fixed i the ε_{ij}'s are independent; (c) for a fixed j the correlation between ε_{ij} and ε_{kj} is $\rho^{|i-k|}$. The assumptions (a) and (b) imply that $s_i^2 = is^2$, so $S_{\text{ind}}^2 = n(n+1)s^2/2$. Replacing sums by integrals one gets the approximation $S_{\text{dep}}^2 \approx (4/9)((n+1/2)^{3/2} - (1/2)^{3/2})^2 s^2$. Therefore, asymptotically $S_{\text{dep}}^2/S_{\text{ind}}^2 \sim 8n/9$. Using the results of Example 4.1, one can show that $S^2 = \{n(n+1)(1-\rho^2)/2 - 2n\rho + 2\rho^2(1-\rho^{n-1})/(1-\rho)\}s^2/(1-\rho)^2$. It follows that in this case also, asymptotically $S^2/S_{\text{ind}}^2 \sim (1+\rho)/(1-\rho)$. As in Example 4.1, S_{ind}^2 is much closer to the true value than S_{dep}^2. $\diamond$

In many cases, the variance of the sum of the ε_i's can be approximated by the AR(1) model of Example 4.1 as well as by a constant correlation model. Define

$$S_{\text{AR}}^2(\varphi) = \sum_{i,j=1}^{n} \varphi^{|i-j|} s_i s_j. \tag{4.3}$$

Since $S_{\text{AR}}^2(0) = S_{\text{ind}}^2$ and $S_{\text{AR}}^2(1) = S_{\text{dep}}^2$, there is also a value $\varphi = \varphi^*$ such that $S_{\text{AR}}^2(\varphi^*) = S^2$, if the average correlation is nonnegative. Similarly, first define $\rho_{ij}(\delta) = 1$ for $i = j$ and $\rho_{ij}(\delta) = \delta$ for $i \neq j$. Then define

$$S_{\text{CC}}^2 = \sum_{i,j=1}^{n} \rho_{ij}(\delta) s_i s_j. \tag{4.4}$$

It follows that there is a $\delta = \delta^*$ such that $S_{\text{CC}}^2(\delta^*) = S^2$, if the average correlation is nonnegative. Then, the correct variance can be obtained using either an AR(1) or a constant correlation assumption. In fact, both representations can also be used for some cases in which the average correlation (4.2) is negative, if it is not too large in absolute value. Note that if the true model is $S^2 = S_{\text{CC}}^2(\delta)$, then $\bar{\rho} = \delta$.

4.2. *Scaled Model for Error*

The preceding discussion may serve as a motivation for a class of relatively simple stochastic models that are capable of approximating a wide variety of error structures. The models are designed to handle errors of nonstationary processes applicable to demographic forecasts. Recalling the problems that derive from limiting the forecast error with fixed bounds (see Complement 25, Chapter 7; Keilman 2002 provides a formulation that uses fixed bounds but does not suffer from a similar defect), we provide a way to limit the errors stochastically. The following description is adapted from Alho and Spencer (1997) and Alho (1998).

We will first show how up to time $T \leq \infty$ the expected error may be determined by a model based assessment. Then we indicate how for longer term forecasting ($t \geq T$) we may specify a subjective structure that continues smoothly from the earlier part, but remains bounded *ad infinitum*. The choice of T will depend on the series. If the forecast errors increase to levels that are considered implausible by expert demographers, then we may want to switch to a subjective specification that incorporates such judgment.

Consider error processes $X(j, t)$, where $j = 1, \ldots, J$ may refer to age or region, for example, and $t > 0$ is the forecast year. It can always be written in the form

$X(j,t) = \varepsilon(j,1) + \cdots + \varepsilon(j,t)$. To define the process further, we have in mind that $X(j,t)$ could be a random walk with a drift (in t), for example. We consider a more general case, however, and suppose that the error increments are of the form

$$\varepsilon(j,t) = S(j,t)(\eta_j + \delta(j,t)). \tag{4.5}$$

Here, the $S(j,t) > 0$ are known weights whose specification will be discussed shortly. Assume that for each j, (a) the variables $\delta(j,t)$ are independent over time $t = 1, 2, \ldots$; (b) the variables $\{\delta(j,t) | j = 1, \ldots, J; t = 1, 2, \ldots\}$ are independent of the variables $\{\eta_j | j = 1, \ldots, J\}$; and (c) that

$$\eta_j \sim N(0, \kappa_j), \quad \delta(j,t) \sim N(0, 1 - \kappa_j), \tag{4.6}$$

where $0 < \kappa_j < 1$ are known. Thus, if the scales would not depend on t (or, $S(j,t) \equiv S(j)$), then we would have a random walk with a random drift for every j.

As discussed in abstract terms in the previous section we may assume that $\mathrm{Corr}(\eta_i, \eta_j) = \rho_\eta^{|i-j|}$, or $\mathrm{Corr}(\eta_i, \eta_j) = \rho_\eta$, for some $|\rho_\eta| \leq 1$. Similarly, $\mathrm{Corr}(\delta(i,t), \delta(j,t)) = \rho_\delta^{|i-j|}$, or $\mathrm{Corr}(\delta(i,t), \delta(j,t)) = \rho_\delta$ for some $|\rho_\delta| \leq 1$. Since the increments are scaled by the $S(j,t)$, or $\mathrm{Var}(\varepsilon(j,t)) = S(j,t)^2$, we call this a *scaled model* for error. Intuitively, allowing the scales to vary with t provides a way to account for changing volatility (Section 4.1, Chapter 7). The role of the correlation parameters is to represent the phenomenon that forecast errors of vital rates in close ages tend to be similar, but in distant ages they may be quite different.

Example 4.3. Autoregressive Model for Correlations Across Age. We considered the logarithms of the age-specific fertility rates for the white U.S. population in 1921–1988 in ages 14, 15, ..., 46. We studied the crosscorrelations of the first and second differences of the series for lag = 0. The correlations involving the youngest and the oldest ages deviated from the rest, so we will use medians to describe typical correlations. The median crosscorrelations between ages that are one year apart was 0.97 for the first differences (0.94 for the second differences); for ages 5 years apart 0.84 (0.82); for ages 10 years apart 0.63 (0.62); for ages 20 years apart 0.33 (0.31). We see that an autoregressive model (over age) with the first autocorrelation ≈ 0.95 gives a reasonable description of the typical correlations. ◊

Note that $\kappa_j = \mathrm{Corr}(\varepsilon(j,t), \varepsilon(j,t+h))$ for all $h \neq 0$. Therefore, κ_j can be interpreted as a constant correlation between the error increments. Under a random walk model the error increments would be uncorrelated, with $\kappa_j = 0$. Suppose that we have an increasing sequence of error variances $\sigma(j,1)^2 < \sigma(j,2)^2 < \cdots < \sigma(j,T)^2$ available with $\mathrm{Var}(X(j,t)) = \sigma(j,t)^2$. One can show with some algebra that we can estimate the corresponding increment variances by taking $S(j,1)^2 = \sigma(j,1)^2$ and

$$S(j,t) = -\kappa_j s(j;t-1) + \left[\kappa_j^2 s(j;t-1)^2 + \sigma(j,t)^2 - \sigma(j,t-1)^2\right]^{1/2}, \tag{4.7}$$

for $t > 1$, where $s(j;t-1) = S(j,1) + \cdots + S(j,t-1)$. Note that in the case $\kappa_j = 0$, (4.7) simplifies to $S(j,t)^2 = \sigma(j,t)^2 - \sigma(j,t-1)^2$.

The key properties of the above model are the following. First, since the choice of the scales $S(j,t)$ is unrestricted, any sequence of non-decreasing error variances can be matched. Second, any sequence of cross-correlations can be majorized using either of the two correlational models (because at $\varphi = 1$ or $\rho = 1$ the sums they represent reduce to S^2_{dep}). Third, any sequence of autocorrelations for the error increments can be majorized. This means that we can always find a conservative approximation to any covariance structure using the model we have introduced.

The scaled model (4.5) can be used to simulate forecast errors. Both empirical estimates and judgmental factors are, in practice, used to determine the parameters of the model (Alho 1998). In particular, the scaled model may provide an approximation to the errors of an ARIMA forecast. One first derives the covariance structure of the forecast error as given in (2.13) of Chapter 7, and then finds a suitable approximating sequences of scales S and appropriate correlation parameters κ. At the other end of the spectrum, purely judgmental forecasts can also be accommodated.

Example 4.4. Specifying a Linear Process to Match Judgment. Suppose we have judgmental forecasts for n successive years and an associated sequence of standard deviations $0 < S_1 < S_2 < \cdots < S_n$. Suppose the process being forecasted is nonstationary. We may then look for a once-integrated process that would have forecast errors similar to the ones specified. Write $\Psi_k \equiv \psi_0 + \cdots + \psi_k$, $k = 0, 1, \ldots$. Based on (2.13) of Chapter 7 we can write $\mathrm{Var}(E_{(k)}) = \sigma_\varepsilon^2(\Psi_0^2 + \cdots + \Psi_{k-1}^2)$. Equating $\mathrm{Var}(E_{(k)}) = S_k^2$ for $k = 1, \ldots, n$ yields first $\sigma_\varepsilon^2 = S_1^2$, and then the estimates $\sigma_\varepsilon^2 \Psi_{k-1}^2 = S_k^2 - S_{k-1}^2$ for $k = 2, \ldots, n-1$. This gives us $\Psi_{k-1} = (S_k^2 - S_{k-1}^2)^{1/2}/\sigma_\varepsilon$. Knowing the Ψ_k's yields the ψ_j's via $\psi_k = \Psi_k - \Psi_{k-1}$. For a given n there are infinitely many linear processes for which the first n ψ_j values agree with the ones obtained, and the judgmental standard deviations are compatible with any one of them. In any case, ARIMA models can be used to simulate realizations of these errors. Presenting these to the judge one can try to determine if the judgmental specification is really as intended. Once a resolution has been found, the scaled model can be used to implement the judgment in error propagation. $\diamondsuit$

We have noted earlier that the usual time-series methods may produce prediction intervals that will eventually be too wide. This may happen if the methods do not incorporate sufficient information about the boundedness of the vital processes. In Alho and Spencer (1997) we proposed to take such additional information into account by allowing for modifications in the error structure so that levels of error that contradict the additional information are excluded. Suppose we judge that the error structure we have specified yields what should be a maximum variance by year T. We may then assume that from T on the error structure will follow an AR(1) process centered around the point forecast that has the standard deviation $\mathrm{Var}(X(j, T))^{1/2}$ and the first autocorrelation $\mathrm{Corr}(X(j, T-1), X(j, T))$. We will consider $X(T)$ as the first value of the AR(1) process, so there is a smooth transition from one process to the next.

To provide a theoretical basis for the eventual AR(1) assumption, it is useful to note that the AR(1) process is the discrete time version of the Ornstein-Uhlenbeck process of diffusion theory. There, the process is obtained from a Brownian motion as subjected to an elastic force towards a mean function (Feller 1971, 99, 335–336).

This notion seems to capture the idea that the errors should be centered around the point forecast and have a bounded variance in the long run.

4.3. Structure of Error in Migration Forecasts

Characterizing the error of migration forecasts in a multistate setting will yield approximations that can be used in the specification of error for net migration for a single state model.

Consider a closed system of J regions with two sexes ($s = 1, 2$), and ages $x = 0, \ldots, \omega$. Define $M_{sij}(x, t)$ = number of those of sex s who are at time t in age x in region j and survive to age $x + 1$ in region i. Then, we can define the in-migrants to region i as

$$M_{si.}(x, t) = \sum_{j \neq i} M_{sij}(x, t), \tag{4.8}$$

the number of out-migrants from region i as

$$M_{s.i}(x, t) = \sum_{j \neq i} M_{sji}(x, t), \tag{4.9}$$

the net number of migrants to region i as

$$N_{si}(x, t) = M_{si.}(x, t) - M_{s.i}(x, t), \tag{4.10}$$

and the gross number of migrants as

$$G_{si}(x, t) = M_{si.}(x, t) + M_{s.i}(x, t). \tag{4.11}$$

Suppose we have forecasts $\hat{M}_{sij}(x, t)$ for the out-migrants. Similar notation will be used for (4.9), (4.10), and (4.11). We assume that the forecast error $\varepsilon_{sij}(x, t)$ is proportional to the forecast, or

$$M_{sij}(x, t) = \hat{M}_{sij}(x, t)(1 + \varepsilon_{sij}(x, t)). \tag{4.12}$$

A possible variance components representation for the error is the following,

$$\varepsilon_{sij}(x, t) = \xi(t) + \eta_i - \eta_j + \theta_{sij}(x, t). \tag{4.13}$$

Here the ξ, η, ζ, and θ terms are assumed to be random and independent of each other. The role of $\xi(t)$ is to represent unexpected error in the overall level of migration for all regions. It has been empirically noted that there are times when migration speeds up, and other times when it slows down. This can be associated with the level of economic activity in the country, with economic growth being associated with fast movement of people. A change in speed can occur without any change in the shares of the regions. In an exaggerated case one can imagine that there is a fixed number of jobs (or places to live, for example); individuals can only move when a job (or house) becomes vacant; and during economic boom many movements occur. The role of η_i is to represent unexpected rise in the economic potential of region i, which influences the outflow from region i negatively and the inflow positively. The terms $\theta_{sij}(x, t)$ represent uncorrelated residual error.

To approximate the error of the net migration forecast, let us set the terms θ to zero. Summing over both sexes we may write the total net migration to region i in age x during year t as

$$N_{.i}(x,t) = \hat{N}_{.i}(x,t) + \hat{N}_{.i}(x,t)\xi(t) + \hat{G}_{.i}(x,t)\{\eta_i - \bar{\eta}_i\}, \tag{4.14}$$

where

$$\bar{\eta}_i = \sum_{j \neq i} \frac{\hat{M}_{.ji}(x,t) + \hat{M}_{.ij}(x,t)}{\hat{G}_{.i}(x,t)} \eta_j \tag{4.15}$$

is a weighted average of the unexpected attractiveness of *all the other regions besides i*. We see that the error in (4.14) consists of two pieces. One is proportional to the forecast of net migration and represents the error in the overall level of migration. The second piece is proportional to gross migration and represents the error in the assumed attractiveness of region i relative to all the other regions. Frequently, net migration is set to zero in forecasts. Thus, even if variations in overall migration were larger than changes in attractiveness, this source may not be as important as the latter when it comes to assessing the uncertainty of forecasting net migration.

5. Analytical Propagation of Error

Initially, analytical propagation of error formulas were derived for population forecasts for computational reasons (e.g., Sykes 1969, Alho and Spencer 1991; Lee and Tuljapurkar 1994). However, with the tremendously increased speed of computers, a primary virtue of analytical propagation of error formulas is that they may help us see "what is going on", i.e., how an error in a particular variable or variables influences other variables of interest. We will consider two cases. The first one shows how the uncertainty of births can be decomposed into a component that is due to the uncertainty of past fertility and current fertility. The second example deals with a general linear growth model.

5.1. Births

Consider a single region female population and assume, for simplicity, that time is discrete and the uncertainty in mortality can be ignored (justification for the assumption is provided in Alho 1992b). Let $B(t) = \exp(b(t))$ be the number of births during year t, let $f(x,t)$ be the log of the fertility rate in age x during year t, and let $s(x,t)$ be the log of the probability of surviving from age 0 to be in age x in the beginning of the year t. In analogy with (5.2) of Chapter 6 we can write

$$b(t) = \log\left(\sum_{x=\alpha}^{\beta} \exp(b(t-x) + f(x,t) + s(x,t))\right). \tag{5.1}$$

Let $B(x, t)$ be the number of children born to women in age x during year t, and define the shares $c(x, t) = B(x, t)/B(t)$. Let $\hat{b}(t)$, $\hat{c}(x, t)$ and $\hat{f}(x, t)$ be the forecasts of $b(t)$, $c(x, t)$ and $f(x, t)$, respectively, and write $b(t) = \hat{b}(t) + \varepsilon_b(t)$ and $f(x, t) = \hat{f}(x, t) + \varepsilon_f(x, t)$. Using a linear Taylor series approximation to the right hand side of (5.1), around the point forecast, we get that (cf., Lee 1974)

$$\varepsilon_b(t) \approx \xi(t) + \sum_{x=\alpha}^{\beta} \hat{c}(x, t)\varepsilon_b(t - x), \tag{5.2}$$

where

$$\xi(t) = \sum_{x=\alpha}^{\beta} \hat{c}(x, t)\varepsilon_f(x, t). \tag{5.3}$$

We see that the errors are (approximately) a linear combination of a current error increment $\xi(t)$ and past errors. In this application the forecast error $\xi(t)$ would be expected to be a highly autocorrelated process. In fact it should behave approximately the same way as the relative error of the total fertility rate.

5.2. General Linear Growth

Consider a double sequence of random vectors $(\mathbf{X}_t, \mathbf{Y}_t)$ for $t = 0, 1, 2, \ldots$, and a differentiable vector-valued function $\mathbf{f}(\ldots)$ such that

$$\mathbf{Y}_{t+1} = \mathbf{f}(\mathbf{X}_t, \mathbf{Y}_t). \tag{5.4}$$

We assume that there are point forecasts $\hat{\mathbf{X}}$ for $\mathbf{X}$ such that $\mathbf{X}_t = \hat{\mathbf{X}}_t + \varepsilon_t$, with $\mathbf{E}[\varepsilon_t] = \mathbf{0}$, and forecasts $\hat{\mathbf{Y}}$ for $\mathbf{Y}$ such that $\hat{\mathbf{Y}}_{t+1} = \mathbf{f}(\hat{\mathbf{Y}}_t, \hat{\mathbf{X}}_t)$ and $\mathbf{Y}_t = \hat{\mathbf{Y}}_t + \eta_t$, where η_t is the error. Define the (matrices of) partial derivatives $\partial\mathbf{f}/\partial\mathbf{X}^T = \mathbf{H}$, and $\partial\mathbf{f}/\partial\mathbf{Y}^T = \mathbf{K}$.

Example 5.1. Representation of a Closed Female Population. Let $\mathbf{Y}_t = \mathbf{V}(t)$ be a vector representing a closed female population (Section 2.1 of Chapter 6), and let $\mathbf{X}_t = (\mathbf{F}(t)^T, \mathbf{S}(t)^T)^T$ be a vector that has the age-specific fertility rates of year t in vector $\mathbf{F}(t)$ and the age-specific survival proportions in vector $\mathbf{S}(t)$. Let $\mathbf{f}$ correspond to multiplication $\mathbf{R}(t)\mathbf{V}(t)$. Then, (5.4) represents the linear growth model (2.2) of Chapter 6.[4] In this case $\mathbf{K} = \mathbf{R}(t)$, for example. ◊

Using a linear Taylor series approximation one can write.

$$\mathbf{Y}_{t+1} \approx \mathbf{f}(\hat{\mathbf{X}}_t, \hat{\mathbf{Y}}_t) + \mathbf{H}_t\varepsilon_t + \mathbf{K}_t\eta_t, \tag{5.5}$$

where $\mathbf{H}_t = \mathbf{H}(\hat{\mathbf{X}}_t, \hat{\mathbf{Y}}_t)$, and $\mathbf{K}_t = \mathbf{K}(\hat{\mathbf{X}}_t, \hat{\mathbf{Y}}_t)$ are the partial derivatives evaluated at the point forecast. It follows that we have the approximate recursion for the error,

$$\eta_{t+1} \approx \mathbf{H}_t\varepsilon_t + \mathbf{K}_t\eta_t. \tag{5.6}$$

[4] Similarly, (5.4) can represent the log of the population vector, or it can incorporate external net migration, as in (3.1) of Chapter 6.

This shows how the error at $t+1$, η_{t+1}, arises from the past error η_t and the current forecast error ε_t. By repeated application of (5.6) one can show that

$$\eta_{t+1} \approx \mathbf{M}_{t,0}\eta_0 + \sum_{i=0}^{t} \mathbf{M}_{t,i+1}\mathbf{H}_i\varepsilon_i, \tag{5.7}$$

where

$$\mathbf{M}_{t,k} = \prod_{i=k}^{t} \mathbf{K}_i, \tag{5.8}$$

and $\mathbf{M}_{t,t+1} = \mathbf{I}$. Note the similarity between (5.7) and (3.2) of Chapter 6, where we opened up the population system to external migration. In both cases there is a component deriving from the initial vector (here $\mathbf{M}_{t,0}\eta_0$), and then increments deriving from each subsequent year t that begin to behave according to the growth equation: in (5.7) the increments are past forecast errors that begin to propagate over time according to (5.4), in Chapter 6 they were net migrants. The intuitive interpretation is that errors are like net migrants!

Formula (5.7) shows how the forecast errors of $\mathbf{X}$ for all earlier years influence the error of $\mathbf{Y}$ for year $t+1$. The errors consist of both biases that are due to the nonlinearity of (5.4) and random error. Assuming that the biases are small enough so they can be ignored, (5.7) provides a direct computational formula for the approximate covariance of the forecast error. Formula (5.6), on the other hand shows how a recursive system of calculations can be set up. We have

$$\begin{aligned} &\mathrm{Cov}(\eta_{t+1}) \approx \\ &\mathbf{H}_t\mathrm{Cov}(\varepsilon_t)\mathbf{H}_t^T + \mathbf{K}_t\mathrm{Cov}(\eta_t)\mathbf{K}_t^T + \mathbf{H}_t\mathrm{Cov}(\varepsilon_t, \eta_t)\mathbf{K}_t^T + \mathbf{K}_t\mathrm{Cov}(\eta_t, \varepsilon_t)\mathbf{H}_t^T, \end{aligned} \tag{5.9}$$

where the two last covariances can be calculated from

$$\mathrm{Cov}(\varepsilon_t, \eta_t) \approx \mathrm{Cov}(\varepsilon_t, \eta_0)\mathbf{M}_{t-1,0}^T + \sum_{i=0}^{t-1} \mathrm{Cov}(\varepsilon_t, \varepsilon_i)\mathbf{H}_i^T\mathbf{M}_{t-1,i+1}. \tag{5.10}$$

Note first, that if (unrealistically) the errors ε_t would be an uncorrelated sequence, then the covariances (5.10) would be zero, and (5.9) would be a relatively simple recursion given that we know the terms $\mathrm{Cov}(\varepsilon_t)$. More generally, (5.10) can be interpreted as a recursive system in the sense that the set of coefficient matrices $\mathbf{M}_{t,k}$ can be obtained from the matrices $\mathbf{M}_{t-1,k}$ by left multiplying them by $\mathbf{K}_t$, and by adding $\mathbf{K}_t$ to the set.

In principle, approximate second moments of the forecast error of a linear growth model can be calculated using (5.9) and (5.10) if the point forecast and the covariance structure of the forecast error of the vital rates is known. However, the apparent simplicity of the formulas hides the fact that the derivatives $\mathbf{H}_t$ and $\mathbf{K}_t$ are complicated functions of the vital rates, so their programming is tedious. Another problem in the numerical use of these formulas is that they are approximate: evaluating the magnitude of the error of approximation is more complicated than the use of the formulas themselves.

6. Simulation Approach and Computer Implementation

Stochastic simulation (or Monte Carlo) methods have a history that goes back, at least, to World War II (Ripley 1987).[5] The simulation approach has three primary advantages over the analytic approach. First, no linearizing approximations are required to derive the moments of the predictive distribution. Second, although distributional assumptions are needed for the description of the uncertainty of the vital rates, no assumption needs to be made concerning the predictive distribution of the future population vector. The empirical distribution of the future population computed with respect to the sample of population paths, serves as the estimate of the predictive distribution of the future population vector. Third, with the simulations it is easy to handle functional forecasts – a sample path of a functional forecast is simply the function evaluated on the sample path, and the predictive distribution is readily estimated by their empirical distribution (as in Sections 2.2.4 of Chapter 4 and 1.6 of Chapter 6).

A drawback is that it may be hard to find out the relative roles of different error components in the final result without rerunning the whole simulation. The transparency of some analytical formulations, such as (5.7) may be of help in such an analysis. Also, while simulation may be used to check the accuracy of moment calculations based on analytic approximations, analytical formulas may be used to look for possible errors in the programs used in simulation. Therefore, we view the two approaches as being complementary.

A brute force way to use simulation as part of the database implementation of a stochastic forecast is to store all simulated sample paths of the population vector on a hard disk. Additional programs are used to produce statistical summaries out of the stored data. This provides the real-time performance required for the database implementation. This approach would not have been feasible as late as the early 1990's. With the availability of fast, inexpensive computers, sample sizes in simulation are no more a liming factor.

We have implemented a simulation based database forecast in a computer program PEP (Program for Error Propagation). It is written in the C++ language, and it is based on the estimation procedures discussed in Chapter 4, the one region two-sex linear growth model of Chapter 6, and the scaled model described in Section 4.2. A systematic description of PEP is available at http://www.joensuu.fi/statistics/juha.html. Here, we will summarize the main features as they appear to the user.

PEP is a menu directed Windows program. The user is required to input such information as the number of simulation rounds, the number of forecast years,

[5] Earlier uses of randomization devices include attempts to determine the value of π by repeatedly throwing a needle of length L on a plane that has parallel lines at distance $A > L$ in the latter part of the 1800's ("Buffon's needle problem"; cf., Gnedenko 1976, 36–37). Perhaps Gossett's empirical derivation of the t-distribution from a collection of several thousand biological measurements around 1908 can also be seen as falling into this category.

the lowest and highest child-bearing ages, the highest age, the sex-ratio at birth, and (if mortality rates from the rectangles of the Lexis diagram are being input) the separation factor for mortality in age 0. Then, the user is prompted for file names giving the jump-off population, point forecasts of age and sex-specific mortality, age-specific fertility, and net-migration by age and sex. In addition to these basic data, the user is prompted to give the parameters required for the specification of the scaled models of error for mortality, fertility, and migration. These are partly given as files (e.g., the scales and the kappas), partly as constants requested by the program menus. To facilitate the preparation of the input data, there is another C++ program BEGIN that produces input files that follow some commonly used approaches for formulating forecast assumptions. PEP checks the input data for consistency. For example, the input files must conform to the given age ranges and forecast period. Once the simulation has been carried out, the user is prompted to specify what kinds of aggregate data he or she might wish to study. In most uses of forecasts the interest centers on selected age-groups. There is a third C++ program COMBINE that produces similar aggregated output after a PEP run. The final statistical processing (summary statistics, graphics) is intended to be carried out by a spreadsheet or statistical program of the user's choice.

Example 6.1. Storage Space Required by the Database. Consider a forecast of a population by single years of age for $T = 50$ years. If the whole population vector has ages $0, 1, 2, \ldots, 99, 100+$, by sex, there are 202 components in the vector. Each sample path of the vector is stored into a file containing a 50×202 matrix. If the number of simulation rounds is, say, $N = 3{,}000$, there will 3,000 such files stored. Together, they take up approximately 300 MB of hard disk space. (The exact amount depends on the allocation unit used by the computer.) These files provide the basic material on which everything else is built. PEP automatically converts the N sample paths into T annual files, each containing a 3000×202 matrix. Each column contains $N = 3{,}000$ samples from the distribution of a given component of the population vector for a given forecast year. Together, the annual files also take about 300 MB of disk space. In a typical run, one is also interested in summary data concerning user defined age-groups. The amount of space the results take is proportional to the number of age-sex-groups. In addition, PEP outputs simulated values for life expectancies. Together with the input files and the programs the space required by the database after the initial run is of the order of 650 MB. Increasing the number of forecast years will lead to a proportional increase in space requirements. For example, a corresponding forecast going 65 years into the future with some added output produced by COMBINE took about 50% more, or some 1,000 MB (or 1 GB) of space. The establishment of the original database as described above takes minutes or less on current machines. $\diamondsuit$

To establish the PEP database in the first place requires a professional demographer or statistician capable of understanding both the demographic detail of usual cohort-component forecast and the specification of the error structure, roughly at the level of this book. If the user is willing to accept values for the parameters of the scaled model of error suggested by BEGIN, the demands are comparable to those of a traditional cohort-component forecast. The retrieval of aggregate data is very simple, but the user must be comfortable with some spreadsheet or statistical program to be able to effectively produce numerical or graphical summaries of the simulated data.

7. Post Processing

Any propagation of error program (such as PEP) must limit the range of available models. The primary limiting factor appears not to be the difficulty of implementing probabilistic models of great generality, but rather the user's difficulty of providing meaningful input data for complex models. Given the restricted scope of the program, it is useful in practice to find ways of inferring, from the available output, results that correspond to alternative specifications. There is no hope that one could find an acceptable approximation for arbitrary alternatives, only for certain restricted types. Suppose a forecast database is available that corresponds to the predictive distribution of the future population. By *post processing* we refer to selective uses of forecast database results.

7.1. Altering a Distributional Form

Consider a population characteristic ξ, whose distribution can be estimated from the database values. In the current version of PEP, life expectancy at birth is stored, for example, so ξ could be the female or male life expectancy during any given future year or, say, the average life expectancy over the forecast years. Even if the desired measure is not stored, a proxy may be available. In Example 7.1 we illustrate the use of the *general fertility rate*, i.e., the ratio of births to person years lived in child bearing ages, as a proxy for the total fertility rate.

Assume that the forecast database is based on N simulation rounds, so we have the values $\xi_1, \ldots, \xi_N$ available. Let the empirical distribution function based on the simulated values be $F(x) = (\text{number of } \xi_i\text{'s} \leq x)/N$. Suppose a user is unhappy with $F(.)$. This could take many forms, but suppose for the sake of illustration that the user is satisfied with the distribution up to the median, but thinks that the upper tail is too long, and wishes that the upper half of the distribution be modified in a gradual manner so instead of the current decile $F^{-1}(0.9) = a$ we would have a distribution taking the value b, or $F^{-1}(0.9) = b < a$, instead. This can be achieved by selectively removing or rejecting simulated sample paths from the output. A possible approach (one among many) is as follows.

We exclude the possibility of ties for simplicity of exposition. Let the ordered data be $\xi_{(1)} < \cdots < \xi_{(N)}$. Define $\lfloor x \rfloor$ = largest integer $\leq x$. Then, the median can be taken to be $F^{-1}(0.5) = \xi_{(\lfloor N/2 \rfloor)}$ and the 9th decile can be taken to be $a = \xi_{(\lfloor 9N/10 \rfloor)}$. Take B to satisfy $\xi_{(B)} \leq b \leq \xi_{(B+1)}$. The simulated values can now be split into three segments $\xi_{(1)} < \cdots < \xi_{(\lfloor N/2 \rfloor)}$; $\xi_{(\lfloor N/2 \rfloor+1)} < \cdots < \xi_{(B)}$; and $\xi_{(B+1)} < \cdots < \xi_{(N)}$. A brute force solution that has the virtue of retaining a maximal number of simulated values is as follows:

(i) Retain simulation rounds corresponding to values $\xi_{(\lfloor N/2\rfloor+1)} < \cdots < \xi_{(B)}$, there are $B - \lfloor N/2\rfloor$ of them;

(ii) Retain the fraction $f = (B - \lfloor N/2\rfloor)/4(N - B)$ of the simulation rounds corresponding to values $\xi_{(B+1)} < \cdots < \xi_{(N)}$, so $(1 - f)(N - B)$ values are deleted;

(iii) Delete a total of $(1 - f)(N - B)$ simulation rounds corresponding to values $\xi_{(1)} < \cdots < \xi_{(\lfloor N/2\rfloor)}$.

The value of f in step (ii) is chosen so that the ratio of the number of retained rounds with values above the desired 9th decile $b(= f(N - B))$, to the number of values above the median but below $b(= (B - \lfloor N/2\rfloor)$, is $0.1/0.4 = 1/4$. Or, $f(N - B)/(B - \lfloor N/2\rfloor) = 1/4$. In the third step the same number are deleted below the median as were deleted above the median to keep it at its current value.

The remaining number of simulations in the *purged database*, i.e., a database that remains after rejection sampling, is thus $N^* = N - 2(1 - f)(N - B)$. Denote distribution function of ξ in the purged database as $F^*(.)$. Any summary statistics from the purged database can be interpreted as being conditional on the assumptions made on the distribution of ξ.

In step (iii), it may be preferable to use systematic sampling (based on the ordered values) to delete the simulated values. (This reduces the role of chance fluctuations, but whether or not it introduces biases depends on the finer details of the user's views.) Systematic random sampling was discussed in Chapter 3, Section 6, and implementation details may be found in texts such as Cochran (1977, 265–266) and Kish (1965, 115–116). A simple version of nonrandom systematic sampling for the current application is the following: Since the fraction to be deleted is $g = (1 - f)(N - B)/\lfloor N/2\rfloor$, we may divide the ordered values into segments of length $\lfloor N/2\rfloor/g$, and delete the observation closest to the middle of the segment from each segment. Rounding to integers complicates nonrandom systematic deletions if the segments are small, but the methods of random systematic sampling with fractional intervals can easily avoid rounding to integers. In step (ii) one may also delete the fraction $1 - f$ systematically.

Example 7.1. Stochastic Forecast Database for Finland. Consider a stochastic forecast database of Finland generated by PEP. The number of simulation rounds was $N = 3{,}000$. Suppose we are interested in the level of fertility. The total fertility rate is not stored by the program, but we can reason as follows. Since fertility in ages under 18 and over 40 is low, let us take ξ = (the number of births)/(the female population in ages 18–40). This gives roughly the average fertility in those 23 ages, and an estimate of total fertility would be $23 \times \xi$. Consider a lead time of 35 years. The median of the simulated values is $\xi_{1500} = 0.0761$ and the 90th percentile is $a = \xi_{2700} = 0.1052$. These would correspond roughly to total fertility rates of 1.75 and 2.42, respectively. The distribution is skewed to the right, and suppose we would like to see the consequence of having the predictive distribution with the 9th decile at 2.20, instead. This corresponds to $b = 2.20/23 = 0.0956522$. The closest smaller value is $\xi_{2434} = 0.095631$. Thus $B = 2434$, so we retain the $2434 - 1500 = 934$ observations that are above the median and below b. These represent 40% of the probability mass (the paths between the median and the 90th percentile), so

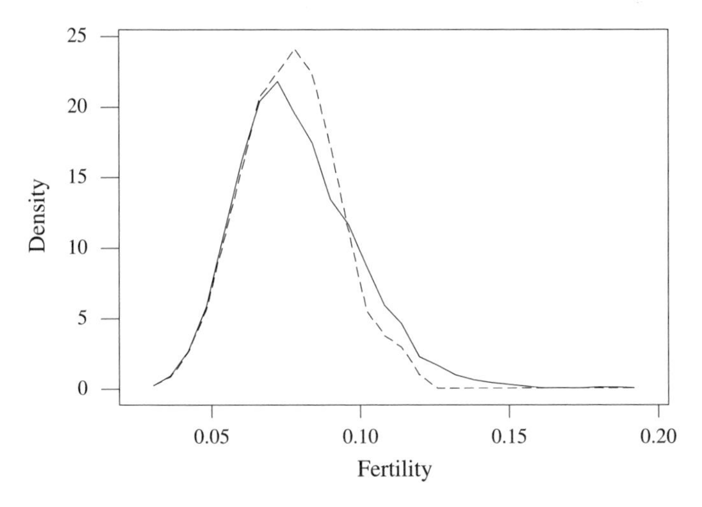

FIGURE 1. Predictive Distribution of a Fertility Measure (Solid) and its Modified Distribution (Dashed).

$N^* = 934/0.40 = 2335$ will be the size of the purged database. Consider step (iii). Below the median a we will retain $\lfloor 2335/2 \rfloor = 1162$ of the original 1500 observations. The deletion of $1500 - 1162 = 338$ observations can be achieved by first deleting, 148 times, every 5th observation from the median down (i.e., 1496, 1491, 1486, ..., 761), and then, 190 times, every 4th observation (i.e., 757, 753, 749, ..., 1), for example. This choice enhances smoothness near b slightly. (Alternatively, using methodology of random systematic sampling we could have randomly chosen a number r between 0 and $k = 1500/338$ and deleted observations $\lfloor r + j \times k \rfloor$, $j = 0, \ldots, 337$ counting from the median and moving down). Consider step (ii). Above b we originally have $3000 - 2434 = 566$ observations, but only $2335 - 1162 - 934 = 239$ can remain, with $566 - 239 = 327$ deleted. This can be achieved by deleting the top 88 observations (for smoothness near b), and then every second observation 239 times. The result can be seen in Figure 1. As an alternative to general fertility, one could use the ratio of observed births to births expected under the point forecast of fertility and observed numbers of women in childbearing ages (cf., (3.6) of Chapter 5). $\Diamond$

If the fractions to be deleted are large, then the method may lead to notable lack of smoothness in $F^*(.)$ at the median and at b. Note that the method as described applies to modify the predictive distribution at one point, but the method may be repeatedly applied to yield modifications at multiple points. More elaborate steps of deletion can be devised by defining a probability of deletion for every value $i = 1, \ldots, N$. This is equivalent to the rejection sampling described above. Using those techniques we may smooth the effect of deletions near the median and b. Depending on the details, this may require a larger number of deletions than the method given above.

Above, we have purged the database according to a fertility related criterion. As long as fertility and mortality forecast errors are independent in the model generating the sample paths, purging will not create dependencies. The situation would be different if fertility and mortality would have been originally somehow dependent. It follows that we may carry out a subsequent purge based on a mortality defined characteristic, such as life expectancy. To compensate for the expected purges, we may use a larger number N of simulation rounds initially.

7.2. Creating Correlated Populations

A full *multistate* representation of a population system becomes quite complex in the stochastic framework. The complexity is caused by the vast number of second order characteristics of the already large numbers of migration (or transition) flows. The difficulties are not connected so much with the technicalities of programming, but rather with the finding parsimonious representations for the covariances and estimating them from the available data. Post processing provides a way to handle the cross-region-correlations, if a single region forecast database is available for each state. We consider two general approaches. One depends on a judicious choice of seeds in random number generation. The latter is a more genuine post-processing technique that depends on sorting the output. We concentrate on the correlations of fertility and mortality, but note that migration can, in principle, be added as a third criterion. However, in practice we have handled migration via other means in multistate settings.

7.2.1. Use of Seeds

Suppose there are J regions. As before, this could be a set of countries (say, the countries of EU), a set of race categories (say, the main racial groups of the U.S.), etc. We are first interested in creating a given level $0 < \rho < 1$ of constant correlation between any pair of the regions, in terms of one characteristic ξ, say life expectancy. For definiteness, let ξ_j be the life expectancy at birth of region $j = 1, \ldots, J$, during a given future year.

Suppose the forecasts of the different regions are exactly equivalent in terms of the number of age-groups, child-bearing ages, and forecast years. Assume that the correlations across age for the errors of the vital processes are the *same* (or at least similar) in all countries. Suppose the generator used to produce the random numbers always produces the same sequence for the same seed. This is true for the congruential generator (Ripley 1987, 20–26) used in PEP, for example. This creates a nearly perfect correlation across countries for the ξ_j's, except that, as in the case of life expectancy, both linear and nonlinear transformations may reduce the correlation to some extent. Suppose too that the desired ρ does not exceed the maximal correlation that can be achieved in this way, say, ρ^*.[6] Let the desired number of simulation rounds be N. Then, the constant correlation ρ can be achieved approximately as follows:

(i) generate $\lfloor N\rho/\rho^* \rfloor$ simulation rounds from the same seed; the internal correlation of these values is (asymptotically) ρ^*;

[6] In the case of life expectancy, values $\rho^* > 0.99$ have been observed in simulation when the same correlation structure was applied within two countries.

(ii) generate an additional number $N - \lfloor N\rho/\rho^* \rfloor$ simulation rounds from randomly chosen seeds; the internal correlation of these values is (asymptotically) 0.

The method works, because both steps produce observations with the same mean and variance, and steps (i) and (ii) cause the fraction ρ/ρ^* of the data to have correlation ρ^* and the fraction $1 - \rho/\rho^*$ to have correlation 0 for any pair of the regions. (Should the countries have highly disparate vital rates, then the nonlinearities induced might make pairwise correlations across countries notably different. In that case the extension immediately following may be of help.)

The simple approach may serve as a building block for more complex, partially hierarchical patterns. For example, suppose there is a subset J' of the regions, for which a higher constant correlation ρ' such that $\rho < \rho' < \rho^*$ is desired. Then, step (ii) would be replaced by the following:

(ii′) generate an additional number $\lfloor N\rho'/\rho^* \rfloor - \lfloor N\rho/\rho^* \rfloor$ simulation rounds from the same seed for $j \in J'$; generate an additional number $\lfloor N\rho'/\rho^* \rfloor - \lfloor N\rho/\rho^* \rfloor$ simulation rounds from randomly chosen seeds for $j \notin J'$;

(iii′) generate an additional number $N - \lfloor N\rho'/\rho^* \rfloor$ simulation rounds from randomly chosen seeds for all j.

The approach outlined is not always as useful as might appear initially, because it creates a corresponding correlation for fertility (and possibly for net-migration), as well. Thus, its use is limited to situations in which one is willing to make broad assumptions about the cross-country correlations of the vital processes in general. Recall, however, that correlation here refers to deviations from the point forecast rather than deviations from the jump-off value (cf., Section 5 of Chapter 8), so the use of the method does not imply that the countries need to have the same overall trends.

7.2.2. Sorting Techniques

We continue in the J region setting of Section 7.2.1. and use life expectancy as the variable ξ_j that we would like to make correlated across regions. Let us assume now that the country-specific simulations are independent (within the practical limits imposed by the selected random number generator). Suppose the goal is to create a constant correlation $0 < \rho < \rho^*$ between any pair of countries, based on N simulation rounds, where ρ^* is the average maximal cross-country correlation that can be achieved by sorting the data from all countries into an increasing sequence. A procedure that accomplishes the task is as follows:

(i) select the number $\lfloor N\rho/\rho^* \rfloor$ of simulation rounds systematically from the ordered data; sort these values into an increasing sequence for each region j; the internal correlation of these values is (asymptotically) ρ^*;

(ii) sort the remaining $N - \lfloor N\rho/\rho^* \rfloor$ simulation rounds into a random order, independently for each country; the internal correlation of these values is (asymptotically) 0.

Replacing the original simulation round labels by the re-ordered labels causes the simulation results of all countries to become correlated in the desired way. This method works for essentially the same reason as the method (i)–(ii) of Section 7.2.1.

Consider a partially hierarchical pattern, in which there is a subset J' of the regions, for which a higher constant correlation $\rho < \rho' < \rho^*$ is desired, with the remaining regions $\{1, \ldots, J\} \backslash J'$ having correlation ρ. Then, step (ii) would be replaced by the following:

(ii′) select an additional number $\lfloor N\rho'/\rho^* \rfloor - \lfloor N\rho/\rho^* \rfloor$ of simulation rounds systematically for $j \in J'$ and sort them into an increasing sequence; select an additional number $\lfloor N\rho'/\rho^* \rfloor - \lfloor N\rho/\rho^* \rfloor$ simulation rounds for $j \notin J'$ and sort them into a random order, independently for each country;

(iii′) sort the remaining $N - \lfloor N\rho'/\rho^* \rfloor$ simulation rounds into a random order for all j.

The sorting method differs from the use of seeds in that sorting depends on the variable of interest only. In our illustration sorting is by life expectancy, for example. Since fertility is independent of mortality within each country, the fertility data will remain uncorrelated across countries, even after sorting. Thus, the set of those simulation rounds whose order has been randomized, can be further sorted by fertility, if desired. In the method (i)–(ii) the remaining fraction is ρ/ρ^*. In the case (i), (ii′), (iii′) it is ρ'/ρ^* for $j \in J'$ and ρ/ρ^* for $j \notin J'$. Having to limit the second round of sorted values to a fraction of the total means that there is an upper bound one can achieve during the second sorting. This is where the ability to use seeds comes into play. Using sorting in combination with the choice of seeds allows us to achieve a wider range of positive correlations for both mortality and fertility.

Exercises and Complements (*)

1. Referring to Section 1, if Törnqvist had used a random walk model with volatility 0.06^2 (as in Chapter 8), would his 80% interval for the short-term forecast have included the realized value of 2.98?
2. Multiplying (2.2) and (2.3) show the exponent of (2.4) is of the form given.
3. Show that if (2.5), (2.6), and (2.7) hold, then (2.4) must be of the form given, and vice versa. (Hint: for the multivariate normal $\mathbf{Y} \sim N(\boldsymbol{\mu}, \boldsymbol{\Sigma})$ the exponent is of the form $-(\mathbf{Y} - \boldsymbol{\mu})^T \boldsymbol{\Sigma}^{-1}(\mathbf{Y} - \boldsymbol{\mu})/2 = -\mathbf{Y}^T \boldsymbol{\Sigma}^{-1}\mathbf{Y}/2 + \mathbf{Y}^T \boldsymbol{\Sigma}^{-1}\boldsymbol{\mu} - \boldsymbol{\mu}^T \boldsymbol{\Sigma}^{-1}\boldsymbol{\mu}/2$.)
4. Show that one can view (2.7) as a weighted average of $\hat{\boldsymbol{\beta}}$ and $\mathbf{b}$, with "weights" $\mathbf{M}(\mathbf{X}_1^T \boldsymbol{\Sigma}_{11}^{-1} \mathbf{X}_1)$ and $\mathbf{M}\mathbf{S}^{-1}$, respectively.
5. Show that (2.9) holds.
6. Show that $S^2 = (1 - \bar{\rho})S^2_{\text{ind}} + \bar{\rho} S^2_{\text{dep}}$.
7. Starting from $S^2 = S^2_{\text{ind}} + \bar{\rho}(S^2_{\text{dep}} - S^2_{\text{ind}})$, show that $S^2 \le S^2_{\text{dep}}$ for all $\bar{\rho}$.

8. Show that $S^2_{\text{dep}} \approx (4/9)((n + 1/2)^{3/2} - (1/2)^{3/2})^2$, when $s_i = i^{1/2}$. (Hint: replace the sums by integrals from 1/2 to $n + 1/2$.)
9. Show that $S^2 = (2\rho^{n+1} - n\rho^2 - 2\rho + n)s^2/(1 - \rho)^2$ in Example 4.1. (Hint: (a) $1 + 2 + \cdots n = n(n + 1)/2$; (b) define $f(x) = 1 + x + x^2 + \cdots + x^{n-1} = (1 - x^n)/(1 - x)$, so $1 + 2x + 3x^2 + \cdots + (n - 1)x^{n-2} = f'(x)$.)
10. Show that $S^2 = \{n(n + 1)(1 - \rho^2)/2 - 2n\rho + 2\rho^2(1 - \rho^{n-1})/(1 - \rho)\}s^2/(1 - \rho)^2$ in Example 4.2. (Hint: use the result of Exercise 9.)
11. Prove (4.7). (Hint: omitting reference to j, note that under (4.6) we have $\text{Var}(X(t)) = \sigma(t)^2 = \kappa s(t)^2 + (1 - \kappa)(S(1)^2 + \cdots + S(t)^2)$, where $s(t) = S(1) + \cdots + S(t)$. Starting from this, derive the equation $S(t)^2 + 2\kappa s(t - 1)S(t) + \sigma(t - 1)^2 = \sigma(t)^2$. Verify that the solution of this is (4.7).)
12. Show that in Example 5.1, $\mathbf{H}$ is an $(\omega + 1) \times (\beta - \alpha + \omega + 2)$ matrix with the following structure. Its first row is $(V(\alpha, t), \ldots, V(\beta, t), 0, \ldots, 0)$. The remaining rows have zeros in the first $\beta - \alpha + 1$ locations. The last column is zero except it has $V(\omega, t)$ as the last element. The remaining $\omega \times \omega$ block is a diagonal matrix with $V(i - 1, t)$ as the ith diagonal element.
13. Note that the function $f(x_1, \ldots, x_n) = \log(\Sigma_i \exp(x_i))$ has the partial derivatives $\partial f/\partial x_j = \exp(x_j)/\Sigma_i \exp(x_i)$, $j = 1, \ldots, n$. Using this, derive (5.2) and (5.3).
14. (a) Derive (5.7). (b) Derive (5.9). (c) Show that (5.10) holds.
15. Consider a simple case of (5.4), where $f(x, y) = ye^x$; $Y_0 = 1$, and $X_t = \hat{X}_t + \varepsilon_t$, where $\varepsilon_t \sim N(0, \sigma^2)$ are i.i.d. What is the bias of the forecast relative to the true mean as a function of t and σ?
16. Show that the methods (i)–(ii) of Sections 7.2.1 and 7.2.2 produce (approximately) the desired correlation.

10
Errors in Census Numbers

1. Introduction

Jump-off population numbers typically are based on population registers if available or on census numbers. Jump-off populations serve not only as the base for forecasts, but also for the denominators of rates whose numerators are based on vital registration, such as mortality rates, fertility rates, marriage rates, etc. Jump-off errors affect vital rates, including mortality rates (Section 2.1), as well as forecasts (Section 2.2). Jump-off populations are also used in evaluation of past forecasts or postcensal estimates, and when the evaluations compare the forecasts to incorrect benchmarks (i.e., census or register results containing error), the evaluations can be biased (Section 2.3).

In this chapter we consider the estimation of accuracy in jump-off populations that are based on censuses. *Gross undercount* occurs when people are missed entirely or not counted in their appropriate geographic area or as members of their demographic subgroups (e.g., age groups, racial or ethnic groups, etc). *Gross overcount* occurs when people are enumerated multiple times, are enumerated where they should not have been, or are enumerated when they should not be. The difference between gross undercount and gross overcount is *net undercount*. Two main sources of information concerning undercount and overcount are demographic analysis (Section 3) and samples (Section 4); Anderson and Fienberg (1999). The information is used to estimate the net error in the census counts for various geographic and demographic groups and to adjust the census counts. Whether adjustment improves accuracy depends both on the accuracy of the adjustments as well as on the accuracy of the census counts (Mulry and Spencer 1993, Brown et al. 1999b, and Anderson et al. 2000).

A *postcensal estimate* refers to a time point since the last census but, unlike a forecast, the time point is in the past. An *intercensal estimate* refers to a time point in between two censuses. Postcensal estimates of population can be interpreted as the sum of the latest census number plus an estimate of change since the census. At the national level, the estimate of change may be explicitly estimated as births − deaths + net migration. At the subnational level, a variety of methods are used. For example, the *housing unit method* multiplies an estimate of the current number of housing units by an estimate of the average number of persons per housing unit to estimate the population total (e.g., Smith and Mandell 1984). A variety of regression methods are in use to estimate postcensal change in a local area's share of national (or other larger aggregate) population by a function of changes in administrative data, such as school enrollments, births, automobile registrations, etc. Methods for allowing the model to change over time have been developed using data from ongoing large scale sample surveys, where available (Ericksen 1974, Bryan 2004). The form of the postcensal estimators is such that errors in the last census numbers directly affect the postcensal estimates (Panel on Small Area Estimates of Population and Income 1980).

The accuracy of postcensal estimates typically is evaluated by comparison to census numbers for the same time points, or to intercensal estimates for a time point close to the census (e.g., Moltchanov, Kuulasmaa, and Torppa 1999). The errors in the consecutive census numbers may affect both the postcensal estimates and the difference between the postcensal estimate and the concurrent census number, as in the case of forecasts discussed in Sections 3 and 4, below.

The difficulty in assessing the accuracy of postcensal estimates is that a census itself must be used as the standard. As we know, censuses have their own errors, and the analyses of accuracy of postcensal estimates have tended to ignore the errors in the census. Furthermore, national censuses are taken only infrequently, limiting the information about accuracy for alternative postcensal intervals. (Complement 1.) In the U.S., special subnational censuses are occasionally conducted for specific cities or areas, but only when requested and paid for by the localities. Since local areas will request a special census only when they think they have something to gain, e.g., when the area is growing rapidly or the official postcensal estimate is believed to be too low, the evaluations of postcensal estimates based on special censuses may understate accuracy.

2. Effects of Errors on Estimates and Forecasts

2.1. *Effects on Mortality Rates*

Consider an estimated one-year mortality rate of the form (1.3) in Chapter 4. Let n be the actual number of persons in the population to begin with, let m denote the number of deaths, let K be the person-years lived in the year by the m persons who die during the year, and for simplicity ignore migration or births. If we knew n, we would calculate the rate

$$\hat{\mu} = \frac{m}{K + n - m}. \tag{2.1}$$

If errors from census undercount or postcensal estimation cause us to estimate n by $\tilde{n} = (1 - \lambda)n$, we would instead calculate $\tilde{\mu} = m/(K + \tilde{n} - m)$. Notice that

$\tilde{\mu} - \hat{\mu}$ has the same sign as λ. To a first approximation, the relative error in $\tilde{\mu}$ compared to $\hat{\mu}$ is $\lambda n/(K + n - m)$, and this exceeds $|\lambda|$ in absolute value (Exercise 2).

Error in the past population can also have a second-order effect on postcensal population estimates that use mortality rates in adjusting for population change (Exercise 3). If there is net undercount, or $\lambda > 0$, not only is the base too low but the estimated mortality rate is too high. Thus, the estimated probability of survival is too low, and the effects of the two estimates are compounded, making the estimate of the surviving population too small. This bias would tend to be present in ages where mortality is an important component of population change, and it could be dominated by other factors for other ages.

2.2. *Effects on Forecasts*

Errors in jump-off populations directly affect population forecasts. Consider the linear growth model (2.7) of Chapter 6. The population vector is $\mathbf{V}(T) = \mathbf{R}_T\mathbf{V}(0)$, where we have written $\mathbf{R}_T = \mathbf{R}(T-1)\mathbf{R}(T-2)\cdots\mathbf{R}(0)$, for short. The effect of multiplying the jump-off population vector by a disturbance matrix $\mathbf{A}$ is to change the representation to $\mathbf{V}(T) = \mathbf{R}_T\mathbf{A}\mathbf{V}(0)$, and hence to change the future population by $\mathbf{R}_T(\mathbf{I} - \mathbf{A})\mathbf{V}(0)$. The matrix $\mathbf{A}$ need not be diagonal, for example, age misreporting could move counts of people from one age to another. In the special case where $\mathbf{A} = (1 - \lambda)\mathbf{I}$, the effect is to reduce the forecasted population by the proportion λ.

Errors in past populations also affect the forecasts through the errors in past vital rates, which may lead to biases in forecasts of future vital rates. The effect of errors in past vital rates on forecasts of vital rates may be complex, but generally if a time series of age-specific vital rates is uniformly off by a factor, the forecasts of rates will be off by the same factor.

2.3. *Effects on Evaluation of Past Population Forecasts*

Errors in benchmarks adversely affect estimates of accuracy of past population forecasts. Let $\varepsilon(t)$ denote the actual error in the forecast for time t and let $\beta(t)$ denote the error in the benchmark for t. The natural estimate of $\varepsilon(t)$ is $\hat{\varepsilon}(t)$ defined as forecast minus benchmark. To see the effect of error in the benchmark, observe that $\hat{\varepsilon}(t) = \varepsilon(t) - \beta(t)$. A similar effect holds for estimates of error in the forecast of change since jump-off time 0, say $\Delta(t)$. The error typically is estimated by the difference between the forecasted change and the change in the benchmarks, say $\hat{\Delta}(t)$. Observe that $\hat{\Delta}(t) = \Delta(t) + \beta(t) - \beta(0)$, so that the estimate of $\Delta(t)$ is confounded by change in the errors in the benchmark. The effect of errors in the benchmarks can be nontrivial in some situations, such as the evaluation of the U.N. forecasts that was considered in Section 4.3 of Chapter 8. (Also see Complement 4.)

3. Use of Demographic Analysis to Assess Error in U.S. Censuses

Information about the national extent of net undercount by age, race, and sex is provided by the method of demographic analysis (DA), which uses administrative and vital statistics data to construct an alternative estimate of population. In the United States, selective service registration in October 1940 included more men of draft age (21-35) than did the 1940 census, and the discrepancy varied by region and race (Price 1947, Anderson and Fienberg 1999, 29). The first systematic analysis in the U.S. was the development by Coale (1955) of alternative estimates of the 1950 population by age, sex, and race (whites and nonwhites). Coale's methodology was complex, as it had to cope with incompleteness in birth and death registrations, especially for the older population. Subtracting the census count from Coale's numbers yields estimates of net undercount in the 1950 census by age and race. For the 1990 census, Robinson et al. (1993) calculated births − deaths + immigration − emigration by age, race, and sex using data since 1935 to estimate the population under age 55 in 1990. Administrative data (aggregate Medicare enrollments, adjusted by estimates for incompleteness of enrollment) were used to estimate the population 65 and over. To account for incompleteness in earlier vital statistics, more complex methods were used to estimate the population aged at least 55 and under 65, although this extra step was not needed for the 2000 census.

The main weakness of DA in many applications is the inaccuracy of emigration and, especially, immigration numbers. Immigration is the sum of authorized immigration and unauthorized immigration. The Census Bureau estimates numbers of unauthorized migrants by comparing statistics on legal immigration with numbers of foreign born reported to the census, interpreting an excess of the latter as an approximation to the extent of unauthorized immigration (Citro, Cork, and Norwood 2004, 165). When a small quantity is estimated by the difference between two large quantities, even small relative errors in the latter can cause large error in the estimate. For example, preliminary DA estimates implied a net overcount of 0.4% in the 1980 census (ESCAP 2001, 18; Passel, Siegel, and Robinson 1982) but when later revised for undocumented immigration, they showed a net undercount of 1.2% (Robinson et al. 1993). Redfern (2001) has developed a variant of DA which avoids relying on migration data. In his application to the 1991 census of England and Wales, he estimates the number of emigrants by numbers of persons recorded in censuses of other countries as having been born in England and Wales, and he estimates the number of immigrants from the England and Wales census. The method is subject to a variety of errors, whose effect he estimates in a simulation study.

An important contribution from DA is the provision of sex ratios (numbers of males over numbers of females) by age and race. The DA sex ratios are believed to be more accurate than the DA numbers of persons (because relative errors in numbers of males and females are believed highly positively correlated) and the ratios provide an important tool for assessing the accuracy of the census and alternative estimates of population. In the U.S. sex ratios have tended to be lower in the census than in DA for persons 20 and over, particularly for blacks (Bell 1993, 1110, Table 2; Robinson 2001). This indicates a larger net undercount for males than for females (cf., Example 2.2 of Chapter 2). The sex ratios from the census for blacks aged 20–29, 30–44, and 45–64 were only about 90 percent as large as the ratios from DA. DA estimates are only available at the national level, unless very strong assumptions are invoked to account for internal migration (as in Siegel et al. 1977). We regard the accuracy of DA estimates as unknown territory, although some beginnings of assessment have been made by Robinson et al. (1993), Clogg and Himes (1993), Passel (1993), and Fay (1974).

4. Assessment of Dual System Estimates of Population Size

In a census, an enumeration is correct if it is not a duplicate enumeration, if it refers to an actual person who was in-scope (met the inclusion criteria for the census), and if it classified the person in the correct geographic area. An enumeration is erroneous if it is not a correct enumeration. An individual person may either be enumerated or unenumerated (missed) by the census.[1]

Information for estimating gross overcount (the number of erroneous enumerations) and, to a rough approximation, gross undercount (the number of persons unenumerated) and for obtaining geographic breakdowns is also provided by a *post-enumeration survey* or *PES*. In the U.S. in 1990 a PES was taken (Hogan 1993) and subjected to a careful though controversial analysis of error; Mulry and Spencer (1993), Anderson and Fienberg (1999), Brown et al. (1999b), Anderson et al. (2000), and references therein. A PES was taken again in the U.S. in 2000 (U.S. Census Bureau 2001). We will provide here a schematic discussion of the PES and the evaluation of accuracy. Although detailed discussion would occupy a separate book, a schematic analysis at an intermediate detail is of interest. We include it here for two reasons. First, it is similar in spirit to the study of propagation of error in forecasts, and is a natural component of statistical demography. Second, the assessment of jump-off error and error in benchmarks is important for evaluation of forecast accuracy.

Example 4.1. Post Enumeration Surveys in the 1990 and 2000 U.S. Censuses. The PES consists of an E sample, used primarily to estimate erroneous enumerations, and a P sample, used to estimate census omissions. The E sample is a sample of census enumerations and the P sample is like a second census conducted on a sample of blocks, independently of the original census. The 1990 U.S. PES was based on a stratified random sample of 5,290 "block clusters". (A block is

[1] The "out-of-scope" enumeration category includes people who according to residency rules should not be counted and fictitious enumerations, which are essentially fabrications by enumerators.

a connected space enclosed by streets or natural boundaries, e.g., a city block. A block cluster is one or more contiguous blocks.) All the housing units in the block clusters were to be included, except for the largest block clusters, which were divided into contiguous groups of housing units and were subsampled. In all, 165,000 housing units were selected. The 1990 PES sample was restricted to the noninstitutional population, and excluded people living in jails, nursing homes, and other institutions, military personnel living in barracks or on ships, and homeless people (Hogan 1993). The PES for 2000 was called the Accuracy and Coverage Evaluation (A.C.E.) and it was based on a larger and more efficient sample. A stratified sample of slightly more than 300,000 housing units was selected from a two-phase stratified sample of more than 11,300 block clusters (Sands and Navarro 2001). The universe for the 2000 A.C.E. included only the population living in housing units, unlike the 1990 PES which included noninstitutional group quarters (e.g., college dormitories).

The *E sample* consists of census enumerations within the sampled block clusters (or subsamples of the enumerations in the larger blocks), although certain enumerations are excluded. Excluded are enumerations whose characteristics were wholly imputed ("whole person imputations" or "WPI" cases) and enumerations that were processed too late for inclusion in the E sample ("late additions" or "LA" cases). Out of the 273.6 million enumerations in the household population in the 2000 census, there were 5.8 million people WPIs and 2.4 million LAs; in 1990 the latter two numbers were about 1.9 million and 0.3 million, respectively (Citro, Cork, and Norwood 2004, 161–162). All enumerations in the E sample were checked to see whether they are erroneous or not. In checking whether E-sample enumerations were erroneous or correct, the Census Bureau attempted to match every E-sample enumeration to the P-sample interview data for cases within the search area. If a match was made, the person was considered a correct enumeration. Some enumerations, called "KE" cases, had too little information to permit a match, i.e., the person's name was missing or incomplete or the person had a valid name but fewer than two other characteristics (on a specified list) reported. If a match was not made, a follow-up interview was conducted to obtain additional information to determine enumeration status (correct or erroneous); if the follow-up interview could not be conducted, interviewers attempted to collect information from neighbors.

The ratio of erroneous enumerations to total number of enumerations in the E sample is only a very approximate estimate of the erroneous enumeration rate for the E sample. The problem is that an estimated 4.8 million enumerations were KE cases (Feldpausch 2002, 4) and that although they were all classified as erroneous, some if not the vast majority of the KE cases were really valid enumerations. That issue aside, the enumeration rate for the E sample is still not the same as the erroneous enumeration rate for the general population (noninstitutional for 1990 and household in 2000) because the WPI and LA cases are not included, and their fraction erroneous could well be larger than the general population.

The *P-sample* survey was conducted in the sample blocks, independently of the census operations. To ensure the operational independence, a new listing of housing units was prepared for each P-sample block independently of the listing used in the census. Very large blocks were subsampled, as in the E sample, and all other blocks had 100 percent of their listed housing units included in the P sample. Interviews were to be conducted in every sampled housing unit to enumerate all persons living in the housing units at the time of the interview (typically within four months of the census) and to obtain sufficient information to ascertain whether they were enumerated in the census. Attempts were made to match each P-sample enumeration to a census enumeration. Aside from operational errors in assessing matches, a match will be made if the P-sample person was enumerated in the census with sufficient information to allow matching (i.e., neither a LA nor WPI nor KE case). Thus, the non-match rate estimates the fraction of people who were either omitted from the census, were added late, or were enumerated but with insufficient information for matching. Although such cases prevent accurate assessment of the gross numbers of erroneous enumerations and census omissions, they should in theory cancel out so that the net number of omissions can be ascertained.

The Census Bureau constructed a DSE separately for each poststratum in the PES (Chapter 3, Examples 3.3 and 7.2) as the product of three factors: the correct enumeration rate as estimated from the E sample, the reciprocal of the match rate in the P sample, and the census count excluding WPI and LA cases. For rationale, note that Census enumerations are of two kinds, (a) correct enumerations of people in the population, with sufficient information for matching to the P sample, and (b) all others (WPI cases, LA cases, KE cases, and other erroneous enumerations). Denote the size of the population by N and the number of type (a) enumerations by N_a. If we knew N_a and if we knew the ratio $\rho = N_a/N$, we could estimate N by N_a/ρ. If we estimate N_a by the number of non-WPI and non-LA enumerations $\times$ (1− erroneous-enumeration rate) from the E sample and estimate ρ by the match rate from the P sample, we get what is essentially the estimator used by the Census Bureau. ◊

Example 4.2. Post Enumeration Survey in the U.K. in 2001. A PES was also taken in the U.K. in 2001 (Brown et al. 1999a). In contrast to the U.S., the 2001 PES in the U.K. used a two-stage sample, with a stratified sample of Enumeration Districts (areas) and then a subsample of Postcodes (smaller areas) within each selected Enumeration District; all housing units within each selected Postcode were surveyed. ◊

We will emphasize the accounting of alternative sources of error rather than the empirical estimation of the moments of the various error components. We will also simplify in ignoring certain aspects whose complexity outweighs their importance, such as treatment of movers or the possibility of the so-called contamination bias (see Complement 16). For treatments that do not make these simplifications, see Mulry and Spencer (2001, 1993, 1991)[2] and Spencer (2000b).

---

[2] The quantitative estimates of error in Mulry and Spencer (2001) were incorrect due to data problems not understood at that time.

An alternative to the DSE/PES methodology is the *reverse record check* that has been used in Canada and the United Kingdom. The idea in the reverse record check is to partition the current population into four sampling frames: (i) enumerations in the previous census, (ii) registered births since the last census, (iii) immigrants since the last census, and (iv) persons missed in the previous census; Statistics Canada (1999), Fellegi (1980), Germain and Julien (1993), Burgess (1988). Frame (iv) is conceptual and practically may be constructed from the results of the reverse record check for the previous census. Estimation of the size of frame (i) requires tracing forward in time samples of enumerations from the previous census. Such tracing may be difficult, e.g., when the U.S. Census Bureau tested tracing methods in the 1980s, it found that the fraction it could not trace exceeded the net undercount rate (Hogan 1983; Mulry and Dajani 1989). A recent study suggested that use of administrative records could improve the tracing and hence that the reverse record check might be useful for the 2010 or 2020 censuses in the U.S. (Cork, Cohen, and King 2004, 203–204).

5. Decomposition of Error in the Dual System Estimator[3]

5.1. A Probability Model for the Census[4]

Consider a set S_0 that represents potential census enumerations for the population. Suppose a potential enumeration is either correct (C), erroneous (E), or unenumerated (U). For $x = E, C, U$, and $s \in S_0$, define indicators of enumeration status as $\delta_x(s) = 1$ if s has status x, and $\delta_x(s) = 0$ otherwise. Then, the true population is $S = \{s \in S_0 | \delta_C(s) + \delta_U(s) = 1\}$, and the correct population total is

$$N = \sum_{s \in S_0} [\delta_C(s) + \delta_U(s)]. \tag{5.1}$$

Define $S_e = \{s \in S_0 | \delta_C(s) + \delta_E(s) = 1\}$ as the set of actual census enumerations, with size

$$N_e = \sum_{s \in S_0} [\delta_C(s) + \delta_E(s)]. \tag{5.2}$$

It is useful to consider enumeration statuses as random variables, with probability functions $P(\delta_x(s) = 1) = p_x(s)$. Sometimes it is reasonable to consider the true population size fixed at N, in which case the probabilities $p_x(s)$ are conditional on N (and perhaps on subgroup sizes as well). Sometimes it is reasonable

[3] Sections 5 and 6 are technical and may be skipped without loss of continuity.

[4] This section uses a model developed in Haberman, Jiang, and Spencer (1998).

to condition on the attained census count, N_e (and on counts for subgroups). Define

$$A(s) = \frac{p_C(s) + p_U(s)}{p_C(s) + p_E(s)}. \tag{5.3}$$

If $p_x(s)$ is known for each s in S_e, then N may be estimated by (cf., Complements 5 and 6)

$$\sum_{s \in S_e} A(s). \tag{5.4}$$

This is a random variable, since the S_e is a random subset of S_0.

Example 5.1. Artificial Example of Probability Model for a Census. Suppose that a census is expected to include 500 enumerations, of which 400 are people who were enumerated in the correct block as classified by the census, 60 are people who were not enumerated in the correct block (i.e., outside the E-sample search area), 25 are duplicate enumerations of people, and 15 are out-of-scope enumerations. In addition, suppose there are expected to be 50 people who are not enumerated at all by the census. The true population thus includes 510 people, with 400 expected correct enumerations, 50 expected misses, and 60 expected erroneous enumerations due to incorrect classification of area. The expected number of erroneous enumerations is 100: 15 fictitious, 60 enumerations in the wrong block, and 25 duplicates. A variety of probability models may be constructed. For a simple example, take the number of potential enumerations, N_0, to be 610 and set $p_U(s) = 110/610$, $p_C(s) = 400/610$, and $p_E(s) = 100/610$ for each s in S_0. From (5.3), we obtain a constant value of $A(s) = (510/610)/(500/610) = 510/500$. To calculate the expected value of (5.4), multiply $A(s)$ times the expected number of enumerations and obtain 510, the true size of the population. (Complement 7) ◊

Our probability model specifies marginal probabilities for each potential enumeration. To specify joint probabilities, a dependence structure is needed. Census misses may fail to be independent in a variety of ways. For example, an individual could be missed if his or her household were missed or if the household was not missed but the individual was not reported. Clustering of individuals within households thus leads to a (positive) dependence in whether or not they are missed (Cowan and Malec 1986). Similarly, effects of individual enumerators could lead to a positive dependence for individuals in an enumerator's caseload being missed or being erroneously enumerated, postal delivery mixups could lead to positive dependence for events for individuals on the same postal route, etc.

5.2. *Poststratification*

In practice, we use modeling to estimate the probabilities $p_E(s)$, $p_C(s)$, and $p_U(s)$ and the resulting $A(s)$. We treat individuals with similar recorded characteristics as having similar probabilities of being missed, and enumerations with similar recorded characteristics as having similar probabilities of being erroneous.

The U.S. Census Bureau estimates $A(s)$ by assuming they are constant within poststrata although it is known that this is only approximately true. P-sample persons are grouped into poststrata based on age, sex, race/ethnicity, type of metropolitan statistical area, mail return rate, owner or renter, and other variables. Within each poststratum the sample-weighted fraction of P-sample persons who were missed in the census is computed; that fraction is used for $\hat{p}_U(s)$ if person s belongs to the poststratum and if there are no WPI or LA enumerations. (Exercise 8 discusses $\hat{p}_U(s)$ if WPI enumerations are present.)

E-sample enumerations are grouped into poststrata and within each poststratum the estimated fraction of E-sample enumerations that were erroneous is computed; that fraction is used for $\hat{p}_E(s)$ if enumeration s belonged to the poststratum. Recall that by definition, we have that $\delta_C(s) + \delta_E(s) + \delta_U(s) = 1$. Therefore, we can estimate (5.3) by

$$\hat{A}(s) = \frac{1 - \hat{p}_E(s)}{1 - \hat{p}_U(s)}. \tag{5.5}$$

The poststratified DSE can be written in the form

$$\hat{N} = \sum_{s \in S_e} \hat{A}(s). \tag{5.6}$$

Exercise 9 shows that (5.6) may be computed from grouped data. Thus, defining C_h to be the number of census enumerations in poststratum h and defining $\hat{A}_h$ to equal $\hat{A}(s)$ if enumeration s belongs to poststratum h, we may write the poststratified DSE as

$$\hat{N} = \sum_h \hat{A}_h C_h. \tag{5.7}$$

Exercise 10 shows that $\hat{p}_U(s)$ and $\hat{p}_E(s)$ do not need to be based on the same sets of poststrata. Complement 11 discusses logistic regression models as an alternative to poststratification (see also Section 5 of Chapter 5).

5.3. *Overview of Error Components*

The estimates $\hat{A}$ are based on a sample of housing units in a sample of blocks and as such are subject to random error and bias. Random sampling error arises because only a sample is used to develop the estimates. Since the estimates $\hat{A}$ involve ratios of random variables (see, e.g., (5.5)), and the expectation of a ratio typically is not equal to the ratio of the expectations of the numerator and denominator, the estimates are subject to "ratio estimator bias" (see Exercise 14 and Section 2 of Chapter 3). Let $\hat{A}^*(s)$ denote the value of $\hat{A}(s)$ that we would have if there were no ratio-estimator bias and let $\hat{A}^{**}(s)$ denote the value of $\hat{A}(s)$ that we would have if there were no ratio-estimator bias or bias due to data collection or data processing errors. It follows that $\hat{A}(s) - A(s) = (\hat{A}(s) - \hat{A}^*(s)) + (\hat{A}^*(s) - \hat{A}^{**}(s)) + (\hat{A}^{**}(s) - A(s))$, where the last component reflects modeling error that arises from the use of simplified models (cf., Section 3.3 of Chapter 8).

Each potential enumeration in the population may be considered to be a member or nonmember of a subgroup, G, for example, persons in a given region who are male and aged 15–19. Errors can cancel across potential enumerations within any G, but biases may remain on the group level. Alternative additive decompositions are feasible, depending on how the interactions of the various types of errors are handled. We will present a simple analysis.

Consider the true group membership, and define $K_G(s) = 1$ if $s \in G$ and $K_G(s) = 0$ otherwise. An enumeration may be assigned to the correct group or not; define $\hat{K}_G(s) = 1$ if s is enumerated and recorded as belonging to group G and $\hat{K}_G(s) = 0$ otherwise. Consider a population subgroup G with size N_G,

$$N_G = \sum_{s \in S_0} (\delta_C(s) + \delta_U(s)) K_G(s). \tag{5.8}$$

In the following, we will condition on the actual population counts, so (5.8) is not random. The estimator (5.6) for N_G is

$$\hat{N}_G = \sum \delta_e(s) \hat{K}_G(s) \hat{A}(s), \tag{5.9}$$

where the summation is over all potential enumerations $s \in S_0$, and we have written $\delta_e(s) = \delta_E(s) + \delta_C(s)$, for short.

Let $E[.]$ denote expectation with respect to sampling as well as probabilistic aspects of behavior in the census and PES, and define $p_e(s) = E[\delta_e(s)]$ and $p_{e,G}(s) = E[\delta_e(s)\hat{K}_G(s)]$. We decompose estimation error first as

$$\hat{N}_G - N_G = \varepsilon_G + (E[\hat{N}_G] - N_G), \tag{5.10}$$

where

$$\varepsilon_G = \hat{N}_G - E[\hat{N}_G] \tag{5.11}$$

represents *random error* due to sampling and mis-classification. This may be reexpressed as

$$\varepsilon_G = \sum \{\delta_e(s)\hat{K}_G(s)\hat{A}(s) - p_{e,G}(s) E[\hat{A}(s) | \delta_{e,G}(s) = 1]\}. \tag{5.12}$$

Further decompositions of (5.12) can be made to see how randomness in enumeration and randomness in group classification influence the error. Continuing with the second term of (5.10) we define first the *ratio-estimator bias* as

$$b_{R,G} = \sum p_{e,G}(s) b_{R,G}(s) \tag{5.13}$$

with $b_{R,G}(s) = E[\hat{A}(s) - \hat{A}^*(s) | \delta_{e,G}(s) = 1]$. The *bias from data error* involving the estimation of $\hat{A}$ is defined as

$$b_{D,G} = \sum p_{e,G}(s) b_{D,G}(s), \tag{5.14}$$

with $b_{D,G}(s) = E[\hat{A}^*(s) - \hat{A}^{**}(s) | \delta_{e,g}(s) = 1]$. *Model bias* is defined as

$$b_{M,G} = \sum \{p_{e,G}(s) E\big[\hat{A}^{**}(s) | \delta_{e,G}(s) = 1\big] - K_G(s) A(s)\}. \tag{5.15}$$

This reflects both the use of simplified models and the use of biased classification procedures. Notice that $\hat{N}_G - N_G = \varepsilon_G + b_{R,G} + b_{D,G} + b_{M,G}$, so that we can decompose the error in the population estimator for a subgroup G as total error = random error + ratio-estimator bias + data error bias + model bias. If group G were the entire population, then $\hat{K}_G = K_G$, and the various component biases depend on the average biases in the estimates of $A(s)$ for $s \in S_e$. For more general G, the component errors also depend on the observed or true group classifications for the enumerations: sampling error and ratio-estimator bias depend on the actual classifications ($\hat{K}_G$), the data bias reflects misclassifications ($\hat{K}_G \neq K_G$), and the model bias depends on the true classifications (K_G).

A *total error model* is a decomposition of the total error into pieces or components that can be estimated or at least bounded. The decomposition is an algebraic identity, possibly derived under simplifying assumptions, and if the means and variances and covariances of the components can be estimated, the mean and variance of the total error can be approximated with some assumptions about the distributions of the components. Typically, normal distributions have been used for the total error of the DSE (Mulry and Spencer 1991, 1993). The accuracy of the estimates of mean and variance of total error depend on the accuracy of the estimates of the moments of the components.

The variance of the random error (5.11) can be estimated directly by jackknife replication methods (Chapter 3, Section 8), which also provide estimates of the first-order approximation to the ratio-estimator bias (5.13). Data error bias is discussed in Section 5.4, below, and model error is discussed in Section 5.5.

Example 5.2. Error Components in the 1990 PES. Mulry and Spencer (1993, 1082, Tables 1, 2) report estimates of relative error in the national population estimates in the U.S. based on the 1990 DSE, calculated under the assumption that no other errors were present. Ratio-estimator bias was estimated at 0.11 percent (of the true size of the population), the standard error was estimated at 0.19 percent, and the model bias was estimated at −0.29 percent, with a standard error of 0.09 percent. The net bias was estimated at 0.49 percent. The bias from data error was not directly reported, but may be inferred to be about 0.60 or 0.70 percent. The positive bias from data error is thus offset to some extent by downward correlation bias. ◊

Example 5.3. Error Components in the 2000 A.C.E.. The 2000 PES or A.C.E. for the U.S. was difficult to evaluate. The initial evaluation occurred in the first quarter of 2001 to support a decision about whether to adjust the census for undercount for the purposes of redrawing the boundaries of Congressional districts (Chapter 12, Section 4.4.2). Law required a decision some time in March, but the evaluations of data error were not then available. Estimates of data error were made under assumptions of similarity to those of the 1990 PES, with allowances for changes in scale, but later study showed the errors to be quite different. At the time, the Census Bureau (correctly) distrusted those assumptions because the A.C.E. results differed in unusual ways from census and demographic analysis results. For the March 2001 evaluation, ratio-estimator bias was estimated at 0.01 percent (of the true size of the population), the standard error was estimated at 0.13 percent, and the model bias was estimated at −0.39 percent (Mulry and Spencer 2001, 80). Later, additional evaluations based on matching E-sample records to the whole census and to administrative records indicated that the A.C.E. failed to detect millions of duplicate and other erroneous enumerations; the 12.5 million erroneous enumerations estimated from the A.C.E. in March 2001 was increased to 17.2 million in a later revised DSE (known as "A.C.E. Revision II"), and even that figure may be low (Citro, Cork, and Norwood 2004, 214–217). Correspondingly, although the DSE reported in March 2001 indicated a net undercount of 1.18 percent, A.C.E. Revision II indicated a net *over*count of 0.49 percent (Citro, Cork, and Norwood 2004, 229). In fact, the discrepancy was even greater than it seems because the revised DSE incorporated an upward adjustment for correlation bias but the earlier DSE did not. Without the correlation bias adjustment, the revised DSE would have been smaller and so would have indicated of a larger net overcount. ◊

5.4. Data Error Bias

To understand how data error and other nonsampling errors affect the DSE, we consider the representation of the post-stratified DSE in (5.7) and focus on the "adjustment factor" $\hat{A}_h$ for poststratum h. The DSE as computed by the U.S. Census Bureau is quite a complex statistic (U.S. Census Bureau 2004; Kostanich 2003a,b; Citro, Cork, and Norwood 2004), and for insight we will simplify in several ways (Complements 17, 18). Define the following quantities.

- $\hat{E}_{\mathrm{E},h}$ weighted number of erroneous enumerations in in poststratum h in the E sample;
- $\hat{N}_{\mathrm{E},h}$ weighted number of E-sample enumerations in poststratum h
- $N_{\mathrm{I},h}$ number of census enumerations in poststratum h that are whole person imputations or late adds
- C_h number of census enumerations in poststratum h as defined below (5.6)
- $\hat{N}_{\mathrm{P},h}$ sample estimate of number of people who were enumerated in the P-sample in poststratum h
- $\hat{N}_{\mathrm{CP},h}$ weighted number of P-sample matches in poststratum h to the census

The E sample excludes whole person imputations and census enumerations that occurred too late for processing in the PES; thus, $N_{\mathrm{I},h}$ equals the expected value of $C_h - \hat{N}_{\mathrm{E},h}$. Note that $\hat{N}_{\mathrm{P},h}$ refers to people rather than enumerations, and excludes estimated P-sample fabrications and adjusts for missing data in the P sample. For s in poststratum h, estimate $p_E(s)$ by $\hat{E}_{\mathrm{E},h}/\hat{N}_{\mathrm{E},h}$, and $p_C(s) + p_E(s)$ by $\hat{N}_{\mathrm{CP},h}/\hat{N}_{\mathrm{P},h}$. This yields an adjustment factor of the form

$$\hat{A}_h = (1 - \hat{E}_{\mathrm{E},h}/\hat{N}_{\mathrm{E},h})\hat{N}_{\mathrm{P},h}/\hat{N}_{\mathrm{CP},h} \tag{5.16}$$

Nonsampling errors influence these estimates in several ways.

Define $n_{\mathrm{EE},h}$ as the nonsampling error in estimate of number of erroneous enumerations in the census that were included in the E sample in poststratum h. The error $n_{\mathrm{EE},h}$ arises during the data collection and processing of the E sample if respondents are misclassified as to whether they are correctly or erroneously enumerated in the original enumeration. Some cases are immediately classifiable but others, the "unresolved", are assigned an imputed probability being an erroneous enumeration.

Define $n_{\mathrm{P},h}$ as the nonsampling error in the estimated size of the P-sample population in poststratum h. The error arises from whole-household and partial-household fabrications in the P sample, imperfect weighting adjustments for whole-household noninterviews and imperfect imputations for missing data, and misclassification of whether the individual was a resident on Census Day.

Next, define $n_{\mathrm{CP},h}$ to be the nonsampling error in the estimated number of matches between the P sample in poststratum h and the E sample. This error arises from several sources, including misreporting of data, clerical matching error, incorrect imputation for missing data, and imperfect weighting adjustments for household non-interviews.

An additional source of error is inconsistent assignment of postrata in the P sample and in the E sample. This error arises because the P-sample match rate, $\hat{N}_{\mathrm{CP},h}/\hat{N}_{\mathrm{P},h}$ is based on the P-sample data for assigning membership to poststratum h, but the DSE in (5.7) is based on assignment of postrata from the census, which is the same as the E sample. The theory for poststratification rests on an assumption that poststratum membership is defined independently of the sample selection or survey procedure. This assumption is violated, for example, if the poststratum to which a unit is assigned depends on the sample that was selected. Thus, if the E sample and P sample would classify the same person in different poststrata, perhaps as a result of interviewer error or reporting error, a bias may result. Preliminary analysis suggests that this may be a problem for the 2000 Census in the U.S. (Farber 2001; Haberman and Spencer 2001; Complement 19).

If there were no data error, the estimate of the adjustment factor would be $\tilde{A}_h = (1-(\hat{E}_{\mathrm{E},h}-n_{\mathrm{EE},h})/\hat{N}_{\mathrm{E},h})(\hat{N}_{\mathrm{P},h}-n_{\mathrm{P},h})/(\hat{N}_{\mathrm{CP},h}-n_{\mathrm{CP},h})$. Therefore, the net data error bias in $\hat{A}_h$ is $\hat{A}_h - \tilde{A}_h$.

Estimation of the expected values of the data errors is based on reinterview surveys, rematch studies, and other methods for redoing the post-enumeration survey with higher quality but smaller scale, and possibly on administrative data as well. The estimates of the data error biases for individual poststrata typically are imprecise and require some modeling adjustments to reduce variance. The followup studies for evaluation are themselves imperfect, for example, they may be conducted only after a significant lapse of time, and attempts to obtain complete reporting when data in the post-enumeration survey were missing may be largely unsuccessful.

5.5. *Decomposition of Model Bias*

5.5.1. *Synthetic Estimation Bias and Correlation Bias*

Estimators of $A(s)$ depend on at least two kinds of assumptions. One assumption concerns the dependence structure for captures. The usual poststratified DSE, such as (5.16), depends on assumptions of independence. Methods for allowing for certain kinds of dependence have been proposed, but they also depend on explicit or implicit assumptions about the dependence structure (Wolter 1990). Generalizing the usual terminology, we refer to any bias arising from failure of the assumed dependence structure as correlation bias. A second assumption is used when we aggregate data across areas or other domains and base the estimate for a particular area on data from other areas as well. The method of synthetic estimation assigns to smaller groups the same parameter values held by the larger groups to which they belong. Some methods, such as empirical Bayes methods, generalize the assumption of equality of parameters to an assumption that the parameter values are drawn from a distribution that is structurally the same for all subgroups (Rao 2003; Ghosh and Rao 1994). The effect of failure of such assumptions is a form of aggregation bias, but in keeping with the simplest form of modeling, we refer to it as synthetic estimation bias.

To describe the components of model bias more formally, let $\hat{A}^{***}(s)$ denote the value of $\hat{A}^{**}(s)$ if, in addition to lack of data bias and ratio estimator bias, there were no bias due to incorrect allowance for the dependence structure. Let $\hat{A}^{**}_{|G}(s)$ denote the value of $\hat{A}^{***}(s)$ if the estimation were based solely on data for group G. Further details about $\hat{A}^{***}$ and $\hat{A}^{**}_{|G}$ are provided in Section 5.5.2 for the simple poststratified DSE. Define the *correlation bias for subgroup G* as

$$b_{\text{corr},G} = \sum p_{e,G}(s) b_{\text{corr},G}(s), \tag{5.17}$$

with $b_{\text{corr},G}(s) = E[K_G(s)(\hat{A}^{**}(s) - \hat{A}^{***}(s))|\delta_{e,G}(s) = 1]$. Define the *synthetic estimation bias for subgroup G* as

$$b_{\text{syn},G} = \sum p_{e,G}(s) b_{\text{syn},G}(s), \tag{5.18}$$

with $b_{\text{syn},G}(s) = E[K_G(s)(\hat{A}^{***}(s) - \hat{A}^{***}_{|G}(s))|\delta_{e,G}(s) = 1]$. The *residual model bias for subgroup G* is

$$b_{\text{res},G} = \sum \left\{ p_{e,G}(s) E\left[\hat{K}_{e,G}(s)\hat{A}^{***}_{|G}(s)|\delta_{e,G}(s) = 1\right] - K_G(s)A(s) \right\}, \tag{5.19}$$

Observe that the model bias for subgroup G is the sum of correlation bias, synthetic estimation bias, and residual model bias, $b_{M,G} = b_{\text{corr},G} + b_{\text{syn},G} + b_{\text{res},G}$. Notice also that the synthetic estimation bias is defined so as not to include the effect of bias from imperfect modeling of dependence, and the correlation bias is defined apart from bias that could arise from synthetic estimation.

5.5.2. Poststratified Estimator

In this section we derive expressions for correlation bias, synthetic estimation bias, and residual bias for the poststratified DSE derived under an independence model. The results may be useful as guides to other estimators derived under independence models, such as logistic regression estimators, to the extent that such estimators can be approximated by a poststratified DSE. Even when the approximations are not available, the correlation bias and synthetic estimation bias for the poststratified DSE may serve as useful benchmarks. We will see that for the special case of the poststratified DSE derived under the independence assumption, (5.16), residual model bias is zero and correlation bias and synthetic estimation bias have simple representations.

To show these results, let $H(s)$ be the poststratum to which potential enumeration s belongs, let $S_{0h} = \{s \in S_0 | H(s) = h\}$ denote the set of potential enumerations in poststratum h, and let N_{0h} denote the size of S_{0h}. Further, let $S_{0hG} = S_{0h} \cap G$ denote the set of potential enumerations in poststratum h and in group G, and let N_{0hG} denote the size of S_{0hG}. Now let $S_{eh} = \{s \in S_{0h} | \delta_e(s) = 1\}$, let N_{eh} denote the size of S_{eh}, let $S_{ehG} = \{s \in S_{0h} | K_G(s)\delta_e(s) = 1\}$, and let N_{ehG} denote the size of S_{ehG}.

It is useful to extend our probability model to cover the P sample as well. In analogy with the argument of Section 5 in Chapter 5, the set S_0 of potential enumerations is considered to allow for bivariate outcomes, in the census and in the P sample in the post-enumeration survey. As before, we write $p_e(s) = p_C(s) + p_E(s)$ to denote be the probability that potential enumeration $s \in S_0$ is included in the census. Similarly, we let $p_{\mathrm{P}}(s)$ denote the probability of $s \in S_0$ being included in the P sample. Let $p_{e\mathrm{P}}(s)$ denote the probability of s being included in both. For each $x = C, E, U, e, \mathrm{P}, e\mathrm{P}$ define the mean probability of enumeration status x for the part in poststratum h that is in group G, and for the whole postratum h, as

$$\bar{p}_{x,hG} = \sum_{s \in S_{0hG}} p_x(s)/N_{0hG}, \quad \bar{p}_{x,h} = \sum_{s \in S_{0h}} p_x(s)/N_{0h}. \tag{5.20}$$

If S_{0hG} or S_{0h} are empty, their respective mean probabilities will be taken to be zero.

Some simplifying assumptions will be used to analyze the components of model bias.

(i) We condition on true membership in group G, so that $K_G(s)$ is fixed for each potential enumeration s.
(ii) We assume that N_{ehG} is uncorrelated with $\hat{A}^{**}(s)$, $\hat{A}^{***}(s)$, and $\hat{A}^{**}_{|G}(s)$.

Assumption (ii) could hold, for example, if the analysis were conditional on N_{ehG}.

The DSE under consideration is developed as if the two enumerations are independent, or $p_e(s) = p_{e\mathrm{P}}(s)/p_{\mathrm{P}}(s)$ and it assumes that $p_E(s)$, $p_e(s)$, $p_{e\mathrm{P}}(s)$, and $p_{\mathrm{P}}(s)$ each are constant for all s in the same poststratum. As a result, $\hat{A}^{**}(s)$ has a constant value for all $s \in S_{0h}$, say $\hat{A}^{**}_h$, and $E[\hat{A}^{**}_h] = (1 - \bar{p}_{E,h})/(\bar{p}_{e\mathrm{P},h}/\bar{p}_{\mathrm{P},h})$. Similarly, $\hat{A}^{***}(s)$ has a constant value for all $s \in S_{0h}$, say $\hat{A}^{***}_h$. By definition $\hat{A}^{***}$ properly accounts for dependence and so $E[\hat{A}^{***}_h] = (1 - \bar{p}_{E,h})/\bar{p}_{e,h} = A_h$, say. Observe that $E[\hat{A}^{**}_h] - E[\hat{A}^{***}_h] = -(1 - \bar{p}_{E,h})\psi_h$ with $\psi_h = \mathrm{Cov}(p_e, p_{e\mathrm{P}})/(\bar{p}_{e,h}\bar{p}_{e\mathrm{P},h})$ and $\mathrm{Cov}(p_e, p_{e\mathrm{P}}) = \bar{p}_{e\mathrm{P},h} - \bar{p}_{e,h}\bar{p}_{\mathrm{P},h}$. It follows (Exercise 20) that

$$b_{\mathrm{corr},G} = -\sum_h N_{0hG}\bar{p}_{e,hG}(1 - \bar{p}_{E,h})\psi_h. \tag{5.21}$$

If the enumeration probabilities for either the P sample or the census are constant within the poststratum, ψ_h must be zero. Similarly, if there is no dependence, so that $\psi_h = 0$, then there is zero correlation bias. The correlation bias may be interpreted as a negative weighted sum of ψ_h terms, with weights equal to the expected number of correct enumerations from poststratum h for the subgroup G if the probability of an enumeration being correct did not depend on the subgroup. Notice that ψ_h is defined independently from the subgroup.

Turning attention to synthetic estimation error, note that $\hat{A}^{***}_{|G}(s)$ has constant value, say $\hat{A}^{***}_{hG}$, for all $s \in S_{0h}$, and $E[\hat{A}^{***}_{hG}] = (1 - \bar{p}_{E,hG})/\bar{p}_{e,hG} = A_{hG}$, say. It follows (Exercise 20) that the synthetic bias for group G is

$$b_{\text{syn},G} = \sum_h N_{0hG}\bar{p}_{e,hG}(A_h - A_{hG}). \tag{5.22}$$

This expression shows that if the average census-enumeration probability and average erroneous enumeration probability within a poststratum differ for group G versus the rest of the population, synthetic estimation bias will be present. If group G is the whole population, synthetic estimation bias is zero. If we consider a set of disjoint groups whose union is the whole population, the sum of synthetic estimation biases for the groups is zero.

Residual bias $b_{\text{res},G}$ is zero because for $s \in S_{0hG}$ the sample means of $A(s)$ and $\hat{A}^{***}_{hG}$ have the same expected value (Exercise 20).

5.6. Estimation of Correlation Bias in a Poststratified Dual System Estimator

Consider a national population with θ_s persons of sex $s = F, M$. Suppose we have DA estimates Y_s and DSE's X_s for the two population sizes, such that

$$X_s = \theta_s(1 + \beta_{X,s} + \varepsilon_{X,s}), \quad Y_s = \theta_s(1 + \beta_{Y,s} + \varepsilon_{Y,s}), \tag{5.23}$$

where $E[\varepsilon_{X,s}] = E[\varepsilon_{Y,s}] = 0$, and $\beta_{X,s}$ and $\beta_{Y,s}$ are the relative biases. Decompose the bias in the DSE into correlation bias ("c") and other bias ("o"):

$$\beta_{X,s} = \beta_{X,s,c} + \beta_{X,s,o}. \tag{5.24}$$

Let $\hat{b}_{X,s,o}$ be an unbiased estimator of the "other" bias, $E[\hat{b}_{X,s,o}] = \theta_s\beta_{X,s,o}$.

Example 5.4. Estimates of Correlation Bias Based on DA Totals. A simple approach is to estimate the correlation bias in the DSE by the difference between the DSE and DA estimates, $\delta_s = X_s - Y_s$. Observe that $\delta_s = \theta_s(\beta_{X,s} - \beta_{Y,s} + \varepsilon_{X,s} - \varepsilon_{Y,s})$. If there is no error in DA, then Y_s is equal to θ_s and $E[\delta_s] = \theta_s\beta_{X,s} = \theta_s(\beta_{X,s,c} + \beta_{X,s,o})$, and $\text{Var}(\delta_s) = \theta_s^2 \text{ Var}(\varepsilon_{X,s})$. Should the DSE have no "other" bias, the expected value of δ_s would be equal to correlation bias. More generally, allowing for error in DA, we have $E[\delta_s] = \theta_s(\beta_{X,s} - \beta_{Y,s})$ and $\text{Var}(\delta_s) = \text{Var}(X_s - Y_s)$. Wachter and Freedman (1999) adjusted for other biases in the DSE and used $\delta'_s = X_s - \hat{b}_{X,s,o} - Y_s$ to estimate correlation bias in the 1990 U.S. census. If we reexpress this as $\delta'_s = \theta_s(\beta_{X,s} - \theta_s^{-1}\hat{b}_{X,s,o} - \beta_{Y,s} + \varepsilon_{X,s} - \varepsilon_{Y,s})$, we find that $E[\delta'_s] = \theta_s(\beta_{X,s,c} - \beta_{Y,s})$ and $\text{Var}(\delta'_s) = \text{Var}(\ X_s - \hat{b}_{X,s,o} - Y_s)$. $\Diamond$

The estimates of correlation bias in Example 5.4 are subject to large error if the DA estimates of totals are inaccurate. As noted earlier, the DA sex ratios are believed to be more accurate, because errors in DA estimates of male and female populations are believed to be of similar magnitude. Bell (1993) discusses a variety of estimates of correlation bias based on an assumption that some parametric function of the cross-product ratio $\theta = p_{11}p_{22}/(p_{12}p_{21})$ is constant across poststrata, where the population of unknown size N is assumed to be multinomially distributed with p_{11} the probability of being in both the census and the PES, p_{22} the probability of being in neither, p_{12} the probability of being in the census only, and p_{21} the probability of being in the PES only.[5] A special case is what he calls the "two group model", where θ is taken to be 1 for females and θ is estimated for males from DA sex ratios. In this case, the estimate of correlation bias for males is

$$\gamma_M = X_M - R_Y X_F, \tag{5.25}$$

where R_Y is the DA-based estimate of sex ratio. This estimate may be re-expressed as

$$\gamma_M = \theta_M \left[1 + \beta_{X,M} + \varepsilon_{X,M} - (1 + \beta_{X,F} + \varepsilon_{X,F}) \frac{(1 + \beta_{Y,M} + \varepsilon_{Y,M})}{(1 + \beta_{Y,F} + \varepsilon_{Y,F})} \right]. \tag{5.26}$$

If R_Y is equal to R_θ, the true sex ratio, then γ_M has the same distribution as $\theta_M[(\beta_{X,M} - \beta_{X,F}) + (\varepsilon_{X,M} - \varepsilon_{X,F})]$, with variance $\mathrm{Var}(X_M - R_\theta X_F)$ and expected value $\theta_M[(\beta_{X,M,c} - \beta_{X,F,c}) + (\beta_{X,M,o} - \beta_{X,F,o})]$. Thus, if there is no correlation bias for females, the effect of bias other than correlation bias is the same for males and females, and the DA sex ratio is correct, the expected value of γ_M is equal to correlation bias for males. It is possible to adjust this estimate for estimated "other" bias (Complement 21). It is also possible to exploit DA sex ratios using assumptions other than lack of correlation bias for females; see Elliott and Little (2000), Wolter (1986, 1990) and Bell (1993).

Example 5.5. Estimates of Correlation Bias Based on DA Sex Ratios. In preparing its revised estimates of undercount in the 2000 census, the U.S. Census Bureau incorporated adjustments for correlation bias for adult males.[6] The Bureau was fully aware that "alternative models for correlation bias can be used that are equally consistent with the data but that produce different subnational estimates" (U.S. Census Bureau 2003b, 48). Correlation bias was assumed 0 for females and was not estimated for children. The two group model was used to estimate correlation bias separately for Blacks and Non-Blacks aged 18–29, 30–49, and 50 and over. The data were not consistent with an assumption of (negative) correlation bias for Non-Black males 18–29, and the correlation bias was assumed to be zero for that group. The two-group model includes Hispanics with Non-Blacks, which is suspected to slightly overstate the correlation bias for Non-Hispanic Non-Blacks and greatly understate it for Hispanics. The difficulty is that DA sex ratios are not available for Hispanics. A "modified two group" method was developed to assign the estimate of relative correlation bias for Blacks (by age and sex) to Hispanics, and then to derive the correlation bias adjustment for Non-Hispanic Whites so it

---

[5] The selection probabilities for the PES need to be taken into account, but for insight it suffices to consider that the PES was a 100% sample of all blocks.

[6] The revised estimates were developed in December 2002. Previous estimates of undercount in the 2000 census did not adjust for correlation bias, as noted in Example 4.5 of Chapter 12.

TABLE 1. Estimates of Net Undercount in the 2000 U.S. Census with Various Adjustments for Correlation Bias

	Estimated Net Undercount (%)			Estimated Standard Error (%)		
	Correlation Bias Adjustment			Correlation Bias Adjustment		
Race/Ethnicity	None	Two Group	Modified Two Group	None	Two Group	Modified Two Group
Non-Hispanic White	−1.53	−1.13	−1.39	0.20	0.20	0.20
Non-Hispanic Black	−0.53	1.84	1.84	0.41	0.43	0.43
Hispanic	0.42	0.71	3.17	0.44	0.44	0.49

matched the DA sex ratio for Whites (both Hispanic and Non-Hispanic combined). The results, shown in the Table 1, more than double the estimate of net undercount for Hispanics. Data analysis did not quite support the modified two group model (U.S. Census Bureau 2003b, 52), and the Census Bureau was not willing to make an unsupported assumption and use the modified two group model. ◊

5.7. Estimation of Synthetic Estimation Bias in a Poststratified Dual System Estimator

Assessment of synthetic estimation bias (5.22), or more generally (5.18), is difficult when we lack available highly accurate *direct estimates*, with the latter term referring to estimates based solely on data from just the group (or area) in question. Several approaches have been tried to estimate synthetic estimation bias (or SEB for short). One is to estimate the SEB for variables believed correlated at the smallest area level with the ones of interest, so-called *surrogate variables*. A second approach is to compare the synthetic estimates with direct estimates and make an allowance for error in the direct estimates. A third approach is is to fit models of the variable of interest that include "effects" for the groups and use the magnitudes of the estimated effects to assess the SEB. These approaches will be described in turn.

The idea behind the use of surrogate variables is that one can estimate SEB by calculating it directly for variables other than the adjustment factors A_{hG} in (5.22). Such an analysis can be useful if the synthetic error for the surrogate variables is similar to the synthetic error for the variable(s) of interest. The general problem with surrogate variables is that they correlate imperfectly with undercount adjustment factors, and conflicting results (low synthetic estimation bias for some surrogates, and high for others) are not uncommon. The approach cannot tell us whether the error from the failure of the synthetic assumption is appreciable for the variable of actual interest. Although the surrogate variables may show strong correlations with direct estimates of undercount at the larger area level (i.e., ecological correlations), the correlations could be quite different at the block level, where direct estimates of undercount typically are unavailable except for a sparse sample of block clusters.

Example 5.6. Surrogate Variables for Undercount and Overcount in the 2000 U.S. Census. The Census Bureau's use of the same adjustment factors A_h for all enumerations in a poststratum h gives rise to synthetic estimation bias when enumerations from different areas G should have different adjustment factors (Anderson et al. 2000, Brown et al. 1999b, Freedman and Wachter 1994, Fay and Thompson 1993). The Census Bureau has estimated synthetic estimation bias by assuming the distribution of undercount across areas can be derived by models based on the post-enumeration survey and distributions of surrogate variables whose geographic dispersion was perfectly known; see Malec and Griffin (2001, Appendix). Denote the estimated correct-enumeration rate in poststratum h by $f_{\mathrm{CE},h} = \hat{N}_{\mathrm{CE},h}/(\hat{N}_{\mathrm{CE},h} + \hat{E}_{\mathrm{E},h})$. Now estimate the gross undercount for poststratum h by $\hat{N}_{u,h} = C_h(\hat{A}_h - f_{\mathrm{CE},h})$ and the estimate of gross overcount by $\hat{N}_{o,h} = C_h(1 - f_h)$. Let X denote a surrogate for gross undercount and Y a surrogate for gross overcount, with values X_{hG} and Y_{hG} for the part of poststratum h that is in group G and with poststratum totals $X_h = \sum_G X_{hG}$, $Y_h = \sum_G Y_{hG}$. The *artificial population size* for poststratum h in group G is defined as

$$N_{hG} = C_{hG} + (X_{hj}/X_h)\hat{N}_{u,h} - (Y_{hj}/Y_h)\hat{N}_{o,h} \tag{5.27}$$

with C_{hG} denoting the number of census enumerations in poststratum h and group G. The artificial population size for group G is $\sum_h N_{hG}$. The SEB for group G is estimated by

$$\sum_h C_{hG} A_h - \sum_h N_{hG}. \tag{5.28}$$

Malec and Griffin (2001) considered a variety of surrogate variables and chose those having the greatest correlations with a direct estimates of net undercount at the block cluster level (for blocks in the post-enumeration survey). The squared correlations were weak, however (< 0.08). For similar analyses of SEB in the 1990 U.S. census, see Fay and Thompson (1993), Kim, Blodgett and Zaslavsky (1991), and Freedman and Wachter (1994). $\Diamond$

A nonparametric method for estimating SEB compares synthetic estimates S with direct estimates D for the individual groups or areas. If the variance of $D - S$ is estimated by V and if D has negligible bias, then $(D - S)^2 - V$ estimates the squared bias in S (Exercise 22). There are a variety of difficulties with this method. First, the direct estimates D are subject to ratio-estimator bias, measurement error bias, and correlation bias. Even if the latter two are assumed equal in D and S, the ratio estimator bias in D will typically exceed that in S. It is conceivable that DSEs that are corrected for first-order bias (e.g., by jackknife methods) could be used to avoid this problem, although use of simulations to verify its accuracy would seem prudent. A second set of problems in estimating synthetic estimation error for an area is caused by sample size limitations which prevent calculation of A_{hG}. Even if a group G or area is represented in the sample, some poststrata that are present in the population may be empty in the sample or may fail to contain any matches, preventing calculation of A_{hG} for some h. A possible solution might be to restrict attention to poststrata that have a minimum number, say τ, of E-sample members in the area. (Although this may lead to a slightly biased view by emphasizing poststrata that have higher enumeration rates, the differences in enumeration rates are not so large as to make this an overwhelming problem.) The threshold τ should be chosen so that if there are at least τ E-sample members, there will be at least one match. (Complement 23 pursues this idea in more detail.) An additional problem is that even if V and D are unbiased, the estimator $(D - S)^2 - V$ may be subject to large variance. This issue often can be addressed by aggregating results across large numbers of groups G.

An alternative and more fully parametric approach is to fit models of the variable of interest that include "effects" for the areas. For example, in the context of the 1990 census in the U.S., Hengartner and Speed (1993) considered block effects for surrogate variables (the erroneous enumeration rate and gross omissions rate) in the 1990 PES. For another example, the poststratified DSE can be derived as a special case of logistic regression models for enumeration and for erroneous enumeration; each observation is associated with a single predictor variable, namely the poststratum (Haberman, Jiang, and Spencer 1998). This model may be expanded to allow for synthetic estimation bias by including an additional parameter or parameters for each area or group of interest G. A minimal model would include an extra intercept for each area. Alternatively, the model could include an extra intercept for each major grouping of poststrata, where the major groupings could be defined (for example) by sex, tenure, age, and minority status. The logistic models with and without the extra parameters directly give predictions of A_h and A_{hG}. These may be substituted into (5.22), along with N_{ehG} to estimate $N_{0hG}\bar{p}_{e,hG}$, to provide an estimate of the SEB.

6. Assessment of Error in Functions of Dual System Estimators and Functions of Census Counts

For many purposes, accuracy of functions of the population vector may be more important than accuracy of the population sizes. For example, as will be discussed in Section 3 of Chapter 12, allocation of benefits such as funds or political representation may be proportional to the fraction of the population that is associated with a given group, or what is known as the group's *share*. The necessary components for the estimation of accuracy were discussed in Section 5. Here we sketch how to combine the components to estimate the bias and variance for functions of the DSE and how to estimate the bias and MSE of functions of the census. Section 6.1 provides a detailed overview and Section 6.2 discusses computational methods.

6.1. Overview

The MSE is the sum of the variance and the squared bias (cf., (4.1) of Chapter 8). To estimate the squared bias, we need to allow for the variance in the estimate of bias, because the expectation of the square of a random quantity is equal to its variance plus the square of its mean. For simplicity of exposition, we will use a slightly different notation in this section than in earlier sections.

Let $\boldsymbol{\theta}$ denote the vector of true populations for all groups, let $\mathbf{D}$ denote the vector of DSE's, $\mathbf{N}$ the vector of census counts, $\mathbf{B} = E[\mathbf{D}] - \boldsymbol{\theta}$ the bias in $\mathbf{D}$, and $\hat{\mathbf{B}}$ the estimate of bias in $\mathbf{D}$. Define $\mathbf{T} = \mathbf{D} - \hat{\mathbf{B}}$, so that $\mathbf{T}$ adjusts for known bias in $\mathbf{D}$. Let f be a real-valued function of the vectors, e.g., $f(\mathbf{N})$ could be the population share of a given area. Estimates of covariance matrices are needed for $\mathbf{D}$ and $\mathbf{T}$ jointly. Under distributional assumptions we can use simulation to estimate the variances of $f(\mathbf{T})$, $f(\mathbf{D})$, and $f(\mathbf{D}) - f(\mathbf{T})$; denote these estimates by $\hat{V}_{f(\mathbf{T})}$, $\hat{V}_{f(\mathbf{D})}$, and $\hat{V}_{f(\mathbf{D})-f(\mathbf{T})}$, respectively, and denote the estimated covariance of $f(\mathbf{T})$ and $f(\mathbf{D})$ by $\hat{C}_{f(\mathbf{D}),f(\mathbf{T})}$. In practice, a multivariate normal distribution is used; Mulry and Spencer (1991) provide some empirical justification.

The squared error of $f(\mathbf{N})$ is $(f(\mathbf{N}) - f(\boldsymbol{\theta}))^2$. Assuming $\mathbf{T}$ is unbiased for $\boldsymbol{\theta}$, we can estimate the squared error by $(f(\mathbf{N}) - f(\mathbf{T}))^2 - \hat{V}_{f(\mathbf{T})}$. The MSE of $f(\mathbf{D})$ is $(E[f(\mathbf{D})] - f(\boldsymbol{\theta}))^2 +$ Var$(f(\mathbf{D}))$. We estimate it correspondingly by $(f(\mathbf{D}) - f(\mathbf{T}))^2 + \hat{V}_{f(\mathbf{D})} - \hat{V}_{f(\mathbf{D})-f(\mathbf{T})}$.

In the U.S., considerable interest has centered on whether the DSE is more accurate than the census. Denote the excess MSE of the census over the DSE by Δ. For a single area, Δ is equal to $(f(\mathbf{N}) - f(\boldsymbol{\theta}))^2 - (E[f(\mathbf{D})] - f(\boldsymbol{\theta}))^2 - V_{f(\mathbf{D})}$. If $\Delta < 0$, the census is more accurate for the area, and if $\Delta > 0$ then the DSE is more accurate. In the case of a single area, we estimate Δ by $\hat{\Delta} = (f(\mathbf{N}) - f(\mathbf{T}))^2 - (f(\mathbf{D}) - f(\mathbf{T}))^2 - \hat{V}_{f(\mathbf{T})} - \hat{V}_{f(\mathbf{D})} + \hat{V}_{f(\mathbf{D})-f(\mathbf{T})}$. This may be more simply written as $(f(\mathbf{N}) - f(\mathbf{T}))^2 - (f(\mathbf{D}) - f(\mathbf{T}))^2 - 2\hat{C}_{f(\mathbf{D}),f(\mathbf{T})}$. If the dependence between $\hat{\mathbf{B}}$ and $\mathbf{D}$ is negligible, we can further simplify the estimate of Δ to $(f(\mathbf{N}) - f(\mathbf{T}))^2 - (f(\mathbf{D}) - f(\mathbf{T}))^2 - 2\hat{V}_{f(\mathbf{D})}$, which does not involve any correction for the variance of $\hat{\mathbf{B}}$. The precision of $\hat{\Delta}$ is still affected by that of $\hat{\mathbf{B}}$, however.

6.2. Computation

We now describe the calculations in more detail. We use "area" as a general concept, and some care may be needed in practice. For example, if focusing on numbers in a demographic group in an area, the area should be taken to exclude the groups not of interest. We use the term "share" to denote a population share or more generally any function of the population vector.

The following variables are inputs to the calculation process, and are derived as discussed in Section 5. The census count for area i is denoted by N_i. The census count for area i, poststratum $h = 1, \ldots, H$ is denoted by C_{ih} and the vector of counts for poststrata is $\mathbf{C}_i = (C_{i1}, \ldots, C_{iH})^T$. The adjustment factor for poststratum h is A_h and the vector of adjustment factors for poststrata is $\mathbf{A} = (A_1, \ldots, A)^T$. The estimated bias in A_h is denoted by b_h and the vector of bias estimates across poststrata is $\mathbf{b} = (b_1, \ldots, b_H)^T$. The estimated bias in the DSE for area i is $B_i = \mathbf{b}^T\mathbf{C}_i$. The estimated $H \times H$ covariance matrix of $\mathbf{A}$ is $\boldsymbol{\Sigma}_\mathbf{A}$, the estimated covariance matrix of $\mathbf{b}$ is $\boldsymbol{\Sigma}_\mathbf{b}$, and the estimated covariance matrix of $\mathbf{A}$ and $\mathbf{b}$ is $\boldsymbol{\Sigma}_{\mathbf{Ab}}$. The full covariance matrix for $(\mathbf{A}^T, \mathbf{b}^T)^T$ is estimated by

$$\boldsymbol{\Sigma} = \begin{pmatrix} \boldsymbol{\Sigma}_\mathbf{A} & \boldsymbol{\Sigma}_{\mathbf{Ab}} \\ \boldsymbol{\Sigma}_{\mathbf{Ab}} & \boldsymbol{\Sigma}_\mathbf{b} \end{pmatrix}. \tag{6.1}$$

The adjusted estimate for area i is $A_i = \mathbf{A}^T\mathbf{C}_i$. If we adjust the adjustment factors for estimated bias and multiply by the census counts, we obtain the "target" estimate for area i, $T_i = (\mathbf{A} - \mathbf{b})^T\mathbf{C}_i = A_i - B_i$. The "target" adjusts the adjustment factor for estimated bias – the adjustment need not improve accuracy. Finally, the estimated covariance matrix of $\mathbf{A} - \mathbf{b}$ is $\boldsymbol{\Sigma}_{\mathbf{A}-\mathbf{b}} = \boldsymbol{\Sigma}_{\mathbf{A}} + \boldsymbol{\Sigma}_{\mathbf{b}} - 2\boldsymbol{\Sigma}_{\mathbf{Ab}}$.

The estimate of the MSE for C_i, say $M_{C,i}$, is equal to $(N_i - T_i)^2 - \mathbf{C}_i^T\boldsymbol{\Sigma}_{\mathbf{A}-\mathbf{b}}\mathbf{C}_i$. The MSE of the DSE D_i is estimated by $M_{D,i} = B_i^2 + \mathbf{C}_i^T\boldsymbol{\Sigma}_{\mathbf{A}}\mathbf{C}_i - \mathbf{C}_i^T\boldsymbol{\Sigma}_{\mathbf{b}}\mathbf{C}_i$. To estimate the MSE's for nonlinear functions of the population vectors we will use simulation. As an example, we describe the procedure for population shares.

Consider area i's share of G, where G is a union of areas. The census share is $N_i' = N_i / \sum_{j\in G} N_j$. The DSE share is $D_i' = D_i / \sum_{j\in G} D_j$. The target share is $T_i' = T_i / \sum_{j\in G} T_j$. The bias of the DSE share is estimated by $B_i' = D_i' - T_i'$.

For estimating variances of shares, we will generate a set of R replicates of $\mathbf{A}$ and $\mathbf{b}$. The value of R might be initially set at 1,000, and simulation variance can be calculated to assess whether R needs to be increased.

Generate independently $2H\times 1$ vectors $\mathbf{z}^{(r)} = ((\mathbf{x}^{(r)})^T, (\mathbf{y}^{(r)})^T)^T$, $r = 1, \ldots, R$, from $N(\mathbf{0}, \boldsymbol{\Sigma})$, where the covariance matrix is given in (6.1). Observe that $\mathbf{x}^{(r)}$ is distributed as the random error in $\mathbf{A}$, $\mathbf{y}^{(r)}$ is distributed as the random error in $\mathbf{b}$, and the covariance between $\mathbf{x}^{(r)}$ and $\mathbf{y}^{(r)}$ is $\boldsymbol{\Sigma}_{\mathbf{Ab}}$. Define replicates of the adjustment factors $\mathbf{A}$ and bias estimates $\mathbf{b}$ by $\mathbf{A}^{(r)} = \mathbf{A} + \mathbf{x}^{(r)}$ and $\mathbf{b}^{(r)} = \mathbf{b} + \mathbf{y}^{(r)}$ for $r = 1, \ldots, R$. Replicates of adjusted estimates of shares and target values of shares are based on $\mathbf{A}^{(r)}$ and $\mathbf{b}^{(r)}$, and the variances and covariances can be derived from the replicates.

Specifically, notice that the adjusted share for area i may be written as $D_i' = \mathbf{A}^T\mathbf{C}_i / \sum_{j\in G} \mathbf{A}^T\mathbf{C}_j$. The rth replicate of the adjusted share is

$$A_i'^{(r)} = \mathbf{A}^{(r)T}\mathbf{C}_i \Big/ \sum_{j\in G} \mathbf{A}^{(r)T}\mathbf{C}_j. \tag{6.2}$$

The sample variance among the R values of $D_i'^{(r)}$ is used to estimate the variance of D_i'. Denote the variance estimate by $\hat{V}_{D_i'}$.

Similarly, the rth replicate of the target share is defined by

$$T_i'^{(r)} = (\mathbf{A}^{(r)} - \mathbf{b}^{(r)})^T\mathbf{C}_i \Big/ \sum_{j\in G} (\mathbf{A}^{(r)} - \mathbf{b}^{(r)})^T\mathbf{C}_j. \tag{6.3}$$

The sample variance among the R values of $T_i'^{(r)}$ is used to estimate the variance of T_i'. Denote the variance estimate by $\hat{V}_{T_i'}$.

The rth replicate of the bias in the adjusted share is defined by $B_i'^{(r)} = D_i'^{(r)} - T_i'^{(r)}$. The sample variance among the R values of $B_i'^{(r)}$ is used to estimate the variance of B_i'. Denote the variance estimate by $\hat{V}_{B_i'}$.

The MSE for the census share for area i is estimated by $M_{C_i'} = (N_i' - T_i')^2 - \hat{V}_{T_i'}$. The MSE for the DSE share for area i is estimated by $M_{A_i'} = B_i'^2 + \hat{V}_{A_i'} - \hat{V}_{B_i'}$. (If there is negligible correlation between $\mathbf{A}$ and $\mathbf{b}$, then in $M_{C_i'}$ we may replace $\hat{V}_{T_i'}$ by the sum, $\hat{V}_{D_i'} + \hat{V}_{B_i'}$.)

Exercises and Complements (*)

*1. *Accuracy of Postcensal Estimates.* Some empirically based conjectures about accuracy of U.S. postcensal estimates are that postcensal increase or decrease is understated, accuracy degrades as the postcensal period increases, and larger areas tend to be more accurately estimated (in percentage terms) than smaller areas, whether because of better data sources or some cancellation of nonrandom but haphazard errors (e.g., Panel on Small-Area Estimates of Population and Income 1980, Davis 1994).

2. As in Section 2.1, define $\hat{\mu} = m/(K + n - m)$ and $\tilde{\mu} = m/(K + (1 - \lambda)n - m)$. Observe that $(\tilde{\mu} - \hat{\mu})/\hat{\mu}$ equals $\lambda n/(K + n - m - \lambda n)$, which has a first-order Taylor approximation at $\lambda = 0$, say, $b = \lambda n/(K + n - m)$. Show that $K \leq m$ and hence $|b| = |\lambda| n/(K + n - m) \geq |\lambda|$.

3. Consider the survival of an age-group during a unit time. Suppose the hazard of mortality is constant (cf., Section 1 of Chapter 4). Using the notation of Exercise 2, denote the postcensal estimate of population, adjusted for mortality (only), by $(1 - \lambda)n \times \exp(-\tilde{\mu})$. Note that if there were no error in the mortality rate due to the base population, the postcensal estimate would be $(1 - \lambda)n \times \exp(-\hat{\mu})$. The difference between the latter and the estimate obtained with $\lambda = 0$ is $\delta \approx -\lambda n(1 - \hat{\mu})$. Show that in general, the error in the postcensal estimate (when nonzero λ causes a bias in the mortality rate) is, say, $\varepsilon \approx \delta - (\tilde{\mu} - \hat{\mu})(1 - \lambda)n$. Show that $|\varepsilon| \geq |\delta|$ with equality only when $\lambda = 0$.

*4. Let $V(t)$ denote the true population at time t, let $\hat{V}(t)$ denote the forecast (or postcensal estimate), and let $\tilde{V}(t)$ denote the benchmark, e.g., a census number. Denote the error in the benchmark by $\beta(t) = \tilde{V}(t) - V(t)$, the forecast error by $\varepsilon(t) = \hat{V}(t) - V(t)$, and the error in the forecast of change by $\Delta(t) = (\hat{V}(t) - \hat{V}(0)) - (V(t) - V(0))$. Let $\hat{\mu}(t) = \hat{V}(t) - \tilde{V}(t)$ and $\hat{\Delta}(t) = (\hat{V}(t) - \hat{V}(0)) - (\tilde{V}(t) - \tilde{V}(0))$ denote estimates of $\varepsilon(t)$ and $\Delta(t)$. (i) Show that $\hat{\varepsilon}(t) = \varepsilon(t) - \beta(t)$. (ii) Show that $\hat{\Delta}(t) = \Delta(t) - \beta(t) + \beta(0)$. (iii) Denote the undercoverage rates in the benchmarks by $u(t) = (V(t) - \tilde{V}(t))/V(t)$ and show that $\hat{\Delta}(t) = \Delta(t) + u(0)(V(t) - V(0)) + V(t)(u(t) - u(0))$. Thus, even if the benchmarks at times t and 0 have the same relative error, the estimate of $\Delta(t)$ can be affected (Spencer 1980b).

*5. Note from (5.3) that $A(s) = (1 - p_E(s))/(1 - p_U(s))$. Our model should be considered to apply only to those individuals with $p_U(s) < 1$. Wachter and Freedman (1999) speculate that some individuals are impossible to count, i.e., have $p(s) = 1$. In this case $A(s)$ is undefined.

*6. If $p_U(s) < 1$ for each s, expression (5.4) has expectation N. Let N_U denote the number of population members s for which $p_U(s) = 1$. Note that the expectation of (5.4) equals $N - N_U$.

*7. In Example 5.1, assume that the status assignments occur independently with probabilities p_E, p_C, and p_U. Define N_E = erroneous enumerations, N_C = correct enumerations, and $N_U = N_0 - N_E - N_C$. Then, $(N_E, N_C, N_U) \sim \text{Mult}(N^*; p_E, p_C, p_U)$ where $N^* = N_E + N_C + N_U$.

8. Referring to Section 5.2, show that if there are N_{I} LA or WPI enumerations out of the C census enumerations in the poststratum, the estimator used by the Census Bureau in Example 4.1 is equal to the product of (5.5) and $(C - N_{\mathrm{I}})/C$. Show that an alternative way to derive the Census Bureau's estimator is to use (5.5) but with $\hat{p}_U(s)$ in the denominator replaced by $(C\hat{p}_U(s) - N_{\mathrm{I}})/(C - N_{\mathrm{I}})$.
9. Let $\hat{N}_{C,h}$ denote the number of census enumerations in poststratum h and let $\hat{A}(s) = \hat{A}_h$ if enumeration s belongs to poststratum h. Show that the poststratified DSE in (5.6) may be written as (5.7).
10. Using the same poststrata for computing $\hat{p}_U(s)$ and $\hat{p}_E(s)$ can reduce variability of estimates, for better or for worse. People who are enumerated in the wrong place are treated as both erroneous enumerations in their E-sample poststrata and as census misses in their P-sample poststrata. If the poststrata are assigned consistently (Complement 19), the errors net out and do not affect the adjustment factor for the poststratum. On the other hand, it is conceivable that one could increase homogeneity of $\hat{p}_U(s)$ using a set of poststrata indexed by l, say and homogeneity of $\hat{p}_E(s)$ using a different set of postrata, indexed by k. Define $\hat{p}_{Ul}$ to equal $\hat{p}_U(s)$ if enumeration s belongs to poststratum l and define $\hat{p}_{Ek}$ to equal $\hat{p}_E(s)$ if enumeration s belongs to poststratum k. Define $\hat{A}_{kl} = (1 - \hat{p}_{Ek})/(1 - \hat{p}_{Ul})$. Let $\hat{N}_{Ckl}$ equal the number of census enumerations in the intersection of poststrata k and l and show that the poststratified DSE in (5.6) may be written as

$$\hat{N} = \sum_{k,l} \hat{A}_{kl}\hat{N}_{Ckl}.$$

Such treatment can lead to increased variability of adjustments for areas. Thus, the U.S. Census Bureau (2003b, 25, Table 10) found that the DSE was at least 10% under the census count for more than 5% of the jurisdictions known as "places" (which tend to be small in size and represent about 57% of the population).

*11. *Poststratification and logistic regression.* In Example 5.1, the probabilities were constant for all potential enumerations $s \in S_0$. More generally, one may group enumerations into poststrata within which the probabilities are assumed constant (Section 5.2). An alternative is to model the probabilities with logistic regression, which allows covariates to enter additively without high order interactions and more easily accommodates continuous covariates (Alho et al. 1993, Haberman, Jiang, and Spencer 1998). Through use of dummy variables, a poststratification model may be treated as a special case of a logistic regression. Logistic regression can also be used to define poststrata by assigning individuals with similar predicted enumeration probabilities into the same poststratum (R.J.A. Little, quoted in National Research Council 2001, 28).

*12. Logistic regression models can account for the two-stage process by which the census misses an individual, e.g., the person's housing unit could be missed, or the housing unit might be included but not the individual (Cowan and Malec 1986). Show that if the conditional probability that a person is enumerated given that the person's household was not missed is constant throughout the poststratum, then the unconditional probability is constant only if the probability of being missed is constant for each household containing the members of the poststratum. If households contain members of diverse poststrata, constant probabilities of household misses may be a very strong condition.

*13. The poststratified DSE (5.7) uses as poststratification variables only those that are recorded both in the census and in the P-sample. This may lead to correlation bias if the miss rates vary with a poststratification variable X observed just in the P sample. Marks (1979) suggested a modification allowing adjustment for variables recorded in the P-sample but not in the census (or E-sample). Consider a single poststratum and for simplicity ignore erroneous enumerations. Let X be a categorical variable defined for P-sample respondents and taking values $1, \ldots, k$. Let f_i denote an estimate of the fraction of the census enumerations for which $X = i$. Although X may not be measured outside the P sample, one could, e.g., use P-sample data to estimate the f_i's. Let C denote the census count (or weighted E-sample count, as the case may be). Let P_i denote the weighted size of the P-sample population with $X = i$, and let $C_i = f_i C$ denote the estimated number of enumerations with $X = i$. The number of matches between the P-sample and the census for persons with $X = i$ is denoted by M_i. Define the modified poststratified DSE by

$$\sum_{i=1}^{k} C_i P_i / M_i.$$

14. *Ratio estimator bias.* Consider estimating a ratio $R = Y/X$ by $\hat{R} = \hat{Y}/\hat{X}$, with $E[\hat{X}] = X$ and $E[\hat{Y}] = Y < \infty$. Assume that $\hat{X} > c > 0$ and hence that $E[\hat{R}]$ is defined and finite. Show that $E[\hat{R} - R] = -E[\hat{R}(\hat{X} - X)]/X = -\operatorname{Cov}(\hat{R}, \hat{X})/X$. Show that the bias is not large relative to the variance provided the CV of $\hat{X}$ is small. (Hint: show that $|E(\hat{R} - R)|/V(\hat{R})^{1/2} \geq CV$. Hartley and Ross 1954).

*15 *Estimating the variance of the DSE.* Notice from (5.7) that $\hat{N}$ is the sum of products $\hat{A}_h C_h$. An estimate ν_{hk} of the covariance between $\hat{A}_h$ and $\hat{A}_k$ (or the variance, if $h = k$) is obtained, e.g., by the jackknife or delta method, and $\operatorname{Var}(\hat{N})$ is estimated by, say,

$$\hat{\nu} = \sum_h \sum_k \nu_{jk} C_h C_k.$$

Any randomness in the census counts C_h arising from the probabilistic nature of the census (Section 5.1) is treated inconsistently by this method: the randomness is reflected in covariance estimates ν, but once ν is obtained C_h's are treated as fixed in the computation of $\hat{\nu}$. Thus, $\hat{\nu}$ may differ from variance estimates based directly on $\hat{N}$, e.g., jackknife estimates.

*16. *Contamination bias.* "Contamination" arises if the selection of a sampling unit into the PES alters the way the census is conducted there. Let $A_0(s)$ denote the value of $A(s)$ that would pertain if the post-enumeration survey sample had included the housing unit containing potential enumeration s. In the absence of contamination, $A_0(s) = A(s)$. Also, for persons whose housing units were sampled in post-enumeration survey, $A(s) = A_0(s)$. The *contamination bias* is defined as

$$E\left[\sum_{s \in S_e \cap G} A_0(s)\right] - E\left[\sum_{s \in S_e \cap G} A(s)\right].$$

Contamination bias was investigated but not detected in the 1990 and 2000 post-enumeration surveys in the U.S.. Show that if $A(s)$ is constant within each poststratum, that the ratio

$$E\left[\sum_{s \in S_e \cap G} A_0(s)\right] \Big/ E\left[\sum_{s \in S_e \cap G} A(s)\right]$$

can be estimated by the ratio of the census count for group G to the sample-weighted number of census enumerations for group G in the post-enumeration survey sample. This result leads to the decision that if the latter ratio is sufficiently close to 1, contamination bias is not present. Contamination bias was believed small in the U.S. A.C.E. in 2000 (Bench 2002a; Hogan 2001, 50–51).

*17. *Coping with unresolved enumeration, match, and residency status due to incomplete data.* In the U.S. PES, E-sample cases that cannot with certainty be classified as erroneous or not are assigned imputed probabilities of being erroneous. Similarly, P-sample cases that cannot be resolved as matches and as residents are assigned imputed probabilities of match and residency. The imputations are based on models fitted to cases that can be classified with certainty. (To be classified with certainty implies there was no ambiguity, but does not imply that the classification was correct.) Imputation error has two parts. One is an increase in variance due to use of existing sample data to impute an unobserved missing value. The imputation is conditional on an implicit or explicit model, and if the model is incorrect a bias is induced as well. The U.S. Census Bureau's variance estimation methods, based on jackknife replication, account for the variance due to imputation. To estimate the bias is more difficult, because the model they use is the one they think is the best. Some information about bias can be gleaned from subsamples of unresolved cases where additional information was obtained to avoid the need for imputation. The sample sizes are not large enough to support direct estimation of the vector of biases. A model-based approach to the problem is to consider the bias in the vector of poststratum adjustment factors to be chosen at random from a population of possible bias vectors, independently of the sample. For simplicity, suppose the population is approximately multivariate normal with some covariance matrix, say Σ. A possible way to estimate Σ is to conduct a sensitivity analysis and to look at the empirical covariance matrix of the resulting vectors of adjustment factors as the imputation model is varied. This raises the question of how does one decide what alternatives to use in the sensitivity analysis. The Census Bureau developed a set of 128 "reasonable alternatives" that were available for use in the sensitivity analysis, which led to an estimate of 531,751 for the standard deviation of the March 2001 A.C.E. estimate of national population, where the standard deviation reflects uncertainty of the choice of imputation model (Kearney 2002) for unresolved erroneous enumeration status (E sample) and match status and residency status (P sample). For comparison, the standard error from sampling was 378,222. The question is whether this set of reasonable alternatives is either too narrow or too wide to provide realistic variance estimates for imputation model uncertainty. Partial information about this is provided by followup surveys that provided additional data for a large part of a subsample of the cases with unresolved erroneous enumeration status. Analysis of that data indicated that empirical standard deviation for the erroneous enumeration rate when calculated on the 128 reasonable alternatives should be increased by a factor of 1.3. The followup survey did not provide sufficient information about the unresolved match status, and so the 1.3 factor was used to multiply the 531,751 estimate. The analysis did not address the question of whether the correlations of the empirical estimate of Σ are realistic.

*18. The description of $\hat{A}$ in Section 5.4 is simplified. (i) We assume that any census enumeration includes enough data to allow for possible matching to the P sample or for imputation of match status when matching results are ambiguous. (ii) We do not consider separate treatment of movers and nonmovers unless they are in separate poststrata.

*19. *Effects of inconsistency of poststratum assignments.* The adjustment factor $\hat{A}_h$ in (5.7) is the product of a factor (≤ 1) for erroneous enumerations and a factor (≥ 1) for gross undercoverage. The latter is based on poststratum assignment in the P sample. The DSE multiplies $\hat{A}_h$ by the number of census enumerations classified in poststratum h, C_h. The same person can be classified in discrepant ways in the P sample and the census if the two surveys record the person's characteristics differently. This may cause a bias in the DSE in (5.7) because the coverage factors for gross undercount are derived for poststrata based on the P sample and are applied to the poststrata based on census enumerations, or what we will call E-sample poststrata.

To understand the bias, it is useful to consider that each person enumerated in the P-sample could be enumerated both ways and assigned to a poststratum two ways, based on either the P-sample data or the census data. Some imagination is required to describe a person enumerated in the P-sample but not the census, and we must make some assumptions to estimate what the E-sample poststratum is for such a person. Let h and k denote P-sample and E-sample poststrata, respectively, and let $f(h|k)$ denote the proportion of persons enumerated in P-sample poststratum k but not in the census who belong to E-sample poststratum k.

Define $\hat{N}_{\mathrm{P}}(h, k)$ as the number of P-sample persons in E-sample poststratum h and P-sample poststratum k, and define $\hat{N}_{\mathrm{CP}}(h, k)$ as the number of P-sample matches belonging to E-sample poststratum h and P-sample poststratum k. The undercount adjustment adjustment factor for poststratum i is of the form $\sum_h \hat{N}_{\mathrm{P}}(h,i)/\sum_h \hat{N}_{\mathrm{CP}}(h,i)$, whereas the factor should, to be unaffected by bias from inconsistent E-sample and P-sample poststratification, be $\sum_k \hat{N}_{\mathrm{P}}(i,k)/\sum_k \hat{N}_{\mathrm{CP}}(i,k)$. The bias in the undercount adjustment factor for poststratum i from inconsistent poststratification is thus

$$\frac{\sum_h \hat{N}_{\mathrm{P}}(h,i)}{\sum_h \hat{N}_{\mathrm{CP}}(h,i)} - \frac{\sum_k \hat{N}_{\mathrm{P}}(i,k)}{\sum_k \hat{N}_{\mathrm{CP}}(i,k)}.$$

To estimate the bias, first define $\hat{N}_{\mathrm{P,nm}}(h,k) = \hat{N}_{\mathrm{P}}(h,k) - \hat{N}_{\mathrm{CP}}(h,k)$, with the subscript "nm" indicating "nonmatch". Although we do not observe $\hat{N}_{\mathrm{P,nm}}(h,k)$, we do observe $\hat{N}_{\mathrm{CP}}(h,k)$ and $\hat{N}_{\mathrm{P,nm}}(k) = \sum_h \hat{N}_{\mathrm{P,nm}}(h,k)$. Given values of the conditional probabilities $f(h|k)$, we can estimate $\hat{N}_{\mathrm{P,nm}}(h,k)$ by $f(h|k)\hat{N}_{\mathrm{P,nm}}(k)$. Thus, we can estimate the bias in the undercount adjustment factor for poststratum i by

$$\frac{\sum_h \hat{N}_{\mathrm{CP}}(h,i) + f(i|k)\hat{N}_{\mathrm{P,nm}}(k)}{\sum_h \hat{N}_{\mathrm{CP}}(h,i)} - \frac{\sum_k \hat{N}_{\mathrm{P}}(i,k)}{\sum_k \hat{N}_{\mathrm{CP}}(i,k)}.$$

A simple way to estimate $f(h|k)$ is by $\hat{N}_{\mathrm{CP}}(h,k)/\sum_h \hat{N}_{\mathrm{CP}}(h,k)$, using the observed joint distribution of matches, but if the number of poststrata is large then there may be too many conditional probabilities to estimate from even a large sample. Haberman and Spencer (2002) use log-linear models for estimating $f(h|k)$. Either method assumes that the poststratum inconsistencies for P-sample nonmatches are the same as for P-sample matches, which might lead to underestimation of the inconsistency. Analyses carried out for the 2000 A.C.E. detected small biases for some poststrata (Bench 2002b, 2003).

20. To derive (5.21), observe that

$$\begin{aligned}
b_{\mathrm{corr},G} &= E\Big[\sum_h \sum_{S_{0h}} \delta_e(s) K_G(s)\big(\hat{A}^{**}(s) - \hat{A}^{***}(s)\big)\Big] \\
&= E\Big[\sum_h \Big\{\sum_{S_{0h}} \delta_e(s) K_G(s)\Big\}\big(\hat{A}_h^{**} - \hat{A}_h^{***}\big)\Big] \\
&= E\Big[\sum_h N_{ehG}\big(\hat{A}_h^{**} - \hat{A}_h^{***}\big)\Big] \\
&= \sum_h E\big[N_{ehG}\big]E\big[\hat{A}_h^{**} - \hat{A}_h^{***}\big] \\
&= \sum_h N_{0hG}\bar{p}_{e,hG} E\big[\hat{A}_h^{**} - \hat{A}_h^{***}\big], \\
&= -\sum_h N_{0hG}\bar{p}_{e,hG}(1 - \bar{p}_{E,h})\psi_h,
\end{aligned}$$

with the fourth equality following from assumption (ii) in Section 5.5.2. Show by a similar analysis that the synthetic bias for group G is given by (5.22) and that $b_{\text{res},G} = 0$.

*21. It is possible to adjust the estimate of correlation bias (5.21) for "other" bias, for example by $\gamma'_M = (X_M - \hat{b}_{X,M,o}) - R_Y(X_F - \hat{b}_{X,F,o})$. Show that γ'_M/θ_M is equal to

$$1 + \beta_{X,M} + \varepsilon_{X,M} - \theta_M^{-1}\hat{b}_{X,M,o} - \left(1 + \beta_{X,F} + \varepsilon_{X,F} - \theta_F^{-1}\hat{b}_{X,F,o}\right)\frac{(1 + \beta_{Y,M} + \varepsilon_{Y,M})}{(1 + \beta_{Y,F} + \varepsilon_{Y,F})}.$$

Show that if R_Y is equal to R_θ then γ'_M has the same distribution as

$$\theta_M\left[(\beta_{X,M} - \beta_{X,F}) + (\varepsilon_{X,M} - \varepsilon_{X,F})\right] - (\hat{b}_{X,M,o} - R_\theta\hat{b}_{X,F,o})$$

and has expected value $\theta_M(\beta_{X,M,c} - \beta_{X,F,c})$ and variance

$$\text{Var}(X_M - R_\theta X_F - \hat{b}_{X,M,o} + R_\theta\hat{b}_{X,F,o}).$$

Show that γ'_M has the approximate representation

$$\theta_M\left[\beta_{X,M} + \varepsilon_{X,M} - \theta_M^{-1}\hat{b}_{X,M,o} - \beta_{X,F} - \varepsilon_{X,F} + \theta_F^{-1}\hat{b}_{X,F,o} - \beta_{Y,M} - \varepsilon_{Y,M} + \beta_{Y,F} + \varepsilon_{Y,F}\right].$$

and hence that

$$E[\gamma'_M] \approx \theta_M\left[\beta_{X,M,c} - \beta_{X,F,c} - \beta_{Y,M} + \beta_{Y,F}\right]$$

and

$$\text{Var}(\gamma'_M) \approx \text{Var}(X_M - R_\theta X_F - (\hat{b}_{X,M,o} - \hat{b}_{X,F,o}) - (Y_M - R_\theta Y_F)).$$

22. Let D and S estimators such that D is unbiased, S has bias b, and the variance of $D - S$ is unbiasedly estimated by V. Show that the expected value of $(D - S)^2 - V$ is equal to b^2.

*23. Further consideration of nonparametric method of estimating SEB in Section 5.7. Let C_{hG} be the census count and E_{hG} be the (unweighted) E-sample count in poststratum h and group G and let $I_{hG} = 1$ if $E_{hG} \geq \tau$ and $= 0$ otherwise. Let π_{hG} denote the probability that $I_{hG} = 1$. (The calculation of π_{hG} is in principle straightforward, as the census numbers for each poststratum and each block are known, and the E-sample selection probabilities are known for all blocks.) Define the relative synthetic estimation bias of area G by

$$\text{RSEB} = \frac{E\left[\sum_h (A_h - A_{hG})C_{hG}I_{hG}/\pi_{hG}\right]}{E\left[\sum_h C_{hG}I_{hG}/\pi_{hG}\right]},$$

with the convention that if $\pi_{hG} = 0$ we set I_{hG}/π_{hG} equal to zero. Estimate RSEB by substituting in the sample estimates for A_h and A_{hG}, with the additional convention that if the estimate of A_{hG} is undefined we set I_{hG}/π_{hG} equal to zero. (Possibly, the E-sample estimates should receive first-order bias corrections.) Call the resulting estimator $\hat{R}_{1G}$. (A desirable simplification may be used if the probabilities π_{hG} can be taken to be unrelated to the synthetic estimation error; in that case we may ignore the π's in calculating $\hat{R}_{1G}$.) In case $\hat{R}_{1G}$ is be subject to large sampling variance, let B_{1G} and V_{1G} denote jackknife estimates of bias and variance of $\hat{R}_{1G}$ and define an estimate of the mean square relative synthetic estimation bias by

$$\hat{R}_{2G} = (\hat{R}_{1G} - B_{1G})^2 - V_{1G}.$$

The fraction of the population of group G for which synthetic bias can be estimated is $\gamma = \sum_h I_{hG} C_{hG} / \sum_h C_{hG}$. In presenting results, it could be useful to construct a table with the following information for groups for which γ exceeds some threshold: characteristics of the group (e.g., size, region, racial or ethnic composition, etc.), γ, $\hat{R}_{1G}$, B_{1G}, $\sqrt{V_{1G}}$, $\hat{R}_{2G}$. Although $\hat{R}_{1G}$ and $\hat{R}_{2G}$ for a single group G might be imprecise, averages and standard deviations within groupings possibly might be informative. Rao (2003, 46–53) provides further discussion and references.

24. Using the notation of Section 6, show that $E[\hat{\Delta}] - \Delta = 2(E[f(\mathbf{D})] - f(\mathbf{N}))(E[f(\mathbf{T})] - f(\boldsymbol{\theta}))$ if the variance estimates in $\hat{\Delta}$ are unbiased.

11
Financial Applications

Being able to quantify the expected error of demographic forecasts allows us tackle some economic problems from a fresh angle. We concentrate here on pensions, and consider how adaptive policies can be used to alleviate problems caused by demographic uncertainty. We also consider how sustainability and intergenerational equity might be assessed statistically. We will see that changing demographics can have surprising implications for policies seemingly unrelated to demography.

1. Predictive Distribution of Adjustment for Life Expectancy Change[1]

An increase in life expectancy puts a strain on the finances of a defined-benefit pension system, which we introduced in Section 2.2.4 of Chapter 4. In anticipation of future gains in life expectancy, Finland passed a law in 2003 that automatically adjusts pensions if life expectancy changes. The aim was to preserve the present value of future pensions. Thus, the system shifts part of the risk from future workers to future retirees.

1.1. Adjustment Factor for Mortality Change

The revised Finnish pension law (enacted in 2003 and taking effect in 2005) set $\beta = 62$ as the base age for pension, $\rho = 0.02$ as the discount rate such that the value of a euro received in age $\beta + x$ is worth $e^{-\rho x}$ at age β, and $T = 2009$ as the base year of the system. Suppose the pension is paid continuously at the rate of one euro per year. If $p(x)$ is the probability of surviving to age x, the *expected present value* of the pension conditional on survival to age β is (cf., (2.15) of Chapter 4)

$$\xi = \int_{\beta}^{\infty} e^{-\rho(x-\beta)} p(x)\, dx / p(\beta). \tag{1.1}$$

[1] This section is based on Alho (2003b).

The Finnish law stipulated that present values be estimated from data for the past five years. Specifically, consider the cohort of individuals who become β years old during calendar year $t \geq T$. For them, data from the period $[t-6, t-2)$ would be used. Let $D_x(u)$ be the number of deaths and $K_x(u)$ be the number of person years in age x, during year u. The o/e rate (cf., (1.3) of Chapter 4) in age x during $[t-6, t-2)$ is

$$\hat{\mu}_x(t) = \sum_{u=t-6}^{t-2} D_x(u) \Big/ \sum_{u=t-6}^{t-2} K_x(u). \tag{1.2}$$

Using the actuarial estimator (Example 2.9 of Chapter 4) the probabilities of survival could be estimated as

$$\hat{p}(x,t) = \prod_{y=0}^{x-1} \frac{2-\hat{\mu}_y(t)}{2+\hat{\mu}_y(t)}. \tag{1.3}$$

The trapezoid method (cf., (2.10) of Chapter 4) yields the approximate expected present value for year t as

$$\hat{\xi}(t) = 1/2 + \sum_{x=\beta+1}^{\infty} e^{-\rho(x-\beta)} \hat{p}(x,t)/\hat{p}(\beta,t). \tag{1.4}$$

Two technical remarks are in order. First, we know from Examples 2.2 and 2.9 of Chapter 4 that the actuarial estimator is based on a hypothesis that $p(.)$ is linear in the interval $[x, x+1)$. This contradicts the linearity hypothesis underlying the trapezoid method (Section 2.2.1, Chapter 4) that leads to (1.4). In Exercise 1 we give a more accurate formulation. Second, an alternative to (1.2) would be to average the annual rates $D_x(u)/K_x(u)$. We indicate in Complement 2 that the latter estimator has a higher variance.

If mortality changes, then the expected present values may also change. By multiplying the pension payout by the adjustment factor

$$A(t) = \xi(T)/\xi(t) \tag{1.5}$$

for $t \geq T$, the expected present value can be maintained at the level it was in year T. By the Finnish law the same factor was to be used for both females and males.

We conclude with two comments. First, for each new retirement cohort, $A(t)$ is calculated using by the most recent mortality data available. This is a period calculation, i.e., no attempt at forecasting the eventual life expectancy of the cohort is made. This can be justified on practical grounds, since cohort calculations would require forecasts going approximately 40 years into the future. Such forecasts and models underlying them would necessarily be uncertain, and might lead to political and other disagreements among the various parties involved. Second, despite the expediency and political realism of the procedure, the pensions will be paid to actual cohorts. An intriguing problem would be to assess the predictive distribution of the difference between expected present values as calculated based on cohort experience, and expected present values as calculated here.

1.2. Sampling Variation in Pension Adjustment Factors

Even if the underlying forces of mortality were constant over time, the number of deaths would vary from year to year due to Poisson variation. We will first determine how the pension adjustment factors would vary if this were the only source of variation.

As in Section 3.1 of Chapter 5, assume that deaths $D_x(u) \sim \text{Po}(\mu_x(u)K_x(u))$ are independent over x and u. To simplify, assume that neither the rates nor the person years change, so that $D_x(u) \sim \text{Po}(\mu_x K_x)$. In this case,

$$\hat{\mu}_x(t) = \sum_{u=t-6}^{t-2} D_x(u)/5K_x. \tag{1.6}$$

Ignore the Poisson variation in $\xi(T)$, and concentrate on years $t > T + 6$. It is reasonable to treat $\xi(t)$ as fixed because we are only concerned with uncertainty after the base year T.

To define the setting, we used the female life table of 2001 to approximate the combined male and female life table around 2025. Based on the life table we used (1.4) to estimate $\xi(T)$ at 17.50 euros. The age-structure was also taken to correspond to the female life table population of 2001. The size of the population was fixed at 1,000,000, a close approximation of the size of the age-group 62+ (both sexes combined) in Finland in 2002. The expected number of deaths in this population is 43,100. A Poisson count with this expectation has a CV of 0.005 (cf., Section 5 of Chapter 4). Thus, with a probability of approximately 95% the annual deaths are within $\pm 1\%$ of the expected value.

Under unchanging mortality we expect roughly that $A(t) = \xi(T)/\xi(t) \approx 1$. (Since $A(t)$ is a convex function of $\xi(t)$, we have that $E[A(t)] \geq 1$ by Jensen's inequality (Complement 8 of Chapter 4), but the CV of $\xi(t)$ is small enough that the bias is small.) Define the change in the adjustment factor for two calender years that are $k = 1, 2, \ldots$ years apart, as

$$\Delta(k) = A(t + k) - A(t). \tag{1.7}$$

This is the difference in pensions, per unit earned, between cohorts born k years apart. We have $E[\Delta(k)] \approx 0$ for all k, but we will determine how much variation one can expect from year to year.

Based on 1,000 simulations, the results are as follows. First, the standard deviation of $A(t)$ is 0.007. Second, the autocorrelations $\text{Corr}(A(t + k), A(t))$ are 0.80, 0.60, 0.43, 0.22 for $k = 1, 2, 3, 4$. We see from (1.6) that for $k \geq 5$ the theoretical correlation is zero. Third, the standard deviations of the $\Delta(k)$'s are 0.00042, 0.00060, 0.00072, 0.00084, 0.00097 for $k = 1, \ldots, 5$.

We see that for adjustment factors that are $k \geq 5$ years apart, pure sampling variability induces a standard deviation for (1.7) of 0.1%. Due to the use of 5-year data periods, adjustment factors that are closer ($k < 5$) are positively autocorrelated, so their differences have less variance. In particular, pure sampling variability induces a standard deviation of 0.04% for the difference (1.7) of consecutive years.

The distributions are approximately normal, so with a probability of 95%, pure sampling variability induces a difference that does not exceed $\pm 0.08\%$ between adjustment factors of consecutive years.

In summary, Poisson type sampling variability is small, even in a small country like Finland. If single-year data were used instead of data from five consecutive years, then all standard deviations would have to multiplied by approximately $5^{1/2} = 2.24$. The use of a five year period has the dual effect of both reducing variability and smoothing the correction factors over time.

1.3. The Predictive Distribution of the Pension Adjustment Factor

The essential source of uncertainty concerning the future values of the correction factors comes from the fact that the future mortality rates $\mu_x(t)$ are unknown. Empirical estimates derived in Alho (2002b) could be used. Suppose a forecast is made at jump-off time $t_0 < T$. In Alho (2003b) we took $t_0 = 2002$. Assume that for $t > t_0$ the rates are of the form

$$\mu_x(t) = \hat{\mu}_x(t) \exp(Y_x(t - t_0)), \tag{1.8}$$

where $\hat{\mu}_x(t)$ is the point forecast of age-specific mortality $u = t - t_0$ years ahead, and $Y_x(t - t_0)$ represents its relative error. The point forecast was specified by assuming that the rate of decline observed for each age and sex during recent past would continue indefinitely. For life expectancy this implied a median forecast of 86.7 for females and 81.8 for males, in 2050.

For the uncertainty we used the scaled model of Section 4.2 of Chapter 9 (where we wrote $X(j, u)$ for $Y_j(u)$). The scales were estimated from Finnish data from 1900–1994 to match the level of error in naive mortality forecasts (cf., Alho 1998). To a good approximation the standard deviation (SD) of the relative error was $0.032(0.15u^2 + 0.85u)^{1/2}$. At $u = 30$, for example, this implies a CV of 0.41. This indicates a higher level of uncertainty than that of Figure 4 of Chapter 8, but corresponds to the actual Finnish experience. The autocorrelation parameters for the η-terms were 0.945 for males and 0.888 for females. The autocorrelation parameters for the δ-terms were 0.977 for males and 0.979 for females. One can deduce that an 80% prediction interval for life expectancy in 2050 would be [83.3, 90.1] and [76.7, 86.4] for females and males, respectively.

To simplify the numerical analysis we used two approximations. First, we assumed that the cross-correlation of error increments between males and females is equal to 1, instead of the empirically estimated value of 0.795. As discussed in Exercise 3, this will not inflate standard deviations by more than 5%. The assumption implies that the relative errors of the males and females could be taken to be the same. Therefore, sex only needs to be considered when computing the point forecast for the average age-specific mortality of the two sexes. The required sex ratios in ages $x \geq \beta$ were available from the sex ratios of the median forecasts.

We denote the resulting point forecasts by $\hat{\lambda}_x(t)$, and the resulting model for the age-specific rate during $t > t_0$ is

$$\tilde{\lambda}_x(t) = \hat{\lambda}_x(t) \exp\left(\eta_x \sum_{u=1}^{t-t_0} S_x(u) + \sum_{u=1}^{t-t_0} \delta_x(u) S_x(u)\right). \tag{1.9}$$

Second, define $\hat{K}_x(u)$ as the median forecast of the person years in age x during year u. We simplified the calculations by using these values (instead of the random population values) to weight the random rates in the calculation of net present values. After the approximations, the average mortality rate needed in the calculation of the present values for years $t \geq T$ is

$$\tilde{\mu}_x(t) = \sum_{u=t-6}^{t-2} \tilde{\lambda}_x(u) \hat{K}_x(u) \Bigg/ \sum_{u=t-6}^{t-2} \hat{K}_x(u). \tag{1.10}$$

These rates lead to probabilities of survival via

$$\tilde{p}(x, t) = \prod_{y=0}^{x-1} \frac{2 - \tilde{\mu}_y(t)}{2 + \tilde{\mu}_y(t)}, \tag{1.11}$$

to expected present values via

$$\tilde{\xi}(t) = 1/2 + \sum_{x=\beta+1}^{\infty} e^{-\rho(x-\beta)} \tilde{p}(x, t) / \tilde{p}(\beta, t). \tag{1.12}$$

and to adjustment factors for $t > T$ via

$$\tilde{A}(t) = \tilde{\xi}(T) / \tilde{\xi}(t). \tag{1.13}$$

The numerical calculations were carried out by simulation (using Minitab). The median (M), the first and third quartiles (Q_1, Q_3), and the first and ninth deciles (d_1, d_9), for the predictive distribution of the adjustment factors in 2010–2060 were as follows:

year	d_1	Q_1	M	Q_3	d_9
2010	0.990	0.992	0.995	0.998	1.001
2020	0.915	0.933	0.953	0.973	0.995
2030	0.863	0.884	0.918	0.951	0.985
2040	0.814	0.842	0.889	0.936	0.983
2050	0.778	0.811	0.865	0.921	0.982
2060	0.751	0.787	0.843	0.905	0.974

We expect the adjustment factor to decline to about 0.87 in 2050, with an 80% prediction interval [0.78, 0.98]. These intervals are valid provided that the volatility of the trends of mortality during the next 50 years does not exceed the volatility of mortality during 1900–1994. Figure 1 has the corresponding data for 2010–2060. We note here that in the preparation of the legislation a number of alternative

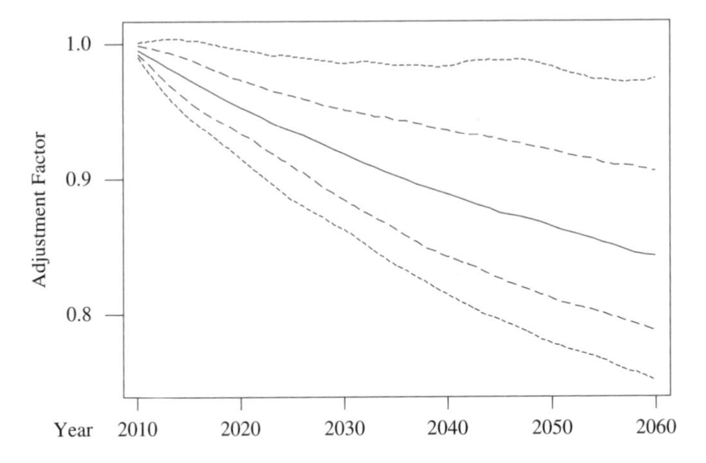

FIGURE 1. Predictive Distribution of the Adjustment Factor in 2010–2060: Median (Solid), First and Third Quartiles (Dashed), and First and Ninth Deciles (Dotted).

scenarios were considered. However, although the scenarios were viewed as spanning a plausible range of future contingencies, they were essentially within the area between the median and the third quartile of Figure 1 (i.e., between the second and third graphs from the top).

2. Fertility Dependent Pension Benefits[2]

The main earnings-related pension system in Finland is based on the Employees' Pensions Act (TEL, a Finnish abbreviation) of the year 1962. TEL is a defined-benefit system, in which the level of future pension evolves according to complex rules that depend on time worked, type of employment, and level of salary. The system is mostly PAYG – recall that PAYG means that current workers pay contributions at time t to cover payments to pensioners at time. From the start, partial prefunding was established, so that part of the contributions are set side into a fund to cover future pensions. By a *pension right* we refer to the level of pension a person is entitled to, for as long as he or she lives, based on his or her work history. The main rule for the accrual of pension rights was (until the reform of 2005) that rights accrue in ages 23–64 at the rate of 1.5% of the wage. Part of this, the rights that accrue in ages 23–54, are prefunded to cover 0.5% of the wage. Specifically, let $p(x, t)$ denote the probability of surviving to age x according to the life table used during period t and let ρ be the interest rate. Then,

$$h(x, t) = 0.005 \int_{65}^{\infty} e^{-\rho(y-x)} p(y, t)\, dy / p(x, t) \tag{2.1}$$

[2] This section is draws heavily from Lassila and Valkonen (1999).

is the prefunding in ages $x \in [23, 55)$ during year t, for a unit wage. In addition to $h(x, t)$, contributions are collected from workers at t to cover the benefits that the pensioners are entitled to at t, but that have not been covered by earlier prefunding. This is the source of a possible inequity across cohorts, because if the cohort of workers is small relative to size of the cohort of pensioners, the workers' contributions may greatly exceed their own future pension returns.

However, suppose that (2.1) were actually a nominal target of prefunding, but the actual prefunding would be more or less: small cohorts would prefund less, large cohorts would prefund more than the target, if we make the actual level depend on cohort size in suitable way.[3] In a stylized fashion, the proposal of Lassila and Valkonen (1999) can be described as follows. Define $B(t)$ as the density of births at time t. The ages that are subject to prefunding are in age $x \in [23, 55)$. Those in age x at time t were born at $t - x$, and we would like to compare $B(t - x)$ to the density of births later, $B(t - y), 0 \leq y < x$, in some way. Define a coefficient

$$\gamma(x, t) = B(t - x) \Bigg/ \int_0^x w(y, x, t) B(t - y)\, dy, \tag{2.2}$$

where the weights $w(y, x, t) \geq 0$ integrate (over y) to 1 for each x and t. The optimal choice of the weights is an open question. One possibility is to choose $w(y, x, t)$ proportional to the expected number of individuals born at $t - y$ who are in contributing age at the time of retirement of those who are in age x at t. In any case, we then have that for large cohorts $\gamma(x, t) > 1$ and for small cohorts $0 < \gamma(x, t) < 1$. The prefunding payment collected would then be $\gamma(x, t)h(x, t)$ per unit wage, instead of $h(x, t)$ given by (2.1).

How would such a rule work in practice? The Finnish Overlapping Generations (*FOG*) model (developed at the Research Institute of the Finnish Economy) is an economic equilibrium model of type proposed by Auerbach and Kotlikoff (1987). From a demographic point of view it is a cohort-component book-keeping system (cf., Chapter 6), in which one keeps track of the work force by age. Each cohort optimizes its involvement in the labor market and leisure time, in different ages, according to a life time plan. The model has markets that clear every period. Enterprises maximize shareholder value by deciding how much capital and labor to use during each period, according to the marginal productivity of labor. The wage rate is determined by equating supply and demand. There is a capital market, in which saving and investment are balanced by the domestic interest rate. The model also includes a government with an intertemporal budget constraint.

In a defined-benefit pension system of the Finnish type, pension depends on earlier wages. Prefunding depends on wages, so the pension system can be made a part of the FOG model. Given a set of prefunding rules, the model endogenously determines the time paths of wages, prices, contributions, and replacement rates. Demographic development influences the system via the sizes of the cohorts,

---

[3] This notion of cohort size is similar to relative cohort size in Easterlin's (1961) theory of fertility cycles.

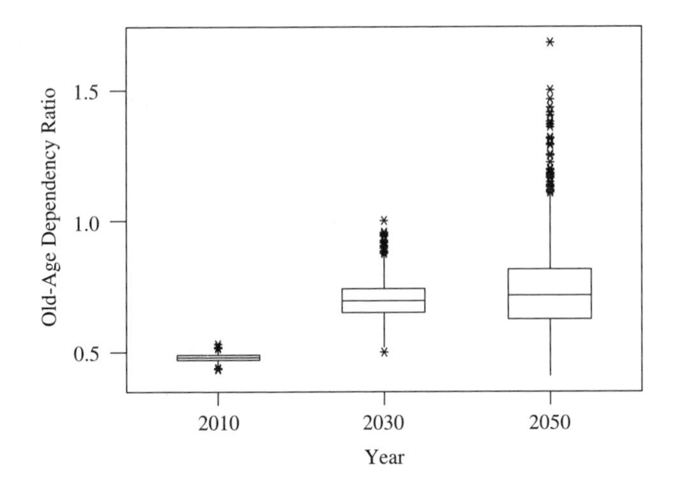

FIGURE 2. Predictive Distribution of Old-Age Dependency Ratio (Ages 60+/Ages 20–59) in Finland in 2010, 2030, and 2050. (based on 1500 sample paths.)

mortality, and net migration. The model calculations are based on a perfect foresight hypothesis.

Lassila and Valkonen (1999) used two carefully chosen sample paths of the population vector produced by PEP (Section 6 of Chapter 9) to provide alternative scenarios to the base calculation that used the median forecast. The sample paths were chosen so that they corresponded to the lower and upper end points of the 80% prediction interval of the *old-age dependency ratio* (in this case the ratio of those in ages 60+ to those in ages 20–59) in 2030. Figure 2 has a boxplot of the predictive distribution of the old-age dependency ratio of Finland based on Alho (1998). The end points of the 80% prediction interval [0.61, 0.78] in 2030 roughly correspond to the midpoints of the whiskers. The number of simulated sample paths was 1,500.

The findings came as a shock, since no-one had earlier contemplated, with any seriousness, alternative population scenarios that would be so different from the median. In particular, the sample path corresponding to the ninth decile of the predictive distribution of the old age dependency ratio (0.78) would have dramatic implications. The possibility of experiencing unexpectedly low fertility with unexpectedly low mortality had not been sufficiently emphasized in the studies of aging.

Figure 3 gives the proportion of total pension contributions of the total wages, for the years 1999–2070, under the high old-age dependency ratio scenario. Initially, current rules lead to a lower contribution level, but in about 50 years, the adaptive rule of Lassila and Valkonen (1999) cuts the extreme levels. In contrast, under the median forecast variant (details not shown) both rules lead to essentially the same contributions that rise roughly linearly to only 30%, at 2070 (for extensions, see Alho, Lassila and Valkonen 2005).

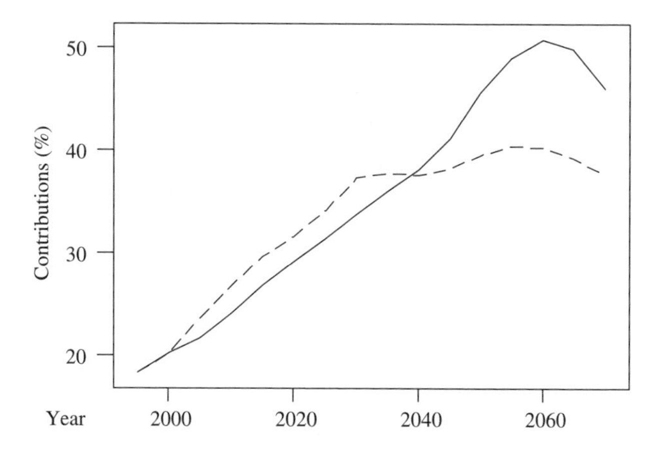

FIGURE 3. Pension Contributions, as Percent of the Total Wages of the Covered Employees, in Finland in 1995–2070, Under Current Rules (Solid) and Under a Fertility Dependent Rule (Dashed), If the Population Follows the High Old-Age Dependency Ratio Variant.

3. Measuring Sustainability[4]

A well-known definition of the sustainability of fiscal policies is the OECD view: "Sustainability is basically about good housekeeping. It is essentially about whether, based on the policy currently on books, a government is headed towards excessive debt accumulation." (Blanchard et al. 1990, 8). This means that the government debt should stay below an acceptable level in the long term (cf., Section 5). In the case of pension systems, sustainability requires contribution rates to remain sufficiently low and replacement rates sufficiently high that the system remains viable. If the likelihoods of the various risks were known, both the current and future pensioners and the current and future workers could prepare for them in a rational manner. Reducing the scope of the unexpected is a natural aspect of sustainability.

Time horizon is also important. Valdés-Prieto (2000) noted that long-term financial stability can be irrelevant if sufficiently large imbalances will cause a political intervention. Thus, neither from contributors' nor pensioners' point of view can sustainability be simply viewed in terms of average contributions and average replacement during any chosen period. It may be that if certain boundaries are crossed, then the individuals affected will feel sufficiently threatened to try to use the political process to change the rules.

Using the Lithuanian forecast mentioned in Section 4.4 of Chapter 8 as a backdrop, we will now consider a proposed reform of the Lithuanian pension system. The existing system consisted of a basic pension and a supplementary component. The system was purely PAYG. The pension was fully indexed to the average wage, so an increase in average wage by p percent was matched by an increase in pensions

---

[4] For a more detailed analysis, see Alho, Lassila and Valkonen (2005).

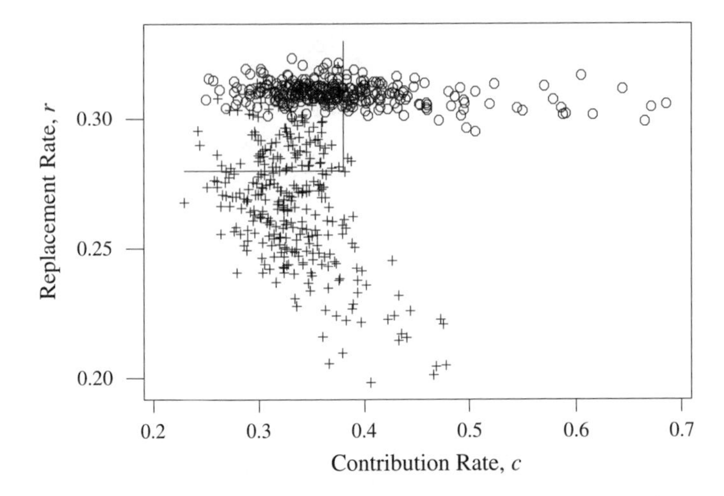

FIGURE 4. Replacement Rate and Contribution Rate Under Full Wages Indexation (o) and Full Wage-Bill Indexation (+), and an Example of Potential Viable Region $\{(c, r)|c \leq 0.38, r \geq 0.28\}$.

of p percent. The forecast of population indicated that the current age-dependency ratio would persist for a decade or so, but then would increase rapidly. The population aging would put a heavy burden on the working population. If pensions could made to depend on wage-bill (i.e., the total sum of wages) instead of average wage, then the burden of the small future work force might be alleviated.

Partial indexing to wage-bill was considered such that the pensions would follow the total wage-bill with weight $0 \leq \alpha \leq 1$, and the average wage with weight $1 - \alpha$. As in the previous section, an overlapping generations (*OLG*) model of the Lithuanian economy (similar to that of the FOG model) was used to predict the consequences of the various policy choices. The stochastics derived from the program PEP produced population paths that were fed to the OLG model. For each path a different optimal wage, consumption, and savings pattern emerged that implied a *contribution rate* c, i.e., the fraction of wages going into the pension system, and a *replacement rate* r, i.e., the ratio of average pension to average wage. How would wage-bill indexation work?

Figure 4 has a plot of the pairs (c, r) corresponding to 300 population paths. We see that under wage indexation ($\alpha = 0.0$) replacement rates are nearly constant (median $= 0.31$, SD $= 0.004$), irrespective of population development, but contribution rates are quite uncertain (median $= 0.36$, SD $= 0.07$). Under wage-bill indexation the replacement rates are reduced and less certain (median $= 0.27$, SD $= 0.02$), but this does lead to contribution rates that are both lower and less uncertain (median $= 0.33$, SD $= 0.04$). Thus, wage-bill indexation does provide the relief promised.

We see that different degrees of wage-bill indexation have an effect on both the mean and variance of the outcomes. Thus, *variance reducing policies* may be popular among a risk averse population. Note that current workers are future pensioners, so in the long term reducing variance should be appealing to everyone.

However, as noted above, studying the moments of the joint distribution of the pairs (c, r) may hide some political realities. Under wage-bill indexation both contribution rates and replacement rates depend on uncertain future demographics. As an approximation to the complex political process, one can postulate an *upper bound* $c^* > 0$ such that contribution rates $c > c^*$ would be considered politically unacceptable and a *lower bound* $r^* > 0$ such that replacement rates $r < r^*$ would be considered politically unacceptable. Thus, the set $\{(c, r)|c \leq c^*, r \geq r^*\}$ is the *viable region* of the policy, outside which a political reform is initiated (cf., Aubin 1991). The upper left hand corner of Figure 4 provides an illustration with $c^* = 0.38$ and $r^* = 0.28$. Although a reform would typically end in a compromise of some sort, the outcome is uncertain for all involved.

Neither c^* nor r^* can be known with any certainty. The value $c^* = 0.38$ represents a 10 percentage points increase in the contribution rate relative to the current level. Many countries face the prospect of such increases in their payroll taxes, and it is exactly these projections that have led the experts and decision-makers to pay attention to the aging problem and seek for ways to avoid such increases. We omit the details here (see Alho, Lassila and Valkonen 2005) but note that, as one would expect, the result depends on what r^* might be. For $r^* = 0.30$, the probability that the pair (c, r) is within the viable region in 2050 is at most 63%, a value that is achieved at $\alpha = 0.0$ (i.e., 63% of the population paths produced by PEP lead to a pair (c, r) that was within the region in 2050). For $r^* = 0.20$ the probability of being within the viable region can be made as high as 90%, if full wage-bill indexation ($\alpha = 1.0$) is adopted. In the case of Figure 4, the probability is 63% of staying in the viable region when $\alpha = 0.0$, 30% when $\alpha = 1.0$, and (not shown) the maximum is 72% when $\alpha = 0.4$.

Other measures for operationalizing sustainability can be developed, based on the minimal r and maximal c over the forecast horizon (2001–2050), for example.

4. State Aid to Municipalities[5]

Substantial amounts of funding from national to local governments are determined by statistical allocation formulas (cf., Chapter 12, Section 3). It has long been recognized that errors in the data can have an appreciable impact on the allocations (Spencer 1980a; U.S. Federal Committee on Statistical Methodology 1978; Louis, Jabine, and Gerstein 2003). Statisticians have paid less attention to the stability of the system, when the true values themselves change over time (cf., Chapter 12, Section 3.1.2).

Finland is divided into approximately 450 municipalities that range from Helsinki with 1/2 million inhabitants to villages of a few hundred people. The median size of a municipality is 5,000 inhabitants. The municipalities are obligated

[5] This section is largely based on Alho and Salo (1998).

by law to provide basic educational, health, and social services to their inhabitants. These are financed in part by municipal taxes collected at a flat rate of 15–20%. In part they are covered by state aid that derives from the national income tax and other taxes and fees collected by the government. The system is entirely PAYG. The aid is allocated by statistical formulas that depend primarily on population size and age structure. We concentrate here on health and social services.

Suppose $V(x, t)$ is the number of people in age x during year t. Suppose $s(x)$ is the amount of state aid allocated for health and social care per inhabitant in age x. In 1997 these amounts (in euros) were the sum of the "Health" and "Social" columns given below:

Age x	Health	Social	Sum
0–6	460	3,775	4,235
7–64	510	255	765
65–74	1,230	460	1,690
75–84	2,315	2,560	4,875
85+	3,895	7,140	11,035

These are nominal allocations that are based on estimates of past age-specific costs. Although they are adjusted in various ways to come up with the final allocations, the basic age-dependence is as indicated: on average a person aged 0–6 or 75+ was allocated about 8 times as much as a person aged 7–74. Define $V_{19-64}(t)$ as the size of population in ages 19–64 during year t. It follows that $C(t) = \Sigma_x s(x)V(x, t)/V_{19-64}(t)$ is a measure of the tax burden posed by the state aid system on tax payers. Using the year 1997 as a base year we can further define a standardized burden as $D(t) = C(t)/C(1997)$.

We can investigate the burden both retrospectively, in terms of *what it would have been, had the same legislation been in force earlier*, and prospectively, by calculating *what it will be if the legislation continues to be in force*. We can use the stochastic forecast of the population to produce a predictive distribution of $D(t)$. Figure 5 shows the past hypothetical development of $D(t)$ in 1940–1997 and prediction intervals for 1998–2050. We see that the past burden would have been within ± 10% of the level in 1997. However, the predictive distribution shows that even the first decile rapidly increases after the year 2010, and the burden may very well increase by a half or more in the coming decades.

The reason for the rapid increase in burden is the strong age-dependence of the allocations. One can calculate what $D(t)$ would look like, if the allocations would have been the same in all ages (i.e., $s(x) \equiv s$ for all x). The resulting curve (not shown) is notably flat in the past, and the median of its predictive distribution only rises to less than 1.1 (instead of 1.5) by 2050. Figure 5 shows that from the perspective of this piece of legislation, the change in population age structure is much more dramatic than anything seen since 1940, even if we use the first decile of the predictive distribution, let alone the median or a higher quantile, as a guide. While it is true that actual costs depend on age, we expect that there will be

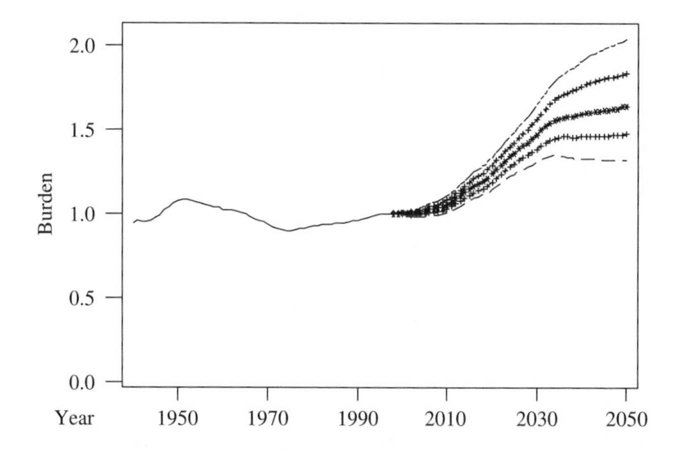

FIGURE 5. Relative Burden of Social and Health Care Allocations in 1940–1997 in Finland, and the Median(*), Quartiles (+), and 1st and 9th Deciles (.) of its Predictive Distribution in 1998–2050.

continuing debate as to the acceptable level of the burden. This has implications that have not been seriously considered.

For example, one can show that if the allocations were to be made less age dependent then the winners would be the urban Helsinki region and northern Lappland (home of the Sami, see Example 1.2 of Chapter 2), an otherwise unlikely sounding alliance! The losers would be the coastal municipalities, many of which have Swedish speakers as a majority, where mortality is low and populations are, therefore, older than elsewhere. Possibly, changing demographics will expand debates on social and health policy to include discussions of regional and ethnic equity.

Alternatively, the government might decrease allocations across the board. This might be politically more acceptable, but the cuts would be borne most heavily by those cohorts that are young (ages 0–6) or old (ages 75+) after 2010, who would be disadvantaged relative to those who were in such ages before 2010. A problem of intergenerational inequity (cf., Auerbach, Kotlikoff, and Leibfritz 1997) would arise. This is the theme of the next section.

5. Public Liabilities[6]

Part of the inheritance we leave to future generations is the positive or negative public net wealth. Stochastic forecasts have been used to analyze the solvency of the Social Security Trust Fund in the United States (cf., Social Security Administration 2004, Burdick, Manchester, and Bang 2003, Lee and Tuljapurkar 1998, Lee 2000,

[6] This section is based on a paper presented at the International Meeting on Age Structure Transitions and Policy Dynamics: The Allocation of Public and Private Resources Across Generations, in Taipei, December 2001, sponsored by the International Union for the Scientific Study of Population. For further details, see Alho and Vanne (2005).

Auerbach and Lee 2001). In the following analysis, we consider the whole of public sector finances. In fact, we define the public sector in a rather broad manner (appropriate in the Finnish context) to include not only states and municipalities but also institutions that manage statutory pensions (such as TEL, discussed in Section 2) and unemployment and disability insurance. In Finland, many of the latter institutions are privately owned.

Intertemporal public liabilities (IPL) can be defined as IPL = (current net debt) + (discounted future entitlements) − (discounted future taxes) (Raffelhüschen 1999). The uncertainty of future demographics influences the IPL via future population size and age-structure. To simplify, we assume that the uncertainty of the future economy influences the IPL via three processes. First, the value of government debt varies according to the real interest rate of government bonds. Second, the value of stocks held by the government varies like a large portfolio of publicly traded stocks.[7] Third, the tax rates and the entitlements follow the productivity of labor. Finally, we assume that the evolution of future demographics is independent of the evolution of the economic variables.

All economic calculations below are inflation adjusted.

5.1. Economic Series

Using the time-series models of Chapter 7, the economic series were analyzed as follows. First, the German real interest rate series from 1955–2000 was used to estimate a model for the future interest rates of the government bonds. The series looks roughly stationary with mean 0.038 (SE = 0.0032). Deviations from the mean can be approximated by an AR(1) process: $\rho_t = 0.62\rho_{t-1} + \varepsilon_{1,t}$, where $\varepsilon_{1,t} \sim N(0, 0.0082^2)$.

The Dow-Jones stock index series from 1949–2000 was used to model stock returns. The series has a mean of 0.054 (0.019), and a standard deviation of 0.134. The series displays some positive autocorrelations, but in general a stock index series such as this should be close to an uncorrelated series. This was assumed.

Productivity was measured via the real Gross Domestic Product (*GDP*) *per capita* series from Finland in 1860–2000 (Hjerppe 1989, Statistics Finland 2001). It reflects both productivity per person, labor force participation, and unemployment rate, for males and females. The series varies around a mean 0.022 (standard error 0.005). It is reasonably well approximated by an MA(1) process $\gamma_t = \varepsilon_{3,t} + 0.32 \times \varepsilon_{3,t-1}$, where $\varepsilon_{3,t} \sim N(0, 0.045^2)$.

5.2. Wealth in Terms of Random Returns and Discounting

Putting the various money flows on an equal footing at jump-off time requires an assumption about the way the government portfolio is managed. Assuming, as we do, that the government maintains a *fixed value portfolio* leads to a solution that

[7] In Finland the government has negative debt that derives from the prefunding of the pension system and from government owned stocks in companies it has helped to found. Another country with negative debt is Norway, due to oil revenues.

is similar in spirit to the assumptions made in generational accounting (Auerbach, Gokhale, and Kotlikoff 1991, Auerbach, Kotlikoff, and Leibfritz 1997). Under this policy, if the stocks change in value, then enough are sold or bought, so that the value of the holdings does not change (inflation adjusted). A meaningful discount rate for any future income or out-payments is the expected real interest rate of bonds, or 0.038.

The wealth W at $t = 2000$ can be written as $W = W_1 + W_2$, where W_1 is the debt owed primarily in terms of bonds, and W_2 represents the wealth owned primarily as stocks. Both can be viewed as a discounted sum of the income and out-payments they generate, if held at current level. As discussed in Alho and Vanne (2005), we can then write $W_1 = -0.323 \times (1 + X_1)$ in the units of the year 2000 GDP, where $E[X_1] = 0$ and $\text{Var}(X_1)^{1/2} = 0.111$. Similarly, $W_2 = 0.913 \times (1.39 + X_2)$, where $E[X_2] = 0$ and $\text{Var}(X_2)^{1/2} = 0.684$. In this example the wealth in stocks is six times more volatile than the debt owed in bonds.

The two components were taken to be independent, and W had an approximate normal distribution with the expectation $E[W] = -0.323 + 1.272 = 0.950$, and standard deviation $\text{Var}(W)^{1/2} = ((0.323 \times 0.111)^2 + (0.913 \times 0.684)^2)^{1/2} = 0.626$. The conclusion is that with probability 95% the value of the current wealth is in the range $[-0.28, 2.18]$, in the units of the GDP of year 2000, under a constant portfolio policy.

5.3. Random Public Liability

To include the future net taxes into the calculation, let $V(x, s, t)$ be the population in age x, who are of sex $s = M, F$, at time t. Let $\hat{S}(x, s)$ and $\hat{T}(x, s)$ be the entitlements and taxes, at $t = 2000$. We assumed that the entitlements and taxes grow with productivity. The difference in the discount rate and the growth rate of the GDP is $0.038 - 0.022 = 0.016$, so the discounted entitlements were of the form

$$S = \sum_{x,s} \sum_{u=1}^{\infty} V(x, s, t+u) \hat{S}(x, s) \exp\left(-0.016u + u(\mu - 0.022) + \sum_{k=1}^{u} \gamma_{t+k}\right). \tag{5.1}$$

Here the role of $\mu \sim N(0.022, 0.005^2)$ is to represent the uncertainty in the mean rate of increase in productivity, and γ is an MA(1) process of the type considered at the end of Section 5.1. Replacing $\hat{S}$ in (5.1) by taxes $\hat{T}$ gives the formula for the discounted taxes.

We can now define the *random public liability* as

$$L = S - T - W. \tag{5.2}$$

This random version of the IPL has a probability distribution that it inherits from S, T, and W.

Combining population sample paths from PEP with simulated paths of the economic processes, simulated values of S, T, and W were obtained. Truncating the calculation of S and T after year 2100, we obtained the following statistics for the predictive distribution,

	Mean	Median	Q_1	Q_3	SD
S − T	1.01	0.70	0.08	1.59	1.50
−W	−0.95	−0.95	−1.36	0.53	0.62
L	0.07	−0.18	−0.95	0.75	1.63

In terms of standard deviations, the uncertainty of the future primary balances $S - T$ is 2.5 times as high as the uncertainty of the current wealth W. The uncertainty of the future liability L approximately equals that of the future primary balances.

In addition, we can decompose the variance in $S - T$ into a component deriving from the uncertainty of future population development and a component deriving from the uncertainty in future productivity. The result is (details omitted) that the share of demographics out of Var(L) is 30%. The rest is due to economics and possible interactions. The share of uncertain future taxes and entitlements is 54%. The remainder, or approximately 15%, is due to uncertainty in W, under the fixed value portfolio policy. Or, contrary to the generally held view, demographics is a major component of uncertainty in the evaluation of the fiscal soundness of the public sector.

Exercises and Complements (*)

*1. Assume that $p(x + y) = p_x + (p_{x+1} - p_x)y$ for $y \in [0, 1)$, so $p(.)$ is linear on $[x, x + 1)$ with $p(x) = p_x$ and $p(x + 1) = p_{x+1}$. By a direct calculation one can show that

$$\int_x^{x+1} p(z)e^{-\rho z}\,dz = e^{-\rho x}\{p_x(1 - e^{-\rho})/\rho + (p_{x+1} - p_x)(1 - \rho e^{-\rho} - e^{-\rho})/\rho^2\}.$$

This leads to an alternative to (1.4). Verify that in the case $\rho = 0.02$, the formula gets the form

$$\xi' = 0.496683 + 1.00003 \sum_{x=\beta+1}^{\infty} \hat{p}(x)e^{-0.02(x-\beta)}/\hat{p}(\beta).$$

*2. Assume that $D(u) \sim \text{Po}(mK(u))$ are independent, and define

$$\hat{m} = \sum_{u=1}^{n} D(u) \Big/ \sum_{u=1}^{n} K(u), \quad \tilde{m} = \frac{1}{n}\sum_{u=1}^{n} D(u)/K(u).$$

Show that,

$$\mathrm{Var}(\hat{m}) = m \Big/ \sum_{u=1}^{n} K(u), \quad \mathrm{Var}(\tilde{m}) = \frac{m}{n^2} \sum_{u=1}^{n} 1/K(u).$$

The first variance is smaller because, by Jensen's inequality, we have that

$$1 \Big/ \frac{1}{n} \sum_{u=1}^{n} K(u) \leq \frac{1}{n} \sum_{u=1}^{n} 1/K(u),$$

with equality only if $K(1) = \cdots = K(n)$. Why is Jensen's inequality applicable here?

3. Suppose $\mathrm{Var}(X_1) = 1$, $\mathrm{Var}(X_2) = c^2$, $0 < c \leq 1$, with correlation $\rho(X_1, X_2) = \rho$. Then, $\mathrm{Var}(X_1 + X_2) = 1 + c^2 + 2\rho c$. If $\rho = 1$, the variance is $(1 + c)^2$. Define $f(c) = (1 + c^2 + 2\rho c)/(1 + c)^2$. Show that the minimum is at $c = 1$, with $f(1) = (1 + \rho)/2$. If $\rho = 0.795$ show that assuming $\rho = 1$ does not inflate the standard deviation of the sum by more than 5%, no matter what c is.

12 Decision Analysis and Small Area Estimates

1. Introduction

Now that we are equipped with methods to assess the accuracy of estimates of current or future demographic quantities, we consider questions related to the importance of accuracy.

Does it matter in practice whether demographic estimates are accurate? Is the accuracy of the estimates acceptable? If not, can the accuracy be improved with the expenditure of more money (or resources more generally) on data, analysis, research, and interpretation?

How might you want to modify your use of the estimates given knowledge of their accuracy? For example, if consequences of underestimates are more severe than overestimates, should one choose estimates with asymmetric error distributions? If forecasts are highly uncertain, should one attempt to make flexible plans (such as those discussed in Sections 1 and 2 of Chapter 11) that can adapt over time?

Should one spend more money (or less money) for development of the estimates? How can one find an optimal or at least satisfactory balance between consequences of errors in estimates and costs of improving accuracy of the estimates?

Statistical decision theory provides a framework for thinking about such questions in a systematic way. To apply the theory one must take careful account of the context of the uses of the data. Some uses are not easily tied to actions, for example uses of demographic data in social science research. Other uses are directly tied to allocations, such as uses of U.S. census data to apportion the House of Representatives and Electoral College and to allocate government moneys among states and local governments (Section 4 of Chapter 11; Louis, Jabine, and Gerstein 2003). Such uses depend on estimates for "small areas", which we will take to be any geographic level below the national level. Allocative uses of data tend to be relatively easy to know, at least in principle, when the allocations are based on a specified formula. Some related uses are difficult to analyze, however, such as drawing of legislative districts with equal population size, where changes in estimates at very small geographic levels lead to changes in the shape of the district, and the mapping between the (high-dimensional) set of population estimates and the set of possible shapes of districts is determined partly by a computer program and partly by back room negotiations. Other uses are intermediate in specificity between scientific inference and allocative uses, and include uses of data for planning purposes (e.g., public and private pension funding, understanding the need for new public works) and public policy interventions (e.g., to change population growth rates from fertility or from net immigration).

In Section 2 we discuss some aspects of estimation for small areas. Such estimates are important for formula-based allocations of public benefits (Section 3). A brief introduction to statistical decision theory (Section 4) allows us to separate the analysis of uncertainty in the estimates (as in prior chapters) from the assessment of consequences of errors in estimates by means of what is called a loss function. Decision theory allows us to jointly consider uncertainty in the estimates and consequences of errors by looking at the expected value of the loss function. We can also compare the expected losses of alternative estimators, specifically estimates of census population based on the census alone or supplemented by estimates of coverage error, so-called "unadjusted" and "adjusted" estimates, respectively, to see which has more accuracy for certain kinds of uses (Section 5). That information can be used in a schematic decision analysis for choosing whether to adjust the census or not (Section 6). Finally, by comparing changes in the expected loss with changes in cost of developing estimates, we can conduct cost-benefit analysis of demographic data (Section 7). This is enormously difficult in practice, but the analysis can provide useful insight nonetheless.

2. Small Area Analysis

Methods for demographic estimation for small areas tend to differ from methods for national estimates, due to differences in data availability and underlying population dynamics. A distinction is made between "direct" and "indirect" estimates. A *direct estimate* for an area depends only on data for the area itself, e.g., fertility estimates depend only on fertility data for the area and not on fertility in other areas. An *indirect estimate* relies on data for other areas, e.g., fertility estimates are based on data aggregated across areas deemed similar. Methods for subnational forecasts are discussed by Smith, Tayman, and Swanson (2001) and methods for postcensal estimates are described by Rao (2003, Chapter 3) and in the U.K. context by Rees, Norma, and Brown (2004).

For example, if the postcensal population estimate for an area is based on a housing unit method (Chapter 10, Section 1), where the average number of persons per housing unit in the area is estimated from the latest census and the total number of housing units is estimated from current data on utility connections, the estimate is a direct estimate. If the average number of persons per housing unit is based on a regression analysis applied to current survey data from an aggregation of areas, the estimate would be an indirect estimate. Other examples of indirect estimation are the use of Poisson regression for the estimation of counts in municipalities and other small areas (Example 3.5 of Chapter 5), and the synthetic estimates that adjust for net undercount in a census (Chapter 10). These examples highlight the fact that some small areas (here, municipalities) are truly "small" in the sense that Poisson type variability is an important concern, and others may actually be quite large (states or large cities).

The important distinction for our purposes is the dependence on a model – indirect estimates depend on assumptions of homogeneity of one sort or another, whereas direct estimates may rely on different assumptions to overcome limitations in the data or else may be subject to greater errors due to data limitations, e.g., increased sampling variability. Sometimes even direct estimates are partly affected by external data, e.g., state estimates may be calibrated to sum to a national estimate, for a so-called "modified direct" estimator (Rao 2003).

From the point of view of accuracy, we need to recognize that either direct or indirect estimates may be affected by sampling variability, by biases in the statistics used to develop the estimates (e.g., biases in administrative data or nonsampling errors in survey statistics), and biases arising from model imperfection. For indirect estimates, some of the latter will arise from aggregation across areas, but model error may well be present in direct estimates as well. Two approaches for evaluating the accuracy of the estimates include consideration of the various sources of error and their propagation, as we have done for national population forecasts in Chapter 9, as well as examination of past performance of the estimates (along the lines of Keyfitz 1981, Stoto 1983, Keilman 1990, 1997, 1998). While the same kinds of methods we have discussed for analyzing accuracy in national estimates may apply to small area estimates, the empirical findings may of course differ. For example, Smith and Tayman (2003) found different age-patterns of errors for national and subnational forecasts in the U.S. Bryan (2004) summarizes some evaluations of postcensal estimates for small areas, including general findings that the magnitude of relative error varies inversely with the size of the area and that population change (positive or negative) tends to be underestimated.

3. Formula-Based Allocations

3.1. Theoretical Construction

Formulas often are used by governments and large organizations to determine the amounts of benefits or costs to be shared among members. Currently, the federal government in the U.S. allocates more than $250 billion to state and local governments via formulas depending on statistics such as population, income, and tax revenues (Louis, Jabine, and Gerstein 2003), and state and local governments allocate yet additional monies via their own formulas, e.g., state educational funding for school districts. Smith, Rice, and Carr-Hill (2001) describe the use of formulas to allocate large amounts of funds in the United Kingdom. In Finland formulas are used to allocate funds to municipalities (cf., Complement 24, Chapter 5 and Section 4, Chapter 11) and to fund universities (Alho and Salo 1998). Formulas are also used by international organizations such as the United Nations

(Suzara 2002) and the European Unition to allocate the organization's costs among its members.

3.1.1. Apportionment of the U.S. House of Representatives

The U.S. Constitution provided for a Congress consisting of a Senate with two members from each state and a House of Representatives with membership "apportioned among the several States which may be included within this Union, according to their respective Numbers". The Constitution included explicit constraints that each state have at least 1 representative and the number of representatives does not exceed 1 for every 30,000 people, thereby setting upper and lower bounds on the number of representatives. The question of who counts as a person (and how much) was settled as a result of political compromise. Originally, the "numbers" included women and children even though they did not vote, excluded Indians not taxed (who were then viewed as outside the scope of U.S. law) and, as a compromise, included only three fifths of the number of slaves (Chapter 2, Example 1.1).[1] The question of whether illegal immigrants should be included in the numbers remains contentious to this day (Citro, Cork, and Norwood 2004, 38).

Given a set of population numbers, there exist many different methods to determine House apportionment. Let h denote the number of seats to be allocated, let p_i denote the population in state i, let p denote the sum of the state populations, and define $a_i = hp_i/p$. If there were no constraints on the apportionments, state i would receive a_i seats in the House. To accommodate the constraints noted above and the implicit constraint that the apportionments be integer, one might consider rounding each a_i up or down to the nearest integer.[2] For example, one might require that any $a_i < 1$ gets rounded up to 1 and remaining a_i's be rounded up or down, perhaps based on ranking the remainder $r_i = a_i - \lfloor a_i \rfloor$ or the relative size r_i/a_i, etc. The former method was advocated by Secretary of the Treasury Alexander Hamilton based on the results from the first U.S. census in 1791. Although Congress passed a law to enact the apportionment, it was vetoed by President Washington. Ultimately a method advocated by Thomas Jefferson of Virginia was

[1] In 1868 the 14th Amendment to the Constitution eliminated the 3/5 treatment of slaves. Also, it stipulated reduction in the count for a state to the degree that voting rights were denied to male inhabitants (females could not vote until 1920). "Representatives shall be apportioned among the several states according to their respective numbers, counting the whole number of persons in each state, excluding Indians not taxed. But when the right to vote at any election for the choice of electors for President and Vice President of the United States, Representatives in Congress, the executive and judicial officers of a state, or the members of the legislature thereof, is denied to any of the male inhabitants of such state, being twenty-one years of age, and citizens of the United States, or in any way abridged, except for participation in rebellion, or other crime, the basis of representation therein shall be reduced in the proportion which the number of such male citizens shall bear to the whole number of male citizens twenty-one years of age in such state."

[2] Randomization provides one statistical solution to the problem: randomly allocate the r_i's among the states so that each state's expected allocation would equal a_i; this does not seem politically acceptable.

chosen[3], with the result that larger states were favored over small (Balinski and Young 1982, 126).

The choice of apportionment method was based partly on fairness considerations, so that the apportionments did have to be proportional to population, but some latitude existed concerning choice of a metric (or loss function, cf., Section 4.4) for measuring proportionality and choice of h. Historically, both have been chosen to benefit the party in power, although h has been fixed by law at 435 since 1935. For example, in 1941 President Franklin D. Roosevelt reported to Congress two alternative apportionments based on the 1940 census, which differed for two states: Arkansas received one more seat than Michigan under the so-called Hill method (or "equal proportions") and one fewer seat under the so-called Webster's method ("method of major fractions"). As Arkansas was a strongly Democratic state and Michigan was viewed as Republican, the Democratic Congress chose Hill's method. Later that year Hill's method was set into law and has been used ever since. Hill's method has the property that it can round by more than a whole seat – the number of seats allocated to state i can differ from a_i by more than 1 – although such a transgression has not occurred yet. Balinski and Young (1982) present a lively history and elegant mathematical treatment Saari (1995, Chapter V) presents a novel point of view.

The U.S. President is not directly elected by the popular vote but by the Electoral College. Each of the 50 states sends as many electors as it has senators and representatives. The District of Columbia sends the minimum number of electors of any state, currently 3. The total number of electors is thus $2 \times 50 + 435 + 3 = 538$. In the 2000 presidential election, Albert Gore won the majority of the popular vote but George W. Bush won the majority in the Electoral College and was elected President. Had Hamilton's method been used to apportion the House, and the popular vote remained unchanged, there would have been a tie in the Electoral College – Gore would have won the Electoral College vote and become President if Jefferson's method had been used (Jenkins 2004).

3.1.2. Rationale Behind Allocation Formulas

One may interpret an allocation formula in various ways. One interpretation holds that the formula specifies legislative intent, which reflects the will of the people. From this perspective, if perfect data are input into the formula, perfect allocations will result. This interpretation ignores the political give and take involved in the development and enactment of the formula.

At the other extreme is a view that the formula is simply a parametric approximation to data points – legislators look at alternative formulas to see the resulting vectors of allocations as shown on computer printouts and choose the formula that best approximates the compromise allocation numbers, subject to constraints of fair or rational appearance. From this perspective, the fact that the formula stays fixed while data change over time (and hence allocations change) may be undesirable;

[3] Jefferson's method is also known as "method of greatest divisors", and the "method of d'Hont". It works by finding a divisor x such that the sum of $h_i = \lfloor p_i/x \rfloor$ equals h, and then taking h_i as the apportionment for state i.

formulas often contain "hold harmless" provisions to limit or even prevent decreases in fund allocations over time as new statistical series are used to compute the allocations (Louis, Jabine, and Gerstein 2003). In fact, correction of errors in the statistics underlying the printouts may also be undesirable to the extent that the resulting allocations depart from the compromise numbers. On the other hand, the legislators may not know the exact results of the formula even when they are voting on it. For example, the emergence of a complicated formula for General Revenue Sharing, a very large fund allocation program in the U.S. in the 1970s, was described by one of the Representatives James C. Corman: "We finally quit, not because we hit on a rational formula, but because we were exhausted. And finally we got one that almost none of us could understand at the moment. We were told that the statistics were not available to run the [computer] print on it. So we adopted it, and it is here for you today"(*Congressional Record*, June 22, 1972 Daily Edition, p. H5949).

The political reality is that the development of a formula allows coalitions to form and achieve political compromise that might not be otherwise achieved (Melnick 2002). Legislative goals behind the formulas are often vaguely enough stated to achieve widespread support, while the votes and compromises focus on the numerical allocations. Appearance is important, and so formulas include proxies for need, capability, and effort – even if the elements cancel each other out and lead to allocations that are essentially per capita, because a pure population formula does not appear to address particular problems (Spencer 1982, 525).

3.2. Effect of Inaccurate Demographic Statistics

Formula-based allocations that depend on demographic statistics are to some degree sensitive to changes in the statistics. The apportionments described in Section 3.1.1 are highly sensitive for certain data configurations – at some point, a change in population count of a single person must lead to a gain or loss of a House seat. The apportionments, being integer, are not continuous functions of the data. Most fund allocations do change continuously with data values, although use of data-based thresholds to determine eligibility may imply discontinuities (Zaslavsky and Schirm 2002). For example, an area may be eligible for funds only if its population size, its growth rate, its computed allocation, etc. exceeds a given threshold.

Some allocation rules are expressed in words in legislation but are so complex that they only have been expressed as iterative algorithms rather than as analytical formulas. For example, the General Revenue Sharing allocations to more than 39,000 local governments were based on an iterative algorithm that was not guaranteed to converge (Spencer 1980a, Ch. 5). Such complexity makes it difficult to apply approximations such as the delta method in error analysis.

When allocation formulas divide a fixed total, such as a fixed number of seats or a fixed sum of money, they often involve population shares, or the ratio of an area's population to the total. If all areas experience the same proportional change in their population, their allocations may remain unchanged. This fact too often is ignored by local governments in calculating how much funding they are shortchanged due to errors in population counts. The most reliable way to assess the sensitivity of the allocation to changes in the data is to compute the allocations under alternative data configurations, as in a sensitivity analysis. Such analysis will show how changes in the data cause changes in the allocations.

To say how changes in data affect *errors* in the allocations is a different matter, because a discussion of error presumes the existence of a target value. Often, one can define the target value as the value of the allocations that would result if error-free data were input into the existing allocation formula (cf., the discussion of estimating equations in Chapter 3). In some cases this principle does not yield an unambiguous definition. For example, as discussed in Chapter 2 there may exist alternative conceptual definitions of population, not to mention different concepts of income or employment status. Therefore, we cannot casually assume that published statistics are *valid*, i.e., that the concept they measure is appropriate for the allocative uses they are put. But, assuming that particular statistics are valid, we may say that error-free statistics arise if there is no sampling error or non-sampling error in the measurements as applied to the operational definitions for the statistics.

Most formulas for allocating monies involve more than just a single data series. Reducing error in one data series is not guaranteed to reduce overall errors in allocations, unless the errors in the different series are independent with zero means. In fact, it is possible that eliminating error in one series may *increase* the error in the allocations.

3.3. Beyond Accuracy

In addition to validity, timeliness and the level of detail are important aspects of data quality. In many countries population estimates are used that are years out of date and, as discussed in Section 4 of Chapter 2, for some types of events it may not be possible to collect timely statistics at all. Level of geographic, economic, or educational detail is an issue of aggregation. In many countries, statistical agencies have a more difficult time producing statistics for smaller areas as compared to larger areas, due to data availability and the inherent unstability of genuinely small domains. Legislators may want more detail than the statistical system can readily handle, which presents both a problem and an opportunity for statistical agencies.

For example, during the development of the General Revenue Sharing formulas, legislative staff needed to decide whether the allocations should be made to states and large cities or to smaller jurisdictions as well. At the time the Census Bureau was not producing postcensal estimates for very small areas (with population sizes of 200 and below), which is a difficult challenge for a statistical agency, but importance of General Revenue Sharing presented an opportunity for increased funding and decreased bureaucratic barriers so that the small area estimation program could be strengthened.

Assuring timeliness in data used for allocations seems generally desirable. This does not mean that the most recent data should always be used as such. Local governments as well as individuals on government pensions need to be able to predict what their allocations will be from one budget period to the next. Therefore, some form of averaging recent past data may be desirable to prevent the allocations from fluctuating too much (e.g., Alho and Salo 1998).

Stability may even be important enough to warrant a decision to not correct the statistical series for errors that are discovered (and corrected in various revisions of data series). For example, the Consumer Price Index (*CPI*), which measures inflation, is composed of sub-indexes measuring inflation in costs of food, housing, etc. While the U.S. government publishes changes from such revisions to the component price indexes, it generally does not publish retroactive changes to the overall CPI because of the significant role the CPI plays in setting cost-of-living increases contracts and fund allocations (U.S. Federal Committee on Statistical Methodology 1982, 2).

4. Decision Theory and Loss Functions

4.1. Introduction

Consider a decision maker who will choose an action based on available information. The action might be to choose how much money to set aside for future pensions (Chapter 11), to choose which set of alternative estimates of current population to publish for apportioning the U.S. House of Representatives (Section 3.1.1), or deciding how much money to spend on research and production of statistical information (Spencer 1980a, Savage 1975, Törnqvist 1948). The consequences of each alternative action may vary in different ways depending on the state of nature, such as the true population size, and the consequences need not be symmetric.

Common methods of estimation such as maximum likelihood do not allow for such distinctions. Although the usual Neyman-Pearson methods for hypothesis testing can account for asymmetrical consequences in hypothesis testing by varying the probabilities of rejecting the null hypothesis when it is true and failing to reject it when false, researchers and decision makers too often simply set the first type of error probability at 0.05 and allow the second type to be whatever it might be (Exercise 1).

Example 4.1. Asymmetric Consequences of Forecast Error. A government agency is forecasting total national population. Suppose the future population size is θ_1, and the forecast is normally distributed with mean θ_1 and variance θ_2. An error in the forecast leads to a cost or loss proportional to the error, but the loss from an underestimate is λ times as severe as the loss from an overestimate. See Muhsam (1956) for a more detailed example in the context of planning for future water supply. $\Diamond$

We consider that a decision maker is to choose among a set of actions, A, called the *action space*. In Example 4.1 the action is a guess at the future population, so $A = \{0, 1, 2, \dots\}$, the set of positive integers. We assume that the action space is pre-specified, i.e., it does not depend on which element in the sample space was realized. This is an "uninformativeness" assumption concerning the action that does not always hold in practice. For example, an action may lead to new insights, a rethinking of strategies, and a formulation of new alternative actions.

The consequences of an action depend on the state of the world, θ, which belongs to a *state space* or *parameter space*, Θ. In Example 4.1, $\Theta = \{(\theta_1, \theta_2)^T | 0 < \theta_1, 0 < \theta_2\}$. Partial information about the state of the world is provided by an *estimate* whose possible values lie in the *sample space, S*. Typically, in statistical decision theory, the set of possible probability distributions over S is specified so it depends on the state space. In Example 4.1, the possible distributions for the estimate were $N(\theta_1, \theta_2)$.

The consequences of an action depend on the state of nature, and the relative valuations of consequences are represented by a real-valued *loss function* l defined on $\Theta \times A$ such that $(\boldsymbol{\theta}, a)$ is at least as good as $(\boldsymbol{\theta}, a')$ if and only if $l(\boldsymbol{\theta}, a) \leq l(\boldsymbol{\theta}, a')$. In Example 4.1, the loss function is of the "linear" form (Raiffa and Schlaifer 1972) and only depends on the error in estimating the first component of $\boldsymbol{\theta}$

$$l(\boldsymbol{\theta}, \hat{\theta}_1) \propto \begin{cases} c(\hat{\theta}_1 - \theta_1) & \text{if } \hat{\theta}_1 > \theta_1 \\ \lambda c(\theta_1 - \hat{\theta}_1) & \text{if } \hat{\theta}_1 \leq \theta_1. \end{cases} \tag{4.1}$$

The proportionality constant $c > 0$ can be interpreted as translating the estimation error to monetary loss, for example.

A *decision rule* $d \colon S \to A$ tells us which action to take after seeing the estimate. The decision $d(s)$ for $s \in S$ is a random variable with a conditional probability distribution given any $\theta \in \Theta$. (It is possible that $\boldsymbol{\theta}$ is a vector.) We define the *risk* of the decision rule as $R(\theta, d) = E_s[l(\theta, d(s))|\theta]$. Suppose that in advance of seeing the estimate we have a probability distribution, or a *prior distribution*, over Θ. Then, we define the *Bayes risk* of the decision rule as the expected risk, or $B(d) = E[l(\theta, d(s))] = E_\theta[R(\theta, d)]$. In the case of Example 4.1, the forecast equals the estimate s, and the decision rule is simply $d(s) = s$ for $s \in S = A$.

The decision-theoretic framework allows one to choose among decision rules. If $R(\theta, d_1) \leq R(\theta, d_2)$ for all θ and $R(\theta', d_1) < R(\theta', d_2)$ for some θ', then we say that d_1 *dominates* d_2. If neither rule dominates the other, then the unknown state of nature θ determines which has a smaller risk. If a prior distribution for θ is available, we can compare the Bayes risks, and choose d_1 over d_2 if $B(d_1) \leq B(d_2)$. A *Bayes rule* minimizes the Bayes risk. For insight concerning the nature of a Bayes rule, consider the *posterior risk* $E_\theta[l(\theta, d(s))|s]$. Observe that the Bayes risk $B(d) = E_s[E_\theta[l(\theta, d(s))|s]]$ is minimized by a decision rule d that minimizes the posterior risk for each s (provided that one exists). This means that once s is observed, i.e., once the estimate is produced, we choose the action $a = d(s)$ that minimizes $E_\theta[l(\theta, a)|s]$. For example, if $l(\theta, a) = (\theta - a)^2$, we have that $E_\theta[l(\theta, a)|s] = E_\theta[(\theta - a)^2|s] = E_\theta[(\theta - E_\theta[\theta|s])^2|s] + (E_\theta[\theta|s] - a)^2 \geq E_\theta[(\theta - E_\theta[\theta|s])^2|s]$; the lower bound is attained when $a = E_\theta[\theta|s]$, the posterior mean of θ.

Example 4.2. Posterior Risk Under Linear Loss. Suppose θ_1 has a posterior probability density function f and posterior c.d.f. F and the loss function (4.1) applies. The posterior risk of the action x is

$$c \int_{-\infty}^{x} (x - t) f(t)\, dt + \lambda c \int_{x}^{\infty} (t - x) f(t)\, dt$$

$$= cxF(x) - c \int_{-\infty}^{x} t f(t)\, dt - \lambda c x(1 - F(x)) + \lambda c \int_{x}^{\infty} t f(t)\, dt. \tag{4.2}$$

The derivative with respect to x is (after simplification) $cF(x) - \lambda c(1 - F(x))$. The second derivative is positive, so a minimum occurs when the first derivative is zero, or when $F(x) = \lambda/(\lambda + 1)$. If $\lambda = 1$, the optimal forecast is the median of the posterior distribution of θ_1. When $\lambda > 1$, so that underestimates are more consequential than overestimates, we see that the optimal forecast is greater than the median. If the posterior distribution is $N(s, \hat{\sigma}^2)$, the optimal forecast x satisfies $\Phi((x - s)/\hat{\sigma}) = \lambda/(\lambda + 1)$, where Φ is the c.d.f. of a $N(0, 1)$ random variable, so $x = s[1 + (\hat{\sigma}/s)\Phi^{-1}(\lambda/(\lambda + 1))]$. The proportional shift in x from s is equal to the product of the CV, $\hat{\sigma}/s$, and $\Phi^{-1}(\lambda/(\lambda + 1))$. If the CV is 0.1, the proportional shift ranges from 4% to 13% as λ ranges from 2 to 10, but if the CV is 0.5, it ranges from 22% to 67% as λ ranges from 2 to 10. The shift can exceed 100% if the CV is large. ◊

4.2. Decision Theory for Statistical Agencies

In many developments of statistical decision theory, the decision maker is assumed to be an individual person with preferences that satisfy certain postulates of "rational behavior" (cf., Section 3.5.1 of Chapter 8), and under those conditions one can show that the decision maker can represent uncertainty with a probability distribution and will seek to minimize the expected value of the loss function l or, equivalently, maximize the expected value of a utility function defined as $-l$ (Savage 1954)[4]. Even for an individual, the postulates of rational behavior often are more normative than descriptive of actual behavior under uncertainty. For example, the postulates do not take into account the cost of information processing and they assume people know their preferences, will choose from a prespecified set of possible actions, and will represent uncertainty by a probability distribution (March 1978).

In many applications of statistical demography, decisions are made not by individuals working alone or representing just themselves, but by committees or agency heads within bureaucracies. As such the decisions are affected by many actors in complex ways. If the decision making process can be represented by a decision rule $d\colon S \to A$, if the decision maker is willing to adopt a particular

[4] Formally, this means that $l(\theta, a'') = pl(\theta, a) + p'l(\theta, a')$, where a'' is the randomized act (in A) equal to a with probability p and a' with probability $p' = 1 - p$.

function l as an optimization criterion, and if uncertainty can be reflected by specifications of probability distributions, then the decision goal can be represented as minimization of the expected value of l and we may sidestep the question of postulates of rational behavior for organizations and the issue of whose preferences are under consideration. This representation is simplistic and certainly not descriptive of much of organizational behavior (e.g., March 1994), but it does have some benefits. By simplifying, it allows one to understand the major components of the decision problem. The representation may thus lead to some useful insights. In fact, *much of the value of the analysis may even be in setting up the representation of the decision problem and identifying how and why errors in statistics may lead to adverse consequences.*

There should be some restrictions on the loss function l used by statistical agencies. Note that comparison of two alternative decision rules on the basis of risk is unaffected if the loss function $l(\theta, a)$ is modified to $l(\theta, a) - c(\theta)$, because the difference in loss for decision $d(s)$ and decision $d'(s)$ is $l(\theta, d(s)) - l(\theta, d'(s))$ in either case. Consideration of $c(\theta) = \inf_a l(\theta, a)$ suggests that there is no loss of generality in assuming that $l(\theta, a) \geq 0$ and that $l(\theta, d) = 0$ if $\inf_a l(\theta, a)$ is attained for some action $a \in A$. In this case the loss function measures the excess loss relative to the optimal attainable (or limit thereof); such a loss function is called a *regret* function. When a statistical agency (e.g., a government agency or other agency concerned with a reputation for honesty) is using statistical decision theory in developing an estimate, it is important that the decision theoretic analysis guide it towards truth. In formal terms, this means that if θ is a quantity being estimated and l is a regret function, then $l(\theta, \theta)$ must equal zero. This requirement is called *Fisher-consistency*. If a loss function is not Fisher-consistent, the optimal value of the estimate will be one with error!

Example 4.3. When Policy Makers Prefer Error to Accuracy. (a) If the national government policy is to move the actual total fertility rate τ to a target τ_0, the top decision maker(s) in the government may view the loss in an estimate $\hat{\tau}$ as having a minimum when $\hat{\tau} = \tau_0$, regardless of the actual value of τ From a formalistic point of view, the optimal estimate would then be $\hat{\tau} = \tau_0$, irrespective of what data analysis indicates. A statistical agency or research group that simply reported $\hat{\tau} = \tau_0$ would lose credibility, however. Such credibility issues are common in forecasts of the national budget surplus or deficit and energy supply and demand, with different parties (and more neutral observers) developing forecasts that differ widely. (b) Jabine and Schwartz (1974) developed a loss function for determining optimal sample size for data collection in the Supplemental Security Income program that distributed U.S. federal government monies to the states according to sample estimates $\hat{\theta}_i$ of population characteristics θ_i of state i. They adopted the federal government's point of view, which they took to "minimize losses from estimates which result in *overpayments* to the States" (p. 104), and they defined a loss function equivalent to (4.1) but with $\lambda = 0$. The optimal estimate is then to take $\hat{\theta}_i$ as small as possible, which violates Fisher-consistency. (To avoid such an implication, they imposed additional constraints of unbiasedness.) ◇

Some developments of decision theory seek to take an objective view and avoid reliance on subjective probabilities and subjective utilities or loss functions. To formulate the loss function, one would evaluate consequences of data error for making wrong decisions, misallocating public benefits, or misguiding research. For example, Wald (1947) suggested that

> "In industrial problems, [the loss function] may be thought of as expressing the financial loss caused by taking the [wrong] action, [with loss equal to zero for] a correct decision.... The determination of these [loss] functions cannot be regarded as a statistical problem. They will be chosen on the basis of practical considerations in each particular problem."

The losses attributable to errors in statistics produced by government agencies are not merely financial but may involve changes in social welfare. Unfortunately, we know but few uses of the estimates; even known allocative uses – Congressional apportionment is one example – resist empirical estimation of the *consequences* of perturbations in allocations that arise from errors in the statistics, let alone objective valuation of alternative consequences. Thus, such an approach is not feasible for multi-purpose large scale demographic statistics.

A practical approach is to view a loss function as an abstract optimization criterion and to select a criterion whose use in other, similar situations has resulted in good practices. The loss function typically will involve as its action either choice of an estimator (as in apportionment or allocation problems, decisions about whether to adjust for census undercount, and which point forecast to feature as the middle variant) or aspects of the design of a data program (such as how much money to spend on a census or survey and how to allocate a sample). In either situation, attention will be paid to estimation of a population characteristic or parameter, say $\boldsymbol{\theta}$. Although abstract, the criterion should be generally related to the uses of the statistics and one should define $\boldsymbol{\theta}$ to represent those uses in a general way. Consider a statistical agency. For planning purposes and general information, including the development of denominators for demographic rates, population totals are important. Thus, $\boldsymbol{\theta}$ could be defined in terms of total population of areas or groups. For many allocative uses, $\boldsymbol{\theta}$ should reflect population shares. For example, to reflect uses of population estimates to apportion the U.S. House of Representatives, it would be appropriate to set $\boldsymbol{\theta}$ to be the proportion of the U.S. population that is in each state – the state "share" of the population – instead of setting $\boldsymbol{\theta}$ to be the (integer) allocation of seats that would occur if the true population were known; analysis of the latter is more difficult technically than the former. For another example, consider the allocations that would occur if a fixed sum of money were allocated to areas or subgroups proportionally to population. If the fixed total is allocated nationally, $\boldsymbol{\theta}$ should refer to population share relative to the nation; if the allocation is within a state, then the population share is relative to the state population; if the allocation is to all units of a given type, the share is relative to the sum for all units.

The matter of representing uncertainty with probability distributions is another concern for statistical agencies. We have discussed the penchant of forecasters to avoid specifying forecast uncertainty with probability distributions and instead to produce alternative scenario-based "projections" (Chapters 7, 8), and we have shown how the intervals constructed from the latter do not have consistent probabilistic interpretations (Chapter 9). The postulates of rational behavior imply that if probabilities are unknown, people will formulate subjective probabilities and act according to those subjective probabilities. Some experiments of Ellsberg (1961) presented people with a set of choices between gambles with known probabilities and gambles with unknown probabilities and found that people showed a preference for the lotteries with known probabilities and a preference against the "ambiguous" lotteries with undetermined probabilities. The results of the experiment if taken at face value show violations of postulates of rational behavior (Complement 2).[5] The issue is compounded in use of decision theory by a statistical agency, where it is unclear whose preferences or subjective probabilities dominate. An actor in a bureaucracy might find the information compelling enough to justify a personal decision (indeed, given sufficiently strong preferences, the person might hardly need any information), but if the decision must be justified to others then stronger information may be needed (Boruch 1990). We have observed situations where statistical agencies are unwilling to act on the basis of statistical models whose assumptions cannot be validated, even though lack of action may correspond to implicit use of a statistical model demonstrated to be false.

Example 4.4. Non-Adjustment of Undercount Estimates for Correlation Bias. As discussed in Chapter 10, Section 5.6, DA sex ratios may be used to provide a variety of adjustments for correlation bias. The dual systems estimates of undercount in the 1990 U.S. census and the March 2001 estimates of undercount in the 2000 U.S. census did not adjust for correlation bias, in large part because there was no unambiguous way to choose one method over another and the impact would be appreciable. Although not a statistical argument, such considerations are quite important for public confidence in the government. Another major consideration was a desire that the estimates of net undercount have the correct sign, even if the estimates were pulled in towards zero (U.S. Census Bureau 2003a, 50). This attitude is consistent with the notion that a sin of commission is worse than a sin of omission, or "do no harm". ◊

Example 4.5. Adjustment for Correlation Bias for Hispanics in the 2000 U.S. Census. In March 2003 the U.S. Census Bureau published its A.C.E. Revision II estimates of net undercount in the 2000 census, which unlike previous estimates *included* adjustments for correlation bias. Whereas the March 2001 estimate of net undercount for the 2000 census was 1.18%, the March 2003 estimate was a net undercount of −0.49% (i.e., a net overcount), and if a correlation bias adjustment had not been made the estimated net undercount would have been yet more negative, at −1.12% (U.S. Census Bureau 2003a, 2, 5). As discussed in

[5] Daniel Ellsberg became famous for leaking the Pentagon Papers to the New York Times in opposition to the U.S. involvement in war in Vietnam. Kadane (1992) suggests that the subjects in the experiments did not trust the experimenter, and is unconvinced that the postulates of rational behavior were violated.

Example 5.5 of Chapter 10, the Census Bureau did not have demographic data to support a correlation bias adjustment for Hispanics that was different than the adjustment for Non-Blacks as a group. Ethnographic studies have supported a view that Hispanics would be more prone to correlation bias than other Non-Blacks, but there is no straightforward statistical basis to support any one of the alternative estimates of correlation bias for Hispanics. ◊

Example 4.6. Alternative Estimates of Population. Despite evidence of very large errors in the 2000 census – 17. 2 million erroneous enumerations and 15.9 million census omissions for a census count of 281.4 million (National Academy of Sciences 2004, 253), the U.S. Census Bureau ultimately decided not to adjust the 2000 census for undercount. The Census Bureau noted that although the adjusted numbers might have been more accurate "on average", it could not "be confident of improvements in accuracy at the levels of geography for which estimates are produced" (U.S. Census Bureau 2003a, 2). The Bureau identified four particular sources of concern, notably the uncertainty in the estimates of correlation bias, and an independent review by the National Academy of Sciences (Citro, Cork, and Norwood, 2004, 256–258) shared the Bureau's concern for correlation bias, among others. In our view, it is quite plausible that one could have several alternative estimates of population, each derived under a different model for correlation bias and each more accurate than the census count, yet because there would be no clear basis for choosing one over the other, the Census Bureau would reject them all in deference to the census counts, despite the lower accuracy of the latter. ◊

4.3. Loss Functions for Small Area Estimates

To develop a loss function for small area estimates, we proceed in two steps, first developing component loss functions for individual areas and then forming a summary of the components. If $\hat{\theta}_i$ denotes an estimate of the population characteristic θ_i for area or subgroup i, we will consider the component loss function $l_i(\theta_i, \hat{\theta}_i)$ or more generally $l_i(\boldsymbol{\theta}, \hat{\boldsymbol{\theta}})$ to measure the discrepancy between $\hat{\theta}_i$ and θ_i. The expected value of the component loss function will be called the *component risk function*. Examples of widely used measures of risk in this context are standard deviation, RMSE, CV, and their squares. The measure of risk provides a summary of the accuracy of a statistic or, more generally, of the statistical program giving rise to the statistic. In some cases, the measures are most useful for comparative purposes, i.e., for comparing the accuracy of one program or statistic with another.

Moving from a measure of a statistic's accuracy for a single area or group to a summary measure of accuracy for a set of areas or groups requires some value judgements concerning the importance of one set relative to another (Spencer 1980c). We define an *aggregate loss function* as a weighted sum of component loss functions,

$$l(\boldsymbol{\theta}, \hat{\boldsymbol{\theta}}) = \sum_i w_i l_i(\boldsymbol{\theta}, \hat{\boldsymbol{\theta}}), \tag{4.3}$$

with $w_i > 0$. Value judgments cannot, of course, be derived completely on technical grounds. Furthermore, value judgements may vary over time and among different people. The loss function, which serves to rank alternative choices of multivariate distributions of $\hat{\boldsymbol{\theta}}$, represents a social *convention* (Keyfitz 1979). Thus, there is an advantage to having flexibility in how the aggregate loss function is constructed, and there is an advantage to using several loss functions, e.g., with alternative choices of w_i in (4.3). If the alternative loss functions give different rankings among alternative decisions (e.g., sample allocations or estimators), then the choice will depend on values. We find that clarity is advanced when we can separate technical issues from political judgements, and the decision-theoretic framework provides a means for doing so. We do not regard the results of a formal decision analysis as necessarily providing "the" answer to a complicated question, but rather we find that the exercise of carrying out the analysis leads to greater understanding of the issues.

Example 4.7. Value Judgements in Sample Allocation. For example, in the context of designing a stratified sample, suppose sample allocation A leads to a CV of 1% for every local government, and allocation B leads to a CV of 0.1% for every local government with more than 5000 population and a CV of 3% for the others. The latter provides more accurate estimates for most of the people, and the former provides more accurate estimates for most of the local governments. Thus, neither allocation is uniformly more accurate for all areas. To say that one allocation (or sample design, more generally) estimator is, overall, more accurate than the other reflects a value judgement concerning what kinds of accuracy are more important. Suppose further that the small areas each have 1,000 population and the large areas each have 35,000, and suppose 70% of the areas are small and the rest are large.

Type of Area	Population per Area	Percent of All Areas	CV (%) for Allocation	
			A	B
Large	35,000	30%	0.1%	1.0%
Small	1,000	70%	3.0%	1.0%

If the expected aggregate loss is the average of the coefficients of variation, the expected loss for sample allocation A is 0.02 and the expected loss for B is 0.01, and so B is more accurate. If, on the other hand, the expected aggregate loss is the weighted sum of the coefficients of variation, with weights equal to population size, then the expected losses for A and B are proportional to 31.5 and 112, respectively, and so A is more accurate.

To see how this is reflected by alternative weightings of loss functions, let λ denote the weight given to the loss function for small areas, and $1 - \lambda$ the weight for the loss function for large areas, with $0 \leq \lambda \leq 1$. The overall expected loss function for sample allocation A is $0.001(1 - \lambda) + 0.03\lambda$ and the overall expected loss function for B is 0.01 $(= 0.01(1 - \lambda) + 0.01\lambda)$. If $\lambda < 0.31$ then the expected loss is smaller for A than B, and so A is more accurate. If $\lambda > 0.31$ then the expected loss is larger for A than B, and so B is more accurate. The choice of λ represents a value judgement. Such issues played a role in the A.C.E. sample design for 2000 Census in the U.S. (Schindler 1998), where the goal was to achieve CVs for the dual system estimates of 0.5% in all states and standard errors of about 60,000 in the larger states. $\Diamond$

Aggregate loss functions commonly used in statistical decision problems involving small areas include a weighted sum (U.S. Federal Committee on Statistical Methodology 1978)

$$l(\boldsymbol{\theta}, \hat{\boldsymbol{\theta}}) = \sum_i w_i |\hat{\theta}_i - \theta_i|^\alpha, \tag{4.4}$$

where $\alpha > 0$. Choosing $\alpha = 2$ leads to the commonly used *squared error* loss function. Consider θ_i as the population share or population size of state i. Setting w_i*'s* constant gives unweighted squared error loss; setting w_i inversely proportional to θ_i gives the so-called *weighted squared error loss*, and setting w_i inverse to the square of θ_i gives the so-called *relative squared error loss.* In application, θ_i is unknown and empirical comparisons based on (4.4) will be based on an estimate. If the various loss functions give different rankings among the alternative sample allocations, then neither allocation is clearly superior to the other.

Loss functions of the form (4.4) are based on a sum of individual "losses" to the areas or groups, consistent with a utilitarian view of social welfare measurement (Spencer 1985b, 816–817). Depending on its notion of fairness, an agency may require that minimization of aggregate expected loss not be attained at the cost an excessively large component loss (Rawls 1971, Efron and Morris 1971). It is possible that the sum of the losses is smaller for one choice of $\hat{\boldsymbol{\theta}}$ but the expected loss for a particular unit is quite large and larger than under an alternative choice of $\hat{\boldsymbol{\theta}}$. In contrast, loss functions of the form $\max_i\{l_i(\boldsymbol{\theta},\hat{\boldsymbol{\theta}})\}$are concerned with the maximum component loss. The expected values of the maximum are difficult to estimate, however, and a more tractable criterion to check that no individual component loss is excessive is

$$\max_i \{E[l_i(\boldsymbol{\theta}, \hat{\boldsymbol{\theta}})]\}. \tag{4.5}$$

It seems reasonable that in practice aggregate loss functions of the form (4.3) or (4.4) should be used subject to the restriction that (4.5) is not excessively large.

4.4. Loss Functions for Apportionment and Redistricting

4.4.1. Apportionment

If $\theta_i = hp_i$ is the number of House seats times the population share of state i and $\hat{\theta}_i$ is the integer apportionment, then the apportionments that minimize (4.4) with $w_i = 1$ are obtained from Hamilton's method of apportionment (Section 3.1.1; Birkhoff 1976). Websters's method and Hill's method minimize (4.4) for $\alpha = 2$ when w_i equals $1/\theta_i$ and $1/\hat{\theta}_i$, respectively. Adams's method and Jefferson's method minimize (4.5) when $l_i(\boldsymbol{\theta}, \hat{\boldsymbol{\theta}})$ equals $\theta_i/\hat{\theta}_i$ and $\hat{\theta}_i/\theta_i$, respectively. (Balinski and Young 1982)

4.4.2. Redistricting

Consider that the population of a large area such as a nation is partitioned into mutually exclusive states, and an integer number n_i of districts is set for each state. Each state is to be partitioned into n_i districts such that each district is of equal size. In the U.S., the number of Congressional districts equals the number of House seats determined in the apportionment (Section 3.1.1) and law requires that districts be constructed so all districts in any state i have the same estimated share of the state population, $1/n_i$. Ideally, the *true* district shares would all be equal within each state.

A simple measure of inequality is the average within-state variance among the true population sizes of the districts. Let P_{ij} denote the true population size of congressional district $j = 1, \ldots, n_i$ in state i, and define $n = \Sigma_i n_i$. Let $P_i = \Sigma_j P_{ij}$ be the total population of state i, let $\bar{P}_i = P_i/n_i$ be the average district size in state i, let $\theta_{ij} = P_{ij}/P_i$ be the fraction of state i population that is in district j, and let $\bar{\theta}_i = \sum_j \theta_{ij}/n_i = 1/n_i$ be the average fraction. The within-state variance among the true population sizes of the districts is $\sigma_i^2 = \sum_j (P_{ij} - \bar{P}_i)^2/n$, and the overall mean squared deviation from the state means is $\sigma^2 = \sum_i \sum_j (P_{ij} - \bar{P}_i)^2/n = \sum_i (n_i/n)\sigma_i^2$.

Let $\hat{P}_{ij}$ denote the estimate of P_{ij} on which the district boundaries were constructed, $\hat{P}_i = \sum_j \hat{P}_{ij}$, and $\hat{\theta}_{ij} = \hat{P}_{ij}/\hat{P}_i$. By construction, $\hat{\theta}_{ij} = 1/n_i = \bar{\theta}_{ij}$, so one can show (Exercise 3) that

$$\begin{aligned} \sigma^2 &= \sum_i P_i^2 \sum_j (\theta_{ij} - \bar{\theta}_i)^2/n \\ &= \sum_i P_i^2 \sum_j (\hat{\theta}_{ij} - \theta_{ij})^2/n. \end{aligned} \tag{4.6}$$

This is of the form (4.4), where the summation runs over all n districts and the weights are P_i^2/n for any district j in state i. Given an estimate $\hat{V}_{ij}$ of the mean squared error of the estimated population share $\hat{\theta}_{ij}$ of district j in state i, we can estimate σ^2 by $\hat{\sigma}^2 = \sum_i \tilde{P}_i^2 \sum_j \hat{V}_{ij}/n$, where $\tilde{P}_i$ is some estimate of P_i. (If we are comparing $\hat{\sigma}^2$ for alternative estimators, we will want to use the same $\tilde{P}_i$'s for each; otherwise we may take $\tilde{P}_i = \hat{P}_i$.)

Example 4.8. Expected Loss of Adjusted and Unadjusted 2000 U.S. Census for Redistricting. The U.S. Census Bureau was prohibited by law from using the sample-based DSE to adjust the 2000 census for undercount, but it had the option of providing adjusted numbers in March of 2001 for use in redistricting. The Census Bureau developed estimates of the difference between the expected values of (4.6) for adjusted and unadjusted estimates population based on the 2000 U.S. census (Mulry and Spencer 2001), using the methods described in Chapter 10, Section 6. Although the formal analysis suggested that adjustment would improve accuracy of redistricting (Hogan 2001, Table 6; Navarro and Asiala 2001, Table 1.A), time pressures for the decision analysis meant that much of the evaluation data for estimating non-sampling errors in the DSE were not yet available, and the analysis assumed that the patterns of error were similar to those in 1990. That assumption turned out to be incorrect. The Census Bureau, concerned that errors were not being accurately estimated (Chapter 10, Example 5.3), decided not to adjust at that time. To permit additional time for developing evaluations of the non-sampling errors in future DSE's, the National Academy of Sciences has recommended possible extension of the deadline for producing population numbers for redistricting (Citro, Cork, and Norwood 2004, 346–347). ◊

4.5. *Loss Functions and Allocation of Funds*

The choice of a loss function for the effect of data error on allocation of funds should depend on the decision problem at hand. If one really views an allocation formula as optimal in some sense, then one could consider quantifying the social benefits accruing from more accurate statistics in the formula. In such a case, for example, if demographic statistics were part of a formula for allocating government monies for education, one might consider the increase in test scores that would arise from use of more accurate statistics to determine the fund allocations. Such an analysis seems hopelessly difficult. In light of the discussion in Section 4.2, we thus urge that transparency and tractability should be strong considerations. If the problem involves choice of one set of estimators rather than another, a loss function of the form (4.4) with $\alpha = 2$, i.e., squared error, should suffice. Use of other choices of α presents analytical problems for estimating moments of $\hat{\theta}_k$, particularly when biases are estimated with non-negligible variances; Mulry and Spencer (1993, 1084) and Fay (1992) discuss difficulties with $\alpha = 1$ in this context.

If the decision problem at hand concerns spending more money or less money on statistics for use in allocating funds, we recommend a loss function of the form (4.4) with $\alpha = 1$. Such loss functions are simple to interpret (Section 4.5.1). Also, such loss functions lead to decisions to spend less money on data when the allocation formulas are very far from optimal, if the nonoptimality is acknowledged (Section 4.5.2).

4.5.1. *Effects of Over- and Under-Allocation*

To a good approximation, many allocation formulas can be viewed as distributing a fixed amount of funds (or other benefits). First consider the formula's allocations if there were no error in the statistics used to compute them, and let θ_i denote the amount for recipient i. When imperfect data are used, the amount for recipient i is, say, $\hat{\theta}_i$. If the error in the allocation, $\hat{\theta}_i - \theta$, is positive (an overpayment), the recipient is really receiving a benefit! Thus, consider relaxing our requirement that the component losses be non-negative, so that if the error is positive, $\hat{\theta}_i - \theta > 0$, the loss to the recipient is negative, and if the error is negative (an underpayment) then the loss is positive. Taking an underpayment to be more significant than an overpayment of the same magnitude and taking the loss to be linear leads us to the component loss function $l_i(\boldsymbol{\theta}, \hat{\boldsymbol{\theta}}) = a(\theta_i - \hat{\theta}_i)^+ - b(\hat{\theta}_i - \theta_i)^+$, with $a, b > 0$ and where we define $(x)^+ = \max\{0, x\}$. Summing l_i over recipients then yields the aggregate loss function (4.3). If $\boldsymbol{\theta} = \hat{\boldsymbol{\theta}}$ then $l(\boldsymbol{\theta}, \hat{\boldsymbol{\theta}}) = 0$. The requirement of Fisher-consistency implies, however, that if $\boldsymbol{\theta} \neq \hat{\boldsymbol{\theta}}$ then $l(\boldsymbol{\theta}, \hat{\boldsymbol{\theta}}) > 0$, and that in turn implies (Exercise 4)

$$1 \leq \frac{\max_i\{w_i\}}{\min_i\{w_i\}} < \frac{a}{b}. \tag{4.7}$$

If we were to take w_i to be some measure of the size of recipient i, we would typically find that ratio a/b will be much closer to 1 than (4.7) would allow. As a result, it is reasonable to take $w_i = 1$ in (4.3). In the (common) case of allocating a fixed amount, $\sum_i(\theta_i - \hat{\theta}_i) = 0$, and (4.3) then simplifies to

$$\sum_i c|\hat{\theta}_i - \theta_i| \tag{4.8}$$

with $c = (a - b)/2 > 0$. Spencer (1980a, Chapter 1 and 1980c) provides further discussion.

The assumption of piecewise linear loss can be derived from more general considerations. If we take the component loss from an overallocation to be a constant times the component loss from an underallocation, we obtain the component loss function $l_i(\boldsymbol{\theta}, \hat{\boldsymbol{\theta}}) = ag((\theta_i - \hat{\theta}_i)^+) - bg((\hat{\theta}_i - \theta_i)^+)$, where $g(0) = 0$ and it is natural to take g to be increasing. Considering the aggregate loss function to be given by (4.3), requiring Fisher-consistency, i.e., $l(\boldsymbol{\theta}, \hat{\boldsymbol{\theta}}) \geq l(\boldsymbol{\theta}, \boldsymbol{\theta})$ to hold, and considering both the number n of areas or recipients to be arbitrary and allowing all possible vectors of errors in allocations $\hat{\boldsymbol{\theta}} - \boldsymbol{\theta}$ such that $\sum_i(\hat{\theta}_i - \theta_i) = 0$, we may show that there exist constants A and B not depending on n such that $B < g(x)/x < A$ for all $x \geq 1$ (Spencer 1980a, 43–45). This implies that $g(.)$ cannot be far from linear for large misallocations x.

4.5.2. *Formula Nonoptimality*

It is only realistic to recognize that many allocation formulas are imperfect, both for reasons of political compromise (Section 3.1.2) and the fact that the statistics used as inputs into the allocation formulas may be only crude proxies for the concepts of need, capability, and effort that are reflected in so many allocation formulas (Louis, Jabine, and Gerstein 2003, 35–39, Spencer 1982b). Here we will consider modifying the loss function to account for the nonoptimality while at the same time preserving Fisher-consistency. Let $\hat{\boldsymbol{\theta}}$ denote the vector of allocations and $\boldsymbol{\theta}$ denote the vector of allocations if the statistics used to compute the allocations were error-free. In addition, let $\boldsymbol{\theta}^*$ be the vector of "ideal allocations" that would occur if the formula were optimal and data were perfect and define $\boldsymbol{\eta} = \boldsymbol{\theta} - \boldsymbol{\theta}^*$, so that η_i denotes the amount of nonoptimality in the formula's allocation for recipient i. To reflect the non-optimality in the loss function, we modify (4.4) to

$$l(\boldsymbol{\theta}, \hat{\boldsymbol{\theta}}) = \sum_i w_i |\hat{\theta}_i - \theta_i + \eta_i|^{\alpha}. \tag{4.9}$$

If the η_i's were known, then (4.9) would not lead one to estimate θ_i by $\hat{\theta}_i$. To preserve Fisher-consistency, we will consider the η_i's to be random variables symmetrically distributed about 0 and independent of $\hat{\boldsymbol{\theta}}$ given $\boldsymbol{\theta}$ (Spencer 1985a). This implies that the loss $l(\boldsymbol{\theta}, \hat{\boldsymbol{\theta}})$ is random even for fixed $\hat{\boldsymbol{\theta}}$ and $\boldsymbol{\theta}$. We will consider implications of (4.9) in Example 7.1, below.

4.5.3. Optimal Data Quality with Multiple Statistics and Uses

Statistics produced by government agencies are multi-purpose. Even among allocative uses of statistics, the same population estimates may be used in a myriad of programs and in quite different ways. For some, an area will receive more money if its population estimate increases whereas for others, an area with declining population will receive more money if its population decreases (and its population decline thus becomes larger). In principle, one could develop a separate loss function for each allocation program that uses the data and then construct an overall loss function by summing the loss functions for the individual programs.[6,7] (Attention would need to be paid to scale factors for the loss functions for individual programs.) Kish (1987, 228–229) discusses such formal analysis in the context of the design of the World Fertility Survey.

Moving beyond uses for allocations to consider uses for decisions and research makes construction of a loss function far more difficult although not completely impossible (Panel on Methodology for Statistical Priorities 1976, Spencer 1982a). Still, we are lacking a convincing analysis of whether accuracy of demographic statistics for allocation should outweigh the need for accuracy for research and other less superficially obvious uses. Comparing the importance of allocative uses with non-allocative uses is challenging (Keyfitz 1979).

5. Comparing Risks of Adjusted and Unadjusted Census Estimates

In several countries, including the U.S., Australia, and the U.K., considerable attention has focused on whether the census population numbers should be left as reported or whether attempts should be make to "correct" them for estimates of

---

[6] One might think that the allocations to an area from the diverse allocation programs should be summed, and the loss function based on the net allocation. Such an analysis is not necessarily better, however, because the allocation programs often are targeted within areas, and because such an analysis would be enormously complex.

[7] Alternatively, one could look at a sample of allocation programs and estimate the overall loss function from the sample's loss functions.

net undercount, leading to so-called "adjusted" estimates. A key consideration is whether adjustment will improve accuracy. Given the large number of areas and subgroups, there is a question of how to measure accuracy for such highly multivariate statistics. Judgments concerning relative importance of errors in different statistics may be far from unanimous. We believe that decision making will be more transparent if such concerns, which are inherently political, can be separated from technical questions of error assessment. The political priorities concerning different statistics can be reflected in a weighted aggregate loss function, as discussed in Section 4.3. Once the form of the loss function, or loss functions if several are to be used, is set, attention can focus separately on estimates of bias and variance of the alternative estimates of population.

Whether the change in expected loss from adjustment is positive or negative depends critically on the biases and variances of the adjusted and unadjusted estimates. The adjusted estimate has both bias and variance resulting from the sources of error discussed in Chapter 10, Section 5. The net effect of the sources of error is estimated in a process called total error modeling (e.g., Chapter 10, Section 5.3). The unadjusted census is viewed as having bias (net undercount) but negligible variance. The bias can be estimated by comparing the census to the adjusted estimates, and allowing for bias and variance in the adjusted estimates.

Here we discuss estimation of the difference in expected loss based on (4.4) with $\alpha = 2$, $\theta_i =$ population size or share in area i, and $\hat{\theta}_i$ equal to the estimate of θ_i, based on the census, say $\hat{\theta}_{i,c}$, or adjusted census numbers, say $\hat{\theta}_{i,a}$. In the rest of Section 5, let us *condition on the realized census results* $\hat{\theta}_{i,c}$, so θ_i and $\hat{\theta}_{i,c}$ are fixed and only $\hat{\theta}_{i,a}$ are random. To estimate the difference in expected loss we will need to estimate $\Delta_i = (\hat{\theta}_{i,c} - \theta_i)^2 - E[(\hat{\theta}_{i,a} - \theta_i)^2]$. We may interpret Δ_i as the excess MSE of the census over the adjusted numbers. The difference in expected losses based on (4.4) is thus a weighted sum of Δ_i, say

$$\Delta = \sum_i w_i \Delta_i. \tag{5.1}$$

5.1. Accounting for Variances of Bias Estimates

Estimation of Δ_i is not entirely straightforward because the components of error are estimated with error. If the first two moments of the estimates of error components are known, they can be taken into account, as we now show.

Let us write $U_i = \theta_i - \hat{\theta}_{i,c}$, for short. We estimate it by $\hat{U}_i = \hat{\theta}_{i,a} - \hat{\theta}_{i,c}$whose expectation is $E[\hat{U}_i] = U_i + B_i$, where B_i denotes the bias in $\hat{\theta}_{i,a}$, and whose variance is $V_{U,i} = \mathrm{Var}(\hat{U}_i)$. Suppose we have an estimator of the bias $\hat{B}_i$ with variance $V_{B,i} = \mathrm{Var}(\hat{B}_i)$, and covariance $\mathrm{Cov}(\hat{B}_i, \hat{U}_i) = C_{BU,i}$. Next we assume that we have available estimators of the second moments $\hat{V}_{U,i}$, $\hat{V}_{B,i}$, and $\hat{C}_{BU,i}$. Observe that $E[(\hat{\theta}_{i,a} - \theta_i)^2] = B_i^2 + V_{U,i}$, so that $\Delta_i = U_i^2 - (B_i^2 + V_{U,i})$. Suppose for the moment that $\hat{B}_i$, $\hat{V}_{U,i}$, $\hat{V}_{B,i}$, and $\hat{C}_{BU,i}$ are all unbiased. It follows (cf., (4.1) of Chapter 8) that $(\hat{U}_i - \hat{B}_i)^2 - (\hat{V}_{B,i} + \hat{V}_{U,i} - 2\hat{C}_{BU,i})$ is an unbiased estimator of $U_i{}^2$. Similarly, $(\hat{B}_i{}^2 - \hat{V}_{B,i}) + \hat{V}_{U,i}$ is an unbiased estimator of $B_i{}^2 + V_{U,i}$. By subtracting the latter from the former we get that $\hat{\Delta}_i = (\hat{U}_i - \hat{B}_i)^2 - \hat{B}_i{}^2 - 2\hat{V}_{Ui} + 2\hat{C}_{BU,i}$

is an unbiased estimator of Δ_i. Differences in expected loss based on (4.4) with $\alpha = 2$ can be estimated by weighted sums, $\hat{\Delta} = \sum_i w_i \hat{\Delta}_i$.

5.2. *Effect of Unmeasured Biases on Comparisons of Accuracy*

A problem with the preceding analysis is that the estimate of bias in $\hat{U}_i$, namely $\hat{B}_i$, may itself be biased. For example, estimates of correlation bias in the U.S. PES can be flawed (e.g., Alho et al. 1993 show that there is residual heterogeneity in poststrata that are treated as being homogeneous), and estimates of synthetic estimation error are speculative (Chapter 10, Section 5.7). To allow for the possibility that $\hat{B}_i$ is biased, write $E[\hat{B}_i] = B_i - \beta_i$. The unmeasured bias β_i affects the estimates of accuracy of both the adjusted and the unadjusted estimates, and it can cause the comparison of expected loss, $\hat{\Delta}$, to be biased upward or downward. The bias in $\hat{\Delta}_i$ can be shown to be $E[\hat{\Delta}_i - \Delta_i] = 2\beta_i E[\hat{U}_i]$, and hence the bias in $\hat{\Delta}$ is

$$E[\hat{\Delta} - \Delta] = 2 \sum_i w_i \beta_i E[\hat{U}_i]. \tag{5.2}$$

This shows formally the fact that omitting components of bias in the adjusted estimate from the loss function analysis can be to tilt the analysis in favor of adjustment or non-adjustment, here depending on the sign of the right hand side of (5.2). This explains why analyses of synthetic estimation bias based on surrogate variables (as discussed in Chapter 10, Example 5.6) can give conflicting results. For different choices of surrogate variables, $E[\hat{U}_i]$ will be fixed but β_i will vary, and the sign of the right hand side of (5.2) can vary. Thus, some choices of surrogate variables (or other assumed models for synthetic estimation bias) suggest that the effect of omitting synthetic estimation bias from $\hat{B}$ increases $\hat{\Delta}$ and other choices suggest that the omission decreases $\hat{\Delta}$.

6. Decision Analysis of Adjustment for Census Undercount

Censuses contain error and post-enumeration surveys and demographic analysis provide information about the census error. A fundamental statistical question is whether and how to use the additional information to adjust the census numbers in such a way as to improve their accuracy. Comparisons of expected loss provide quantitative summaries concerning the relative accuracy of the alternative estimates. The estimates of expected loss are imperfect, as they depend on estimates of bias and variance that are themselves imperfect to a greater or lesser degree.

Example 6.1. Expected Loss of Adjusted and Unadjusted 1990 U.S. Census. Mulry and Spencer (1993) present estimates of the expected values of (4.4) for adjusted and unadjusted estimates of state shares of total population based on the 1990 U.S. census. The weights w_i were based on the unadjusted census counts in each case. The loss function with $w_i = 1$ and $\alpha = 1$ is called absolute error loss.

Expected Loss for State Shares	Unadjusted Est.	Adjusted Est.
Squared Error ($\times$ 10^6)	7.2	0.5
Weighted Squared Error ($\times$ 10^{-4})	1.2	0.1
Absolute Error ($\times$ 10^3)	9.7	2.8
Expected sum of errors in apportionment of House seats	1.8	0.5

Estimation of the expected value of absolute error loss is difficult due to uncertainty in the bias estimates and variance estimates for the adjusted and unadjusted estimates (Mulry and Spencer 1993, 1084). Estimates of the expected value of squared error loss tend to be inflated, but the inflation is constant for each estimator and the differences in expected values across the two sets of estimators are unaffected (Section 5.1). The estimates of variance of the adjusted estimate are understated to some degree, and biases due to synthetic estimation error are not taken into account. The latter omission biases the comparison of expected losses but not in a known direction (Section 5.2). For a critique of the analysis, see Brown et al. (1999b) and Anderson et al. (2000) and references therein. $\Diamond$

Example 6.2. Expected Loss of Adjusted and Unadjusted 2000 U.S. Census, A.C.E. Revision II. Mulry and ZuWallack (2003) report estimates of expected loss for the adjusted and unadjusted census estimates based on (4.4) with $\alpha = 2$ and $w_i = 1/\theta_i$. For comparison purposes, in computations w_i was set proportional to the reciprocal of the census count for unit i. A total of 10 loss functions were considered, 5 with θ_i based on population size (or "level") and 5 based on population share, and the 5 variants within each group considered geographic breakouts, e.g., states, counties, and places of various sizes. With the exception of the loss function based on population level for places with more than 100,000 population, the expected loss was smaller for the adjusted estimates. Sensitivity analyses were run with various assumptions about errors, including synthetic estimation error (Griffin 2002) and the results were essentially unchanged. A significant limitation of the analysis was that, unlike previous DSE's, the A.C.E. Revision II estimates incorporated adjustments for most of the nonsampling biases, including correlation bias. Estimates of the remaining bias after those adjustments, e.g., adjustments for correlation bias and for coding error, were not available and were not included in the estimates of expected loss nor in the sensitivity analyses. The U.S. Census Bureau (2003b) decided that the apparent improvement in accuracy as shown in the loss function analyses could not be trusted, a decision affirmed by the National Academy of Sciences (Citro, Cork, and Norwood 2004, 254–258). $\Diamond$

Given sufficient time and resources, it is possible at least in principle to develop confidence intervals (or regions, more generally) or other prediction intervals for the difference in expected loss. The sensitivity analyses mentioned in Example 6.2 are a step in that direction. Given that the results of total error modeling, and their summarization in terms of comparison of expected loss for alternative estimators, are imperfect and potentially misleading, should the analysis still be done? If one wants to quantify the accuracy of the estimates, we do not see any good alternative. It may well be, however, that even if the evidence is clear cut that adjustment improves accuracy, a statistical agency might decide not to adjust because assumptions lacked an unambiguous basis (as discussed in Example 4.6).

7. Cost-Benefit Analysis of Demographic Data

In a formal cost-benefit analysis for a data program, we compare the costs of alternative data programs (e.g., censuses with different levels of effort to reduce differential net undercoverage among areas) with differences in their benefits. In this context, benefit can be quantified as the negative of the Bayes risk, when the loss functions reflect data uses. To compare costs and benefits most directly, it is useful to quantify benefits in the same units as costs, e.g., dollars. When such a comparison is not feasible, one should not try to force the issue but instead one should simply prepare summaries showing what benefits are attainable at what costs.

Censuses are expensive programs and, other things being equal, data quality can increase or decrease if the expenditure on the census increases or decreases. For example, in the U.S. the decennial census costs, on a per household basis and adjusted for inflation, increased from $13 in 1970 to $24 in 1980, $32 in 1990, and $56 in 2000; total census costs in billions increased even more, from $0.9 in 1970 to $2.2 in 1980, $3.3 in 1990, and $6.6 in 2000 (Citro, Cork, and Norwood 2004).

Example 7.1. Decennial Census. Spencer (1980a) carried out a cost-benefit analysis of two alternative versions of the 1970 U.S. census: the census as conducted and an improved version that incorporated a coverage improvement program. The coverage improvement program was projected to cost an additional $12.7 million (in 1970 dollars) and to multiply undercount differentials by 0.65. This would have reduced the expected sum of errors in allocations under a large government program, General Revenue Sharing (GRS), by $224 million. Let θ_i in (4.8) denote the GRS allocation to recipient i if all statistics used by the allocation formula were perfect, and let $\hat{\theta}_i$ denote the allocations under the actual data. Thus, if the loss function is (4.8) with $c = 1$, the expected loss is $244 million. Was the extra cost justified? The answer depends on the magnitude of c and on how much of the additional census cost needs to be justified by the GRS program. A choice of $c = 0.01$ seemed reasonable (Spencer 1980a, 33), which meant that GRS could justify about 18% of the cost of the improved census data. Given the extent of other uses, the data improvement program appeared to be justified in cost-benefit terms. Similar explicit cost-benefit analyses were not conducted for the subsequent decennial censuses. For sensitivity analysis of the cost-benefit comparison, see Spencer (1994, 22–24).

We now consider some implications of the parameter α in (4.4) and the effect of the presence of errors in many other statistics used to compute the GRS allocations.

The estimate of a $224 million reduction of errors in allocations took into account the presence of errors in many other statistics used to compute the GRS allocations. If those errors had not been present, the expected reduction would have been much greater than $224 million, and a more expensive data improvement program could have been justified. Alternatively, if we had used $\alpha = 2$ in (4.4) instead of $\alpha = 1$, we would find that the optimal expenditure on the census would be unrelated to the extent of the other errors if they had mean zero and were independent of the census errors. For insight, suppose we model the effect of those other errors by η in (4.9), which we take to be independent of the errors in the census. Spencer (1985a) showed (under regularity conditions met by the example at hand) that if $\alpha < 2$, as the variance of η increases, the optimal expenditure on data decrease. If $\alpha = 2$, the optimal expenditure is unaffected by the variance of η or the presence of other data errors. If $\alpha > 2$, the presence of other data errors implies that the optimal expenditure on data needs to *increase* to compensate for those errors. In determining the needed data quality for allocation uses, we think that $\alpha = 1$ gives the most plausible relationship between optimal data expenditure and the imperfections of the allocations due to other statistical error or to the approximate nature of the allocation formula. ◊

Example 7.2. Mid-Decade Census. In the U.S., legislation requiring a mid-decade census has been in effect since 1976, but funding has not been provided to carry out a mid-decade census in 1985, 1995, or 2005 (Edmonston and Schultze 1995, 163–164). Was the decision not to fund it rational? A schematic cost-benefit analysis of a mid-decade census in the U.K. was carried out by Redfern (1974). To compare benefits and costs, he assumed that an allocation program was to distribute money to areas in proportion to their population and that the cost of the mid-decade census was to be funded by the allocation program itself. To allow for uses of the data other than for allocations, we could specify that if the mid-decade cost increased by x, the amount to be distributed would decrease by λx, where λ reflects the portion of the costs to be borne by the allocation program. Redfern's analysis did not appear to support a mid-decade census (Spencer 1980a, 13–17). ◊

We do not believe that the optimal expenditure on a data program can be determined from a cost-benefit analysis, and we caution against over-reliance on formal analyses. The loss functions in use are simply too approximate. We do believe that carrying out such analyses can improve decision making for the design of demographic data programs, however (Törnqvist 1948, 265). Conducting cost-benefit analyses focuses attention on how the data are used and what aspects of data quality are important or not important.

Exercises and Complements (*)

1. Consider the statistical problem of testing the null hypothesis that the means of two groups are equal. Suppose that we have an estimate $\hat{\delta}$ of the true difference, δ, between the means, and $\hat{\delta} \sim N(\delta, \sigma^2/n)$, where n is a measure of the sample

size and for we assume the n and σ^2 are known. Consider a test that rejects the null hypothesis if $|\hat{\delta}/(\sigma/\sqrt{n})| > c$, where c is chosen so that the significance level, i.e., the probability of rejecting the null hypothesis when $\delta = 0$, is α. Let β denote the probability of a failing to reject the null hypothesis, when $\delta \neq 0$. One can show that when the true difference in the means is δ, we have the approximate relation $1 \approx \frac{\sigma^2/n}{\delta^2}(z_{1-\beta} + z_{1-\alpha/2})^2$, where z_p denotes the pth quantile of a $N(0, 1)$ distribution. Thus, any four of the quantities $\sigma^2, \delta, n, \alpha$, and β determine the fifth. This implies that choosing $\alpha = 0.05$ can lead to large values of β when n is small, for any fixed σ^2 and δ. (Snedecor and Cochran 1967, 111–113)

2. Following Ellsberg (1961), consider two urns containing 100 balls each, with each ball colored red or black. The composition of urn I is unknown but urn II has 50 red and 50 black balls. Consider four gambles:
   A: a ball is picked at random from urn I and if it comes out red you win 100$;
   B: a ball is picked at random from urn I and if it comes out black you win 100$;
   C: a ball is picked at random from urn II and if it comes out red you win 100$;
   D: a ball is picked at random from urn II and if it comes out black you win 100$.

   A person is inquired about his or her preferences as to which gamble to choose:
   (i) Do you prefer A or B or are you indifferent?
   (ii) Do you prefer C or D or are you indifferent?
   (iii) Do you prefer A or C or are you indifferent?
   (iv) Do you prefer B or D or are you indifferent?

   Suppose that a person is indifferent between A and B in (i) and between C and D in (ii). Suppose further that the person does have a preference, one way or the other, in (iii) or in (iv). Show that the person's preferences are inconsistent with existence of a prior probability, p, for the probability of selection of a red ball from Urn I. (Hint: Observe that if, say, C is preferred in (iii) then $p < 1/2$.) What do you conclude if the same urn is preferred in (iii) and (iv)?
3. Show that (4.6) holds.
4. Show that (4.7) holds. (Hint: assume the maximum and minimum w_i's are for $i = 1$ and $i = 2$ consider errors in allocations $x = (\hat{\theta}_1 - \theta_1) = -(\hat{\theta}_2 - \theta_2) > 0$ and $\hat{\theta}_i = \theta_i = 0, i > 2$. Notice that $0 < l(\boldsymbol{\theta}, \hat{\boldsymbol{\theta}}) = -b \max_i \{w_i\}|x| + a \min_i \{w_i\}|x|$.)

*5. In Section 4.5.2, we considered the loss function to be a random variable. Sometimes it is reasonable to view the use of the data as a random variable. For example, consider the design of a study for evaluation of a government program to affect the total fertility rate. If the measured effect exceeds a threshold τ, the program is considered a success and will be continued, and otherwise funding will be cut. The measurement accuracy can be improved if more money is spent on the evaluation, but the optimal amount to spend on the evaluation depends on τ. Although the optimal threshold τ can be developed using statistical decision theory (e.g., reflecting social losses from wrong decisions of either kind), the actual threshold is not optimal but reflects transient priorities and political compromises. How should one decide how much money to spend on the evaluation study, and hence how accurate it will be? One possibility is to consider τ to be a random variable with a prior distribution, and then to derive the Bayes rule for optimal expenditure on the study. The optimal expenditure is not necessarily greatest when the optimal threshold is used (Spencer and Moses 1990).

6 To derive (5.2), note that

$$\begin{aligned} E[\hat{\Delta}_i - \Delta_i] &= E\big[(\hat{U}_i - \hat{B}_i)^2 - \hat{B}_i{}^2 - 2\hat{V}_{U,i} + 2\hat{C}_{BU,i}\big] - \big(U_i{}^2 - B_i{}^2 - V_{U,i}\big) \\ &= E[\hat{U}_i - \hat{B}_i]^2 + \mathrm{Var}(\hat{U}_i - \hat{B}_i) - E[\hat{B}_i]^2 - \mathrm{Var}(\hat{B}_i) - 2V_{U,i} \\ &\quad + 2C_{BU,i} - U_i{}^2 + B_i{}^2 + V_{U,i} \\ &= (U_i + \beta_i)^2 - (B_i - \beta_i)^2 - U_i{}^2 + B_i{}^2 \\ &= 2\beta_i U_i + 2\beta_i B_i \\ &= 2\beta_i E(\hat{U}_i). \end{aligned}$$

Show that if $\sum_i \hat{U}_i$ is fixed, as in the case of estimating population shares, that $E[\hat{\Delta} - \Delta]$ is proportional to the cross-area correlation between β_i and $E[\hat{U}_i]$.

References

Aalen O. (1976) Nonparametric inference in connection with multiple decrement models. *Scandinavian Journal of Statistics* 3, 15–27.

Afifi A.A. and Azen S.P. (1979) *Statistical analysis, 2nd ed.* New York: Academic Press.

Ahlburg D.A. and Vaupel J.W. (1990) Alternative projections of the U.S. population. *Demography* 27, 639–652.

Alho J.M. (1989) Relating changes in life expectancy to changes in mortality. *Demography* 26, 705–709.

Alho J.M. (1990a) Adjusting for nonresponse bias using logistic regression. *Biometrika* 77, 617–624.

Alho J.M. (1990b) Logistic regression in capture-recapture models. *Biometrics* 46, 623–635.

Alho J.M. (1990c) Stochastic methods in population forecasting. *International Journal of Forecasting* 6, 521–530.

Alho J.M. (1991) Effect of aggregation on the estimation of trend in mortality. *Mathematical Population Studies* 3, 53–67.

Alho J.M. (1992a) Estimating the strength of expert judgment: the case of U.S. mortality forecasts. *J. Forecasting* 11, 157–167.

Alho J.M. (1992b) The magnitude of error due to different vital processes in population forecasts. *International Journal of Forecasting* 8, 301–314.

Alho J.M. (1992c) On prevalence, incidence, and duration in general stable populations. *Biometrics* 48, 587–592.

Alho J.M. (1994) Analysis of sample-based capture-recapture experiments. *Journal of Official Statistics*, 10, 245–256.

Alho J.M. (1997) Scenarios, uncertainty and conditional forecasts of the world population. *Journal of the Royal Statistical Society. Series A* 160, 71–85.

Alho J.M. (1998) A stochastic forecast of the population of Finland. *Reviews 1998/4.* Helsinki: Statistics Finland.

Alho J.M. (Ed.) (1999) *Statistics, registries, and science. Experiences from Finland.* Helsinki: Statistics Finland.

Alho J.M. (2000) A statistical look at Modeen's forecast of the population of Finland in 1934. *Yearbook of Population Research in Finland* XXXVI, 107–120.

Alho J.M. (2002a) *Stochastic forecast of the Lithuanian population 2001–2050.* Project Report. Helsinki: The Research Institute of the Finnish Economy.

Alho J.M. (2002b) *The population of Finland in 2050 and beyond*. Discussion Paper 826. Helsinki: The Research Institute of Finnish Economy.

Alho J.M. (2003a) *Duration-dependent lifetables with applications to nuptiality*. Paper presented at the Annual Meeting of Population Association of America, May 2003, Minneapolis.

Alho J.M. (2003b) Predictive distribution of adjustment for life expectancy change. *Working Papers 3*. Helsinki: Finnish Centre for Pensions.

Alho J.M., Kangas J. and Kolehmainen O. (1996) Uncertainty in expert predictions of the ecological consequences of forest plans. *Applied Statistics* 45, 1–14.

Alho J.M., Lassila J. and Valkonen T. (2005) Demographic uncertainty and evaluation of sustainability of pension systems. Forthcoming in Holzmann R. and E. Palmer (Eds.) (2005, in press) *Pension reform issues and prospects for non-financial defined contribution (NDC) schemes*. Washington, D.C.: The World Bank.

Alho J.M., Mulry M.H., Wurdeman K. and Kim J. (1993) Estimating heterogeneity in the probabilities of enumeration for dual-system estimation. *Journal of the American Statistical Association* 88, 1130–1136.

Alho J.M. and Nyblom J. (1997) Mixed estimation of old-age mortality. *Mathematical Population Studies* 6, 319–330.

Alho J.M., Saari M. and Juolevi A. (2000) A competing risks approach to the two-sex problem. *Mathematical Population Studies* 8, 73–90.

Alho, J.M. and Salo, M.A. (1998). *Kuntien valtionosuuden epävarma kohtaanto*. Studies in Social Policy 1. Joensuu: University of Joensuu.

Alho J.M. and Salo, M.A. (2000) Merit rating and formula-based resource allocation. *International Journal of Educational Management* 14, 95–100.

Alho J.M. and Spencer B.D. (1985) Uncertain population forecasting. *Journal of the American Statistical Association* 80, 306–314.

Alho J.M. and Spencer B.D. (1990a) Effects of targets and aggregation on the propagation of error in mortality forecasts. *Mathematical Population Studies* 2, 209–227.

Alho J.M. and Spencer B.D. (1990b) Error models for official mortality forecasts. *Journal of the American Statistical Association* 85, 609–616.

Alho J.M. and Spencer B.D. (1991) A population forecast as a database: implementing the stochastic propagation of error. *Journal of Official Statistics* 7, 295–310.

Alho J.M. and Spencer B.D. (1997) The practical specification of the expected error of population forecasts. *Journal of Official Statistics* 13, 203–225.

Alho J.M. and Vanne R. (2005) On stochastic generational accounting. To appear in Gauthier A., Chu C., and Tuljapurkar S. (eds.) (2005, in press) *Riding the age-waves: allocating public and private resources across generations*. Dordrecht: Kluwer.

Andersen E.B. (1980) *Discrete statistical models with social science applications*. Amsterdam: North-Holland.

Andersen P.K. (1986) *Time-dependent covariates and Markov processes*. Pp. 82–103 in Moolgavkar and Prentice (1986).

Andersen P.K., Borgan Ø, Gill R.D. and Keiding N. (1993) *Statistical models based on counting processes*. New York: Springer.

Anderson M., Daponte, B.O., Fienberg S.E., Kadane J.B., Spencer B.D. and Steffey D.L. (2000) Sampling-based adjustment of the 2000 census – a balanced perspective. *Jurimetrics* 40, 341–356.

Anderson M.A. and Fienberg, S.E. (1999) *Who counts? The politics of census-taking in contemporary America*. New York: Russell Sage Foundation.

Anderson R.N. and Rosenberg H.M. (1998) Age standardization of death rates: implementation of the year 2000 standard. *National Vital Statistics Reports, vol. 47, no. 3.* Hyattsville: National Center for Health Statistics.

Anderson R.N., Minino A.M., Hoyert D.L. and Rosenberg H.M. (2001) Comparability of cause of death between ICD-9 and ICD-10: preliminary estimates. *National Vital Statistics Reports, vol. 49, no. 2.* Hyattsville: Center for Health Statistics.

Andrews G.H. and Beekman J.A. (1987) *Actuarial projections for the old-age, survivors, and disability insurance program of social security in the United States of America.* Itasca IL: Actuarial Education and Research Fund.

Arrow K.J. (1971) *Essays in the theory of risk-bearing.* Chicago: Markham Publishing Co.

Aubin J. (1991) *Viability theory.* Boston: Birkhäuser.

Auerbach A.J. and Kotlikoff J.L. (1987). *Dynamic fiscal policy.* Cambridge: Cambridge University Press.

Auerbach A.J., Gokhale J. and Kotlikoff L.J. (1991) Generational accounts – a meaningful alternative to deficit accounting. Pp. 55–110 in D. Bradford (1991) *Tax policy and the economy. Vol. 5.* Cambridge MA: MIT Press.

Auerbach A.J., Kotlikoff, L.J. and Leibfritz, W. (1997). *Generational accounting around the world.* Chicago: The University of Chicago Press.

Auerbach A.J. and Lee R.D. (Eds.) (2001) *Demographic change and fiscal policy.* Cambridge: Cambridge University Press.

Avery R.B., Elliehausen G.E., and Kennickell A.B. (1986) *Measuring wealth with survey data: an evaluation of the 1983 Survey of Consumer Finances.* Washington D.C.: Board of Governors of the Federal Reserve System. Last revision April 1988. Available from http://www.federalreserve.gov/pubs/oss/oss2/method.html

Azzalini A. (1996) *Statistical inference.* London: Chapman & Hall.

Bachelier L. (1900) Théorie de la spéculation. *Ann. Sci. École Normale Sup.* III-17, 21–86. (An English translation is available: pp. 17–78 in Cootner P.H. (Ed.) (1964) *The random character of stock market prices.* Cambridge MA: MIT Press.)

Bailey R. A. (1982) Confounding. Pp. 128–134 in Kotz, Johnson, and Read (1982), *vol. 9.*

Balinski M.L. and Young H.P. (1982) *Fair representation.* New Haven: Yale University Press.

Beale C. (2004) Reflections on 50+ years as a federal demographic statistician. *Amstat News*, April 2004, 6–8. Alexandria VA: American Statistical Association.

Bell W.R. (1992) ARIMA and principal component models in forecasting age-specific fertility. Chapter 10 of Keilman N. and Cruijsen H. (1992) *National population forecasting in industrialized countries.* Amsterdam: Swets and Zeitlinger.

Bell W.R. (1993) Using information from demographic analysis in post-enumeration survey estimation. *Journal of the American Statistical Association* 88, 1106–1118.

Bell W.R. (1997) Comparing and assessing time series methods for forecasting age-specific fertility and mortality rates. *Journal of Official Statistics* 13, 279–303.

Bench K. (2002a) *Contamination of Census 2000 data collected in Accuracy and Coverage Evaluation block clusters.* Census 2000 Evaluation N.1. August 22, 2002. Washington, D.C.: U.S. Census Bureau.

Bench K. (2002b) *P-sample match rate corrected for error due to inconsistent post-stratification variables.* DSSD A.C.E. Revision II Memorandum Series #PP-46. Washington, D.C.: U.S. Census Bureau.

Bench K. (2003) P-sample match rate corrected for error due to inconsistent post-stratification variables. *Joint Statistical Meetings – Section on Survey Research Methods Proceedings*, 514–519. Alexandria, VA: American Statistical Association.

Bernardelli H. (1941) Population waves. *Journal of Burma Research Society* 31, 1–18.

Bernstein P.L. (1998) *Against the gods. The remarkable story of risk.* New York: Wiley.

Berger J.O. (1980) *Statistical decision theory, foundations, concepts, and methods.* New York: Springer.

Best N.K., Cowles M.K. and Vines K. (1995) *CODA: Convergence diagnosis and output analysis software for Gibbs sampling output. Version 3.0.* Technical Report. NRC Biostatistics Unit, University of Cambridge.

Bickel, P.J. and Doksum, K.A. (2001) *Mathematical statistics: basic ideas and selected topics, Vol I., 2nd ed.* Upper Saddle River, NJ: Prentice Hall.

Bickel, P.J. and Freedman, D.A. (1984) Asymptotic normality and the bootstrap in stratified sampling. *The Annals of Statistics* 12, 470–482.

Bienen H. and Van de Walle N. (1991) *Of time and power, leadership duration in the modern world.* Stanford: Stanford University Press.

Binder D.A. (1983) On the variances of asymptotically normal estimators from complex surveys. *International Statistical Review* 51, 279–292.

Binder D.A. and Roberts R.R. (2003) Design-based and model-based methods for estimating model parameters. Pp. 29–48 in Chambers and Skinner (2003).

Birkhoff G. (1976) House monotone apportionment schemes. *Proceedings of the National Academy of Sciences, U.S.A.,* 73, 684–686.

Bishop Y.M.M., Fienberg S.E. and Holland P.W. (1975) *Discrete multivariate analysis.* Cambridge MA: The MIT Press.

Blanchard, O., Chouraqui J.-C., Hagemann R.P. and Sartor N. (1990) The sustainability of fiscal policy: new answers to an old question. *OECD Economic Studies No. 15, Autumn 1990.*

Bollerslev T., Chou R.Y. and Kroner K.F. (1992) ARCH modeling in finance. *Journal of Econometrics* 52, 5–59.

Bongaarts J. and Bulatao R.A. (Eds.) (2000) *Beyond six billion.* Panel on Population Projections, National Research Council. Washington D.C.: National Academy Press.

Bongaarts J. and Feeney G. (1998) On the quantum and tempo of fertility. *Population and Development Review* 24, 271–291.

Boruch R.F. (1990) Research on the use of statistical data. *Proceedings of the Social Statistics Section,* 52–57. Alexandria, VA: American Statistical Association.

Bowley A.L. (1924) Births and population of Great Britain. *Journal of the Royal Economic Society* 34, 188–192.

Box G.E.P. and Jenkins G.M. (1976) *Time series analysis, revised ed.* San Francisco: Holden-Day.

Box G.E.P. and Tiao G.C. (1973) *Bayesian inference in statistical analysis.* Reading MA: Addison-Wesley.

Bozik J.E. and Bell W.R. (1987) Forecasting age-specific fertility using principal components. *Proceedings of the American Statistical Association, Social Statistics Section,* 396–401.

Bozon M. and Heran F. (1989) Finding a spouse. A survey of how French couples meet. *Population, English Edition* 44, 90–121.

Breslow N.E. (1974) Covariance analysis of censored survival data. *Biometrics* 30, 89–100.

Breslow N.E. and Day N.E. (1980) *Statistical methods in cancer research, vol. I – The analysis of case-control studies.* Lyon: IARC.

Breslow N.E. and Day N.E. (1987) *Statistical methods in cancer research, vol. II – The design and analysis of cohort studies.* Lyon: IARC.

Brown J.J., Diamond D.D., Chambers R.L., Buckner L.J. and Teague A.D. (1999a) A methodological strategy for a one-number census in the UK. *Journal of the Royal Statistical Society A* 162, 247–267.

Brown L.D., Eaton M.L., Freedman D.A., Klein S.P., Olshen R.A., Wachter K.W., Wells M.T. and Ylvisaker D. (1999b) Statistical controversies in census 2000. *Jurimetrics* 39, 347–375, Summer 1999.

Bryan, T. (2004) Population estimates. Pp. 523–560 in J.S. Siegel and D.A. Swanson (2004).

Burdick C., Manchester J. and Bang E. (2003) Stochastic models of the Social Security Trust Funds. *Research and Statistics Note No. 2003-01.* Office of Policy, Social Security Administration, Washington, D.C.

Burgard S. (2002) Race and children's height in Brazil and South Africa. *Demography* 39, 763–790.

Burgdörfer F. (1932) *Volk ohne Jugend.* Berlin: Kurt Vowinkel.

Burgess R. (1988) Evaluation of the reverse record check estimates of undercoverage in the Canadian census of population. *Survey Methodology* 14, 137–156.

Burke J. and Rust K. (1995) On the performance of jackknife variance estimation for systematic samples with small numbers of primary sampling units. *Proceedings of the American Statistical Association, Survey Research Section*, 321–326.

Butz W.P. and Ward M.P. (1979) The emergence of countercyclical U.S. fertility. *American Economic Review* 69, 318–328.

Cannan E. (1895) The probability of a cessation of the growth of population of England and Wales during the next century. *The Economic Journal* 5, 505–515.

Carroll G.R. and Hannan M.T. (2000) *The demography of corporations and industries.* Princeton: Princeton University Press.

Carvalho A. and Spencer B. (2001) Survival models for leadership duration. Unpublished manuscript.

Caswell H. (2001) *Matrix population models, 2nd ed.* Sunderland: Sinauer.

Chamberlain G. (1982) The general equivalence of Granger and Sims causality. *Econometrica* 50, 569–580.

Chambers R.L. and Skinner C.J. (Eds.) (2003) *Analysis of survey data.* New York: Wiley.

Chandra Sekar C. and Deming E. (1949) On a method of estimating birth and death rates and the extent of registration. *Journal of the American Statistical Association* 44, 101–115.

Chatfield C. (1996) *The analysis of time series, 5th ed.* London: Chapman & Hall.

Chiang C.L. (1968) *Introduction to stochastic processes in biostatistics.* New York: Wiley.

Chiang C.L. (1984) *The life table and its applications.* Malabar: Krieger.

Chib S. and Greenberg E. (1994) Bayes inference in regression models with ARMA(p,q) errors. *Journal of Econometrics* 64, 183–206.

Chung K.L. (1974) *A course in probability theory, 2nd ed.* New York: Academic Press.

Çinlar E. (1975) *Introduction to stochastic processes.* Englewood Cliffs: Prentice Hall.

Citro C.F., Cork D.L. and Norwood J.L. (Eds.) (2004) *The 2000 census: counting under adversity.* Panel to Review the 2000 Census, National Research Council. Washington, D.C.: National Academy Press.

Clayton D. and Schiffers E. (1987a) Models for temporal variation in cancer rates I: age-period and age-cohort models. *Statistics in Medicine* 6, 449–467.

Clayton D. and Schiffers E. (1987b) Models for temporal variation in cancer rates II: age-period-cohort models. *Statistics in Medicine* 6, 469–481.

Clemen R.T. (1989) Combining forecasts: a review and annotated bibliography. *International Journal of Forecasting* 5, 559–583.

Clogg C.C. and Himes C.L. (1993) Comment. *Journal of the American Statistical Association* 88, 1072–1074.

Coale A.J. (1955) The population of the United States in 1950 classified by age, sex, and color – a revision of census figures. *Journal of the American Statistical Association* 50, 16–54.

Coale A. (1972) *The growth and structure of human populations*. Princeton University Press.

Cochran W.G. (1954) Some methods of strengthening the common χ^2 tests. *Biometrics* 10, 417–451.

Cochran W.G. (1977) *Sampling techniques*, 3rd *ed.* New York: Wiley.

Cohen J. (1977) Ergodicity of age structure in populations with Markovian vital rates III: finite state moments and growth rate; an illustration. *Advances in Applied Probability* 9, 462–475.

Cohen J. (1986) Population forecasts and confidence intervals for Sweden: a comparison of model-based and empirical approaches. *Demography* 23, 105–126.

Coleman J.S. (1997) Constructed social networks for the study of diffusion. Pp. 180–193 in Spencer (1997).

Conover W.J. (1980) *Practical nonparametric statistics, 2nd ed.* New York: Wiley.

Cork D.L., Cohen M. L., and King B.F. (Eds.) (2004) *Reengineering the 2010 Census: Risks and Challenges*. Panel on Research on Future Census Methods, National Research Council. Washington, D.C.: National Academy Press.

Cormack R.M. (1968) The statistics of capture-recapture methods. *Oceanography and Marine Biology. Annual Review* 6, 455–506.

Cowan C. and Malec D. (1986) Capture-recapture models when both sources have clustered observations. *Journal of the American Statistical Association* 81, 347–353.

Cox D.R. (1972) Regression models and life-tables. *Journal of the Royal Statistical Society. Series B* 34, 187–220.

Cox D.R. (1975) Partial likelihood. *Biometrika* 62, 269–276.

Cox D.R. and Hinkley D.V. (1974) *Theoretical statistics*. London: Chapman & Hall.

Cox D.R. and Oakes D. (1984) *Analysis of survival data*. London: Chapman & Hall.

Cressie N. (1993) *Statistics for spatial data*. New York: Wiley.

Crouch E.A.C. and Wilson R. (1981) Reply to comments on the regulation of carcinogens. *Risk Analysis* 1, 107–111.

Daponte B.O., Kadane J.B. and Wolfson, L.J. (1997) Bayesian demography: projecting the Iraqi Kurdish population, 1997–1990. *Journal of the American Statistical Association* 92, 1256–67.

Darroch J.N., Fienberg S.E., Glonek G.F.V. and Junker B.W. (1993) A three-sample multiple-recapture approach to census population estimation with heterogeneous catchability. *Journal of the American Statistical Association* 88, 1137–1148.

David I.P. and Sukhatme B.V. (1974) On the bias and mean square error of the ratio estimator. *Journal of the American Statistical Association* 69, 464–466.

Davis S.T. (1994) Evaluation of postcensal county estimates for the 1980s. *Population Division Working Paper No. 5*. Washington, D.C. U.S. Census Bureau.

Day J.C. (1993) Population projections of the United States by age, sex, race, and Hispanic origin: 1993 to 2050. *Current Population Reports, Series P-25, No. 1018*. Washington, D.C.: U.S. Census Bureau.

DeBeer J. (1997) The effects of uncertainty of migration on national population forecasts: the case of the Netherlands. *Journal of Official Statistics* 13, 227–243.

DeBeer J. and Alders M. (1999) Uncertainty of population forecasts: a stochastic approach. *Netherlands Official Statistics* 14, 19–25. Voorburg: Statistics Netherlands.

De Finetti B. (1931) Sul significato soggetivo della probabilità. *Fundamenta Mathematicae* 17, 298–329.

De Finetti B. (1937) La prévision: ses lois logiques, ses sources subjectives. *Annales de l'Institut Henri Poincaré* 7, 1–68.

De Finetti B. (1974) *Theory of probability, Vols. 1, 2*. New York: Wiley.

De Morgan A. (1847) *Formal logic or the calculus of inference, necessary and probable.* London: Taylor and Walton.

DeGans H.A. (1999) *Population forecasting 1895–1945*. Dordrecht: Kluwer.

DeGroot M.H. (1987) *Probability and statistics, 2nd ed.* Reading: Addison-Wesley.

Deming W.E. (1964) *Statistical adjustment of data*. New York: Dover.

Deville J.C. and Särndal C.-E. (1992) Calibration estimators in survey sampling. *Journal of the American Statistical Association* 87, 376–382.

Deville J.C., Särndal C.-E. and Sautory O. (1993) Generalized raking procedures in survey sampling. *Journal of the American Statistical Association* 88, 1013–1020.

Dickey D.A. and Fuller W.A. (1981) Likelihood ratio statistics for autoregressive time series with a unit root. *Econometrica,* 49, 1057–1072.

Diggle P.J. (1983) *Statistical analysis of spatial point patterns*. London: Academic Press.

Doll R. and Hill A.B. (1950) Smoking and carcinoma of the lung. preliminary report. *British Medical Journal ii*, 739–748.

Doll R. and Peto R. (1976) Mortality in relation to smoking: 20 years' observations on male British doctors. *British Medical Journal ii*, 1525–1536.

Doob J.L. (1953) *Stochastic processes*. New York: Wiley.

Dorn H. (1950) Pitfalls in population forecasts and projections. *Journal of the American Statistical Association* 45, 311–334.

dos Santos Silva I. (1999) *Cancer epidemiology: principles and methods*. Lyon: International Agency for Research on Cancer.

Draper D. (1995) Assessment and propagation of model uncertainty (with discussion). *Journal of the Royal Statistical Society. Series B* 57, 45–70, with discussion 71–97.

DuMouchel W.H. and Duncan G.J. (1983) Using sample survey weights in multiple regression analyses of stratified samples. *Journal of the American Statistical Association* 78, 535–543.

Durbin J. (1953) Some results in sampling theory when the units are selected with unequal probabilities. *Journal of the Royal Statistical Society. Series B* 15, 262–269.

Durbin J. and Koopman S.J. (2000) Time series analysis of non-Gaussian observations based on state space models from both classical and Bayesian perspectives. *Journal of the Royal Statistical Society. Series B* 62, 3–56.

Durkheim E. (1937) Les règles de la méthode sociologique. Paris: Presses Universitaires de France.

Easterlin R.A. (1961). The American baby boom in historical perspective. *American Economic Review* 51, 860–911.

Edmonston B. and Schultze C. (Eds.) (1995) *Modernizing the U.S. census*. Panel on Census Requirements in the Year 2000 and Beyond, National Research Council. Washington, D.C.: The National Academy Press.

Edwards W. (1982) *Conservatism in human information processing*. Pp. 359–369 in Kahneman et al. (1982).

Efron B. and Morris C. (1971) Limiting the risk of Bayes and empirical Bayes estimators – part 1: the Bayes case. *Journal of the American Statistical Association* 66, 807–815.

Efron B. and Tibshirani R.J. (1993) *An introduction to the bootstrap*. New York: Chapman & Hall.

Ekamper P. and Keilman N. (1993) Sensitivity analysis in a multidimensional demographic projection model with a two-sex algorithm. *Mathematical Population Studies* 3, 21–36.

Ekert O. (1986) Effets et limits des aides financières aux familles. *Population* 41, 327–348.

Eklund K. (1995) *Kuolevuuden mallittaminen ja ennustaminen vanhusväestön keskuudessa.* Pro Gradu Thesis, Dept. of Statistics, University of Joensuu.

Elliott M.R. and Little R.J.A. (2000). A Bayesian approach to combine information from a census, a coverage measurement survey and demographic analysis. *Journal of the American Statistical Association* 95, 351–363.

Ellsberg D. (1961) Risk, ambiguity, and the Savage axioms. *The Quarterly Journal of Economics* 75, 643–669.

Engle R.F. (1982) Autoregressive conditional heteroscedasticity with estimates of the variance of U.K. inflation. *Econometrica* 50, 987–1008.

Erdös P. and Renyi A. (1959) On the central limit theorem for samples from a finite population. *Publications of the Mathematical Institute of the Hungarian Academy of Sciences* 4, 49–61.

Ericksen E.P. (1974) A regression method for estimating population changes of local areas. *Journal of the American Statistical Association* 69, 867–875.

ESCAP (2001) *Recommendation concerning the methodology to be used in producing the tabulations of population reported to states and localities pursuant to 13 U.S.C. 141(c).* Report of the Executive Steering Committee for Accuracy and Coverage Evaluation Policy. March 1, 2001. Washington, D.C.: Census Bureau.

Espenshade T.J. (1995) Using INS border apprehension data to measure the flow of undocumented migrants crossing the U.S.-Mexico frontier. *International Migration Journal* 29, 545–565.

Espenshade T.J. (Ed.)(1997) *Keys to successful migration.* Washington, D.C.: The Urban Institute Press.

Eurostat (2004) *Population statistics. Theme 3: Population and social conditions.* Luxembourg: European Commission.

Farber J. (2001) *Accuracy and Coverage Evaluation: consistency of post-stratification variables, quality indicators of census 2000 and the Accuracy and Coverage Evaluation.* DSSD Census 2000 Procedures and Operations Memorandum Series B-10. Washington, D.C.: Census Bureau.

Fay R.E. (1974) *Statistical considerations in estimating the current population of the United States.* Ph.D. Dissertation, Department of Statistics, University of Chicago.

Fay R.E. (1992) Absolute values in the PES loss function analysis. Memorandum for the Committee on Adjustment of Postcensal Estimates (CAPE). June 29, 1992. Washington, D.C.: U.S. Census Bureau.

Fay R.E. (1996) Alternative paradigms for the analysis of imputed survey data. *Journal of the American Statistical Association* 91, 490–498, with discussion 507–520.

Fay R.E., Passel J.S., Robinson, J.G. and Cowan C.D. (1988). *The coverage of population in the 1980 census.* Washington, D.C.: U.S. Census Bureau.

Fay R.E. and Thompson J.H. (1993) The 1990 post enumeration survey: statistical lessons, in hindsight. *Proceedings of the 1993 Annual Research Conference of the Bureau of the Census*, 71–91. Washington, D.C.: U.S. Census Bureau.

Feeney G.M. (1970) Stable age by region distributions. *Demography* 6, 341–348.

Feinstein A.R. (1985) Experimental requirements and scientific principles in case-control Studies. *Journal of Chronic Disease* 38, 127–133.

Feldpausch R. (2002) *ESCAP II: E-Sample erroneous enumerations*. Executive Steering Committee for A. C. E. Policy II, Report No. 5, March 13, 2000 (revised). Washington, D.C.: U.S. Census Bureau.

Fellegi I. (1980) Evaluation programme of the 1976 census of population and housing – A Sampling. *The Statistician* 29, 275–312.

Feller W. (1968) *An introduction to probability theory and its applications, vol. I, 3rd ed.* New York: Wiley.

Feller W. (1971) *An introduction to probability theory and its applications, vol. II*. New York: Wiley.

Fieller E.C. (1932) The distribution of the index in a normal bivariate population. *Biometrika* 24, 428–440.

Fienberg S.E. (1971) Randomization and social affairs: the 1970 draft lottery. *Science,* 171, 255–261.

Fine T.L. (1973) *Theories of probability*. New York: Academic Press.

Finney D.J. (1952) *Probit analysis, 2nd ed.* Cambridge: Cambridge University Press.

Flanders W.D., Dersimonian R. and Rhodes P. (1990) Estimation of risk ratios in case-base studies with competing risks. *Statistics in Medicine* 9, 423–435.

Fleischhacker J., DeGans H. and Burch T. (Eds.)(2003) *Populations, projections and politics*. Amsterdam: Rozenberg.

Florens J. and Mouchart M. (1982) A note on non-causality. *Econometrica* 50, 583–591.

Fowlkes E.B. (1987) Some diagnostics for binary regression via smoothing. *Biometrika* 74, 503–515.

Fougstedt G. (1977) Trends and factors of fertility in Finland. *Commentationes Scientiarum Socialium 7/1977*. Societas Scientiarum Fennica, Helsinki.

Francisco C.A. and Fuller W.A. (1991) Quantile estimation with a complex survey design. *Annals of Statistics* 19, 454–469.

Frankel M.R. (1983) Sampling theory. Pp. 21–67 in Rossi et al. (1983).

Frankel M. and Kennickell A. (1995) Toward the development of an optimal stratification paradigm for the Survey of Consumer Finances. *Proceedings of the American Statistical Association, Survey Research Section*, 638–643.

Freedman D. and Wachter K. (1994) Heterogeneity and census adjustment for the intercensal base. *Statistical Science* 9, 476–485.

Fuller W.A. (1984) Least squares and related analyses for complex survey designs. *Survey Methodology* 10, 97–118.

Fuller W.A. (1987) *Measurement error models*. New York: Wiley.

Gabler S., Haeder S. and Lahiri P. (1999) A model based justification of Kish's formula for design effects for weighting and clustering. *Survey Methodology* 25, 105–106.

Gail M.H. (1986) Adjusting for covariates that have the same distribution in exposed and unexposed cohorts. Pp. 3–18 in Moolgavkar and Prentice (1986).

Gantmacher F.R. (1959) *The theory of matrices*, Vols. I–II. New York: Chelsea Publishing.

Gavrilov L.A. and Gavrilova N.S. (1991) *The biology of life span: A quantitative approach*. Chur: Harwood Academic Publishers.

Gelman A., Carlin J.B., Stern H.S. and Rubin D.B. (1995) *Bayesian data analysis*. London: Chapman & Hall.

Geng Z., Guo J. and Fung W. (2002) Criteria for confounders in epidemiological studies. *Journal of the Royal Statistical Society. Series B* 64, 3–15.

Germain M.-F. and Julien C. (1993) Results of the 1991 census coverage error measurement program. Pp. 55–70 in *Proceedings of Seventh Annual Research Conference of the Bureau of the Census.* Washington, D.C.: U. S. Census Bureau.

Ghosh M. and Rao J.N.K. (1994) Small area estimation: an appraisal. *Statistical Science* 9, 55–93.

Gilks W.R., Richardson S. and Spiegelhalter D. (1995) *Practical Markov chain Monte Carlo*. New York: Chapman & Hall.

Gill R.D. and Keilman N. (1990) On the estimation of multidimensional demographic models with population registration data. *Mathematical Population Studies* 2, 119–143.

Gini C. (1930) Calcolo di previsione della popolazione italiana del 1921 al 1961. *Notiziario demografico* 1930, 8–9.

Girosi F. and King G. (2003) *Demographic forecasting*. Manuscript.

Gissler M. (1999) *Routinely collected registers in Finnish health care*. Pp. 241–254 in Alho (1999).

Gnedenko B. (1976) *The theory of probability*. Moscow: Mir Publishers.

Goldstein H. (2003) *Multilevel statistical models, 3rd ed.* London: Arnold.

Goodman L. (1967) On the age-sex composition of the population that would result from given fertility and mortality conditions. *Demography* 14, 423–441.

Goodman L. (1968) Stochastic models for the population growth of the sexes. *Biometrika* 55, 469–487.

Goodman L. (1991) Measures, models, and graphical displays in the analysis of cross-classified data (with discussion). *Journal of the American Statistical Association*, 86, 1085–1138.

Gove W.R. (1973) Sex, marital status, and mortality. *American Journal of Sociology* 79, 45–67.

Gower J.C. and Hand D.J. (1996) *Biplots*. London: Chapman & Hall.

Granger C.W.J. (1969) Investigating causal relations by econometric models and cross-spectral methods. *Econometrica* 37, 424–438.

Granger C.W.J. and Teräsvirta T. (1993) *Modelling nonlinear economic relationships*. Oxford: Oxford University Press.

Green P.J. and Silverman B.W. (1994) *Nonparametric regression and generalized linear models*. London: Chapman & Hall.

Greenacre M. (1984) *Theory and applications of correspondence analysis*. London: Academic Press.

Griffin R. (2002) A.C.E. Revision II analysis of the synthetic assumption. *DSSD A.C.E. Revision II Memorandum Series #PP-49r*. Washington, D.C.: U.S. Bureau of the Census.

Griffith D.A. (1988) *Advanced spatial statistics*. Dordrecht: Kluwer.

Groves R.M., Dillman D.A., Eltinge J.L. and Little R.J.A. (Eds.) (2001) *Survey nonresponse*. New York: Wiley.

Haberman S. (1999) *Actuarial models for disability insurance*. Boca Raton FL: Chapman-Hall.

Haberman S.J. (1978) *Analysis of qualitative data, vol. 1*. New York: Academic Press.

Haberman S.J. (1979) *Analysis of qualitative data, vol. 2*. New York: Academic Press.

Haberman S.J. (1984) Adjustment by minimum discriminant information. *The Annals of Statistics* 12, 971–988.

Haberman S.J., Jiang W. and Spencer B.D. (1998) *Activity 7: develop methodology for evaluating model-based estimates of the population size for states*. Final Report under contract 50-YABC-2-66023 for the Bureau of the Census. Chicago: National Opinion Research Center.

Haberman S.J. and Spencer B.D. (2001) *Estimation of inconsistent poststratication in the 2000 A. C. E.* Report for the Bureau of the Census. July 18, 2001. Abt Associates: Cambridge, MA.

Haberman S.J. and Spencer B.D. (2002) *Estimation of inconsistent poststratification in the 2000 A.C.E.* Report prepared for the Bureau of the Census under Contract No. 46-YABC-7-66020, January 18, 2002. Evanston, IL: Statistics Department, Northwestern University.

Hájek J. (1960) Limiting distributions in simple random sampling from a finite population. *Publications of the Mathematical Institute of the Hungarian Academy of Sciences* 5, 361–374.

Hájek J. (1964) Asymptotic theory of rejective sampling with varying probabilities from a finite population. *The Annals of Mathematical Statistics* 35, 1491–1523.

Hájek J. (1981) *Sampling from a finite population.* New York: Dekker.

Hanika A., Lutz W. and Scherbov S. (1997) Ein probabilistischer Ansatz zur Bevölkerungsvorausscätzung für Österreich. *Statistische Nachrichten*, 984–988. Vienna: Statistics Austria.

Hanski I., Alho, J. and Moilanen, A. (2000) Estimating the parameters of survival and migration of individuals in metapopulations. *Ecology* 81, 239–251.

Harala R. and Tammilehto-Luode M. (1999) *GIS and register-based population census.* Pp. 55–72 in Alho (1999).

Härdle W. (1990) *Applied nonparametric regression.* Cambridge: Cambridge University Press.

Hartley H.O. and Ross A. (1954) Unbiased ratio estimates. *Nature* 174, 270–271.

Harvey A.C. (1989) *Forecasting, structural time series models and the Kalman filter.* Cambridge: Cambridge University Press.

Hausman J.A. (1978) Specification tests in econometrics. *Econometrica*, 46, 1251–1271.

Heligman L. and Pollard J.H. (1980) The age pattern of mortality. *Journal of the Institute of Actuaries* 107, 49–80.

Hendershot G.E. and Placek P.J. (Eds.) (1981) *Predicting fertility.* Lexington: Lexington Books.

Hengartner N. and Speed T.P. (1993) Assessing between-block heterogeneity within the post-strata of the 1990 Post-Enumeration Survey. *Journal of the American Statistical Association* 88, 1119–1125.

Henry L. (1972) *Démographie. Analyse et modèles.* Paris: Larousse.

Hjerppe R. (1989) *The Finnish economy 1860–1985, growth and structural change.* Bank of Finland Publications, Studies on Finland's Economic Growth no. 13. Helsinki: Government Printing Centre.

Hobbs F. (2004) Age and sex composition. Pp.125–173 in Siegel and Swanson (2004).

Hoel D. G. (1985) *The impact of occupational exposure patterns on quantitative risk estimation.* Pp. 105–118 in Hoel D.G., Merrill R.A. and Perera F.P. (1985) *Risk quantitation and regulatory policy.* Banbury Report 19. Cold Spring Harbor NY: Cold Spring Harbor Laboratory.

Hoem J. (1970) *Grunnbegreper i formell befolkningslære.* Oslo: Universitetsforlaget.

Hoem J. (1973) Levels of error in population forecasts. *Artikler 61.* Oslo: Central Bureau of Statistics.

Hoem J. (1987) Statistical analysis of a multiplicative model and its application to the standardization of vital rates. *International Statistical Review*, 55, 119–152.

Hoem J.M. and Funck Jensen U. (1982) Multistate life table methodology: a probabilist critique. Pp. 155–264 in Land K.C. and Rogers A. (1982).

Hogan H. (1983) The forward trace study: its purpose and design. *Proceedings of the American Statistical Association, Survey Research Methods Section*, 168–172.

Hogan H. (1992) The 1990 Post-Enumeration Survey: an overview. *The American Statistician*, 46, 261–269.

Hogan H. (1993) The 1990 Post-Enumeration Survey: operations and results. *Journal of the American Statistical Association* 88, 1047–1060.

Hogan H. (2001) *Accuracy and Coverage Evaluation: data and analysis to inform the ESCAP Report.* DSSD Census 2000 Procedures and Operations Memorandum Series B-1. Washington D.C.: U.S. Census Bureau.

Horowitz J.L. (1994) Bootstrap-based critical values for the information matrix test. *Journal of Econometrics* 61, 395–411.

Hosmer D.W. and Lemeshow S. (2000) *Applied logistic regression, 2nd ed.* New York: Wiley.

Howson C. and Urbach P. (1993) *Scientific reasoning: the Bayesian approach.* 2nd ed. Chicago: Open Court.

Hu Y. and Goldman N. (1990) Mortality differentials by marital status: an international comparison. *Demography* 27, 233–250.

Huggins R.M. (1989) On the statistical analysis of capture experiments. *Biometrika* 76, 133–140.

Hull H.F., Bettinger C.J., Gallaher M.M., Keller M.N., Wilson J., and Mertz G.J. (1988) Comparison of HIV-antibody testing in an STD clinic. *Journal of the American Medical Association* 260, 935–938.

Ilmakunnas P., Laaksonen S., and Maliranta M. (1999) *Enterprise demography and job flows.* Pp. 73–88 in Alho (1999).

I.N.E.D. (1976) *Natalité et politique démograhique*. Paris: Presses Universitaires de France.

Jabine T. and Schwartz R.E. (1974) Use of loss functions to determine sample size in the Social Security Administration. *Proceedings of the Social Statistics Section*, 103–110. Alexandria, VA: American Statistical Association.

Jeffrey R. C. (1983) *The logic of decision, 2nd ed.* Chicago: University of Chicago Press.

Jenkins J. (2004) Apportionment matters: fair representation in the House and Electoral College. Presented at the Institute for Policy Research, May 4, 2004. Department of Political Science, Northwestern University, Evanston, IL, USA.

Jennings D. (1986) Judging inference adequacy in logistic regression. *Journal of the American Statistical Association* 81, 471–476.

Johansen S. (1995) *Likelihood-based inference in cointegrated vector autoregressive models*. Oxford: Oxford University Press.

Johnson N.L. and Kotz S. (1969) *Distributions in statistics, discrete distributions*. New York: Houghton Mifflin.

Kadane J.B. (1992) Healthy scepticism as an expected-utility explanation of the phenomena of Allais and Ellsberg, in *Decision Making under Risk and Uncertainty: New Models and Empirical Findings,* J. Geweke, editor, 11–16 and in Theory and Decision, 32, 57–64.

Kadane J.B. and Wolfson L. (1998) Experiences in elicitation. *The Statistician* 47, 3–19.

Kahn H. (1962) *Thinking about the unthinkable*. New York: Avon Books.

Kahneman D. and Tversky A. (1982) *On the study of statistical intuitions*. Pp. 493–508 in Kahneman et al. (1982).

Kahneman D., Slovic P. and Tversky A. (Eds.) (1982) *Judgment under uncertainty: heuristics and biases*. Cambridge: Cambridge University Press.

Kalton G. and Flores-Cervantes I. (2003) Weighting methods. *Journal of Official Statistics* 19, 81–97.

Kannisto V. (1994) *Development of the oldest-old mortality, 1959–1990.* Odense: Odense University Press.

Kannisto V. (1996) *The advancing frontier of survival*. Odense: Odense University Press.

Kannisto V. and Nieminen M. (1996) *Revised life tables for Finland.* Population 1996:2. Helsinki: Statistics Finland.

Kaplan E.L. and Meier P. (1958) Non-parametric estimation from incomplete observations. *Journal of the American Statistical Association* 53, 457–481, 562–563.

Karlin S. and Lessard S. (1986) *Theoretical studies on sex ratio evolution*. Princeton, N.J.: Princeton University Press.

Karlin S. and Taylor H.M. (1975) *A first course in stochastic processes*, *2nd ed.* New York: Academic Press.

Kass R.E. and Wasserman L. (1996) The selection of prior distributions by formal rules. *Journal of the American Statistical Association* 91, 1343–1370.

Kearney A.T. (2002) *A.C.E. Revision II missing data evaluation.* DSSD A.C.E. Revision II Memorandum Series # PP-48. Washington, D.C.: U.S. Census Bureau.

Keyfitz N. (1979) Information and allocation: two uses of the 1980 census. *The American Statistician* 33, 45–50.

Keiding N. and Hoem J. (1976) Stochastic stable population theory with continuous time I. *Scandinavian Actuarial Journal*, 150–175.

Keilman N. (1990) *Uncertainty in national population forecasting: issues, backgrounds, analyses, recommendations*. Amsterdam: Swets and Zeitlinger.

Keilman N. (1997) Ex-post errors in official population forecasts in industrialized countries. *Journal of Official Statistics* 13, 245–277.

Keilman N. (1998) How accurate are the United Nations world population projections? *Population and Development Review* 24, Supplement: Frontiers of Population Forecasting, 15–41.

Keilman N. (2002) TFR predictions based on Brownian motion theory. *Yearbook of Population Research in Finland* XXXVIII, 207–219.

Keilman N. and Kučera T (1991) The impact of forecasting methodology on the accuracy of national population forecasts: evidence from the Netherlands and Czechoslovakia. *Journal of Forecasting* 10, 371–398.

Keilman N., Pham D.Q. and Hetland A. (2002) *Norway's uncertain demographic future*. Social and Economic Studies 105. Oslo: Statistics Norway.

Keinänen A. (2002) *Informaatiosektorille työllistymiseen vaikuttavat tekijät*. Master's Thesis, University of Joensuu.

Keyfitz N. (1977) *Introduction to the mathematics of population, with revisions*. Reading: Addison-Wesley.

Keyfitz N. (1981) The limits of population forecasting. *Population and Development Review* 7, 579–593.

Keyfitz N. (1982) Can knowledge improve forecasts? *Population and Development Review* 8, 719–751.

Keyfitz N. (1985) *Applied mathematical demography, 2nd ed.* New York: Springer.

Keyfitz N. and Beekman J.A. (1984) *Demography through problems*. New York: Springer.

Kim J. (1991, July 11) *1990 PES evaluation project P12: evaluation of synthetic assumption.* 1990 Coverage Studies and Evaluation Memorandum Series #N-4. Washington, D.C.: Census Bureau.

Kim J., Blodgett R. and Zaslavsky A. (1991) Evaluation of the synthetic assumption – 1990 Post-Enumeration Survey. *Proceedings of the American Statistical Association, Survey Research Section* 254–259.

King G. and Zeng L. (2002) Estimating risk and rate levels, ratios, and differences in case-control studies. *Statistics in Medicine* 21, 1409–1427.

Kish L. (1965) *Survey sampling*. New York: Wiley.

Kish L. (1987) *Statistical Design for Research*. Wiley: New York.

Kish L. (1992) Weighting for unequal P_i. *Journal of Official Statistics* 8, 183–200.

Kish L. (1995) Methods for design effects. *Journal of Official Statistics* 11, 55–78.

Klein J.P. and Moeschberger M.L. (1997) *Survival analysis. Techniques for censored and truncated data*. New York: Springer.

Kleinbaum D.G., Kupper L.L. and Morgenstern H. (1982) *Epidemiologic research*. New York: Van Nostrand.

Kohler H.P. and Philipov D. (2001) Variance effects in the Bongaarts-Feeney formula. *Demography* 38, 1–16.

Korn E.L. and Graubard B.I. (1995) Analysis of health surveys: accounting for the sampling design. *Journal of the Royal Statistical Society. Series A* 158, 263–295.

Korn E.L. and Graubard B.I. (1999) *Analysis of health surveys*. New York: Wiley.

Kostanich D. (2003a) A.C.E. Revision II: design and methodology. *DSSD A.C.E. Revision II Memorandum Series #PP-30*. Washington, D.C.: U.S. Census Bureau.

Kostanich D. (2003b) A.C.E. Revision II: summary of methodology. *DSSD A.C.E. Revision II Memorandum Series #PP-35*. Washington, D.C.: U.S. Census Bureau.

Kotz S., Johnson N. K., and Read C. B. (Eds.) (1982) *Encyclopedia of statistical sciences*. New York: Wiley.

Krewski D. and Rao J.N.K. (1981) Inference from stratified samples: properties of the linearization, jackknife and balanced repeated replication methods. *Annals of Statistics* 9, 1010–1019.

Kruskal W.H. and Mosteller F. (1979a). Representative sampling. I. Scientific literature. *International Statistical Review* 47, 13–24.

Kruskal W.H. and Mosteller F. (1979b). Representative sampling. II. Scientific literature, excluding statistics. *International Statistical Review* 47, 111–128.

Kruskal W.H. and Mosteller F. (1979c). Representative sampling. III. Scientific literature, current statistical literature. *International Statistical Review* 47, 245–265.

Kruskal W.H. and Mosteller F. (1980). Representative sampling. IV. The history of the concept in statistics, 1895–1939. *International Statistical Review* 48, 169–195.

Kyburg H.E. (1970) *Probability and inductive logic*. London: Macmillan.

Land K.C. (1986) Methods for national population forecasts: a review. *Journal of the American Statistical Association* 81, 888–901.

Land K.C. and Rogers, A. (Eds.) (1982) *Multiregional mathematical demography*. New York: Academic Press.

Landwehr J.M., Pregibon D. and Shoemaker A.C. (1984) Graphical methods for assessing logistic regression models. *Journal of the American Statistical Association* 79, 61–71.

Lassila J. and Valkonen T. (1999). *Eläkerahastot ja väestön ikääntyminen*. ETLA Tutkimuksia B 158. Helsinki: ETLA.

LeBras H. (1977) Une formulation générale de la dynamique des populations. *Population* 32, 261–293.

Ledent J. and Rogers A. (1988) Stable growth in native-dependent multistate population dynamics. *Mathematical Population Studies* 1, 157–171.

Ledermann S. and Breas J. (1959) Les dimensions de la mortalité. *Population* 14, 637–682.

Lee R.D. (1974) Forecasting births in post-transition populations: stochastic renewal with serially correlated fertility. *Journal of the American Statistical Association* 69, 607–617.

Lee R.D. (1980) Aiming at a moving target: period fertility and changing reproductive goals. *Population Studies* 34, 205–226.

Lee R.D. (2000). The Lee-Carter method for forecasting mortality, with various extensions and applications. *North American Actuarial Journal,* 4, 80–93.

Lee R.D. and Carter L.R. (1992) Modeling and forecasting the time series of U.S. mortality. *Journal of the American Statistical Association* 87, 659–671.

Lee R.D. and Miller T. (2001) Estimating the performance of the Lee-Carter method for forecasting mortality. *Demography* 38, 537–549.

Lee R.D. and Tuljapurkar S. (1994) Stochastic population forecasts for the United States: beyond high, medium, and low. *Journal of the American Statistical Association* 89, 1175–1189.

Lee R.D. and Tuljapurkar, S. (1998). Uncertain economic futures and social security finances. *American Economic Review,* May, 237–241.

Lee Y. and Nelder J.A. (1996) Hierarchical generalized linear models. *Journal of the Royal Statistical Society. Series B* 58, 619–678.

Lee Y. and Nelder J.A. (2001) Hierarchical generalised linear models: a synthesis of genralised linear models, random effects models and structured dispersions. *Biometrika* 88, 987–1006.

Lehmann E.L. (1983) *Theory of point estimation.* New York: Wiley.

Lehmann E.L. (1986) *Testing statistical hypotheses, 2nd ed.* New York: Wiley.

Lehtonen R. and Pahkinen E. (2004) *Practical methods for design and analysis of complex surveys, 2nd ed.* Chichester: Wiley.

Leslie P.H. (1945) On the use of matrices in certain population mathematics. *Biometrika* 33, 183–212.

Levy P.S. and Lemeshow S. (1999) *Sampling of populations, 3rd ed.* New York: Wiley.

Lewis E.G. (1942) On the generation and growth of a population. *Sankhya* 6, 93–96.

Lexis W. (1875) *Einleitung in die Theorie der Bevölkerungs-Statistik.* Strasbourg: Trubner.

Lillard L.A. and Panis C.W.A. (1996) Marital status and mortality: the role of health. *Demography* 33, 313–327.

Lin G. (1999) Assessing structural change in the U.S. migration patterns: A log-rate modeling approach. *Mathematical Population Studies* 7, 217–238.

Lindsey J.K. (1996) *Parametric statistical inference.* Oxford: Clarendon Press.

Liu J.S. (2001) *Monte Carlo strategies in scientific computing.* New York: Springer.

Lohr S.L. (1999) *Sampling: design and analysis.* Pacific Grove, CA: Duxbury.

Lohr S. and Rao J.N.K. (1997) Jackknife variance estimation in multiple frame surveys. *Proceedings of the American Statistical Association, Survey Research Section,* 552–557.

Louis T.A., Jabine T.B. and Gerstein M.A. (Eds.) (2003) *Statistical issues in allocating funds by formula.* Panel on Formula Allocations, National Research Council. Washington, D.C.: The National Academy Press.

Lutz W. (Ed.) (1994) *The future population of the world.* London: Earthscan.

Lutz W. (Ed.) (1996) *The future population of the world: what can we assume today? Revised edition.* London: Earthscan.

Makridakis S., Andersen A., Carbone R., Fildes R., Hibon M., Lewandowski R., Newton J., Parzen E. and Winkler R. (1984) *The forecasting accuracy of major time series methods.* New York: Wiley.

Malec D.J. and Griffin R.A. (2001, February 28) *Accuracy and Coverage Evaluation: assessment of synthetic assumptions.* DSSD Census 2000 Procedures and Operations Memorandum Series B-2. Washington, D.C.: U.S. Census Bureau.

Mantel N. and Haenszel W. (1959) Statistical aspects of the analysis of data from retrospective studies of disease. *Journal of the National Cancer Institute,* 22(4), 719–748.

March J.G. (1978) Bounded rationality, ambiguity, and the engineering of choice. *Bell Journal of Economics* 9, 587–608.

March J.G. (1994) *A primer on decision making; how decisions happen.* New York: The Free Press.

Marks E.S., Seltzer W. and Krotki K.J. (1974) *Population growth estimation: a handbook of vital statistics measurement.* New York: The Population Council.

Marks E.S. (1979) The role of dual system estimation in census evaluation. P. 56–188 in Karol Krotki (Ed.) (1979) *Recent developments in DSE/PGE.* Alberta: University of Alberta Press.

McCullagh P. and Nelder J.A. (1989) *Generalized linear models, 2nd ed.* London: Chapman & Hall.

McDonald J. (1979) A time series approach to forecasting Australian total live-births. *Demography* 16, 575–601.

McDonald J. (1980) Birth time series, models and structural relationships. *Journal of the American Statistical Association* 75, 39–41.

McDonald J. (1981) Modeling demographic relationships: an analysis of forecast functions for Australian births. *Journal of the American Statistical Association* 76, 782–792.

McFarland D.D. (1972) Comparison of alternative marriage models. Pp. 89–106 in Greville T.N.E. (Ed.) (1972) *Population dynamics.* New York: Academic Press.

McKeown T. (1976) *The modern rise of population.* New York: Academic Press.

Melnick D. (2002) The legislative process and the use of indicators in formula allocations. *Journal of Official Statistics,* 18, 353–369.

Metropolis N., Rosenbluth A.W., Teller A.H. (1953) Equation of state calculations by fast computing machines. *Journal of Chemical Physics* 21, 1087–1092.

Mode C.J. (1985) *Stochastic processes in demography and their computer implementation.* Berlin: Springer.

Modeen G. (1934) Suomen väkiluvun tuleva kehitys ja sen taloudelliset seuraukset. *Kansantaloudellinen aikakauskirja* VI, 351–378.

Moltchanov V., Kuulasmaa K. and Torppa J. (1999) *Quality Assessment of Demographic Data in the WHO MONICA Project.* World Health Organization. http://www.ktl.fi/publications/monica/demoqa/demoqa.htm. URN:NBN:fi-fe19991073.

Moolgavkar S.H. and Prentice R.L. (1986) *Modern statistical methods in chronic disease epidemiology.* New York: Wiley-Interscience.

Moses L.E. (1986) Statistical concepts fundamental to investigations. Pp. 3–26 in Bailar J.C. and Mosteller F. (Eds.) (1986) *Medical uses of statistics.* New England Journal of Medicine Books: Waltham, MA.

Muhsam H.V. (1956) The utilization of alternative population forecasts in planning. *Bulletin of the Research Council of Israel* 5C, 133–146.

Muller T. and Espenshade T.J. (1985) *The fourth wave.* Washington, D.C.: The Urban Institute Press.

Mulry M.H. and Dajani A. (1989) The forward trace study. *Proceedings of the American Statistical Association, Survey Research Methods,* 675–680.

Mulry M.H. and Spencer B.D. (1991) Total error in PES estimates of population: the dress rehearsal census of 1988. *Journal of the American Statistical Association* 86, 839–854 with discussion 855–863.

Mulry M.H. and Spencer B.D. (1993) Accuracy of the 1990 census and undercount adjustments. *Journal of the American Statistical Association* 88, 1080–1091.

Mulry M.H. and Spencer B.D. (2001, February 28) *Accuracy and Coverage Evaluation: overview of total error modeling and loss function analysis.* DSSD Census 2000 Procedures and Operations Memorandum Series B-19*. Washington, D.C.: U.S. Census Bureau. http://www.census.gov/ dmd/www/pdf/Fr19.pdf

Mulry M.H. and ZuWallack R. (2003) Confidence intervals and loss function analysis. *DSSD A.C.E. Revision II Memorandum Series #PP-42.* Washington, D.C.: U.S. Census Bureau.

Muth J.F. (1960) Optimal properties of exponentially weighted forecasts of time series with permanent and transitory components. *Journal of the American Statistical Association* 55, 299.

Myrdal A. and Myrdal G. (1934) *Kris i befolkningsfrågan.* Bonniers, Stockholm.

National Research Council (2001) *Proceedings, first workshop of Panel to Review the 2000 Census (October 6, 1999).* Committee on National Statistics. Washington D.C.: National Academy Press.

Navarro A. and Asiala M. (2001) Accuracy and Coverage Evaluation: comparing accuracy. *DSSD Census 2000 Procedures and Operations Memorandum Series* B-13. February 28, 2001. Washington, D.C.: U.S. Bureau of the Census.

Nelder J.A. and Wedderburn R.W.M. (1972) Generalised linear models. *Journal of the Royal Statistical Society. Series A* 135, 370–384.

Nelson W. (1969) Hazard plotting for incomplete failure data. *Journal of Quality Technology,* 1, 27–52.

Neutra R.R. and Drolette M.E. (1978) Estimating exposure-specific disease rates from case-control studies using Bayes theorem. *American Journal of Epidemiology* 108, 214–222.

Nieminen, K. (1996) Liikenneonnettomuudet Helsingissä vuonna 1985. *Helsingin kaupunkisuunnitteluviraston julkaisuja 1996:16.* Helsinki.

Nieminen M. and Markelin P. (1974) Suomen väestökirjanpito ja väkiluvun laskeminen. *Muistio no. 27.* Helsinki: Tilastokeskus.

Nordberg L. (1989) Generalized linear modeling of sample survey data. *Journal of Official Statistics* 5, 223–239.

Nordberg L. (1999) *Leo Törnqvist – the "grandfather" of Finnish statistics.* Pp. 163–176 in Alho J. (1999).

Ogata Y., Katsura K., Keiding N., Holst C. and Green A. (2000) Empirical Bayes age-period-cohort analysis of retrospective incidence data. *Scandinavian Journal of Statistics* 27, 415–432.

Öller L.E. and Barot B. (2000) The accuracy of European growth and inflation forecasts. *International Journal of Forecasting* 16, 293–315.

Panel on Methodology for Statistical Priorities (1976) *Setting statistical priorities.* National Research Council. Washington, D.C.: The National Academy Press.

Panel on Small-Area Estimates of Population and Income (1980) *Estimating population and income of small areas.* National Research Council. Washington, D.C.: National Academy Press.

Passel J.S. (1993) Comment. *Journal of the American Statistical Association* 88, 1074–1077.

Passel J.S., Siegel J.S. and Robinson J.G. (1982) *Coverage of the national population in the 1980 census, by age, sex, and race: preliminary estimates by demographic analysis.* Current Population Reports, Special Studies, P-23, No. 115. Washington, D.C.: U.S. Census Bureau.

Pearl R. and Reed L.J. (1920) On the rate of growth of the United States since 1790 and its mathematical representation. *Proceedings of the National Academy of Sciences* 6, 275–288.

Pfeffermann D. (1993) The role of sampling weights when modeling survey data. *International Statistical Review* 61, 317–337.

Pflaumer P. (1992) Forecasting U.S. population totals with the Box-Jenkins approach. *International Journal of Forecasting* 8, 329–338.

Pitkänen K. (1977) The reliability of the registration of births and deaths in Finland in the eighteenth and nineteenth centuries: some examples. *The Scandinavian Economic History Review*, 25, 138–159.

Pitkänen K. (1986) Viime vuosisadan vaihteen väestötilasto ja suomalaisten miesten kuolleisuus. *Sosiaalilääketieteellinen Aikakauslehti* 23, 375–382.

Pitkänen K.J. and Laakso M. (1999) The reliability of the Finnish mortality statistics: a historical review. Pp. 15–37 in Alho (1999).

Pollard J.H. (1968) A note on multi type Galton-Watson processes with random branching probabilities. *Biometrika* 55, 589–590.

Pollard J.H. (1973) *Mathematical models for the growth of human populations*. London: Cambridge University Press.

Pollard J.H. (1975) Modelling human populations for projection purposes – some of the problems and challenges. *Australian Journal of Statistics* 17, 63–76.

Portnoy S. (1988) Asymptotic behavior of likelihood methods for exponential families when the number of parameters tends to infinity. *Annals of Statistics* 16, 356–366.

Potter F.J. (1990) A study of procedures to identify and trim extreme sampling weights. *Proceedings of the American Statistical Association, Survey Research Section*, 225–230.

Poulain M. (1993) Confrontation des statistiques de migrations intra-Européennes: vers plus d'harmonisation? *European Journal of Population* 9, 353–381.

Pregibon D. (1981) Logistic regression diagnostics. *Annals of Statistics* 9, 705–724.

Prentice R.L., Self S.G. and Mason M.W. (1986) Design options for sampling within a cohort. Pp. 50–62 in Moolgavkar and Prentice (1986).

Press W.H., Flannery B.P., Teukolsky S.A. and Vetterling W.T. (1992) *Numerical recipes in C, 2nd ed.*. Cambridge: Cambridge University Press.

Pressat R. (1972) *Demographic analysis: methods, results, applications.* Chicago: Aldine-Atherton.

Price (1947) A check on underenumeration in the 1940 census. *American Sociological Review* XII, 44–49.

Qian J. and Spencer B.D. (1994) Optimally weighted means in stratified sampling. *Proceedings of the American Statistical Association, Survey Research Section*, 863–866.

Raffelhüschen B. (1999) Generational accounting: method, data and limitations, in EU generational accounting in Europe. *European Economy, Reports and Studies No 1999:6.* Office for the Official Publications of the EC, Luxembourg.

Raiffa H. and Schlaifer R. (1972) *Applied statistical decision theory*. Cambridge MA.: MIT Press.

Ramsey F.P. (1926) *Truth and probability*. Pp. 61–93 in Kyburg H.E. and Smokler H.K. (Eds.) (1964) *Studies in subjective probability*. New York: Wiley.

Rao C.R. (1973) *Linear statistical inference and its applications, 2nd ed.* New York: Wiley.

Rao J.N.K.(1996) On variance estimation with imputed survey data. *Journal of the American Statistical Association* 91, 499–506, with discussion 507–520.

Rao J.N.K. (2003) *Small area estimation.* New York: Wiley.

Rao J.N.K. and Shao J. (1992) Jackknife variance estimation with survey data under hot deck imputation. *Biometrika* 79, 811–822.

Rao J.N.K. and Wu C.F.J. (1985) Inference from stratified samples: second-order analysis of three methods for nonlinear statistics. *Journal of the American Statistical Association* 80, 620–630.

Rao J.N.K. and Wu C.F.J. (1988) Resampling inference with complex survey data. *Journal of the American Statistical Association* 83, 231–241.

Rao P.S.R.S. and Rao J.N.K. (1971) Small sample results for ratio estimators. *Biometrika* 58, 625–630.

Rawls J. (1971) *A theory of justice*. Cambridge: Harvard University Press.

Redfern P. (1974) The different roles of population censuses and interview surveys, particularly in the U.K. context. *International Statistics Review* 42, 131–146.

Redfern P. (2001) A Bayesian model for estimating census undercount, taking emigration data from foreign censuses. *International Statistical Review* 69, 277–301, with addendum 30 July 2001.

Redfern P. (2004) An alternative view of the 2001 census and future census taking. *Journal of the Royal Statistical Society.* Series A 209–248, with discussion 249–274.

Rees P. and Wilson A.G. (1977) *Spatial population analysis*. London: Arnold.

Rees P., Norma P. and Brown D. (2004) A framework for progressively improving small area population estimates. *Journal of the Royal Statistical Society. Series A* 167, 5–36.

Rice J.A. (1995) *Mathematical statistics and data analysis, 2nd ed.* Belmont, California: Duxbury.

Ripley B.D. (1981) *Spatial statistics*. New York: Wiley.

Ripley B.D. (1987) *Stochastic simulation*. New York: Wiley.

Robert C.P. and Casella G. (1999) *Monte Carlo statistical methods*. New York: Springer.

Roberts L., Lafta R., Garfield R., Khudhairi J. and Burnham G. (2004) Mortality before and after the 2003 invasion of Iraq: cluster sample survey. *Lancet* 364, 1857–64.

Robinson J.G. (2001) *ESCAP II: Demographic analysis results.* Executive Steering Committee for A. C. E. Policy II, Report No. 1, October 13, 2001. Washington, D.C.: U.S. Census Bureau.

Robinson J.G., Adlakha A. and West, K.K. (2002) *Coverage of population in Census 2000. Results from demographic analysis*. Paper presented at the Annual Meeting of the Population Association of America, Atlanta, Georgia, May 2002.

Robinson J.G., Ahmed B., Das Gupta P. and Woodrow K. (1993) Estimation of population coverage in the 1990 United States census based on demographic analysis. *Journal of the American Statistical Association*, 88, 1047–1060.

Rogers A. (1975) *Introduction to multiregional mathematical demography*. New York: Wiley.

Rogers A. (1986) Parametrized multistate population dynamics and projections. *Journal of the American Statistical Association* 81, 48–61.

Rogers A. (1995) *Multiregional demography. Principles, methods and extensions.* New York: Wiley.

Rogers A. and Ledent J. (1976) Increment-decrement life tables a: a comment. *Demography* 13, 287–290.

Rosén, B. (1972) Asymptotic theory for successive sampling with varying probabilities without replacement, I and II. *The Annals of Mathematical Statistics* 43, 373–397, 748–776.

Rosenblatt J.R. and Filliben J.J. (1971) Randomization and the draft lottery. *Science* 171, 306–308.

Rossi P., Fisher G.A. and Willis G. (1986) *The condition of the homeless of Chicago.* Chicago: NORC.

Rossi P.H., Wright J.D. and Anderson A.B. (1983) *Handbook of survey research.* New York: Academic Press.

Rothman K.J. (1986) *Modern epidemiology.* Boston: Little, Brown and Company.

Rubin D.B. (1987) *Multiple imputation for nonresponse in surveys.* New York: Wiley.

Rubin D.B. (1996) Multiple imputation after 18+ years. *Journal of the American Statistical Association* 91, 434–489, with discussion 507–520.

Rust K.F. and Rao J.N.K. (1996) Variance estimation for complex surveys using replication techniques. *Statistical Methods in Medical Research* 5, 283–310.

Ryder N. (1956) Problems of trend determination during transition in fertility. *Milbank Memorial Fund Quarterly* 34, 5–21.

Saari D.G. (1995) *Basic geometry of voting.* Berlin: Springer.

Saboia J.L.M. (1974) Modeling and forecasting population time series. *Demography* 11, 483–492.

Saboia J.L.M. (1977) Autoregressive integrated moving average (ARIMA) models for birth forecasting. *Journal of the American Statistical Association* 61, 706–719.

Sacks J. and Ylvisaker D. (1978) Linear estimation for approximately linear models. *Ann. Statist.* 6, 1122–1137.

Sands R.D. and Navarro A. (2001) 2000 census Accuracy and Coverage Evaluation survey variance estimates. *Proceedings of the American Statistical Association.* Joint Statistical Meetings 2001.

Särndal C.-E., Swensson B. and Wretman J. (1992) *Model assisted survey sampling.* New York: Springer.

Sauvy A. (1932) Calculs démographiques sur la population française jusqu'en 1980. *Journal de la société de statistique de Paris* 1932, 338–347.

Savage I.R. (1975) Cost benefit analysis of demographic data. *Advances in Applied Probability – Supplement* 7, 62–71.

Savage I.R. (1982) Who counts? *American Statistician* 36, 195–200, with discussion 200–207.

Savage L.J. (1954) *The foundations of statistics.* New York: Wiley.

Schindler E. (1998) Allocation of the ICM sample to the states for Census 2000. *Proceedings of the Survey Research Methods Section*, pp. 491–496. Alexandria, VA: American Statistical Association.

Schoen R. (1988) *Modeling multigroup populations.* New York: Plenum Press.

Schultz T.P. (1981) *Economics of population.* Reading: Addison-Wesley.

Searle S.R. (1971) *Linear models.* New York: Wiley.

Sen P.K. (1988) Asymptotics in finite population sampling. Pp. 291–331 in P.R. Krishnaiah and C. R. Rao (eds) *Sampling.* Handbook of Statistics 6. Amsterdam: North-Holland.

Severini T. (2000) *Likelihood methods in statistics.* Oxford: Oxford University Press.

Shao J. and Tu D. (1995) *The jackknife and bootstrap.* New York: Springer.

Sheps M.C. and Menken J.A. (1973) *Mathematical models of conception and birth.* Chicago: University of Chicago Press.

Shryock H.S., Siegel J.S. and Associates (1976) *Methods and materials of demography*, Condensed Edition by Stockwell E.G. New York: Academic Press.

Siegel J.S., Passel R.N.W. and Robinson J. G. (1977) Developmental estimates of the coverage of the population of states in the 1970 census: demographic analysis. *Current Population Reports Series P-23*, No. 65. Washington, D.C.: U.S. Census Bureau.

Siegel J.S. and Swanson D.A. (Eds.) (2004) *The methods and materials of demography, 2nd ed.* New York: Elsevier Academic Press.

Simon J. (1977) *The economics of population growth*. Princeton: Princeton University Press.

Sitter R.R. (1992) A resampling procedure for complex survey data. *Journal of the American Statistical Association* 87, 755–765.

Skinner C.J., Holt D. and Smith T.M.F. (Eds.) (1989) *Analysis of complex surveys*. Chichester: Wiley.

Slutsky E. (1927) The summation of random causes as the source of cyclic processes, translated in *Econometrica* 5, 105–146 (1937).

Smith D.P. (1992) *Formal demography*. New York: Plenum Press.

Smith P.C., Rice N. and Carr-Hill R. (2001) Capitation funding in the public sector. *Journal of the Royal Statistical Society. Series A* 164, 217–257.

Smith S.K. and Mandell M. (1984) A comparison of population estimation methods: housing unit versus component II, ratio correlation, and administrative records. *Journal of the American Statistical Association* 79, 282–289.

Smith S.K. and Tayman J. (2003) An evaluation of population projections by age. *Demography* 40, 741–757.

Smith T.M.F. (1983) On the validity of inferences from non-random samples. *Journal of the Royal Statistical Society. Series A* 146, 394–403.

Smith S.K., Tayman J. and Swanson D.A. (2001) *State and local population projections: methodology and analysis*. New York: Kluwer Academic/Plenum Publishers.

Snedecor G.W. and Cochran W.G. (1967) *Statistical methods, 6th ed.* Ames: Iowa State University Press.

Sommers P. (2003) The writing on the wall. *Chance* 16, 35–38.

Spencer B.D. (1980a) *Benefit-cost analysis of data used to allocate funds*. New York: Springer.

Spencer B.D. (1980b) Effects of biases in census estimates on evaluation of postcensal estimates. Pp. 232–236 in Panel on Small-Area Estimates of Population and Income (1980).

Spencer B.D. (1980c) Implications of equity and accuracy for undercount adjustment: a decision-theoretic approach. *Proceedings of the 1980 conference on census undercount*, 204–216. Washington, D.C.: U.S. Census Bureau.

Spencer B.D. (1982a) Technical issues in allocation formula design. *Public Administration Review* 42, 524–529.

Spencer B.D. (1982b) Feasibility of benefit-cost analysis of data programs. *Evaluation Review* 6, 649–672.

Spencer B.D. (1982c) Technical issues in allocation formula design. *Public Administration Review* 42, 524–529.

Spencer B.D. (1985a) Optimal data quality. *Journal of the American Statistical Assocation*, 80, 564–573.

Spencer B.D. (1985b) Statistical aspects of equitable apportionment. *Journal of the American Statistical Association* 80, 815–822.

Spencer B.D. (1997a) A note on random walks and leadership duration. Unpublished manuscript.

Spencer B.D. (1997b) *Statistics and public policy.* Oxford: Clarendon Press.

Spencer, B. D. (1994) Sensitivity of benefit-cost analysis of data programs to monotone misspecification. *Journal of Statistical Planning and Inference* 39, 19–31.

Spencer B.D. (2000a) An approximate design effect for unequal weighting when measurements may correlate with selection probabilities. *Survey Methodology* 26, 137–138.

Spencer B.D. (2000b) *Total error model for Census 2000: how components of error can be estimated from the Bureau's planned evaluation studies.* Final Report, May 11, 2000. Activity 12 under contract 50-YABC-7-66020 for the Bureau of the Census. Cambridge MA: Abt Associates.

Spencer B.D. and Foran W. (1991) Sampling probabilities for aggregations, with applications to NELS:88 and other educational longitudinal surveys. *Journal of Educational Statistics* 16, 21–34.

Spencer B.D., Frankel M.R., Ingels S.J., Rasinski K.A. and Tourangeau R. (1990) *National Education Longitudinal Study of 1988 base year sample design report.* NCES Technical Report 90463. Washington, D.C.: National Center for Education Statistics.

Spencer B.D. and Moses L.E. (1990) Needed data expenditure for an ambiguous decision problem. *Journal of the American Statistical Association* 85, 1099–1104.

Starmer C. (2000) Developments in non-expected utility theory: the hunt for a descriptive theory of choice under risk. *Journal of Economic Literature* 38, 332–382.

Statistics Canada (1999) Coverage. *1996 Technical Reports*, cat. no. 92-370-XIE, available at http://www.statcan.ca/english/freepub/92-370-XIE/free.htm.

Statistics Finland (2001) Database.

Statistics Finland (2002) Employment Statistics 1999–2000. *Population 2002:3.* Helsinki 2002.

Statistisches Reichsamt (1930) Volkszählung. *Statistik des Deutschen Reichs, Band 401,II.* Berlin.

Stigler S.M. (1986) *The history of statistics.* Cambridge MA: Harvard University Press.

Stoto M. (1983) Accuracy of population projections. *Journal of the American Statistical Association*, 78, 13–20.

Stukel D.M., Hidiroglou M.A. and Särndal C.-E. (1996) Variance estimation for calibration estimators: a comparison of jackknifing versus Taylor linearization. *Survey Methodology* 22, 117–125.

Suzara F.B. (2002) A study on the formulation of an assessment scale methodology: the United Nations experience in allocating budget expenditures among member states. *Journal of Official Statistics* 18, 481–510.

Sykes Z. (1969) Some stochastic versions of the matrix model for population dynamics. *Journal of the American Statistical Association* 64, 111–130.

Tabeau E., van den Berg Jeths A. and Heathcote C. (2001) *Forecasting mortality in developed countries.* Dordrecht: Kluwer Academic.

Talousneuvoston aluejaosto (1972) *Kokonaistaloudellinen alueellinen tarkastelukehikko, liiteosa 1-2.* Helsinki: Valtion painatuskeskus.

Taqqu M. (2001) Bachelier and his times: a conversation with Bernard Bru. *Finance and Stochastics* 5, 3–32.

Teppo L. and Hakulinen T. (1999) *Finnish cancer registry – producing statistics and doing research.* Pp. 225–240 in Alho (1999).

Ter Heide H. and Willekens F.J. (Eds.) (1984) *Demographic research and spatial policy.* London: Academic Press.

Teräsvirta T. (1987) How we got the data. Pp. 1–7 in Pukkila T. and Puntanen S. (1987) *The Second International Tampere Conference in Statistics*. Tampere: University of Tampere.

Theil H. (1966) *Applied econometric forecasting*. Chicago: Rand-McNally.

Theil H. and Goldberger A.S. (1961) On pure and mixed statistical estimation in economics. *International Economic Review* 2, 65–78.

Thisted R.A. (1988) *Elements of statistical computing*. New York: Chapman and Hall.

Thomas A., Speigelhalter D.J. and Gilks W.R. (1992) *BUGS: a program to perform Bayesian inference using Gibbs sampling*. Pp. 837–842 in Bernardo J.M., Berger J.O., Dawid A.P., Smith (Eds.) (1992) *Bayesian statistics 4*. Oxford: Clarendon Press.

Thompson M.E. (1997) *Theory of sample surveys*. London: Chapman & Hall.

Thompson W.S. and Whelpton P.K. (1933) *Population trends in the United States*. New York: McGraw-Hill.

Tukey J.W. (1995) Discussion of the paper by Draper. P. 78 in Draper (1995).

Turner C.F., Lessler J.T. and Gfroerer J.C. (Eds.) (1992) *Survey measurement of drug use. Methodological studies*. DHHS Publication No. (ADM) 92-1929. Washington D.C.: National Institute on Drug Abuse.

Turpeinen O. (1978) Fertility and mortality in Finland since 1750. *Population Studies* 33, 101–114.

Törnqvist L. (1948) An attempt to analyze the problem of an economical production of statistical data. *Nordisk tidskrift for teknisk økonomi*, 265–274. Reprinted as pp. 385–394 in Törnqvist L. (1981) *Collected Scientific Papers of Leo Törnqvist*. Series A7. Helsinki: The Research Institute of the Finnish Economy.

Törnqvist L. (1949) Näkökohdat, jotka ovat määränneet primääristen prognoosioletta-musten valinnan. Pp. 68–74 in Hyppölä J., Tunkelo A. and Törnqvist L. (1949) *Suomen väestöä, sen uusiutumista ja tulevaa kehitystä koskevia laskelmia*. Tilastollisia tiedontantoja 38. Helsinki: Tilastokeskus.

United Nations (1983) *Indirect techniques for demographic estimation. Manual X*. New York: United Nations.

United Nations (1987) *Recommendations for the 1990 censuses of population and housing in the ECE region*. Statistical Standards and Studies No. 40. New York: United Nations.

United Nations (1993) *World population prospects: the 1992 revision*. New York: United Nations.

United Nations (2001) *World population prospects: the 2000 revision. Volume I: comprehensive tables*. New York: United Nations.

U.S. Census Bureau (1949) *Forecasts of population and school enrollment in the United States: 1948 to 1960*. Current Population Reports, Series P-25, No. 18. Washington, D.C.: U.S. Census Bureau.

U.S. Census Bureau (1958) *Illustrative projections of the population of the United States, by age and sex, 1960 to 1980*. Current Population Reports, Series P-25, No. 187. Washington, D.C.: U.S. Census Bureau.

U.S. Census Bureau (1964) *Projections of the population of the United States, by age and sex, 1964 to 1985*. Current Population Reports, Series P-25, No. 286. Washington, D.C.: U.S. Census Bureau.

U.S. Census Bureau (1984) *Projections of the population of the United States, by age, sex, and race: 1983 to 2080*. Current Population Reports. Series P-25, No. 952. Washington, D.C.: U.S. Census Bureau.

U.S. Census Bureau (1992) *Projections of the population of the United States, by age, sex, race and Hispanic origin: 1992 to 2050*. Current Population Reports, Series P-25, No. 1092. Washington, D.C.: U.S. Census Bureau.

U.S. Census Bureau (2000) Racial and ethnic classifications used in census 2000 and beyond. Washington, D.C. Retrieved February 7, 2002 from the World Wide Web: http://www.census.gov/population/www/socdemo/race/racefactcb.html

U.S. Census Bureau (2001) *Census 2000 A.C.E. methodology, vols. 1–4*. Washington, D.C.: author. available at this time at http://www.census.gov/dmd/www/pdf/Vol-1.pdf

U.S. Census Bureau (2002) *Current Population Survey design and methodology.* Technical Paper 63RV. March 2002 revision. Washington, D.C.: U.S. Census Bureau.

U.S. Census Bureau (2003a) *Decision on intercensal population estimates*. March 12, 2003. Washington, D.C.: U.S. Census Bureau.

U.S. Census Bureau (2003b) *Technical assessment of the A.C.E. Revision II estimates.* Washington D.C.: U.S. Census Bureau.

U.S. Census Bureau (2004) *Accuracy and Coverage Evaluation of Census 2000: design and methodology*. Washington, D.C.: U.S. Census Bureau.

U.S. Federal Committee on Statistical Methodology (1978) *Report on statistics for allocation of funds.* Statistical Policy Working Paper 1. Washington D.C.: U.S. Department of Commerce.

U.S. Federal Committee on Statistical Methodology (1982) *An interagency review of time-series revision policies.* Statistical Policy Working Paper 7. Washington D.C.: U.S. Office of Management and Budget.

U.S. Federal Committee on Statistical Methodology (2001) *Measuring and reporting sources of error in surveys.* Statistical Policy Working Paper 31–July 2001. Washington D.C.

Valdés-Prieto S. (2000): The financial stability of notional account pensions. *Scandinavian Journal of Economics* 102, 395–417.

Valkonen T. and Martelin T. (1999) *Social inequality in the face of death – linked registers in mortality research*. Pp. 211–224 in Alho (1999).

Valliant R., Dorfman A.H. and Royall R.M. (2000) *Finite population sampling and inference: a prediction approach*. New York: Wiley.

Van den Berg Jeths A., Hoogenveen R., de Hollander G. and Tabeau E. (2001) *A review of epidemiological approaches to forecasting mortality and morbidity*. Pp. 33–56 in Tabeau et al. (2001).

Van Imhoff E. (2001) On the impossibility of inferring cohort fertility measures from period fertility measures. *Demographic Research* 5-2. http://www.demographic-research.org/.

Van Imhoff E. (1990) The exponential multidimensional demographic projection model. *Mathematical Population Studies* 2, 171–182.

Van Imhoff E. and Keilman N.W. (1991) *Lipro 2.0: an application of a dynamic demographic projection model to household structure in the Netherlands*. Amsterdam: Swets & Zeitlinger.

Van Imhoff E. and Keilman N. (2000) On the quantum and tempo of fertility: comment. *Population and Development Review* 26, 549–553.

Van Imhoff E., van der Gaag N., van Wissen L. and Rees Ph. (1997) The selection of internal migration models for European regions. *International Journal of Population Geography* 3, 137–159.

Vartiainen T., Kartovaara L. and Tuomisto J. (1999) Environmental chemicals and changes in sex ratio. *Environmental Health Perspectives* 107, 813–815.

Vaupel J.W. and Yashin A.I. (1985) Heterogeneity's ruses: some surprising effects of selection on population dynamics. *American Statistician* 39, 176–185.

Verhulst P.F. (1838) Notice sur la loi que la population suit dans son accroissement. *Correspondence Mathématique et Physique Publiée par A. Quételet* (Bruxelles) 10, 113–121.

Vermunt J.K. (1997a) *Log-linear models for event histories*. Thousand Oakes: Sage.

Vermunt J.K. (1997b) *LEM 1.0: A general program for the analysis of categorical data. User's manual.* Tilburg: Tilburg University. (www.kub.nl/mto)

Vinod H.D. and Ullah A. (1981) *Recent advances in regression methods*. New York: Marcel Dekker.

Volinsky C.T., Madigan D., Raftery A.E. and Kronmal R. A. (1997) Bayesian model averaging in proportional hazard models: assessing the risk of stroke. *Applied Statistics*, 46, 433–448.

Väestöennusteryhmä (1973) *Väestöennusteiden laadinnan järjestäminen*. Helsinki: Valtioneuvoston kanslian julkaisu.

Wachter K. and Freedman D. (1999) The fifth cell: correlation bias in U.S. census adjustment. Technical Report Number 570, Department of Statistics, University of California, Berkeley.

Wackernagel H. (1998) *Multivariate geostatistics: an introduction with applications*. Berlin: Springer.

Wade A. (1987) *Social security area population projections: 1987*. Actuarial Study No. 99. Baltimore: U.S. Office of the Actuary, Social Security Administration.

Wade A. (1989) *Social security area population projections 1989*. Actuarial Study No. 105. Baltimore: U.S. Office of the Actuary, Social Security Administration.

Wahba G. and Wold S. (1975) A completely automatic French curve: fitting spline functions by cross validation. *Communications in Statistics* 4, 1–17.

Wahlberg N., Moilanen A. and Hanski I. (1996) Predicting the occurrence of endangered species in fragmented landscapes. *Science* 273, 1536–1538.

Wald A. (1947) *Sequential analysis*. New York: Wiley.

Weisberg S. (1985) *Applied linear regression, 2nd ed.* New York: Wiley.

Welsch R. E. (1983) Leverage. Pp. 610–611 in Kotz, Johnson, and Read (1982) *vol. 4*.

West M., Harrison P.J. and Migon H.S. (1985) Dynamic generalized linear models and Bayesian forecasting (with discussion). *Journal of the American Statistical Association.* 80, 73–97.

Wheeler D. (1984) *Human resource policies, economic growth, and demographic change in developing countries*. Oxford: Clarendon Press.

Whelpton P.K. (1928) Population of the United States, 1925 to 1975. *American Journal of Sociology* 34, 253–270.

Whelpton P.K. (1936) An empirical method of calculating future population. *Journal of the American Statistical Association* 31, 457–473.

Whittle P. (1954) On stationary processes in the plane. *Biometrika* 41, 434–449.

Whelpton P.K., Eldridge H.T. and Siegel J.S. (1947) *Forecasts of the population of the United States 1945–1975*. Washington D.C.: U.S. Census Bureau.

Whelpton P.K. and Kiser C.V. (1946) Social and psychological factors affecting fertility: V. The sampling plan, selection, and the representativeness of couples in the inflated sample. *The Milbank Memorial Fund Quarterly*, 24, 71.

Whelpton P.K. and Kiser C.V. (1947) Social and psychological factors affecting fertility: VI. The planning of fertility. *The Milbank Memorial Fund Quarterly*, 25, 63–111.

Wicksell S.D. (1926) Sveriges framtida befolkning under olika förutsättningar. *Ekonomisk Tidskrift*, 28, 91–123.

Wiebols G.A.H. (1925) *De toekomstige bevolkningsgrootte in Nederland*. Ph.D. thesis, Handelshogeschool Rotterdam. Vlaardingen.

Wiener N. (1956) *The theory of prediction*. Pp. 156–190 in Beckenbach E. (Ed.) (1956) *Modern mathematics for the engineer*. New York: McGraw-Hill.

Willekens F. (1999) Modeling approaches to the indirect estimation of migration flows: from entropy to EM. *Mathematical Population Studies* 7, 239–278.

Williams D. (2001) *Weighing the odds*. Cambridge: Cambridge University Press.

Williams D.A. (1982) Extra-binomial variation in logistic linear models. *Applied Statistics*, 31, 144–148.

Wolf D.A. (1988) The multistate lifetable with duration-dependence. *Mathematical Population Studies* 1, 217–245.

Wolter K.M. (1985) *Introduction to variance estimation.* New York: Springer.

Wolter K.M. (1986) Some coverage error models for census data. *Journal of the American Statistical Association*, 81, 338–346.

Wolter K.M. (1990) Capture-recapture estimation in the presence of a known sex ratio. *Biometrics* 46, 157–162.

Woodruff R. S. (1952) Confidence intervals for medians and other position measures. *Journal of the American Statistical Association* 47, 635–646.

Woodward M. (1999) *Epidemiology. Study design and data analysis*. Boca Raton: Chapman-Hall.

World Bank (1992) *World population projections, 1992–1993.* Baltimore: Johns Hopkins University Press.

Yule G.U. (1925) The growth of population and the factors which control it. *Journal of the Royal Statistical Society* 86, 1–58.

Yamaguchi K. (1989) A formal theory for male-preferring stopping rules of childbearing: sex differences in birth order and in the number of siblings. *Demography* 26, 451–465.

Zaslavsky A. M., Schenker N. and Belin T. R. (2001), Downweighting influential clusters in surveys: application to the 1990 Post-Enumeration Survey. *Journal of the American Statistical Association* 96, 858–869.

Zaslavsky A.M. and Schirm A.L. (2002) Interactions between survey estimates and federal funding formulas. *Journal of Official Statistics*, 18, 371–391.

Zaslavsky A.M. and Wolfgang G.S. (1993) Triple system modeling of the census, post-enumeration survey, and administrative list data. *Journal of Business and Economic Statistics* 11, 279–288.

Zeger S.L. and Karim M.R. (1991) Generalized linear models with random effects: A Gibbs sampling approach. *Journal of the American Statistical Association* 86, 79–86.

Author Index

Subject Index